Computational Noncommutative Algebra and Applications

NATO Science Series

A Series presenting the results of scientific meetings supported under the NATO Science Programme.

The Series is published by IOS Press, Amsterdam, and Kluwer Academic Publishers in conjunction with the NATO Scientific Affairs Division

Sub-Series

I. **Life and Behavioural Sciences**	IOS Press
II. **Mathematics, Physics and Chemistry**	Kluwer Academic Publishers
III. **Computer and Systems Science**	IOS Press
IV. **Earth and Environmental Sciences**	Kluwer Academic Publishers
V. **Science and Technology Policy**	IOS Press

The NATO Science Series continues the series of books published formerly as the NATO ASI Series.

The NATO Science Programme offers support for collaboration in civil science between scientists of countries of the Euro-Atlantic Partnership Council. The types of scientific meeting generally supported are "Advanced Study Institutes" and "Advanced Research Workshops", although other types of meeting are supported from time to time. The NATO Science Series collects together the results of these meetings. The meetings are co-organized bij scientists from NATO countries and scientists from NATO's Partner countries – countries of the CIS and Central and Eastern Europe.

Advanced Study Institutes are high-level tutorial courses offering in-depth study of latest advances in a field.
Advanced Research Workshops are expert meetings aimed at critical assessment of a field, and identification of directions for future action.

As a consequence of the restructuring of the NATO Science Programme in 1999, the NATO Science Series has been re-organised and there are currently Five Sub-series as noted above. Please consult the following web sites for information on previous volumes published in the Series, as well as details of earlier Sub-series.

http://www.nato.int/science
http://www.wkap.nl
http://www.iospress.nl
http://www.wtv-books.de/nato-pco.htm

Series II: Mathematics, Physics and Chemistry – Vol. 136

Computational Noncommutative Algebra and Applications

edited by

Jim Byrnes
Prometheus Inc.,
Newport, Rhode Island, U.S.A.

and

Gerald Ostheimer
Prometheus Inc.,
Newport, Rhode Island, U.S.A.

Kluwer Academic Publishers

Dordrecht / Boston / London

Published in cooperation with NATO Scientific Affairs Division

Proceedings of the NATO Advanced Study Institute on
Computational Noncommutative Algebra and Applications
Il Ciocco
6–19 July 2003

A C.I.P. Catalogue record for this book is available from the Library of Congress.

ISBN 1-4020-1983-1 (PB)
ISBN 1-4020-1982-3 (HB)
ISBN 1-4020-2307-3 (e-book)

Published by Kluwer Academic Publishers,
P.O. Box 17, 3300 AA Dordrecht, The Netherlands.

Sold and distributed in North, Central and South America
by Kluwer Academic Publishers,
101 Philip Drive, Norwell, MA 02061, U.S.A.

In all other countries, sold and distributed
by Kluwer Academic Publishers,
P.O. Box 322, 3300 AH Dordrecht, The Netherlands.

Printed on acid-free paper

This book is dedicated
to the cherished
memory of Edward
Joseph Sanchas, a most
decent and trusted
human being whom we
always counted on to
take charge of any and
all problems. He was a
great man. We miss
him deeply.

Contents

Preface

The chapters in this volume were presented at the July 2003 NATO
Advanced Study Institute on *Computational Noncommutative Algebra
and Applications*. The conference was held at the beautiful Il Ciocco re-
sort near Lucca, in the glorious Tuscany region of northern Italy. Once
again we gathered at this idyllic spot to explore and extend the reci-
procity between mathematics and engineering. The dynamic interaction
between world-renowned scientists from the usually disparate communities of pure mathematicians and applied scientists, which occurred at
our 1989, 1991, 1992, 1998, and 2000 ASI's, continued at this meeting.

The fusion of algebra, analysis and geometry, and their application
to real world problems, have been dominant themes underlying mathematics for over a century. Geometric algebras, introduced and classified
by Clifford in the late 19th century, have played a prominent role in
this effort, as seen in the mathematical work of Cartan, Brauer, Weyl,
Chevelley, Atiyah, and Bott, and in applications to physics in the work
of Pauli, Dirac and others. One of the most important applications of
geometric algebras to geometry is to the representation of groups of Euclidean and Minkowski rotations. This aspect and its direct relation to
robotics and vision were discussed by several of the Principal Lecturers,
and are covered in this book.

Moreover, group theory, beginning with the work of Burnside, Frobenius and Schur, has been influenced by even more general problems. As
a result, general group actions have provided the setting for powerful
methods within group theory and for the use of groups in applications
to physics, chemistry, molecular biology, and signal processing. These
aspects, too, are covered in what follows.

With the rapidly growing importance of, and ever expanding conceptual and computational demands on signal and image processing in

remote sensing, computer vision, medical image processing, and biological signal processing, and on neural and quantum computing, geometric algebras, and computational group harmonic analysis, the topics of the following chapters have emerged as key tools. The authors include many of the world's leading experts in the development of new algebraic modeling and signal representation methodologies, novel Fourier-based and geometric transforms, and computational algorithms required for realizing the potential of these new application fields.

The ASI brought together these world leaders from both academia and industry, with extensive multidisciplinary backgrounds evidenced by their research and participation in numerous workshops and conferences. This created an interactive forum for initiating new and intensifying existing efforts aimed at creating a unified computational noncommutative algebra for advancing the broad range of applications indicated above. The forum provided opportunities for young scientists and engineers to learn more about these problem areas, and the vital role played by new mathematical insights, from the recognized experts in this vital and growing area of both pure and applied science.

The talks and the following chapters were designed to address an audience consisting of a broad spectrum of scientists, engineers, and mathematicians involved in these fields. Participants had the opportunity to interact with those individuals who have been on the forefront of the ongoing explosion of work in computational noncommutative algebra, to learn firsthand the details and subtleties of this exciting area, and to hear these experts discuss in accessible terms their contributions and ideas for future research. This volume offers these insights to those who were unable to attend.

The cooperation of many individuals and organizations was required in order to make the conference the success that it was. First and foremost I wish to thank NATO, and especially Dr. F. Pedrazzini and his most able assistant, Ms. Alison Trapp, for the initial grant and subsequent help. Financial support was also received from the Air Force Office of Scientific Research (Dr. Jon Sjogren), the Defense Advanced Research Projects Agency (Dr. Douglas Cochran), the National Science Foundation (Dr. Sylvia Wiegand), USAF Rome Laboratories (Drs. Michael Wicks, Braham Himed, and Gerard Genello), US Army SMDC (Drs. Pete Kirkland and Robert McMillan), EOARD (Dr. Chris Reuter), the Raytheon Company (Dr. David N. Martin), Bielefeld University (Professor Andreas Dress), Bonn University (Professor Michael Clausen), Melbourne University, and Prometheus Inc. This additional support is gratefully acknowledged.

I wish to express my sincere appreciation to my assistant Marcia Byrnes, to my co-editor and co-organizer, Gerald Ostheimer, and to the co-director, Valeriy Labunets, for their invaluable aid. I am also grateful to Kathryn Hargreaves, our TEXnician, for her superlative work in preparing this volume. Finally, my heartfelt thanks to the Il Ciocco staff, especially Bruno Giannasi and Alberto Suffredini, for offering an ideal setting, not to mention the magnificent meals, that promoted the productive interaction between the participants of the conference. All of the above, the speakers, and the remaining conferees, made it possible for our Advanced Study Institute, and this volume, to fulfill the stated NATO objectives of disseminating advanced knowledge and fostering international scientific contacts.

September 22, 2003   *Jim Byrnes*, Newport, Rhode Island

Acknowledgments

We wish to thank the following for their contribution to the success of this conference: NATO Scientific & Environmental Affairs Division; DARPA Defense Sciences Office; U. S. Air Force AFOSR; U.S. Air Force EOARD; U.S. Air Force Rome Laboratories Sensors Directorate; U.S. Army Space and Missile Defense Command; U.S. National Science Foundation; Raytheon Company, U.S.A.; Universität Bielefeld, Germany; Universität Bonn, Germany; Melbourne University, Australia; and Prometheus Inc., U.S.A.

CLIFFORD GEOMETRIC ALGEBRAS
IN MULTILINEAR ALGEBRA
AND NON-EUCLIDEAN GEOMETRIES

Garret Sobczyk*
Universidad de las Américas
Departamento de Físico-Matemáticas
Apartado Postal #100, Santa Catarina Mártir
72820 Cholula, Pue., México
sobczyk@mail.udlap.mx

Abstract   Given a quadratic form on a vector space, the geometric algebra of
the corresponding pseudo-euclidean space is defined in terms of a simple set of rules which characterizes the geometric product of vectors.
We develop geometric algebra in such a way that it augments, but remains fully compatible with, the more traditional tools of matrix algebra. Indeed, matrix multiplication arises naturally from the geometric
multiplication of vectors by introducing a spectral basis of mutually
annihiliating idempotents in the geometric algebra. With the help of
a few more algebraic identities, and given the proper geometric interpretation, the geometric algebra can be applied to the study of affine,
projective, conformal and other geometries. The advantage of geometric
algebra is that it provides a single algebraic framework with a comprehensive, but flexible, geometric interpretation. For example, the affine
plane of rays is obtained from the euclidean plane of points by adding
a single anti-commuting vector to the underlying vector space. The key
to the study of noneuclidean geometries is the definition of the operations of meet and join, in terms of which incidence relationships are
expressed. The horosphere provides a homogeneous model of euclidean
space, and is obtained by adding a second anti-commuting vector to the
underlying vector space of the affine plane. Linear orthogonal transformations on the higher dimensional vector space correspond to conformal or Möbius transformations on the horosphere. The horosphere
was first constructed by F.A. Wachter (1792–1817), but has only recently attracted attention by offering a host of new computational tools

---

*I gratefully acknowledge the support given by INIP of the Universidad de las Américas. The
author is a member of SNI, Exp. 14587.

1

in projective and hyperbolic geometries when formulated in terms of geometric algebra.

Keywords: affine geometry, Clifford algebra, conformal geometry, conformal group, euclidean geometry, geometric algebra, horosphere, Möbius transformation, non-euclidean geometry, projective geometry, spectral decomposition.

1.   Geometric algebra

A Geometric algebra is generated by taking linear combinations of geometric products of vectors in a vector space taken together with a specified bilinear form. Here we shall study the geometric algebras of the *pseudo-euclidean* vector spaces $\mathcal{G}_{p,q} := \mathcal{G}_{p,q}(I\!\!R^{p,q})$ for which we have the indefinite metric

$$x \cdot y = \sum_{i=1}^{p} x_i y_i - \sum_{j=p+1}^{p+q} x_j y_j$$

for $x = \begin{pmatrix} x_1 & \cdots & x_{p+q} \end{pmatrix}$ and $y = \begin{pmatrix} y_1 & \cdots & y_{p+q} \end{pmatrix}$ in $I\!\!R^{p,q}$. We first study the geometric algebra of the more familiar Euclidean space.

1.1   Geometric algebra of Euclidean n-space

We begin by introducing the geometric algebra $\mathcal{G}_n := \mathcal{G}(I\!\!R^n)$ of the familiar Euclidean n-space

$$I\!\!R^n = \{x|\ x = \begin{pmatrix} x_1 & \cdots & x_n \end{pmatrix} \text{ for } x_i \in I\!\!R\}.$$

Recall the dual interpretations of each element $x \in I\!\!R^n$, both as a point of $I\!\!R^n$ with the coordinates $\begin{pmatrix} x_1 & \cdots & x_n \end{pmatrix}$ and as the position vector or *directed line segment* from the origin to the point. We can thus express each vector $x \in I\!\!R^n$ as a linear combination of the *standard orthonormal basis vectors* $\{e_1, e_2, \cdots, e_n\}$ where $e_i = \begin{pmatrix} 0 & \cdots & 0 & 1_i & 0 & \cdots & 0 \end{pmatrix}$, namely

$$x = \sum_{i=1}^{n} x_i e_i.$$

The vectors of $I\!\!R^n$ are added and multiplied by scalars in the usual way, and the positive definite *inner product* of the vectors x and $y = \begin{pmatrix} y_1 & \cdots & y_n \end{pmatrix}$ is given by

$$x \cdot y = \sum_{i=1}^{n} x_i y_i. \tag{1}$$

The geometric algebra $\mathcal{G}_n$ is generated by the *geometric multiplication* and addition of vectors in $I\!R^n$. In order to efficiently introduce the geometric product of vectors, we note that the resulting geometric algebra $\mathcal{G}_n$ is isomorphic to an appropriate matrix algebra under addition and geometric multiplication. Thus, like matrix algebra, $\mathcal{G}_n$ is an associative, but non-commutative algebra, but unlike matrix algebra the elements of $\mathcal{G}_n$ are assigned a comprehensive geometric interpretation. The two fundamental rules governing geometric multiplication and its interpretation are:

- For each vector $x \in I\!R^n$,

$$x^2 = xx = |x|^2 = \sum_{i=1}^{n} x_i^2 \qquad (2)$$

  where $|x|$ is the usual *Euclidean norm* of the vector x.

- If $a_1, a_2, \ldots, a_k \in I\!R^n$ are k mutually orthogonal vectors, then the product

$$A_k = a_1 a_2 \ldots a_k \qquad (3)$$

  is totally antisymmetric and has the geometric interpretation of a *simple k-vector* or a *directed k-plane* .[1]

Let us explore some of the many consequences of these two basic rules. Applying the first rule (2) to the sum $a + b$ of the vectors $a, b \in I\!R^2$, we get

$$(a + b)^2 = a^2 + ab + ba + b^2,$$

or

$$a \cdot b := \frac{1}{2}(ab + ba) = \frac{1}{2}(|a + b|^2 - |a|^2 - |b|^2)$$

which is a statement of the famous *law of cosines* . In the special case when the vectors a and b are orthogonal, and therefore anticommutative by the second rule (3), we have $ab = -ba$ and $a \cdot b = 0$.

If we multiply the orthonormal basis vectors $e_{12} := e_1 e_2$, we get the 2-vector or *bivector* e_{12}, pictured as the *directed plane segment* in Figure 1. Note that the *orientation* of the bivector e_{12} is counterclockwise, and that the bivector $e_{21} := e_2 e_1 = -e_1 e_2 = -e_{12}$ has the opposite or clockwise orientation.

---

[1]This means that the product changes its sign under the interchange of any two of the orthogonal vectors in its argument.

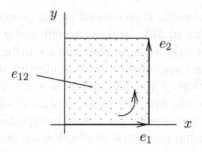

Figure 1.   The directed plane segment $e_{12} = e_1 e_2$.

We can now write down an orthonormal basis for the geometric algebra $\mathcal{G}_n$, generated by the orthonormal basis vectors $\{e_i | 1 \leq i \leq n\}$. In terms of the modified cartesian-like product, $\times_{i=1}^{n}(1, e_i) :=$

$$\{1, e_1, \ldots, e_n, e_{12}, \ldots, e_{(n-1)n}, \ldots, \ldots, e_{1\cdots(n-1)}, \ldots, e_{2\cdots n}, e_{1\ldots n}\}.$$

There are

$$\binom{n}{0} + \binom{n}{1} + \binom{n}{2} + \cdots + \binom{n}{n-1} + \binom{n}{n} = 2^n$$

linearly independent elements in the standard orthonormal basis of $\mathcal{G}_n$. Any *multivector* or *geometric number* $g \in \mathcal{G}_n$ can be expressed as a sum of its homogeneous k-vector parts,

$$g = g_0 + \cdots + g_k + \cdots + g_n$$

where $g_k :=< g >_k= \sum_{\sigma} \alpha_{\sigma} e_{\sigma}$ where $\sigma = \sigma_1 \cdots \sigma_k$ for $1 \leq \sigma_1 < \cdots < \sigma_k \leq n$, and $\alpha_{\sigma} \in \mathbb{R}$. The real part $g_0 :=< g >_0= \alpha_0 e_0 = \alpha_0$ of the geometric number g is just a real number, since $e_0 := 1$. By *definition*, any k-vector can be written as a linear combination of simple k-vectors or k-*blades* , [8, p.4].

Given two vectors $a, b \in \mathbb{R}^n$, we can decompose the vector a into components parallel and perpendicular to b, $a = a_{\parallel} + a_{\perp}$, where

$$a_{\parallel} = (a \cdot b)\frac{b}{|b|^2} = (a \cdot b)b^{-1},$$

and $a_{\perp} := a - a_{\parallel}$, see Figure 2.

With the help of (3), we now calculate the geometric product of the vectors a and b, getting

$$ab = (a_{\parallel} + a_{\perp})b = a_{\parallel} \cdot b + a_{\perp} \wedge b = \frac{1}{2}(ab + ba) + \frac{1}{2}(ab - ba) \quad (4)$$

$$a_{\parallel} = (a \cdot b)\frac{b}{|b|^2}$$
$$= (a \cdot b)b^{-1}.$$

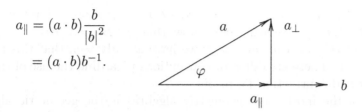

Figure 2.   Decomposition of a into parallel and perpendicular parts.

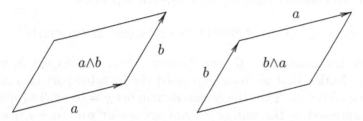

Figure 3.   The bivectors $a{\wedge}b$ and $b{\wedge}a$.

where the *outer product* $a{\wedge}b := \frac{1}{2}(ab - ba) = a_{\perp}b = -ba_{\perp} = -b{\wedge}a$ is the bivector shown in Figure 3. The basic formula (4) shows that the geometric product ab is the sum of a scalar and a bivector part which characterizes the relative directions of a and b. If we make the assumption that a and b lie in the plane of the bivector e_{12}, then we can write

$$ab = |a||b|(\cos\varphi + I\sin\varphi) = |a||b|e^{I\varphi}, \tag{5}$$

where $I := e_{12} = e_1 e_2$ has the familiar property that

$$I^2 = e_1 e_2 e_1 e_2 = -e_1 e_2 e_2 e_1 = -e_1^2 e_2^2 = -1.$$

Equation (5) is the *Euler formula* for the geometric multiplication of vectors.

The definition of the inner product $a \cdot b$ and outer product $a{\wedge}b$ can be easily extended to $a \cdot B_r$ and $a{\wedge}B_r$, respectively, where $r \geq 0$ denotes the *grade* of the r-vector B_r:

DEFINITION 1 *The inner product or contraction $a \cdot B_r$ of a vector a with an r-vector B_r is determined by*

$$a \cdot B_r = \frac{1}{2}(aB_r + (-1)^{r+1}B_r a) = (-1)^{r+1}B_r \cdot a.$$

DEFINITION 2 *The outer product $a{\wedge}B_r$ of a vector a with an r-vector B_r is determined by*

$$a{\wedge}B_r = \frac{1}{2}(aB_r - (-1)^{r+1}B_r a) = -(-1)^{r+1}B_r{\wedge}a.$$

Note that $a \cdot \beta = \beta \cdot a = 0$ and $a \wedge \beta = \beta \wedge a = \beta a$ for the scalar $\beta \in \mathbb{R}$. Indeed, we will soon show that $a \cdot B_r = <aB_r>_{r-1}$ and $a \wedge B_r = <aB_r>_{r+1}$ for all $r \geq 1$; we have already seen that this is true when $r = 1$. There are different conventions regarding the use of the dot product and contraction [5, p. 35].

One of the most basic geometric algebras is the geometric algebra $\mathcal{G}_3$ of 3 dimensional Euclidean space which we live in. The complete standard orthonormal basis of this geometric algebra is

$$\mathcal{G}_3 = \times_{i=1}^{3}(1, e_i) = span\{1, e_1, e_2, e_3, e_{12}, e_{13}, e_{23}, e_{123}\}.$$

Any geometric number $g \in \mathcal{G}_3$ has the form $g = \alpha + v_1 + iv_2 + \beta i$ where $i := e_{123}$. Notice that we have expressed the bivector part of g as the *dual* of the vector v_2. Thus the geometric number $g = (\alpha + i\beta) + (v_1 + iv_2)$ can be expressed as the sum of its *complex scalar part* $(\alpha + i\beta)$ and a *complex vector part* $(v_1 + iv_2)$. Note that the complex scalar part has all the properties of an ordinary complex number $z = x + iy$. This follows easily from the fact that the pseudoscalar $i = e_{123}$ satisfies $i^2 = e_{123}e_{123} = e_{23}e_{23} = -1$.

We can use the Euler form (5) to see that

$$a = a(b^{-1}b) = (ab^{-1})b = \left(\frac{ab}{b^2}\right)b,$$

so the geometric quantity $ab/|b|^2$ *rotates and dilates* the vector b into the vector a *when multiplied by* b *on the left*. Similarly, multiplying b on right by $\frac{ba}{b^2}$ also rotates and dilates the vector b into the vector a. By re-expressing this result in terms of the Euler angle φ, letting $I = ie_3$, and assuming that $|a| = |b|$, we can write $a = \exp(ie_3\varphi)b = b\exp(-ie_3\varphi)$. Even more powerfully, and more generally, we can write

$$a = \exp(I\varphi/2)b\exp(-I\varphi/2),$$

which expresses the $\frac{1}{2}$-angle formula for rotating the vector $b \in \mathbb{R}^n$ in the plane of the simple bivector I through the angle φ. There are many more formulas for expressing reflexions and rotations in $\mathbb{R}^n$, or in the pseudo-euclidean spaces $\mathbb{R}^{p,q}$, [8], [10].

1.2   Basic algebraic identities

One of the most difficult aspects of learning geometric algebra is coming to terms with a host of unfamiliar algebraic identities. These important identities can be quickly mastered if they are established in a careful systematic way. The most important of these identities follows easily from the following two trivial algebraic identities involving the vectors a and b and an r-blade B_r where $r \geq 0$:

$$abB_r + bB_r a \equiv (ab + ba)B_r - b(aB_r - B_r a), \tag{6}$$

and

$$baB_r - aB_r b \equiv (ba + ab)B_r - a(bB_r + B_r b). \tag{7}$$

Whereas these identities are valid for general r-vectors, we state them here only for a simple r-vector B_r, the more general case following by linear superposition.

In proving the identities below, we use the fact that

$$bB_r = < bB_r >_{r-1} + < bB_r >_{r+1} . \tag{8}$$

This is easily seen to be true if B_r is an r-blade, in which case $B_r = b_1 \cdots b_r$ for r orthogonal, and therefore anticommuting, vectors $b_1, \ldots, b_r$. We then simply decompose the vector $b = b_{\parallel} + b_{\perp}$ into parts parallel and perpendicular to the subspace of $\mathbb{R}^n$ spanned by the vectors $b_1, \ldots, b_r$, and use the anticommutivity of the $b's$ to show that $b \cdot B_r = b_{\parallel} B_r = < b_{\parallel} B_r >_{r-1}$ and $b \wedge B_r = b_{\perp} B_r = < b_{\perp} B_r >_{r+1}$. This also shows the useful result that $b \cdot B_r$ and $b \wedge B_r$ are blades whenever B_r is a blade.

The following basic identity relates the inner and outer products:

$$a \cdot (b \wedge B_r) = (a \cdot b)B_r - b \wedge (a \cdot B_r), \tag{9}$$

for all $r \geq 0$. If $r = 0$, (9) follows from what has already been established. If $r \geq 2$ and even, (6) and definition (2) implies that

$$2a \cdot (bB_r) = 2(a \cdot b)B_r - 2b(a \cdot B_r).$$

Taking the r-vector part of this equation gives (9). If $r \geq 1$ and odd, (7) implies that

$$2b \cdot (aB_r) = 2(a \cdot b)B_r - 2a(b \cdot B_r)$$

which implies (9) by again taking the r-vector part of this equation and simplifying. By iterating (9), we get the important identity for *contraction*

$$a \cdot (b_1 \wedge \cdots \wedge b_n) = \sum_{i=1}^{n} (-1)^{i+1} (a \cdot b_i) \, b_1 \wedge \cdots \hat{i} \cdots \wedge b_n.$$

Let $I = e_{12 \cdots n}$ be the unit pseudoscalar element of the geometric algebra $\mathcal{G}_{p,q} = \mathcal{G}(\mathbb{R}^{p,q})$. We give here a number of important identities relating the inner and outer products which will be used later in the contexts of projective geometry. For an r-blade A_r, the $(p + q - r)$-blade $A_r^* := A_r I^{-1}$ is called the *dual* of A_r in $\mathcal{G}_{p,q}$ with respect to the pseudoscalar I. Note that it follows that $I^* = II^{-1} = 1$ and $1^* = I^{-1}$. For $r + s \leq p + q$, we have the important identity

$$(A_r \wedge B_s)^* = (A_r \wedge B_s)I^{-1} = A_r \cdot (B_s I^{-1}) = A_r \cdot B_s^* = (-1)^{s(p+q-s)} A_r^* \cdot B_s$$

(10)

1.3   Geometric algebras of psuedoeuclidean spaces

All of the algebraic identities discussed for the geometric algebra $\mathcal{G}_n$ hold in the geometric algebra with indefinite signature $\mathcal{G}_{p,q}$. However, some care must be taken with respect to the existence of non-zero *null vectors* . A non-zero vector $n \in \mathbb{R}^{p,q}$ is said to be a null vector if $n^2 = n \cdot n = 0$. The inverse of a non-null vector v is $v^{-1} = \frac{v}{v^2}$, so clearly a null vector has no inverse. The spacetime algebra $\mathcal{G}_{1,3}$ of $\mathbb{R}^{1,3}$, also called the *Dirac algebra*, has many applications in the study of the Lorentz transformations used in the special theory of relativity. Whereas non-zero null vectors do not exist in $\mathbb{R}^n$, there are many non-zero geometric numbers $g \in \mathcal{G}_n$ which are null. For example, let $g = e_1 + e_{12} \in \mathcal{G}_3$, then

$$g^2 = (e_1 + e_{12})(e_1 + e_{12}) = e_1^2 + e_{12}^2 = 1 - 1 = 0.$$

Let us consider in more detail the spacetime algebra $\mathcal{G}_{1,3}$ of $\mathbb{R}^{1,3}$. The standard orthonormal basis of $\mathbb{R}^{1,3}$ are the vectors $\{e_1, e_2, e_3, e_4\}$, where $e_1^2 = e_2^2 = e_3^2 = -1 = -e_4^2$. The standard basis of the bivectors $\mathcal{G}_{1,3}^2$ of $\mathcal{G}_{1,3}$ are $e_{14}, e_{24}, e_{34}, ie_{14}, ie_{24}, ie_{34}$, where $i = e_{1234}$ is the *pseudoscalar* of $\mathcal{G}_{1,3}$. Note that the first 3 of these bivectors have square $+1$, where as the *duals* of these basis bivectors have square -1. Indeed, the subalgebra of $\mathcal{G}_{1,3}$ generated by $E_1 = e_{41}, E_2 = e_{42}, E_3 = e_{43}$ is algebraically isomorphic to the geometric algebra $\mathcal{G}_3$ of space. This key relationship makes possible the efficient expression of electromagnetism and the theory of special relativity in one and the same formalisms.

An important class of pseudo-euclidean spaces consists of those that have *neutral signature*, $\mathcal{G}_{n,n} = \mathcal{G}_{n,n}(\mathbb{R}^{n,n})$. The simplest such algebra is $\mathcal{G}_{1,1}$ with the standard basis $\{1, e_1, e_2, e_{12}\}$, where $e_1^2 = 1 = -e_2^2$ and $e_{12}^2 = 1$. We shall shortly see that $\mathcal{G}_{1,1}$ is the basic building block for extending the applications of geometric algebra to affine and projective and other non-euclidean geometries, and for exploring the structure of geometric algebras in terms of matrices.

1.4 Spectral basis and matrices of geometric algebras

Until now we have only discussed the standard basis of a geometric algebra $\mathcal{G}$. The standard basis is very useful for presenting the basic rules of the algebra and its geometric interpretation as a graded algebra of multivectors of different grades, a k-blade characterizing the direction of a k-dimensional subspace. There is another basis for a geometric algebra, called a *spectral basis*, that is very useful for relating the structure of a geometric algebra to corresponding isomorphic matrix algebras [15]. Another term that has been applied is *spinor basis*, but I prefer the term "spectral basis" because of its deep roots in linear algebra [17].

The key to constructing a spectral basis for any geometric algebra $\mathcal{G}$ is to pick out any two elements $u, v \in \mathcal{G}$ such that $u^2 = 1$, v^{-1} exists, and $uv = -vu \neq 0$. We then define the *idempotents* $u_+ = \frac{1}{2}(1 + u)$ and $u_- = \frac{1}{2}(1 - u)$ in $\mathcal{G}$, and note that

$$u_+^2 = u_+, u_-^2 = u_-, u_+u_- = u_-u_+ = 0, \text{ and } u_+ + u_- = 1.$$

We say that $u_{\pm}$ are *mutually annihilating idempotents which partition 1*. Also $vu_+ = u_-v$, from which it follows that $v^{-1}u_+ = u_-v^{-1}$.

Using these simple algebraic properties, we can now *factor out* a 2×2 matrix algebra from $\mathcal{G}$. Adopting matrix notation, first note that

$$\begin{pmatrix} 1 & v \end{pmatrix} u_+ \begin{pmatrix} 1 \\ v^{-1} \end{pmatrix} = \begin{pmatrix} u_+ & vu_+ \end{pmatrix} \begin{pmatrix} 1 \\ v^{-1} \end{pmatrix} = u_+ + u_- = 1.$$

For any element $g \in \mathcal{G}$, we have

$$g = \begin{pmatrix} 1 & v \end{pmatrix} u_+ \begin{pmatrix} 1 \\ v^{-1} \end{pmatrix} g \begin{pmatrix} 1 & v \end{pmatrix} u_+ \begin{pmatrix} 1 \\ v^{-1} \end{pmatrix}$$

$$= \begin{pmatrix} 1 & v \end{pmatrix} u_+ \begin{pmatrix} g & gv \\ v^{-1}g & v^{-1}gv \end{pmatrix} u_+ \begin{pmatrix} 1 \\ v^{-1} \end{pmatrix}. \tag{11}$$

The expression $[g] := u_+ \begin{pmatrix} g & gv \\ v^{-1}g & v^{-1}gv \end{pmatrix} u_+$ is called the *matrix decomposition* of g with respect to the elements $\{u, v\} \subset \mathcal{G}$. To see that the mapping $g \mapsto [g]$ gives a *matrix isomorphism*, in the sense that $[g + h] = [g] + [h]$ and $[gh] = [g][h]$ for all $g, h \in \mathcal{G}$, it is obvious that we only need to check the multiplicative property. We find that

$$[g][h] = u_+ \begin{pmatrix} g & gv \\ v^{-1}g & v^{-1}gv \end{pmatrix} u_+ \begin{pmatrix} h & hv \\ v^{-1}h & v^{-1}hv \end{pmatrix} u_+$$

$$= u_+ \begin{pmatrix} gu_+h + gvu_+v^{-1}h & gu_+hv + gvu_+v^{-1}hv \\ v^{-1}gu_+h + v^{-1}gvu_+v^{-1}h & v^{-1}gu_+hv + v^{-1}gvu_+v^{-1}hv \end{pmatrix} u_+$$

$$= u_+ \begin{pmatrix} gh & ghv \\ v^{-1}gh & v^{-1}ghv \end{pmatrix} u_+ = [gh].$$

To fully understand the nature of this matrix isomorphism, we need to know about the nature of the entries of $[g]$ in (11). We will analyse the special case where $v^2 \in \mathbb{R}$, although the relationship is valid for more general v. In this case, the entries of $[g]$ can be decomposed in terms of *conjugations* with respect to the elements u and v. Let $a \in \mathcal{G}$ for which a^{-1} exists. The *a-conjugate* $\bar{g}^a$ of the element $g \in \mathcal{G}$ is defined by $\bar{g}^a = aga^{-1}$.

We shall use u- and v-conjugates to decompose any element g into the form

$$g = G_1 + uG_2 + v(G_3 + uG_4) \tag{12}$$

where $G_i \in C_{\mathcal{G}}(\{u, v\})$, the subalgebra of all elements of $\mathcal{G}$ which commute with the subalgebra generated by $\{u, v\}$. It follows that $\mathcal{G} = C_{\mathcal{G}}(\{u, v\}) \otimes \{u, v\}$ or $\mathcal{G} \equiv \mathcal{M}_2(C_{\mathcal{G}}(\{u, v\}))$. This means that $\mathcal{G}$ is isomorphic to a 2×2 matrix algebra over $C_{\mathcal{G}}(\{u, v\})$.

We first decompose g into the form

$$g = \frac{1}{2}(g + \bar{g}^u) + v[\frac{v^{-1}}{2}(g - \bar{g}^u)] = g_1 + vg_2$$

where $g_1, g_2 \in C_{\mathcal{G}}(u)$, the geometric subalgebra of $\mathcal{G}$ of all elements which commute with the element u. By further decomposing g_1 and g_2 with respect to the v-conjugate, we obtain the decomposition

$$g = G_1 + uG_2 + v(G_3 + uG_4)$$

where each $G_i \in C_{\mathcal{G}}(\{u, v\})$. Specifically, we have

$$G_1 = \frac{1}{2}(g_1 + \bar{g}_1^v) = \frac{1}{4}(g + ugu + vgv^{-1} + vuguv^{-1})$$

$$G_2 = \frac{u}{2}(g_1 - \bar{g}_1^v) = \frac{u}{4}(g + ugu - vgv^{-1} - vuguv^{-1})$$

$$G_3 = \frac{1}{2}(g_2 + \bar{g}_2^v) = \frac{v^{-1}}{4}(g - ugu + vgv^{-1} - vuguv^{-1})$$

$$G_4 = \frac{u}{2}(g_2 - \bar{g}_2^v) = \frac{uv^{-1}}{4}(g - ugu - vgv^{-1} + vuguv^{-1}).$$

Using the decomposition (12) of g, we find the 2×2 matrix decomposition (11) of g over the module $C_{\mathcal{G}}(\{u, v\})$,

$$[g] := u_+ \begin{pmatrix} g & gv \\ v^{-1}g & v^{-1}gv \end{pmatrix} u_+ = u_+ \begin{pmatrix} G_1 + G_2 & v^2(G_3 - G_4) \\ G_3 + G_4 & G_1 - G_2 \end{pmatrix} \tag{13}$$

where $G_i \in C_{\mathcal{G}}(\{u, v\})$ for $1 \le i \le 4$.

For example, for $g \in \mathcal{G}_3$ and $u = e_1$, $v = e_{12} = -v^{-1}$, we write $g = (z_1 + uz_2) + v(z_3 + uz_4)$, where $z_j = x_j + iy_j$ for $i = e_{123}$ and $1 \le j \le 4$. Noting that $u_{\pm}u = uu_{\pm} = \pm u_{\pm}$ and $u_{\pm}v = vu_{\mp}$, and substituting this complex form of g into the above equation gives

$$g = \begin{pmatrix} 1 & v \end{pmatrix} u_+ \begin{pmatrix} g & gv \\ v^{-1}g & v^{-1}gv \end{pmatrix} u_+ \begin{pmatrix} 1 \\ v^{-1} \end{pmatrix}$$

$$= \begin{pmatrix} 1 & v \end{pmatrix} u_+ \begin{pmatrix} z_1 + z_2 & z_4 - z_3 \\ z_4 + z_3 & z_1 - z_2 \end{pmatrix} \begin{pmatrix} 1 \\ v^{-1} \end{pmatrix}.$$

We say that

$$[g] = u_+ \begin{pmatrix} z_1 + z_2 & z_4 - z_3 \\ z_4 + z_3 & z_1 - z_2 \end{pmatrix}$$

is the matrix decomposition of $g \in \mathcal{G}_3$ over the complex numbers, $\mathbb{C} = \{x + iy\}$ where $i = e_{123}$. It follows that

$$\mathcal{G}_3 \equiv \mathcal{M}_2(\mathbb{C}).$$

There are many decompositions of Clifford geometric algebras into isomorphic matrix algebras. As shown in the example above, a matrix decomposition of geometric algebra is equivalent to selecting a *spectral basis*, in this case

$$\begin{pmatrix} 1 \\ v \end{pmatrix} u_+ \begin{pmatrix} 1 & v^{-1} \end{pmatrix} = \begin{pmatrix} u_+ & v^{-1}u_- \\ vu_+ & u_- \end{pmatrix},$$

as opposed to the standard basis for the algebra. The relative position of the elements in the spectral basis, written as a matrix above, gives the isomorphism between the geometric algebra and the matrix algebra.

There is a matrix decomposition of the geometric algebra $\mathcal{G}_{p+1,q+1}$ that is very useful. For this decomposition we let $u = e_{p+1}e_{p+q+2}$, so that the bivector u has the property that $u^2 = 1$, and let $v = e_{p+1}$. We then have the idempotents $u_{\pm} = \frac{1}{2}(1 \pm u)$, satisfying $vu_{\pm} = u_{\mp}v$, and giving the decomposition

$$\mathcal{G}_{p+1,q+1} = \mathcal{G}_{1,1} \otimes \mathcal{G}_{p,q} \equiv \mathcal{M}_2(\mathcal{G}_{p,q}) \tag{14}$$

for $\mathcal{G}_{1,1} = gen\{e_{p+1}, e_{p+q+2}\}$ and $\mathcal{G}_{p,q} = gen\{e_1, \ldots, e_p, e_{p+2}, \ldots, e_{p+q+1}\}$.

2. Projective Geometries

Leonardo da Vinci (1452–1519) was one of the first to consider the problems of projective geometry. However, projective geometry was not formally developed until the work "Traité des propriés projectives des figure" of the French mathematician Poncelet (1788-1867), published in 1822. The extrordinary generality and simplicity of projective geometry led the English mathematician Cayley to exclaim: "Projective Geometry is all of geometry" [18].

The projective plane is almost identical to the Euclidean plane, except for the addition of ideal points and an ideal line at infinity. It seems natural, therefore, that in the study of analytic projective geometry the coordinate systems of Euclidean plane geometry should be almost sufficient. It is also required that these ideal objects at infinity should be indistinquishable from their corresponding ordinary objects, in this case ordinary points and ordinary lines. The solution to this problem is the introduction of "homogeneous coordinates", [6, p. 71]. The introduction of the tools of homogeneous coordinates is accomplished in a very efficient way using geometric algebra [9]. While the definition of geometric algebra does indeed involve a metric, that fact in no way prevents it from being used as a powerful tool to solve the metric-free results of projective geometry. Indeed, once the objects of projective geometry are identified with the corresponding objects of linear algebra, the whole of the machinery of geometric algebra applied to linear algebra can be carried over to projective geometry.

Let $I\!R^{n+1}$ be an $(n + 1)$-dimensional euclidean space and let $\mathcal{G}_{n+1}$ be the corresponding geometric algebra. The *directions* or *rays* of non-zero vectors in $I\!R^{n+1}$ are identified with the points of the n-dimensional projective plane Π^n. More precisely, we write

$$\Pi^n \equiv I\!R^{n+1}/I\!R^*$$

where $I\!R^* = I\!R - \{0\}$. We thus identify *points*, *lines*, *planes*, and higher dimensional *k-planes* in Π^n with 1, 2, 3, and $(k + 1)$-dimensional subspaces $\mathcal{S}^{k+1}$ of $I\!R^{n+1}$, where $k \leq n$. To effectively apply the tools of geometric algebra, we need to introduce the basic operations of *meet* and *join*.

2.1   The Meet and Join Operations

The *meet* and *join* operations of projective geometry are most easily defined in terms of the *intersection* and *union* of the linear subspaces which name the objects in Π^n. Each r-dimensional subspace $\mathcal{A}^r$ is described by a non-zero r-blade $A_r \in \mathcal{G}(I\!R^{n+1})$. We say that an r-blade $A_r \neq 0$ *represents*, or is a *representant* of an r-subspace $\mathcal{A}^r$ of $I\!R^{n+1}$ if and only if

$$\mathcal{A}^r = \{x \in I\!R^{n+1}|\ x \wedge A_r = 0\}. \tag{15}$$

The equivalence class of all nonzero r-blades $A_r \in \mathcal{G}(I\!\!R^{n+1})$ which define the subspace $\mathcal{A}^r$ is denoted by

$$\{A_r\}_{ray} := \{tA_r \mid t \in I\!\!R, \ t \neq 0\}. \tag{16}$$

Evidently, every r-blade in $\{A_r\}_{ray}$ is a representant of the subspace $\mathcal{A}^r$. With these definitions, the problem of finding the meet and join is reduced to the problem of finding the corresponding *meet* and *join* of the $(r+1)$- and $(s+1)$-blades in the geometric algebra $\mathcal{G}(I\!\!R^{n+1})$ which represent these subspaces.

Let A_r, B_s and C_t be non-zero blades representing the three subspaces $\mathcal{A}^r$, $\mathcal{B}^s$ and $\mathcal{C}^t$, respectively. Following [15], we say that

DEFINITION 3 *The t-blade $C_t = A_r \cap B_s$ is the meet of A_r and B_s if there exists a complementary $(r-t)$-blade A_c and a complementary $(s-t)$-blade B_c with the property that $A_r = A_c \wedge C_t$, $B_s = C_t \wedge B_c$, and $A_c \wedge B_c \neq 0$.*

It is important to note that the t-blade $C_t \in \{C_t\}_{ray}$ is not unique and is defined only up to a non-zero scalar factor, which we choose at our own convenience. The existence of the t-blade C_t (and the corresponding complementary blades A_c and B_c) is an expression of the basic relationships that exists between linear subspaces.

DEFINITION 4 *The $(r+s-t)$-blade $D = A_r \cup B_s$, called the join of A_r and B_s is defined by $D = A_r \cup B_s = A_r \wedge B_c$.*

Alternatively, since the join $A_r \cup B_s$ is defined only up to a non-zero scalar factor, we could equally well define D by $D = A_c \wedge B_s$. We use the symbols $\cap$ *intersection* and $\cup$ *union* from set theory to mark this unusual state of affairs. The problem of "meet" and "join" has thus been solved by finding the direct sum and intersection of linear subspaces and their $(r+s-t)$-blade and t-blade representants.

Note that it is only in the special case when $A_r \cap B_s = 0$ that the join can be considered to *reduce* to the outer product. That is

$$A_r \cap B_s = 0 \quad \Leftrightarrow \quad A_r \cup B_s = A_r \wedge B_s.$$

In any case, once the join $J := A_r \cup B_s$ has been found, it can be used to find the meet

$$A_r \cap B_s = A_r \cdot [B_s \cdot J] = [JJA_r] \cdot [B_s \cdot J] = [(A_r \cdot J) \wedge (B_s \cdot J)] \cdot J \tag{17}$$

In the case that $J = I^{-1}$, we can express this last relationship in terms of the operation of duality defined in (10), $A_r \cap B_s = (A_r^* \wedge B_s^*)^* =$

$(A_r^* \cup B_s^*)^*$ which is DeMorgan's formula. It must always be remembered that the "equalities" in these formulas only mean "up to a non-zero real number". While the positive definite metric of $I\!\!R^{n+1}$ is irrelevant to the definition of the meet and join of subspaces, the formula (17) holds only in $I\!\!R^{n+1}$.

A slightly modified version of this formula will hold in any non-degenerate pseudo-euclidean space $I\!\!R^{p,q}$ and its corresponding geometric algebra $\mathcal{G}_{p,q} := \mathcal{G}(I\!\!R^{p,q})$, where $p + q = n + 1$. In this case, after we have found the join $J = A_r \cup B_s$, we first find any blade $\overline{J}$ of the same step which satisfies the property that $\overline{J} \cdot J = 1$. The blade $\overline{J}$ is called a *reciprocal blade* of the blade J in the geometric algebra $\mathcal{G}_{p,q}$. The meet $A_r \cap B_s$ may then be defined by

$$A_r \cap B_s = A_r \cdot [B_s \cdot \overline{J}] = [(A_r \cdot \overline{J}) \cdot J] \cdot [B_s \cdot \overline{J}] = \{[(A_r \cdot \overline{J}) \wedge (B_s \cdot \overline{J})]\} \cdot J \quad (18)$$

The meet and join operations formulated in geometric algebra can be used to efficiently prove the many famous theorems of projective geometry [9]. See also *Geometric-Affine-Projective Computing* at the website [12].

2.2   Incidence, Projectivity and Colineation

Let $J \in \mathcal{G}_{p,q}^{k+1}$ be a $(k+1)$-blade representing a projective k-dimensional subplane in Π^n where $k \le n$. A point (ray) $x \in \Pi^n$ is said to be *incident* to J if and only if $x \wedge J = 0$. Since J is a $(k + 1)$-blade, we can find vectors $a_1, \ldots, a_{k+1} \in I\!\!R^{p,q}$ such that $J = a_1 \wedge \cdots \wedge a_{k+1}$. Projectively speaking, this means we can find $k + 1$ non-co$(k - 1)$planar points a_i in the k-projective plane J.

Now let $\overline{J}$ be a reciprocal blade to J with the property that $J \cdot \overline{J} = 1$. With the help of $\overline{J}$, we can define a *determinant function* or *bracket* $[\cdots]_{\overline{J}}$ on the projective k-plane J. Let $b_1, \ldots, b_{k+1}$ be $(k + 1)$ points incident to J,

$$[b_1, \cdots, b_{k+1}]_{\overline{J}} := (b_1 \wedge \cdots \wedge b_{k+1}) \cdot \overline{J}. \quad (19)$$

The bracket $[b_1, \cdots, b_{k+1}]_{\overline{J}} \neq 0$ iff the points b_i are not co-$(k - 1)$planar.

We now give the definitions necessary to complete the translation of real projective geometry into the language of multilinear algebra as formulated in geometric algebra.

DEFINITION 5 *A central perspectivity is a transformation of the points of a line onto the points of a line for which each pair of corresponding points is collinear with a fixed point called the center of perspectivity. See Figure 4.*

The key idea in the analytic expression of projective geometry in geometric algebra is that to each projectivity in Π^n there corresponds a non-singular linear transformation[2] $T : \mathbb{R}^{p,q} \longrightarrow \mathbb{R}^{p,q}$. It is clear that each projectivity of *points* on Π^n induces a corresponding *projective collineation* of lines, of planes, and higher dimensional projective k-planes. The corresponding extension of the linear transformation T from $\mathbb{R}^{p,q}$ to the whole geometric algebra $\mathcal{G}_{p,q}$ which accomplishes this is called the *outermorphism* $\mathbf{T} : \mathcal{G}_{p,q} \longrightarrow \mathcal{G}_{p,q}$, which is defined in terms of T by the properties:

$$\mathbf{T}(1) := 1, \quad \mathbf{T}(x) = T(x), \quad \mathbf{T}(x_1 \wedge \cdots \wedge x_k) := T(x_1) \wedge \cdots \wedge T(x_k) \quad (20)$$

for each $2 \leq k \leq p + q$, and then extended linearly to all elements of $\mathcal{G}_{p,q}$. Outermorphisms in geometric algebra, first studied in [16], provide the backbone for the application of geometric algebra to linear algebra. Since in everything that follows we will be using the outermorphism $\mathbf{T}$ defined by T, we will drop the boldface notation and simply use the same symbol T for both the linear transformation and its extension to an outermorphism $\mathbf{T}$.

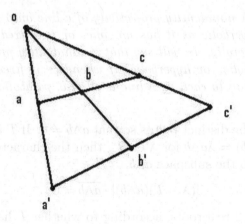

Figure 4. A central perspectivity from the point **o**.

---

[2]Unique up to a non-zero scalar factor.

DEFINITION 6 *A projective transformation or projectivity is a transformation of points of a line onto the points of a line which may be expressed as a finite product of central perspectivities.*

We can now easily prove

THEOREM 1 *There is a one-one correspondence between non-singular outermorphisms* $T : \mathcal{G}_{p,q} \longrightarrow \mathcal{G}_{p,q}$, *and projective collineations on* Π^n *taking* $n + 1$ *non-co(n − 1)planar points in* Π^n *into* $n + 1$ *non-co(n-1)planar points in* Π^n.

Proof: Let $a_1, \ldots, a_{n+1} \in \Pi^n$ be $n+1$ non-co(n-1)planar points. Since they are non-co(n-1)planar, it follows that $a_1 \wedge \cdots \wedge a_{n+1} \neq 0$. Suppose that $b_i = T(a_i)$ is a projective transformation between these points for $1 \leq i \leq n + 1$. The corresponding non-singular outermorphism is defined by considering T to be a linear transformation on the basis vectors $a_1, \ldots, a_{n+1}$ of $I\!\!R^{p,q}$. Conversely, if a non-singular outermorphism is specified on $I\!\!R^{p,q}$ it clearly defines a unique projective collineation on Π^n, which we denote by the same symbol T.

All of the theorems on harmonic points and cross ratios of points on a projective line follow easily from the above definitions and properties [9], but we will not prove them here. For what follows, we will need two more definitions:

DEFINITION 7 *A nonidentity projectivity of a line onto itself is elliptic, parabolic, or hyperbolic as it has no, one, or two fixed points, respectively. More generally, we will say that a nonidentity projectivity of* Π^n *is elliptic, parabolic, or hyperbolic, if whenever it fixes a line in* Π^n, *then the restriction to each such line is elliptic, parabolic, or hyperbolic, respectively.*

Let $a, b \in \Pi^n$ be distinct points so that $a \wedge b \neq 0$. If T is a projectivity of Π^n and $T(a \wedge b) = \lambda a \wedge b$ for $\lambda \in I\!\!R^*$, then the characteristic equation of T restricted to the subspace $a \wedge b$,

$$[(\lambda - T)(a \wedge b)] \cdot \overline{a \wedge b} = 0 \qquad (21)$$

will have 0, 1 or 2 real roots, according to whether T has 0, 1 or 2 real eigenvectors, [15], [8, p.73], which correspond directly to fixed points.

DEFINITION 8 *A nonidentity projective transformation T of a line onto itself is an involution if $T^2 = $ identity.*

2.3   Conics and Polars

Let $a_1, a_2, \ldots, a_{n+1} \in I\!\!R^{p,q}$ represent $n + 1 = p + q$ linearly independent vectors in $I\!\!R^{p,q}$. This means that $I = a_1 \wedge a_2 \wedge \cdots \wedge a_{n+1} \neq 0$. As an element in the projective space Π^n, I represents the projective n-plane determined by the $n + 1$ non-co$(n - 1)$planar points $a_1, \ldots, a_{n+1}$. Representing the points of Π^n by homogeneous vectors $x \in I\!R^{p,q}$, makes it easy to study the *quadric* hypersurface (conic) Q in Π^n defined by

$$Q := \{x| \ x \in I\!R^{p+q}, x \neq 0, \text{ and } x^2 = 0\}. \tag{22}$$

DEFINITION 9 *The polar of the k-blade $A \in \mathcal{G}_{p,q}^k$ is the $(n + 1 - k)$-blade $Pol_Q(A)$ defined by*

$$Pol_Q(A) := AI^{-1} = A^* \tag{23}$$

where A^* is the dual of A in the geometric algebra $\mathcal{G}_{p,q}$.

The above definition shows that *polarization* in the quadric hypersurface Q and *dualization* in the geometric algebra $\mathcal{G}_{p,q}$ are identical operations.

If $x \in Q$, it follows that

$$x \wedge Pol_Q(x) = x \wedge (xI^{-1}) = x^2 I^{-1} = 0. \tag{24}$$

This tells us that $Pol_Q(x)$ is the hyperplane which is tangent to Q at the point x, [12]. We will meet something very similar when we discuss the *horosphere* in section 5.

3. Affine and other geometries

In this section, we explore in what sense "projective geometry is all of geometry" as exclaimed by Cayley. In order to keep the discussion as simple as possible, we will discuss the relationship of the 2 dimensional projective plane to other 2 dimensional planar geometries, [6].

We begin with affine geometry of the plane. Let Π^2 be the real projective plane, and $\mathcal{T}(\Pi^2)$ the group of all projective transformations on Π^2. Let a line $\mathcal{L} \in \Pi^2$ (a degenerate conic) be picked out as the *absolute line*, or the *line at infinity*.

DEFINITION 10 *The affine plane $\mathcal{A}^2$ consists of all points of the projective plane Π^2 with the points on the absolute line deleted. The projective transformations leaving $\mathcal{L}$ fixed, restricted to the real affine plane, are real affine transformations. The study of $\mathcal{A}^2$ and the subgroup of real affine transformations $\mathcal{T}\{\mathcal{A}^2\}$ is real plane affine geometry.*

If we now fix an elliptic involution, called the *absolute involution*, on the line $\mathcal{L}$, then a real affine transformation which leaves the absolute involution invariant is called a *similarity* transformation.

DEFINITION 11 *The study of the group of similarity transformations on the affine plane is similarity geometry.*

An affine transformation which leaves the *area* of all triangles invariant is called an equiareal transformation.

DEFINITION 12 *The study of the affine plane under equiareal transformations is equiareal geometry.*

A *euclidean transformation* on the affine plane is an affine transformation which is both a similarity and an equiareal transformation. Finally, we have

DEFINITION 13 *The study of the affine plane under euclidean transformations is euclidean geometry.*

In our representation of Π^n in a geometric algebra $\mathcal{G}_{p,q}$ where $n+1 = p+q$, a projectivity is represented by a non-singular linear transformation. Thus, the group of projectivities becomes just the *general linear group* of all non-singular transformations on $I\!R^{p,q}$ extended to outermorphisms on $\mathcal{G}_{p,q}$. Generally, we may choose to work in a euclidean space $I\!R^{n+1}$, rather than in the pseudo-euclidean space $I\!R^{p,q}$, [9], [12]. We have chosen here to work in the more general geometric algebra $\mathcal{G}_{p,q}$, because of the more direct connection to the study of a particular nondegenerate quadric hypersurface or conic in $I\!R^{p,q}$.

We still have not mentioned two important classes of non-euclidean plane geometries, *hyperbolic plane geometry* and *elliptic plane geometry* , and their relation to projective geometry. Unlike euclidean geometry, where we picked out a degenerate conic called the absolute line or line at infinity, the study of these other types of plane geometries involves the picking out of a nondegenerate conic called the *absolute conic* .

For plane hyperbolic geometry, we pick out a real non-degenerate conic in Π^2, called the *absolute conic*, and define the points interior to the absolute to be *ordinary* , those points on the absolute are *ideal* , and those points exterior to the absolute are *ultraideal* [6, p.230].

DEFINITION 14 *A real projective plane from which the absolute conic and its exterior have been deleted is a hyperbolic plane. The projective collineations leaving the absolute fixed and carrying interior points onto interior points, restricted to the hyperbolic plane, are hyperbolic isometries. The study of the hyperboic plane and hyperbolic isometries is hyperbolic geometry.*

Hyperbolic geometry has been studied extensively in geometric algebra by Hongbo Li in [3, p.61-85], and applied to automatic theorem proving in [3, p.110-119], [5, p.69-90].

For plane elliptic geometry, we pick out an *imaginary* nondegenerate conic in Π^2 as the *absolute conic*. Since there are no real points on this conic, the points of elliptic geometry are the same as the points in the real projective plane Π^2. A projective collineation which leaves the absolute conic fixed (whose points are in the complex projective plane) is called an *elliptic isometry*.

DEFINITION 15 *The real projective plane Π^2 is the elliptic plane. The study of the elliptic plane and elliptic isometries is elliptic geometry.*

In the next section, we return to the study of affine geometries of higher dimensional pseudo-euclidean spaces. However, we shall not study the formal properties of these spaces. Rather, our objective is to efficiently define the *horosphere* of a pseudo-euclidean space, and study some of its properties.

4. Affine Geometry of pseudo-euclidean space

We have seen that a projective space can be considered to be an affine space with idealized points at infinity [18]. Since all the formulas for meet and join remain valid in the pseudo-euclidean space $I\!R^{p,q}$, using (18), we define the $n = (p+q)$-dimensional affine plane $\mathcal{A}_e(I\!R^{p,q})$ of the null vector $e = \frac{1}{2}(\sigma+\eta)$ in the larger pseudo-euclidean space $I\!R^{p+1,q+1} = I\!R^{p,q}\oplus I\!R^{1,1}$, where $I\!R^{1,1} = span\{\sigma,\eta\}$ for $\sigma^2 = 1 = -\eta^2$. Whereas, effectively, we are only extending the euclidean space $I\!R^{p,q}$ by the null vector e, it is advantageous to work in the geometric algebra $\mathcal{G}_{p+1,q+1}$ of the *non-degenerate* pseudo-euclidean space $I\!R^{p+1,q+1}$. We give here the important properties of the reciprocal null vectors $e = \frac{1}{2}(\sigma + \eta)$ and $\bar{e} = \sigma - \eta$ that will be needed later, and their relationship to the *hyperbolic unit bivector* $u := \sigma\eta$.

$$e^2 = \bar{e}^2 = 0, \quad e\cdot\bar{e} = 1, \quad u = \bar{e}\wedge e = \sigma\wedge\eta, \quad u^2 = 1. \tag{25}$$

The affine plane $\mathcal{A}_e^{p,q} := \mathcal{A}_e(I\!R^{p,q})$ is defined by

$$\mathcal{A}_e(I\!R^{p,q}) = \{x_h = x + e \mid x \in I\!R^{p,q}\} \subset I\!R^{p+1,q+1}, \tag{26}$$

for the null vector $e \in I\!R^{1,1}$. The affine plane $\mathcal{A}_e(I\!R^{p,q})$ has the nice property that $x_h^2 = x^2$ for all $x_h \in \mathcal{A}_e(I\!R^{p,q})$, thus preserving the metric structure of $I\!R^{p,q}$. We can restate definition (26) of $\mathcal{A}_e(I\!R^{p,q})$ in the form

$$\mathcal{A}_e(I\!R^{p,q}) = \{y \mid y \in I\!R^{p+1,q+1}, \quad y\cdot\bar{e} = 1 \quad \text{and} \quad y\cdot e = 0\} \subset I\!R^{p+1,q+1}.$$

This form of the definition is interesting because it brings us closer to the definition of the $n = (p + q)$-dimensional *projective plane*.

The projective n-plane Π^n can be defined to be the set of all points of the affine plane $\mathcal{A}_e(I\!R^{p,q})$, taken together with idealized points at infinity. Each point $x_h \in \mathcal{A}_e(I\!\!R^{p,q})$ is called a *homogeneous representant* of the corresponding point in Π^n because it satisfies the property that $x_h \cdot \bar{e} = 1$. To bring these different viewpoints closer together, points in the affine plane $\mathcal{A}_e(I\!\!R^{p,q})$ will also be represented by *rays* in the space

$$\mathcal{A}_e^{rays}(I\!\!R^{p,q}) = \{\{y\}_{ray}|\ y \in I\!\!R^{p+1,q+1}, y \cdot e = 0,\ \ y \cdot \bar{e} \neq 0\ \} \subset I\!\!R^{p+1,q+1}.$$
$$(27)$$

The set of rays $\mathcal{A}_e^{rays}(I\!\!R^{p,q})$ gives another definition of the affine n-plane, because each ray $\{y\}_{ray} \in \mathcal{A}_e^{rays}(I\!\!R^{p,q})$ determines the unique homogeneous point

$$y_h = \frac{y}{y \cdot \bar{e}} \in \mathcal{A}_e(I\!\!R^{p,q}).$$

Conversely, each point $y \in \mathcal{A}_e(I\!\!R^{p,q})$ determines a unique ray $\{y\}_{ray}$ in $\mathcal{A}_e^{rays}(I\!\!R^{p,q})$. Thus, the affine plane of homogeneous points $\mathcal{A}_e(I\!\!R^{p,q})$ is equivalent to the affine plane of rays $\mathcal{A}_e^{rays}(I\!\!R^{p,q})$.

Suppose that we are given k-points $a_1^h, a_2^h, \ldots, a_k^h \in \mathcal{A}_e(I\!\!R^{p,q})$ where each $a_i^h = a_i + e$ for $a_i \in I\!\!R^{p,q}$. Taking the outer product or *join* of these points gives the projective $(k-1)$-plane $A^h \in \Pi^n$. Expanding the outer product gives

$$A^h = a_1^h \wedge a_2^h \wedge \ldots \wedge a_k^h = a_1^h \wedge (a_2^h - a_1^h) \wedge a_3^h \wedge \ldots \wedge a_k^h$$
$$= a_1^h \wedge (a_2^h - a_1^h) \wedge (a_3^h - a_2^h) \wedge a_4^h \wedge \ldots \wedge a_k^h = \ldots$$
$$= a_1^h \wedge (a_2 - a_1) \wedge (a_3 - a_2) \wedge \ldots \wedge (a_k - a_{k-1}),$$

or

$$A^h = a_1^h \wedge a_2^h \wedge \ldots \wedge a_k^h = a_1 \wedge a_2 \wedge \ldots \wedge a_k +$$
$$e \wedge (a_2 - a_1) \wedge (a_3 - a_2) \wedge \ldots \wedge (a_k - a_{k-1}). \qquad (28)$$

Whereas (28) represents a $(k-1)$-plane in Π^n, it also belongs to the affine (p,q)-plane $\mathcal{A}_e^{p,q}$, and thus contains important metrical information. Dotting this equation with $\bar{e}$, we find that

$$\bar{e} \cdot A^h = \bar{e} \cdot (a_1^h \wedge a_2^h \wedge \ldots \wedge a_k^h) = (a_2 - a_1) \wedge (a_3 - a_2) \wedge \ldots \wedge (a_k - a_{k-1}).$$

This result motivates the following

DEFINITION 16 *The directed content of the $(k-1)$-simplex*

$$A^h = a_1^h \wedge a_2^h \wedge \cdots \wedge a_k^h$$

in the affine (p,q)-plane is given by

$$\frac{\bar{e} \cdot A^h}{(k-1)!} = \frac{\bar{e} \cdot (a_1^h \wedge a_2^h \wedge \ldots \wedge a_k^h)}{(k-1)!}$$
$$= \frac{(a_2 - a_1) \wedge (a_3 - a_2) \wedge \ldots \wedge (a_k - a_{k-1})}{(k-1)!}.$$

4.1 Example

Many incidence relations can be expressed in the affine plane $\mathcal{A}_e(I\!\!R^{p,q})$ which are also valid in the projective plane Π^n, [3, pp.263]. We will only give here the simplest example.

Given are 4 coplanar points $a_h, b_h, c_h, d_h \in \mathcal{A}_e(I\!\!R^2)$. The join and meet of the lines $a_h \wedge b_h$ and $c_h \wedge d_h$ are given, respectively, by $(a_h \wedge b_h) \cup (c_h \wedge d_h) = a_h \wedge b_h \wedge c_h$, and using (18),

$$(a_h \wedge b_h) \cap (c_h \wedge d_h) = [\bar{I} \cdot (a_h \wedge b_h)] \cdot (c_h \wedge d_h)$$

where e_1, e_2 are the orthonormal basis vectors of $I\!\!R^2$, and $\bar{I} = e_2 \wedge e_1 \wedge \bar{e}$. Carrying out the calculations for the meet and join in terms of the bracket determinant (19), we find that

$$(a_h \wedge b_h) \cup (c_h \wedge d_h) = [a_h, b_h, c_h]_{\bar{I}} I = \det\{a, b\} I \qquad (29)$$

where $I = e_1 \wedge e_2 \wedge e$ and $\det\{a, b\} := (a \wedge b) \cdot (e_{21})$, and

$$(a_h \wedge b_h) \cap (c_h \wedge d_h) = \det\{c - d, b - c\} a_h + \det\{c - d, c - a\} b_h. \qquad (30)$$

Note that the meet (30) is not, in general, a homogeneous point. Normalizing (30), we find the homogeneous point $p_h \in \mathcal{A}_e(I\!\!R^2)$

$$p_h = \frac{\det\{c - d, b - c\} a_h + \det\{c - d, c - a\} b_h}{\det\{c - d, b - a\}}$$

which is the intersection of the lines $a_h \wedge b_h$ and $c_h \wedge d_h$. The meet can also be solved for directly in the affine plane by noting that

$$p_h = \alpha_p a_h + (1 - \alpha_p) b_h = \beta_p c_h + (1 - \beta_p) d_h$$

and solving to get $\alpha_p = [b_h, c_h, d_h]_{\bar{I}} / [b_h - a_h, c_h, d_h]_{\bar{I}}$. Other simple examples can be found in [15].

5. Conformal Geometry and the Horosphere

The conformal geometry of a pseudo-Euclidean space can be linearized by considering the *horosphere* in a pseudo-Euclidean space of two dimensions higher. We begin by defining the *horosphere* $\mathcal{H}_e^{p,q}$ in $I\!\!R^{p+1,q+1}$ by *moving up* from the affine plane $\mathcal{A}_e^{p,q} := \mathcal{A}_e(I\!\!R^{p,q})$.

5.1 The horosphere

Let $\mathcal{G}_{p+1,q+1} = gen(I\!\!R^{p+1,q+1})$ be the geometric algebra of $I\!\!R^{p+1,q+1}$, and recall the definition (26) of the affine plane $\mathcal{A}_e^{p,q} := \mathcal{A}_e(I\!\!R^{p,q}) \subset$

$I\!\!R^{p+1,q+1}$. Any point $y \in I\!\!R^{p+1,q+1}$ can be written in the form $y = x + \alpha e + \beta \bar{e}$, where $x \in I\!\!R^{p,q}$ and $\alpha, \beta \in I\!\!R$.

The *horosphere* $\mathcal{H}_e^{p,q}$ is most directly defined by

$$\mathcal{H}_e^{p,q} := \{x_c = x_h + \beta \bar{e} \mid x_h \in \mathcal{A}_e^{p,q} \text{ and } x_c^2 = 0.\} \tag{31}$$

With the help of (25), the condition that

$$x_c^2 = (x_h + \beta \bar{e})^2 = x^2 + 2\beta = 0$$

gives us immediately that $\beta := -\frac{x^2}{2}$. Thus each point $x_c \in \mathcal{H}_e^{p,q}$ has the form

$$x_c = x_h - \frac{x_h^2}{2}\bar{e} = x + e - \frac{x^2}{2}\bar{e} = \frac{1}{2}x_h\bar{e}x_h. \tag{32}$$

The last equality on the right follows from

$$\frac{1}{2}x_h\bar{e}x_h = \frac{1}{2}[(x_h\cdot\bar{e})x_h + (x_h\wedge\bar{e})x_h] = x_h - \frac{1}{2}x_h^2\bar{e}.$$

From (32), we easily calculate

$$x_c \cdot y_c = (x + e - \frac{x^2}{2}\bar{e}) \cdot (y + e - \frac{y^2}{2}\bar{e}) =$$

$$x \cdot y - \frac{y^2}{2} - \frac{x^2}{2} = -\frac{1}{2}(x - y)^2,$$

where $(x - y)^2$ is the square of the pseudo-euclidean distance between the conformal representatives x_c and y_c. We see that the pseudo-euclidean structure is preserved in the form of the inner product $x_c \cdot y_c$ on the horosphere.

Just as $x_h \in \mathcal{A}_e^{p,q}$ is called the *homogeneous representant* of $x \in I\!\!R^{p,q}$, the point x_c is called the *conformal representant* of *both* the points $x_h \in \mathcal{A}_e^{p,q}$ and $x \in I\!\!R^{p,q}$. The set of all conformal representatives $\mathcal{H}^{p,q} := c(I\!\!R^{p,q})$ is called the *horosphere* . The horosphere $\mathcal{H}^{p,q}$ is a non-linear model of both the affine plane $\mathcal{A}_e^{p,q}$ and the pseudo-euclidean space $I\!\!R^{p,q}$. The horosphere $\mathcal{H}^n$ for the Euclidean space $I\!\!R^n$ was first introduced by F.A. Wachter, a student of Gauss, [7], and has been recently finding many diverse applications [3], [5].

The set of all null vectors $y \in I\!\!R^{p+1,q+1}$ make up the *null cone*

$$\mathcal{N} := \{y \in I\!\!R^{p+1,q+1} \mid y^2 = 0\}.$$

The subset of $\mathcal{N}$ containing all the representants $y \in \{x_c\}_{ray}$ for any $x \in I\!\!R^{p,q}$ is defined to be the set

$$\mathcal{N}_0 = \{y \in \mathcal{N} \mid y\cdot\bar{e} \neq 0\} = \cup_{x \in I\!\!R^{p,q}} \{x_c\}_{ray},$$

and is called the *restricted null cone*. The conformal representant of a null ray $\{z\}_{ray}$ is the representant $y \in \{z\}_{ray}$ which satisfies $y \cdot \bar{e} = 1$.

The horosphere $\mathcal{H}^{p,q}$ is the parabolic section of the restricted null cone,

$$\mathcal{H}^{p,q} = \{y \in \mathcal{N}_0 \mid y \cdot \bar{e} = 1\},$$

see Figure 5. Thus $\mathcal{H}^{p,q}$ has dimension $n = p + q$. The null cone $\mathcal{N}$ is determined by the condition $y^2 = 0$, which taking differentials gives

$$y \cdot dy = 0 \quad \Rightarrow \quad x_c \cdot dy = 0 , \tag{33}$$

where $\{y\}_{ray} = \{x_c\}_{ray}$. Since $\mathcal{N}_0$ is an $(n + 1)$-dimensional surface, then (33) is a condition necessary and sufficient for a vector v to belong to the tangent space to the restricted null cone $\mathcal{T}(\mathcal{N}_0)$ at the point y

$$v \in \mathcal{T}(\mathcal{N}_0) \quad \Leftrightarrow \quad x_c \cdot v = 0 . \tag{34}$$

It follows that the $(n + 1)$-pseudoscalar I_y of the tangent space to $\mathcal{N}_0$ at the point y can be defined by $I_y = I x_c$ where I is the pseudoscalar of $\mathbb{R}^{p+1,q+1}$. We have

$$x_c \cdot v = 0 \quad \Leftrightarrow \quad 0 = I(x_c \cdot v) = (I x_c) \wedge v = I_y \wedge v, \tag{35}$$

a relationship that we have already met in (24).

5.2   H-twistors

Let us define an *h-twistor* to be a rotor $S_x \in \mathrm{Spin}_{p+1,q+1}$

$$S_x := 1 + \frac{1}{2} x \bar{e} = \exp\left(\frac{1}{2} x \bar{e}\right). \tag{36}$$

An h-twistor is an equivalence class of two "twistor" components from $\mathcal{G}_{p,q}$, that have many twistor-like properties. The point x_c is generated from $0_c = e$ by

$$x_c = S_x e S_x{}^\dagger, \tag{37}$$

and the tangent space to the horosphere at the point x_c is generated from $dx \in \mathbb{R}^{p,q}$ by

$$dx_c = dS_x \, e \, S_x{}^\dagger + S_x \, e \, dS_x{}^\dagger = S_x (\Omega_S \cdot e) S_x{}^\dagger = S_x dx S_x{}^\dagger. \tag{38}$$

It also keeps unchanged the "point at infinity" $\bar{e}$

$$\bar{e} = S_x \bar{e} S_x{}^\dagger.$$

H-twistors were defined and studied in [15], and more details can be found therein.

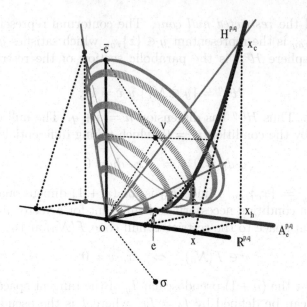

Figure 5.   The restricted null cone and representations of the point x in affine space and on the horosphere.

Since the group of isometries in $\mathcal{N}_0$ is a double covering of the group of conformal transformations $Con_{p,q}$ in $I\!\!R^{p,q}$, and the group $Pin_{p+1,q+1}$ is a double covering of the group of orthogonal transformations $O(p+1, q+1)$, it follows that $Pin_{p+1,q+1}$ is a four-fold covering of $Con_{p,q}$, [10, p.220], [14, p.146].

5.3   Matrix representation

We have seen in (14) that the algebra $\mathcal{G}_{p+1,q+1} = \mathcal{G}_{p,q} \otimes \mathcal{G}_{1,1}$ is isomorphic to a 2×2 matrix algebra over the module $\mathcal{G}_{p,q}$. This identification makes possible a very elegant treatment of the so-called *Vahlen matrices* [10, 11, 4, 14].

Recall in section 1.4, that the idempotents $u_\pm = \frac{1}{2}(1 \pm u)$ of the algebra $\mathcal{G}_{1,1}$ satisfy the properties

$$u_+ + u_- = 1 \; , \;\; u_+ - u_- = u \; , \;\; u_+ u_- = 0 = u_- u_+ \; , \;\; \sigma u_+ = u_- \sigma \; ,$$

where

$$u := \bar{e} \wedge e \; , \;\;\; u_+ = \frac{1}{2}\bar{e}e \; , \;\;\; u_- = \frac{1}{2}e\bar{e} \; ,$$

and

$$u\bar{e} = \bar{e} = -\bar{e}u \; , \;\;\; eu = e = -ue \; , \;\;\; \sigma u_+ = e \; , \;\;\; 2\sigma u_- = \bar{e} \; .$$

Each multivector $G \in \mathcal{G}_{p+1,q+1}$ can be written in the form

$$G = \begin{pmatrix} 1 & \sigma \end{pmatrix} u_+ [G] \begin{pmatrix} 1 \\ \sigma \end{pmatrix} = A u_+ + B u_+ \sigma + C^* u_- \sigma + D^* u_- \qquad (39)$$

where

$$[G] \equiv \begin{pmatrix} A & B \\ C & D \end{pmatrix} \quad \text{for} \quad A, B, C, D \in \mathcal{G}_{p,q}.$$

The matrix $[G]$ denotes the matrix corresponding to the multivector G.

The operation of *reversion* of multivectors translates into the following transpose-like matrix operation:

$$\text{if} \quad [G] = \begin{pmatrix} A & B \\ C & D \end{pmatrix} \quad \text{then} \quad [G]^\dagger := [G^\dagger] = \begin{pmatrix} \overline{D} & \overline{B} \\ \overline{C} & \overline{A} \end{pmatrix}$$

where $\overline{A} = A^{*\dagger}$ is the *Clifford conjugation*, [15].

5.4 Möbius transformations

We have seen in (37) that the point $x_c \in \mathcal{H}_{p,q}$ can be written in the form, $x_c = S_x e S_x^\dagger$. More generally, any conformal transformation $f(x)$ can be represented on the horosphere by

$$f(x)_c = S_{f(x)} e \, S_{f(x)}^\dagger. \qquad (40)$$

Using the matrix representation (39), for a general multivector $G \in \mathcal{G}_{p+1,q+1}$ we find that

$$[GeG^\dagger] = \begin{pmatrix} A & B \\ C & D \end{pmatrix} \begin{pmatrix} 0 & 0 \\ 1 & 0 \end{pmatrix} \begin{pmatrix} \overline{D} & \overline{B} \\ \overline{C} & \overline{A} \end{pmatrix}$$

$$= \begin{pmatrix} B \\ D \end{pmatrix} \begin{pmatrix} \overline{D} & \overline{B} \end{pmatrix} \qquad (41)$$

where

$$[e] = \begin{pmatrix} 0 & 0 \\ 1 & 0 \end{pmatrix}, \quad [G] \equiv \begin{pmatrix} A & B \\ C & D \end{pmatrix}, \quad [G]^\dagger = \begin{pmatrix} \overline{D} & \overline{B} \\ \overline{C} & \overline{A} \end{pmatrix}.$$

The relationship (41) suggests defining the *conformal h-twistor* of the multivector $G \in \mathcal{G}_{p+1,q+1}$ to be

$$[G]_c := \begin{pmatrix} B \\ D \end{pmatrix},$$

which may also be identified with the multivector $G_c := Ge = Bu_+ + D^*e$. The *conjugate* of the conformal h-twistor is then naturally defined by

$$[G]_c^\dagger := (\overline{D} \quad \overline{B}) .$$

Conformal h-twistors give us a powerful tool for manipulating the conformal representant and conformal transformations much more efficiently. For example, since x_c in (37) is generated by the conformal h-twistor $[S_x]_c$, it follows that

$$[x_c] = [S_x]_c[S_x]_c^\dagger = \begin{pmatrix} x \\ 1 \end{pmatrix} \begin{pmatrix} 1 & -x \end{pmatrix} = \begin{pmatrix} x & -x^2 \\ 1 & -x \end{pmatrix} .$$

We can now write the conformal transformation (40) in its spinorial form,

$$[S_{f(x)}]_c = \begin{pmatrix} f(x) \\ 1 \end{pmatrix} .$$

Since $T_x = RS_x$ for the constant vector $R \in Pin_{p+1,q+1}$, its spinorial form is given by

$$[T_x]_c = [R][S_x]_c = \begin{pmatrix} A & B \\ C & D \end{pmatrix} \begin{pmatrix} x \\ 1 \end{pmatrix} = \begin{pmatrix} Ax + B \\ Cx + D \end{pmatrix} = \begin{pmatrix} M \\ N \end{pmatrix} ,$$

where

$$[R] = \begin{pmatrix} A & B \\ C & D \end{pmatrix} , \quad \text{for constants} \quad A, B, C, D \in \mathcal{G}_{p,q}.$$

It follows that

$$[T_x] = \begin{pmatrix} M \\ N \end{pmatrix} = \begin{pmatrix} f(x) \\ 1 \end{pmatrix} H \quad \Rightarrow \quad H = N \quad \text{and} \quad f(x) = MN^{-1}. \quad (42)$$

The beautiful linear fractional expression for the conformal transformation $f(x)$,

$$f(x) = (Ax + B)(Cx + D)^{-1} \quad (43)$$

is a direct consequence of (42), [15].

The linear fractional expression (43) extends to any dimension and signature the well-known Möbius transformations in the complex plane. The components A, B, C, D of $[R]$ are subject to the condition that $R \in Pin_{p+1,q+1}$. Conformal h-twistors are a generalization to any dimension and any signature of the familiar 2-component spinors over the complex numbers, and the 4-component twistors. Penrose's twistor theory [13] has been discussed in the framework of Clifford algebra by a number of authors, for example see [1], [2, pp75-92].

References

[1] R. Ablamowicz, and N. Salingaros, *On the Relationship Between Twistors and Clifford Algebras*, Letters in Mathematical Physics, 9, 149-155, 1985.

[2] *Clifford Algebras and their Applications in Mathematical Physics*, Editors: A. Ablamowicz, and B. Fauser, Birkäuser, Boston (2000).

[3] E. Bayro Corrochano, and G. Sobczyk, Editors, Geometric Algebra with Applications in Science and Engineering, Birkhäuser, Boston 2001.

[4] J. Cnops (1996), *Vahlen Matrices for Non-definite Metrics*, in Clifford Algebras with Numeric and Symbolic Computations, Editors: R. Ablamowicz, P. Lounesto, J.M. Parra, Birkhäuser, Boston.

[5] L. Dorst, C. Doran, J. Lasenby, Editors, *Applications of Geometric Algebra in Computer Science and Engineering*, Birkhäuser, Boston 2002.

[6] W.T. Fishback, *Projective and Euclidean Geometry, 2nd*, John Wiley & Sons, Inc., New York, 1969.

[7] T. Havel, *Geometric Algebra and Möbius Sphere Geometry as a Basis for Euclidean Invariant Theory*, Editor: N.L. White, Invariant Methods in Discrete and Computational Geometry: 245–256, Kluwer 1995.

[8] D. Hestenes and G. Sobczyk, *Clifford Algebra to Geometric Calculus: A Unified Language for Mathematics and Physics*, D. Reidel, Dordrecht, 1984, 1987.

[9] D. Hestenes, and R. Ziegler (1991), *Projective geometry with Clifford algebra*, Acta Applicandae Mathematicae, 23: 25–63.

[10] P. Lounesto, *Clifford Algebras and Spinors*, Cambridge University Press, Cambridge, 1997, 2001.

[11] J. G. Maks, *Modulo (1,1) Periodicity of Clifford Algebras and Generalized Möbius Transformations*, Technical University of Delft (Dissertation), 1989.

[12] A. Naeve, *Geo-Metric-Affine-Projective Computing*, http://kmr.nada.kth.se/papers/gacourse.html, 2003.

[13] R. Penrose and M.A.H. MacCallum, *Twistor Theory: an Approach to Quantization of Fields and Space-time*, Phys. Rep., 6C, 241-316, 1972.

[14] I. R. Porteous, *Clifford Algebras and the Classical Groups*, Cambridge University Press, 1995.

[15] J.M. Pozo and G. Sobczyk, *Geometric Algebra in Linear Algebra and Geometry*, Acta Applicandae Mathematicae 71: 207-244, 2002.

[16] G. Sobczyk, *Mappings of Surfaces in Euclidean Space Using Geometric Algebra*, Arizona State University (Dissertation), 1971.

[17] G. Sobczyk, *The Missing Spectral Basis in Algebra and Number Theory*, The American Mathematical Monthly, **108**, No. 4, (2001) 336-346.

[18] J. W. Young, *Projective Geometry* The Open Court Publishing Company, Chicago, IL, 1930.

CONTENT-BASED INFORMATION RETRIEVAL BY GROUP THEORETICAL METHODS

Michael Clausen
Department of Computer Science III
University of Bonn, Germany[*]
clausen@cs.uni-bonn.de

Frank Kurth
Department of Computer Science III
University of Bonn, Germany
frank@cs.uni-bonn.de

Abstract   This paper presents a general framework for efficiently indexing and searching large collections of multimedia documents by content. Among the multimedia information retrieval scenarios that fit into this framework are music, audio, image and 3D object retrieval. Combining the technique of inverted files with methods from group theory we obtain space efficient indexing structures as well as time efficient search procedures for content-based and fault-tolerant search in multimedia data. Several prototypic applications are discussed demonstrating the capabilities of our new technique.

Keywords: multimedia retrieval, content-based retrieval, music and audio retrieval.

1.   Introduction

The last few years have seen an increasing importance of multimedia databases for a wide range of applications. As one major reason, the availability of affordable high-performance hardware now allows for efficient processing and storage of the huge amounts of data which arise, e.g., in video, image, or audio applications. A key philosophy to accessing data in multimedia databases is *content-based retrieval* ([Yoshitaka

---

[*]This work was supported in part by Deutsche Forschungsgemeinschaft under grant CL 64/3

J. Byrnes (ed.) Computational Noncommutative Algebra and Applications, 29-55.
© 2004 *Kluwer Academic Publishers. Printed in the Netherlands.*

and Ichikawa, 1999]), where the content of the multimedia documents is processed rather than just some textual annotation describing the documents. Hence, a content-based query to an image database asking for all images showing a certain person would basically rely on a suitable feature extraction mechanism to scan images for occurrences of that person. On the other hand, a classical query based on additional textual information would rely on the existence of a suitable textual annotation of the contents of all images. Unfortunately, in most cases such an annotation is neither available nor may it be easily extracted automatically, emphasizing the demand for feasible content-based retrieval methods.

It turns out that many content-based retrieval problems share essential structural properties. We briefly sketch two of those problems.

Let us first consider a problem from music information retrieval. Assume that a music database consists of a collection of scores. That is, each database document is a score representation of a piece of music containing the notes of that piece as well as additional information such as meter or tempo. Now we consider the following database search problem: Given a melody or, more generally, an arbitrary excerpt of a piece of music, we are looking for all occurrences of that *query* or slight variations thereof in the database documents. The result of such a query could for example help music professionals to discover plagiarism. A simpler, yet very active application area is to name a tune that is whistled or hummed into a microphone. As a query result, a user could expect information on title, composer, and consumer information on an audio CD containing the corresponding piece of music.

The second problem is concerned with content-based image retrieval. Consider a database consisting of digital copyrighted images. Assume that we are interested, e.g., for some legal reasons, in finding all web pages on the Internet containing at least one of the copyrighted images or fragments thereof. This problem may be again considered as a database search problem: Given a (query) image taken from some web page, we are looking for an occurrence of that image as a subimage of one of the database images, including the case that the query matches one of those images as a whole. Extensions to this problem include that we are also interested in finding rotated, resized, or lower-quality versions of the original images.

Both of the above problems may be viewed in the following general setting: A query to a database consisting of a collection of multimedia documents has to be answered in the sense that certain transformations (or generalized shift operations) have to be found which transport the query to its location within the database document. In this tutorial paper, we systematically exploit this principle for the case that admissible transformations are taken from a *group* acting on a *set* which in turn constitute the database documents. It turns out that this approach leads to very efficient search algorithms for a large class of content-based search problems, which are in particular applicable to spatial-, temporal-, or spatio-temporal retrieval settings ([Yoshitaka and Ichikawa, 1999]).

We briefly summarize the main contributions of our approach:

- We develop a general framework for retrieval of multimedia documents by example. Our technique's flexibility has been demonstrated by prototypes in various fields (e.g., music, audio, image, and (relational) object retrieval).

- We propose generic algorithms for query evaluation together with efficient algorithms for fault-tolerant retrieval which consequently exploit the structure inherent in the retrieval problems.

- In contrast to previously reported approaches, query evaluation becomes *more efficient* when the complexity of a query increases.

- The concept of partial matches (i.e., a query is only matched to a part of a document) is an integral part of our technique and requires no additional storage.

The proposed technique has been successfully tested on a variety of content-based retrieval problems. We summarize some figures on those prototypes:

- Our PROMS system, for the first time, allowed for efficient polyphonic search in polyphonic scores ([Clausen et al., 2000]). E.g., queries to a database of 12,000 pieces of music containing 33 million notes can be answered in about 50 milliseconds.

- Our system for searching large databases of audio signals allows for both identifying *and* precisely locating short fragments of audio signals w.r.t. the database. The sizes of our search indexes are very small, e.g., only 1:1,000–1:15,000 the size of the original audio data depending on the required retrieval granularity. As an example, a database of 180 GB of audio material can be indexed using about 50 MB only, while still allowing audio signals of only several seconds of length to be located within fractions of seconds ([Ribbrock and Kurth, 2002]).

- Index construction may be performed very efficiently. For the PROMS system, index construction takes only a few minutes, hence allowing for indexing on the fly. Indexing PCM audio data may be performed several times faster than real-time on standard PC hardware.

- In our prototypic image retrieval system containing 3,300 images, exact sub-image queries require about 50ms response time ([Röder, 2002]). The search index is compressed to 1:6 compared to the original (JPEG) data.

The paper is organized as follows. In the next section, we discuss several motivating examples and give an informal overview on the concepts of our approach to content-based multimedia retrieval. Those concepts are introduced formally in Sections 3 and 4. Section 3 deals with a formal specification of documents, queries, the notion of matches, and the derivation of fast retrieval algorithms. Section 4 introduces two general mechanisms to incorporate fault tolerance: mismatches and fuzzy queries. In Section 5 we present prototypic applications from the fields of music-, audio-, image-, and object retrieval. Finally, Section 6 gives a brief overview on related work and suggest some future research directions.

2.   Motivating Examples

The goal of this section is to present three motivating examples that illustrate the desirability of a unified approach to content-based retrieval. The first two are concerned with text retrieval tasks, whereas the third discusses content-based retrieval in score-based music data.

As a first example, suppose we have a collection $\mathcal{T} = (T_1, \ldots, T_N)$ of text documents each consisting of a finite sequence of *words* where each word is contained in a set (i.e., a dictionary) W of all admissible words. Taking a coarse level of granularity, each text document is preprocessed in order to extract a set of *terms* occurring in that document. If T denotes the set of all conceivable terms, then this preprocessing step produces a sequence $\mathcal{D} = (D_1, \ldots, D_N)$ of finite subsets of T, where D_i is the set of terms extracted from the ith document. We may think of the term set T as a reduced version of W, where unimportant words have been left out and verbs, nouns, etc. in W have been replaced by their principal forms, e.g., by means of a stemming algorithm (e.g., *sitting* $\mapsto$ *sit*).

Next, let us think about content-based queries. A usual way of formulating a query is to input a finite set Q of terms to the retrieval system. Then one possible task is to compute the set $H_{\mathcal{D}}(Q)$ of all *exact partial matches*, which is the set of all documents containing all of the terms specified by Q. More formally, with $[1 : N] := \{1, \ldots, N\}$,

$$H_{\mathcal{D}}(Q) := \{i \in [1 : N] \mid Q \subseteq D_i\}. \tag{1}$$

Note that this is a rather strict decision about the Q-relevance of a document: if just one $q \in Q$ does not occur in D_i, then D_i is already considered irrelevant. A more realistic way to estimate relevance would be a ranking of the documents along the quotients $|Q \cap D_i|/|Q| \in [0, 1]$. This quotient equals 1 if and only if Q is a subset of D_i.

From the viewpoint of logic, presenting a set $Q = \{q_1, \ldots, q_n\}$ of terms to the retrieval system is equivalent to the boolean query $q_1 \wedge \ldots \wedge q_n$ which asks for all documents that contain all these terms. More generally, we can form boolean expressions like $(t_1 \wedge t_2) \vee (t_3 \wedge \neg t_4)$, the latter asking for all documents containing both the terms t_1 and t_2 or the term t_3 but not the term t_4.

After this brief introduction to content-based queries and different text retrieval tasks, we are now going to discuss efficient algorithms for text retrieval. For each term $t \in \cup_{i=1}^{N} D_i$ one establishes a so-called inverted file, which is the (linearly ordered) set $H_{\mathcal{D}}(t) := \{i \in [1 : N] \mid t \in D_i\}$. Then, for a query $Q \subseteq T$, the set of all exact partial matches is obtained by intersecting the inverted files of all elements in Q:

$$H_{\mathcal{D}}(Q) = \bigcap_{t \in Q} H_{\mathcal{D}}(t).$$

This generalizes to boolean queries. For example, $H_{\mathcal{D}}((t_1 \wedge t_2) \vee (t_3 \wedge \neg t_4)) = (H_{\mathcal{D}}(t_1) \cap H_{\mathcal{D}}(t_2)) \cup (H_{\mathcal{D}}(t_3) \setminus H_{\mathcal{D}}(t_4))$.

The above discussion immediately leads to the following computational problems: given (linearly ordered) finite sets A and B of integers, compute their intersection, union and difference, again linearly ordered. If a and b denote the cardinality of A and B, respectively, then $A \cap B$, $A \cup B$ and $A \setminus B$ may be computed with at most $a + b$ comparisons using a merge technique. If $a \ll b$ then comparing the elements of A one after another by means of the binary search method (see, e.g., [Cormen et al., 1990]) with the elements of B causes at most $a \log b$ comparisons to solve each of these problems.

So far we have discussed a rather coarse notion of a match. If for example the document ID i is an exact partial match w.r.t. the query Q, then we only know that all terms in Q do also occur in the set of terms extracted from the ith document. However, we do not know *where* these terms occur in the original text document T_i. This in turn is the main goal of full-text retrieval, where questions are allowed that ask, e.g., for a document containing the text passage "to be or not to be." We discuss this kind of retrieval next.

In our second example we consider full-text retrieval where we use a finer level of granularity. For this, we consider the word sequences constituting a text document. If W denotes the set of all words then the ith document is viewed as a sequence $D_i = (w_{i0}, \ldots, w_{in_i})$ over W. To make this example more comparable to the previous one, we will identify this sequence with the set $\{[j, w_{ij}] \mid j \in [0 : n_i]\}$. (Note that replacing sequences by their corresponding sets means a switch from an implicit to an explicit specification of the words' places.) Similarly, a query like "to be or not to be" would then be identified with the set $\{[0, \text{to}], [1, \text{be}], [2, \text{or}], [3, \text{not}], [4, \text{to}], [5, \text{be}]\}$.

In general, for a query $Q = (q_0, \ldots, q_n) \equiv \{[j, q_j] \mid j \in [0 : n]\}$, we are looking for the set $H_{\mathcal{D}}(Q)$ of all pairs (t, i) such that $t + Q := \{[t+j, q_j] \mid j \in [0 : n]\}$ is a subset of D_i. To obtain all those pairs efficiently we use, in analogy to the first example, an inverted file $H_{\mathcal{D}}(w) := \{(j, i) \mid w = w_{ij}\}$, for each word w that occurs in any of the documents. We claim that

$$H_{\mathcal{D}}(Q) = \cap_{k=0}^{n}(H_{\mathcal{D}}(q_k) - k), \tag{2}$$

where $H_{\mathcal{D}}(q_k) - k := \{(j - k, i) \mid (j, i) \in H_{\mathcal{D}}(q_k)\}$. In fact, $(j, i) \in H_{\mathcal{D}}(Q)$ iff $(q_0, \ldots, q_n) = (w_{ij}, \ldots, w_{i,j+n})$. This is equivalent to $(j + k, i) \in H_{\mathcal{D}}(q_k)$, for all $k \in [0 : n]$, i.e., $(j, i) \in H_{\mathcal{D}}(q_k) - k$, for all k, which proves our claim. Note that we use square brackets (such as in $[1, \text{be}]$) to denote the elementary objects constituting our documents in order to avoid confusion with elements of the inverted files (such as (j, i) above) which are denoted by round brackets.

According to this formula, the list of all solutions is again the intersection of all relevant and properly adjusted inverted files. However, for stop words like *the, a, is, to, be*, the lists are rather long. To avoid long lists and to improve the query response time substantially, we consider instead of W the set of all pairs $W2$ as our new universe of elementary data objects. For each pair $[v, w]$ of words we generate the inverted file $H_{\mathcal{D}}(v, w) := \{(j, i) \mid [v, w] = [w_{ij}, w_{i,j+1}]\}$. On the one hand, this simple trick drastically increases the number of inverted files, on the other hand, the new lists are typically much smaller than the original inverted files. The query "to be or not to be" can now be processed by the following intersection (check this!):

$$H_{\mathcal{D}}(to, be) \cap (H_{\mathcal{D}}(or, not) - 2) \cap (H_{\mathcal{D}}(to, be) - 4).$$

Thus instead of intersecting six long lists when working with W, we now have to intersect only three small lists. Obviously, this generalizes to k-tuples of words, for any k.

After these rather classical examples let us now turn to our third example which deals with content-based music information retrieval. The reader should notice the analogy to full-text retrieval.

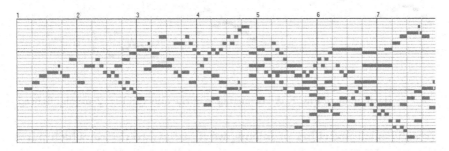

Figure 1. A part of J.S. Bach's Fugue in C major, BWV 846, in the piano roll notation.

Suppose we have a collection $T = (T_1, \ldots, T_N)$ of scores. That is, each T_i is a score representation of a piece of music containing the notes of that piece as well as additional information such as meter, tempo or specifications of dynamics. Musical scores in the conventional staff notation may be visualized using the so-called *piano roll* representation. Fig. 1 shows the piano roll representation of the beginning of Johann Sebastian Bach's Fugue in C major, BWV 846. In this figure, the horizontal axis represents time whereas the vertical axis describes pitches. Each rectangle of width d located (w.r.t. its left lower corner) at coordinates (t, p) represents a note of pitch p, onset time t, and duration d. Such a note will be denoted by the triple $[p, t, d]$. (After a suitable quantization, we can assume w.l.o.g. that p, t and d are integers.)

Figure 2. A query to the database in the piano roll notation.

Now we consider the following music information retrieval task which has already been sketched in the introduction: given a fragment of a melody or, more generally, an arbitrary excerpt of a piece of music, we are looking for all occurrences of that content-based query in the collection. As an example consider the query depicted in Fig. 2. In fact, this query is a part of the fugue's theme. Assume that we are looking for all positions where this theme or a pitch-transposed version thereof occurs. Then, Fig. 3 shows all occurrences of the query within the excerpt of the fugue given in Fig. 1.

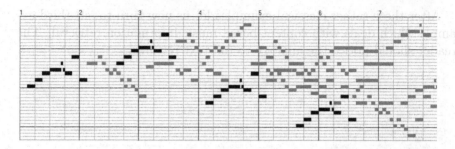

Figure 3.   Bach fugue of Fig. 1. All occurrences of the query in Fig. 2 are highlighted.

To solve such matching problems, we first have to decide which parts of the complete score information are actually needed. By our experience, the most important parameters for recognizing a piece of music are pitch and onset time. In other words, a person will recognize a piece of music even if it is played in staccato throughout, i.e., disregarding the notes' durations. According to this experimental result, the first step is to extract from each score the pitches and onset times. This results in a new collection $\mathcal{D} = (D_1, \ldots, D_N)$ of finite subsets D_i of $\mathbb{Z}^2$. A content-based query is—after a possible preprocessing step—again a finite subset Q of $\mathbb{Z}^2$. The set $H_{\mathcal{D}}(Q)$ of all exact partial matches—when time- and pitch-shifts are allowed—is thus given by

$$H_{\mathcal{D}}(Q) := \{(\pi, \tau, i) \in \mathbb{Z}^2 \times [1:N] \mid (\pi, \tau) + Q \subseteq D_i\},$$

where $(\pi, \tau) + Q := \{[\pi + p, \tau + t] \mid [p, t] \in Q\}$. If we construct for every pair $[p, t] \in \mathbb{Z}^2$ an inverted file $H_{\mathcal{D}}([p, t]) := \{(\pi, \tau, i) \in \mathbb{Z}^2 \times [1:N] \mid [\pi + p, \tau + t] \in D_i\}$, then

$$H_{\mathcal{D}}(Q) = \bigcap_{[p,t] \in Q} H_{\mathcal{D}}([p, t]).$$

As in the case of text retrieval, the set of all exact partial matches is the intersection of all inverted files corresponding to the elements of Q. Obviously, there are infinitely many inverted files. Fortunately, from just *one* list one can recover the other inverted files according to the formula

$$H_{\mathcal{D}}([p, t]) = H_{\mathcal{D}}([0, 0]) - (p, t), \tag{3}$$

where $H_{\mathcal{D}}([0, 0]) - (p, t) := \{(\pi - p, \tau - t, i) \mid (\pi, \tau, i) \in H_{\mathcal{D}}([0, 0])\}$. In fact, $(\pi, \tau, i) \in H_{\mathcal{D}}([p, t])$ iff $[\pi + p, \tau + t] \in D_i$ iff $(\pi + p, \tau + t, i) \in H_{\mathcal{D}}([0, 0])$ iff $(\pi, \tau, i) \in H_{\mathcal{D}}([0, 0]) - (p, t)$, which proves our claim. Due to this formula, we only need to store the inverted file $H_{\mathcal{D}}([0, 0])$. The other lists can easily be computed from it if required. Remarkably, if $H_{\mathcal{D}}([0, 0])$ is linearly ordered, then so is $H_{\mathcal{D}}([0, 0]) - (p, t)$.

Let us analyze the computational cost of computing all exact partial matches w.r.t. the query Q using the formula

$$H_\mathcal{D}(Q) = \bigcap_{[p,t] \in Q} (H_\mathcal{D}([0,0]) - (p,t)). \qquad (4)$$

As the list $H_\mathcal{D}([0,0])$ contains information about *all* notes from *all* scores in the collection, its size is proportional to the total number of notes in the collection. At least for large corpora this leads to unacceptable query response times. So we have to look for alternatives to get shorter lists.

Although Formula (4) looks very similar to Formula (2), there exists a significant difference concerning the number of independent inverted files: in our music retrieval scenario this number is one whereas in the full-text retrieval scenario this number equals the number of different terms in all documents. Thus text retrieval has the advantage of many independent inverted files that are typically small as compared to the storage requirements of the whole collection. Hence it is advantageous to make sure that many independent inverted files exist also in the music retrieval scenario.

There are different ways to achieve this goal. The first way is to reconsider the durations of the notes, resulting in documents D_i and Q over $\mathbb{Z}^3$ where the additional component represents the note duration. Using shifts $(\pi, \tau) + Q := \{[\pi + p, \tau + t, d] \mid [p, t, d] \in Q\}$ and inverted files $H_\mathcal{D}([p, t, d]) := \{(\pi, \tau, i) \in \mathbb{Z}^2 \times [1 : N] \mid [\pi + p, \tau + t, d] \in D_i\}$, we note that now

$$H_\mathcal{D}([p, t, d]) = H_\mathcal{D}([0, 0, d]) - (p, t). \qquad (5)$$

Hence, the number of independent inverted lists equals the number of different durations occuring in the score collection. Besides fundamental difficulties of quantifying the duration, e.g., for staccato or grace notes, there will be extremely long lists for whole, half, quarter and eighth notes and these lists will typically be needed in most queries. So this alternative will generally not be suitable.

A second way of obtaining shorter independent inverted files is to exploit a user's prior knowledge. To avoid technicalities, assume that all pieces of the score collection are in 4/4 meter. Assume furthermore that each meter is subdivided into 16 metrical positions. When a user posing a query knows about the metrical position of the query, e.g., that the query starts at an offbeat, this may be incorporated as follows. We work with modified inverted files of the form

$$H'_\mathcal{D}([\pi, \tau]) := \{(p, k, i) \mid [p + \pi, 16k + \tau] \in D_i\}$$

and note that $H'_{\mathcal{D}}([\pi, \tau]) = H'_{\mathcal{D}}([0, \tau + 16\ell]) - (\pi, -\ell, 0)$ (Check this!). Thus, instead of one independent inverted file we now have 16 independent files

$$H'_{\mathcal{D}}([0, t]) := \{(\pi, k, i) \mid [\pi, 16k + t] \in D_i\}$$

for $t \in [0 : 15]$. For a query Q containing correct indications of the metrical positions, our task is to compute the set $H'_{\mathcal{D}}(Q) := \{(p, k, i) \mid (p, 16k) + Q \subseteq D_i\}$ which is equal to the intersection of all $H'_{\mathcal{D}}(q)$, q ranging over all elements of Q.

However, this only works if the user provides the retrieval system with accurate metrical information. If, in addition, the user knows about the exact pitches, then assuming 128 different pitches (like in the MIDI format), we now have 2048 independent inverted files

$$H''_{\mathcal{D}}([p, t]) := \{(p, k, i) \mid [p, 16k + t] \in D_i\}$$

for $[p, t] \in [0 : 127] \times [0 : 15]$. In this case, a user has to know both the exact pitches and the metrical positions. So this second alternative requires a high degree of user knowledge. One could decrease this requirement by incorporating fault tolerance. We briefly discuss two mechanisms for achieving this: mismatches and fuzzy queries.

The first mechanism considers the case that a query is not completely contained in a document D_i. The elements of $Q \setminus D_i$ are called mismatches. Using a variant of an algorithm (see Section 4) for computing the ratios $|Q \cap D_i|/|Q|$, one may efficiently determine all documents with at most k mismatches.

Another possibility is to allow a user to pose fuzzy queries. In our music scenario, it could be the case that one is unsure about a certain pitch interval or the exact rhythm. Here, a user can specify alternatives for one note. Then a fuzzy query is a sequence $\mathbf{F} = (F_1, \ldots, F_n)$ of n finite sets F_i of alternatives. Such an $\mathbf{F}$ is the shorthand for a family of ordinary queries $Q = \{q_1, \ldots, q_n\}$ where for each i, q_i is allowed to take arbitrary values of F_i. A document D_i is an exact partial match for a fuzzy query $\mathbf{F}$ if $Q \subseteq D_i$, for some Q in this bunch of ordinary queries corresponding to $\mathbf{F}$. If the F_i are pairwise disjoint, then $\mathbf{F}$ consists of $\prod_{i=1}^{n} |F_i|$ ordinary queries. Although this number might grow rapidly, there are efficient algorithms to compute all matches, see Section 4.

Finally, let us discuss a third alternative. Instead of notes $[p, t]$ we now consider *pairs* of notes $([p, t], [p', t'])$ as our basic objects. As a query Q in our score scenario typically refers to a contiguous part of the score, we do not need to store all pairs of $D_i \times D_i$ but only those which are close to another. Noticing that most rows of a score correspond to one particular voice each, the storage complexity is still linear in the total length of the collection.

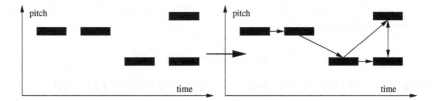

Figure 4. Replacing a document D (left) with a subset of $D2$ (right) consisting of neighboring objects of D. Elements of $D2$ are specified by pairs of notes joined by an arrow.

So documents D_i and queries Q are replaced by suitable subsets of $D_i 2$ and $Q2$, respectively. Figure 4 illustrates this procedure. Considering pairs of notes is advantageous: we now have many independent inverted files: these are indexed by the pairs $([0, 0], [p, t])$ with $p \in [-127 : 127]$ and $t \in \mathbb{Z}$. For details see the next section.

Prepared by these motivating examples, we are going to describe a general concept for multimedia information retrieval. This concept, worked out in detail in the next section, combines classical full-text retrieval techniques with methods from group theory.

3.   General Concept

Let M be a set whose elements model the elementary data objects. A *document* over M is just a finite subset D of M. A *collection* (or *data base*) over M is a finite sequence $\mathcal{D} = (D_1, \ldots, D_N)$ of documents.

In the classical text retrieval example, $M = T$ is the set of all terms, whereas D_i equals the set of terms extracted from the ith text document. In the full-text retrieval scenario, $M = \mathbb{Z} \times W$ is the set of all words together with positional information. The ith text document T_i is viewed as a sequence of words. This in turn may be described by the subset $D_i = \{[j, w_{ij}] \mid j \in [0 : n_i]\}$ of M. Finally, in our score retrieval example, $M = [0 : 127] \times \mathbb{Z}$ or $M = ([0 : 127] \times \mathbb{Z})2$ is the set of all (pairs of) notes. The ith score document is then a finite subset of M. (For technical reasons, in this last example we will replace M by the supersets $\mathbb{Z}^2$ and $(\mathbb{Z}^2)2$, respectively.)

A content-based query—after a possible preprocessing step—is again a finite subset Q of M. Hence, in our three examples, Q is a set of terms, a sequence of words, or a set of (pairs of) notes, respectively.

To define more generally shifted versions of a query Q we use the concept of a group G acting on a set M. Recall that the (multiplicative) group G acts on the set M if there is a map $G \times M \ni (g, m) \mapsto gm \in M$ satisfying for all $g, h \in G$ and $m \in M$: $g(hm) = (gh)m$ and $1_G m = m$.

In this case, M is also called a *G-set*. Such a group action defines an equivalence relation on M:

$$m \sim m' \quad \text{iff} \quad \exists g \in G : gm = m'.$$

The equivalence class containing $m \in M$ is the so-called *G-orbit*

$$Gm := \{gm \mid g \in G\}.$$

Hence the G-set M decomposes into the disjoint union of G-orbits. If R is a *transversal* of the G-orbits, i.e., a set of representatives containing exactly one element of each orbit, then

$$M = \bigsqcup_{r \in R} Gr.$$

By definition, the G-action on M is *transitive* if there is only one G-orbit, otherwise the G-action is *intransitive*.

If G acts on M then G also acts on $P(M) := \{Q \mid Q \subseteq M\}$, the power set of M, via

$$gQ := \{gq \mid q \in Q\},$$

where $g \in G$ and $Q \subseteq M$. As Q and gQ have the same cardinality, G acts on the set of all finite subsets of M, i.e., on the documents and queries over M. Thus the group elements precisely allow us to specify what is meant by a G-shift of a query. Furthermore, as we will see below, we can vary the group G to exploit the user's prior knowledge. With this concept we can define a first information retrieval task.

DEFINITION 1   *(Exact partial $(G, \mathcal{D})$-matches)*
Let $\mathcal{D} = (D_1, \ldots, D_N)$ be a collection over the G-set M. For a query Q over M the set of all exact $(G, \mathcal{D})$-matches is defined as

$$G_{\mathcal{D}}(Q) := \{(g, i) \in G \times [1 : N] \mid gQ \subseteq D_i\}. \quad \bullet$$

In the text retrieval example, G is the trivial group. Hence there is no loss of information if we simply replace $(g, i) \in G_{\mathcal{D}}(Q)$ by i. Thus $G_{\mathcal{D}}(Q) \equiv \{i \in [1 : N] \mid Q \subseteq D_i\}$ equals the set $H_{\mathcal{D}}(Q)$ from equation (1).

In the full-text retrieval example, the additive group $G = \mathbb{Z}$ acts on $\mathbb{Z} \times W$ by $(t, [j, w]) \mapsto [t + j, w]$. In this case, $G_{\mathcal{D}}(Q)$ consists of all pairs (t, i) such that the time-shifted version $t + Q$ of Q is completely contained in D_i.

Finally, in the score retrieval scenario, the additive group $G = \mathbb{Z}^2$ acts on $M = \mathbb{Z}^2$ by vector addition. By the way, this action is transitive. Again, $G_{\mathcal{D}}(Q)$ is the set of all pairs $((p, t), i)$ such that $(p, t) + Q$ is completely contained in the ith document.

So far, in all examples a commutative group is acting. Here is a classical example of a noncommutative group action.

A 2D color image I is modeled as a finite subset of $\mathbb{R}^2 \times C$, where $[(x, y), c] \in I$ describes the color information $c \in C$ of I at position $(x, y) \in \mathbb{R}^2$. Let G denote the group of similarity transformations in $\mathbb{R}^2$. This noncommutative group is generated by rotations, translations and uniform scalings. If $Q \subset \mathbb{R}^2 \times C$ is a fragment of a 2D color image, then G acts on such fragments of images by $gQ := \{[g(x, y), c] \mid [(x, y), c] \in Q\}$. Thus gQ is a rotated, translated and rescaled version of Q.

If $\mathcal{D}$ is a collection over the G-set M then our index will consist of $(G, \mathcal{D})$-*inverted files* or *lists*, defined by

$$G_\mathcal{D}(m) := \{(g, i) \in G \times [1 : N] \mid gm \in D_i\},$$

for $m \in M$. One easily shows that for a query Q the corresponding set of all exact partial $(G, \mathcal{D})$-matches may be computed as the intersection of all $(G, \mathcal{D})$-inverted files specified by Q:

$$G_\mathcal{D}(Q) = \bigcap_{q \in Q} G_\mathcal{D}(q).$$

Thus to obtain all matches, it suffices to have access to all inverted lists $G_\mathcal{D}(q)$. If M or G are infinite or of large finite cardinality, it might be impossible or impractical to store all inverted lists. However, in a number of cases, we can overcome this problem. The crucial observation is that the inverted lists of all elements in one G-orbit are closely related.

LEMMA 2 $G_\mathcal{D}(gm) = G_\mathcal{D}(m)g^{-1} := \{(hg^{-1}, i) \mid (h, i) \in G_\mathcal{D}(m)\}$.

Proof. The following chain of equivalences proves our claim:

$$(h, i) \in G_\mathcal{D}(m) \iff hm \in D_i \iff hg^{-1}(gm) \in D_i$$
$$\iff (hg^{-1}, i) \in G_\mathcal{D}(gm). \quad \bullet$$

Thus if the multiplication in the group G is not too involved, we can quickly recover $G_\mathcal{D}(gm)$ from $G_\mathcal{D}(m)$, which might lead to dramatic storage savings. To be more precise, let R be a transversal of the G-orbits of M. Then every element $m \in M$ can be written as $m = g_m r_m$ with a uniquely determined $r_m \in R$ and an element $g_m \in G$ which is unique modulo $G_m := \{g \in G \mid gm = m\}$, the stabilizer subgroup of m. By the last lemma it is sufficient to store only the inverted lists corresponding to the transversal R. Then the remaining lists can be computed on demand according to the following result.

THEOREM 3 *For the set of all exact partial $(G, \mathcal{D})$-matches w.r.t. the query $Q \subseteq M$ the following formula holds:*

$$G_{\mathcal{D}}(Q) = \bigcap_{q \in Q} G_{\mathcal{D}}(r_q)g_q^{-1}.$$

Proof. Combine $G_{\mathcal{D}}(Q) = \cap_{q \in Q} G_{\mathcal{D}}(q)$ and $q = g_q r_q$ with the last lemma. •

How time and space consuming is the computation of all exact partial $(G, \mathcal{D})$-matches along the formula of the last theorem? Let us first discuss storage requirements.

THEOREM 4 *With the notation of the last theorem suppose that the stabilizer subgroups G_m for all $m \in M$ are trivial. Then the sum of the lengths of all $(G, \mathcal{D})$-inverted lists corresponding to a transversal R of the G-orbits of M equals the sum of the cardinalities of all documents:*

$$\sum_{r \in R} \text{length}(G_{\mathcal{D}}(r)) = \sum_{i \in [1:N]} |D_i|.$$

Proof. As all stabilizers are trivial, each $m \in M$ has a unique decomposition $m = g_m r_m$ with $g_m \in G$ and $r_m \in R$. Thus each $m \in D_i$ contributes exactly one entry to exactly one list: $(g_m, i) \in G_{\mathcal{D}}(r_m)$. •

If $m \in D_i$ has a nontrivial stabilizer G_m and if $m = g_m r$ with $g_m \in G$ and $r \in R$, then m contributes exactly $|G_m|$ entries to the inverted list $G_{\mathcal{D}}(m)$, namely all pairs of the form $(g_m g, i)$ with $g \in G_r$. So if all stabilizers are small ($|G_m| \leq c$, say) then a slight modification of the above reasoning shows that the above equality can be replaced by $\sum_{r \in R} \text{length}(G_{\mathcal{D}}(r)) \leq c \cdot \sum_{i \in [1:N]} |D_i|$. If there are large stabilizers, then the above formula should not be applied directly. Instead, a redesign of the set M of elementary data objects might be helpful to force trivial stabilizers. This trick will be discussed below.

In what follows we concentrate on the case that all stabilizers are trivial. How fast can we compute an intersection like $\cap_{q \in Q} G_{\mathcal{D}}(r_q)g_q^{-1}$? To settle this question let us first discuss the problem of generating the inverted lists. To accelerate the intersection task, we first define a linear ordering on the group G and suppose that we have an algorithm that efficiently decides for $g, h \in G$ whether $g = h$, $g < h$, or $g > h$. For the moment, we also assume that we know a transversal R of the G-orbits of M and that we know an efficient algorithm for computing $M \ni m \mapsto (g_m, r_m) \in G \times R$, where $m = g_m r_m$. Under these assumptions we may, for each element in all the documents, efficiently compute the corresponding inverted file: $m = g_m r_m \in D_i$ yields the entry (g_m, i) in $G_{\mathcal{D}}(r_m)$. Finally, we linearly order the entries in each inverted file according to their first component, which is an element of the linearly ordered group G. Note that not all G-orbits have to be involved in the documents of a particular database $D_1, \ldots, D_N$. Suppose that $R' \subseteq R$ is minimal with $\cup_{r \in R'} Gr \supseteq \cup_{i \in [1:N]} D_i$, then we have a total of $|R'|$ inverted files. If the group operation is compatible with the linear ordering of the group, i.e., $x < y$ implies $xg < yg$, for all $x, y, g \in G$, then the linear ordering of the inverted files $G_{\mathcal{D}}(r)$ for $r \in R$ can be used directly when computing $G_{\mathcal{D}}(Q)$ using the formula $G_{\mathcal{D}}(Q) = \cap_{q \in Q} G_{\mathcal{D}}(r_q) g_q^{-1}$. According to our assumption, each $G_{\mathcal{D}}(r_q)$ is linearly ordered, and hence so is $G_{\mathcal{D}}(r_q) g_q^{-1}$. Consequently, we have to compute the intersection of n linearly ordered lists of length $\ell_1 \leq \ldots \leq \ell_n$, say. If λ_i is the cardinality of the intersection of the first i lists, then computing $G_{\mathcal{D}}(Q)$ with the binary method requires at most $\sum_{i=1}^{n-1} \lambda_i \log \ell_{i+1} \leq (n-1)\ell_1 \log \ell_n$ comparisons in G as well as at most n inversions and $\ell_1 + \ldots + \ell_n$ multiplications in G.

Surprisingly, a similar complexity result can even be obtained when the linear ordering of G is not compatible with the multiplications in G. This is based on the following simple but very useful observation.

LEMMA 5 *Let a, b be elements and A, B subsets of the group G. Then $Aa \perp Bb = (Aab^{-1} \perp B)b = (A \perp Bba^{-1})a$, for $\perp \in \{\cap, \cup, \setminus\}$.*

Proof. Straightforward. •

In the worst case, both Aa and Bb are not linearly ordered in contrast to A and B. Nevertheless, when computing, e.g., the intersection $Aa \cap Bb$ we can use either the linear ordering of A by computing $(A \cap Bba^{-1})a$ or that of B by computing $(Aab^{-1} \cap B)b$.

THEOREM 6 *Let Q be a query of size n, $Q = \{q_1, \ldots, q_n\}$. Suppose that the lists $G_{\mathcal{D}}(q_i)$ are ordered according to their lengths $\ell_1 \leq \ldots \leq \ell_n$, where $\ell_i := |G_{\mathcal{D}}(q_i)|$. For $j \in [1:n]$ let λ_j denote the cardinality of the intersection $\cap_{i \in [1:j]} G_{\mathcal{D}}(q_i)$. Given the linearly ordered lists $G_{\mathcal{D}}(r)$ for all $r \in R$, the list $G_{\mathcal{D}}(Q)$ of all exact partial $(G, \mathcal{D})$-matches of Q can be computed with $\sum_{i=1}^{n-1} \lambda_i \log \ell_{i+1} \leq (n-1)\ell_1 \log \ell_n$ comparisons in G, along with n decompositions $q_i \mapsto (g_i, r_i) \in G \times R$ satisfying $q_i = g_i r_i$, n inversions $g_i \mapsto g_i^{-1}$, $n-1$ multiplications $g_i^{-1} g_{i+1}$, and $\sum_{i=1}^{n} \lambda_i \leq n\ell_1$ multiplications in G.*

Proof. Let $q_i = g_i r_i$ with $g_i \in G$ and $r_i \in R$. Then

$$G_{\mathcal{D}}(Q) = \bigcap_{i=1}^{n} G_{\mathcal{D}}(q_i) = \bigcap_{i=1}^{n} G_{\mathcal{D}}(r_i) g_i^{-1}.$$

According to the last lemma we compute $G_{\mathcal{D}}(Q)$ as follows (here shown for the case $n = 3$ using the shorthand $H_i := G_{\mathcal{D}}(r_i)$):

$$H_1 g_1^{-1} \cap H_2 g_2^{-1} \cap H_3 g_3^{-1} = ((H_1 g_1^{-1} g_2 \cap H_2) g_2^{-1} g_3 \cap H_3) g_3^{-1}.$$

This formula suggests the following procedure: first, compute all decompositions $q_i \mapsto (g_i, r_i)$. Then invert all g_i and compute all $g_i^{-1} g_{i+1}$. Afterwards, inductively compute linearly ordered sets

$$\Lambda_j := \Big(\bigcap_{i \in [1:j]} H_i g_i^{-1} \Big) g_j.$$

Note that $\Lambda_1 = H_1$ is already linearly ordered. As $\Lambda_{j+1} = \Lambda_j g_j^{-1} g_{j+1} \cap H_j$ and H_j is linearly ordered, we can compute this intersection using the binary method. We finish the proof by mentioning that Λ_j has cardinality λ_j and $\Lambda_n g_n^{-1}$ equals $G_{\mathcal{D}}(Q)$. •

Roughly speaking, the last theorem tells us that the number of operations to compute $G_{\mathcal{D}}(Q)$ is at most the product of the cardinality of Q, the length of the shortest list and the logarithm of the longest list among all lists corresponding to Q. Hence it is profitable to have many short lists. But what do we do if G acts transitively on M? In this case one should look at intransitive G-sets, closely related to M. Here are two examples: if $n \geq 2$ then G acts on $\mathcal{P}_{\leq n}(M) := \{X \subseteq M \mid 0 < |X| \leq n\}$ by $gX := \{gx \mid x \in X\}$ and on M^n by $g(m_1, \ldots, m_n) := (gm_1, \ldots, gm_n)$. Both actions are intransitive. The next result shows a close connection between exact partial matches in various G-sets.

THEOREM 7 *Let $\mathcal{D} = (D_1, \ldots, D_N)$ denote a document collection over the G-set M. This induces new collections $\mathcal{D}^n := (D_1^n, \ldots, D_N^n)$. Similarly, for a query $Q \subseteq M$ and $\mathcal{P}_{\leq n}(\mathcal{D}) := (\mathcal{P}_{\leq n}(D_1), \ldots, \mathcal{P}_{\leq n}(D_N))$ we associate the queries $\mathcal{P}_{\leq n}(Q) \subseteq \mathcal{P}_{\leq n}(M)$ and $Q^n \subseteq M^n$. The corresponding sets of exact partial matches are equal:*

$$G_{\mathcal{D}}(Q) = G_{\mathcal{P}_{\leq n}(\mathcal{D})}(\mathcal{P}_{\leq n}(Q)) = G_{\mathcal{D}^n}(Q^n).$$

Proof. For $(g, i) \in G \times [1 : N]$ we have

$$
\begin{aligned}
(g, i) \in G_{\mathcal{D}}(Q) &\iff gQ \subseteq D_i \\
&\iff \forall X \in \mathcal{P}_{\leq n}(gQ) = g\mathcal{P}_{\leq n}(Q) : \ X \in \mathcal{P}_{\leq n}(D_i) \\
&\iff g\mathcal{P}_{\leq n}(Q) \subseteq \mathcal{P}_{\leq n}(D_i) \\
&\iff (g, i) \in G_{\mathcal{P}_{\leq n}(\mathcal{D})}(\mathcal{P}_{\leq n}(Q)).
\end{aligned}
$$

This proves the first equality. The second equality follows in a similar way using $(gQ)^n = gQ^n$. •

It should be clear that the above intransitive G-actions are just two generic constructions which are available in every case. Depending on a particular application one will possibly succeed with smaller G-sets, as, e.g., the action on $\mathcal{P}_n(M) = \{X \subseteq M \mid |X| = n\}$ instead of $\mathcal{P}_{\leq n}(M) = \{X \subseteq M \mid 0 < |X| \leq n\}$.

Here are some specific examples. $G = (\mathbb{R}^3, +)$ acts transitively on $M := \mathbb{R}^3$ by vector addition. The induced actions of G on both $\mathcal{P}_2(M)$ and $M2$ are intransitive. In fact, the G-orbit of $\{x, y\} \in \mathcal{P}_2(M)$ is characterized by $\pm(x - y)$, whereas the G-orbit of $(x, y) \in M2$ is characterized by $x - y$. Thus if the documents consist of finite subsets D_i of $\mathbb{R}^3$ and the group of all translations is the group in question, then it is profitable to view pairs or 2-sets of vectors in $\mathbb{R}^3$ as the new elementary data objects. If, however, the group G_s of all similarity transformation in 3-space (generated by all translations, rotations and uniform scalings) is the group of choice, then $M2$ decomposes only into two G_s-orbits represented by $([0, 0, 0], [0, 0, 0])$ and $([0, 0, 0], [0, 0, 1])$, respectively. Furthermore, G_s acts transitively on $\mathcal{P}_2(M)$. However, $\mathcal{P}_3(M)$ decomposes into many G_s-orbits. These orbits correspond to the classes of congruent triangles in Euclidean 3-space. Thus if $G = G_s$ then one should switch from $M = \mathbb{R}^3$ to $\mathcal{P}_3(M)$. Besides many G_s-orbits by this switch we also obtain small stabilizers. In fact, all stabilizers are isomorphic to a subgroup of the dihedral group of order 6.

A switch from M to $\mathcal{P}_k(M)$ has the drawback that a document $D_i \subseteq M$ with d_i elements has to be replaced by the new document $\mathcal{P}_k(D_i)$ which consists of $\binom{d_i}{k}$ elements. Fortunately, in most applications a query Q will typically be not a random but a structured subset of M. For example, if (M, d) is a metric space and $\theta > 0$ a prescribed constant, a possible property of Q could be its θ-connectedness, i.e., there is an enumeration of Q, $Q = \{q_1, \ldots, q_n\}$, such that $d(q_{i-1}, q_i) \leq \theta$, for all $i \in [2 : n]$. Thus instead of $\mathcal{P}_k(D_i)$ in this case it is sufficient to work with the typically much smaller set $D_{i,k,\theta} := \{X \in \mathcal{P}_k(D_i) \mid X \text{ is } \theta\text{-connected}\}$. A query $Q = \{q_1, \ldots, q_n\}$ with $0 < d(q_{i-1}, q_i) \leq \theta$ should then be replaced by $Q' = \{\{q_{\lambda k+1}, \ldots, q_{(\lambda+1)k}\} \mid 0 \leq \lambda \leq \lfloor n/k \rfloor\} \cup \{\{q_{n-k+1}, \ldots, q_n\}\}$.

4. Fault Tolerance

So far, we have mainly discussed *exact* partial matches. Typically, there are many sources of impreciseness. Think, e.g., of a non professional user humming a melody into a microphone. Another example could be the search in a database of signals where the queries consist of noisy or lossy compressed versions of the original signals. In all of those cases, a certain degree of fault tolerance is required. Now we will discuss two general kinds of fault tolerance: mismatches and fuzzy queries.

We start with mismatches. Let Q be a query and D a document over M. The elements of $Q \cap D$ form the matching part, whereas the elements in Q that do not belong to D are called the *mismatches*. Thus for a fixed non-negative integer k and a query Q the set

$$G_{\mathcal{D}}^k(Q) := \{(g, i) \in G \times [1 : N] \mid |gQ \setminus D_i| \leq k\}$$

specifies all partial $(G, \mathcal{D})$-matches with at most k mismatches. Obviously, $G_{\mathcal{D}}^0(Q)$ equals $G_{\mathcal{D}}(Q)$. For $k > 0$ and $(g, i) \in G_{\mathcal{D}}^k(Q)$ the transformed query gQ is contained in D_i with up to k mismatching elements. Fig. 5 illustrates the concept of mismatches.

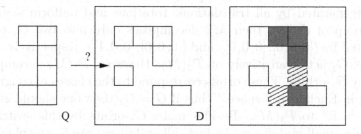

Figure 5. A query Q (left) is matched to a certain location of a document D (middle) using a shift by g. On the right, the matching positions $D \cap gQ$ are represented by gray boxes, whereas the two mismatches $gQ \setminus D$ are marked by dashed boxes.

To determine $G_{\mathcal{D}}^k(Q)$, we use a dynamic programming approach. Let $Q = \{q_1, \ldots, q_n\}$ and define $G_j := G_{\mathcal{D}}(q_j)$ as well as $\Gamma_j := G_1 \cup \ldots \cup G_j$. We inductively define credit functions $C_j : \Gamma_j \to \mathbb{Z}$ as follows. $\Gamma_1(\gamma) := k + 1$, for every $\gamma \in \Gamma_1$. For $2 \leq j \leq n$ we define

$$C_j(\gamma) := \begin{cases} C_{j-1}(\gamma) & \text{if } \gamma \in \Gamma_{j-1} \cap G_j \\ C_{j-1}(\gamma) - 1 & \text{if } \gamma \in \Gamma_{j-1} \setminus G_j \\ k + 2 - j & \text{if } \gamma \in G_j \setminus \Gamma_{j-1}. \end{cases}$$

THEOREM 8 *The elements of Γ_n with a positive credit are just all partial $(G, \mathcal{D})$-matches with at most k mismatches:*

$$G_{\mathcal{D}}^k(Q) = \{\gamma \in \Gamma_n \mid C_n(\gamma) > 0\}.$$

Proof. Let $(g, i) \in \Gamma_j$ and $Q_j := \{q_1, \ldots, q_j\}$. By induction on j we show that $C_j(g, i) = k + 1 - |gQ_j \setminus D_i|$. The start $j = 1$ is clear. To prove the inductive step $(j - 1 \to j)$ we distinguish three cases.

Case 1: $(g, i) \in \Gamma_{j-1} \cap G_j$. As $gq_j \in D_i$, $gQ_j \setminus D_i = gQ_{j-1} \setminus D_i$. Thus $C_j(g, i) := C_{j-1}(g, i) = k + 1 - |gQ_{j-1} \setminus D_i| = k + 1 - |gQ_j \setminus D_i|$.

Case 2: $(g, i) \in \Gamma_{j-1} \setminus G_j$. As $gq_j \notin D_i$, $|gQ_j \setminus D_i| = |gQ_{j-1} \setminus D_i| + 1$. Thus $C_j(g, i) := C_{j-1}(g, i) - 1 = k + 1 - (|gQ_{j-1} \setminus D_i| + 1) = k + 1 - |gQ_j \setminus D_i|$.

Case 3: $(g, i) \in G_j \setminus \Gamma_{j-1}$. Then $gq_j \in D_i$, but $gq_\ell \notin D_i$, for all $\ell \in [1 : j - 1]$. Hence $|gQ_j \setminus D_i| = j - 1$. Thus $C_j(g, i) := k + 2 - j = k + 1 - (j - 1) = k + 1 - |gQ_j \setminus D_i|$.

As $Q_n = Q$, we get $C_n(g, i) = k + 1 - |gQ \setminus D_i|$, for every $(g, i) \in \Gamma_n$. Thus $C_n(g, i) > 0$ iff $|gQ \setminus D_i| \leq k$, i.e., $(g, i) \in G_\mathcal{D}^k(Q)$.   •

To perform a complexity analysis of k-mismatch search, we first note that $C_j(\gamma) \leq C_{j-1}(\gamma)$, for all $\gamma \in \Gamma_{j-1}$. Hence in a practical implementation we can replace Γ_j by $\Gamma^j := \Gamma_j \setminus \{\gamma \in \Gamma_{j-1} \mid C_{j-1}(\gamma) \leq 0\}$. Then $\Gamma_j = \Gamma^j$, for all $j \in [1 : k+1]$ and $\Gamma 1 \subseteq \Gamma 2 \subseteq \ldots \subseteq \Gamma^{k+1} \supseteq \Gamma^{k+2} \supseteq \ldots \supseteq \Gamma^n$. Let ℓ_j and μ_j denote the lengths of G_j and Γ^j, respectively. With Lemma (3.5) and a similar technique as in the proof of Theorem (3.6) we can compute $G_\mathcal{D}^k(Q)$ with at most $\sum_{j=1}^{n-1} \min\{\ell_{j+1} \log \mu_j, \mu_j \log_{j+1}\} \leq L \log \mu_{k+1}$ comparisons, where L denotes the total length of all involved lists. In addition, we have to perform $2n + \sum_{j=1}^{n-1} \min\{\ell_{j+1}, \mu_j\}$ multiplications or inversions in the group.

Let us now turn to fuzzy queries. Recall that a fuzzy query over M consists of a sequence $\mathbf{F} = (F_1, \ldots, F_n)$ of n finite sets F_i of alternatives. We associate to such an $\mathbf{F}$ a family of ordinary queries

$$Q(\mathbf{F}) := \{\{q_1, \ldots, q_n\} \mid \forall i : q_i \in F_i\}.$$

The set $G_\mathcal{D}(\mathbf{F}) := \{(g, i) \mid \exists Q \in Q(\mathbf{F}) : gQ \subseteq D_i\}$ specifies the set of all exact partial $(G, \mathcal{D})$-matches w.r.t. the fuzzy query $\mathbf{F}$. Obviously, $G_\mathcal{D}(\mathbf{F})$ is the union of all $G_\mathcal{D}(Q)$, where Q ranges over all ordinary queries associated to $\mathbf{F}$. As already mentioned, if the F_i are pairwise disjoint, then $\mathbf{F}$ consists of $\prod_{i=1}^n |F_i|$ many ordinary queries. So, the naive algorithm that separately computes sets $G_\mathcal{D}(Q)$ for each $Q \in Q(\mathbf{F})$ and finally merges those sets is rather inefficient. Fortunately, the following result indicates an efficient algorithm to compute $G_\mathcal{D}(\mathbf{F})$.

THEOREM 9 *Let* $\mathcal{D} = (D_1, \ldots, D_N)$ *be a collection of documents over the G-set M. If* $\mathbf{F} = (F_1, \ldots, F_n)$ *is a sequence of subsets of M then the set $G_\mathcal{D}(\mathbf{F})$ of all exact partial $(G, \mathcal{D})$-matches w.r.t. the fuzzy query $\mathbf{F}$ may be computed using the following formula*

$$G_\mathcal{D}(\mathbf{F}) = \bigcap_{j \in [1:n]} \left(\bigcup_{q \in F_j} G_\mathcal{D}(q) \right) = \bigcap_{j \in [1:n]} \left(\bigcup_{q \in F_j} G_\mathcal{D}(r_q) g_q^{-1} \right).$$

Proof. $(g, i) \in G_\mathcal{D}(\mathbf{F})$ iff $gQ \subseteq D_i$, for some $Q \in Q(\mathbf{F})$. Equivalently, there exists an element $q_j \in F_j$ satisfying $gq_j \in D_i$, for all $j \in [1 : n]$, i.e., $(g, i) \in \cup_{q_j \in F_j} G_\mathcal{D}(q_j)$, for all j, proving our claim.   •

The complexity analysis for computing $G_\mathcal{D}(\mathbf{F})$ is straightforward and left to the reader. It turns out that $G_\mathcal{D}(\mathbf{F})$ may be computed with a number of comparisons which is—modulo logarithmic factors—linear in the total number of entries in all involved lists.

To incorporate prior user knowledge, we consider subgroups $U < G$. Then, for a transversal $R \ni 1_G$ of U's cosets in G we have $G = \sqcup_{r \in R} Ur$. Using this decomposition, G- and U-inverted lists are connected by

$$G_\mathcal{D}(m) = \bigsqcup_{r \in R} U_\mathcal{D}(rm)r,$$

for each $m \in M$. Then, a speed-up in query processing when restricting ourselves to subgroups $U < G$ may result from the fundamental property

$$G_\mathcal{D}(Q) = \bigsqcup_{r \in R} U_\mathcal{D}(rQ)r,$$

for all $Q \subseteq M$. Hence, $G_\mathcal{D}(Q)$ consists of many lists of the form $U_\mathcal{D}(rQ)r$. Now, assuming prior knowledge about the coset of a match $g \in G$ w.r.t. U, as in the above example where a user knows about the exact metrical position of a query, we only need to determine the list for the case $r = 1_G$, i.e., $U_\mathcal{D}(Q)$.

5.   Applications, Prototypes, and Test Results

In this section we give an overview on several applications of the proposed indexing and search technique. A more detailed treatment for the case of music retrieval may be found in our related work ([Clausen and Kurth, 2003]). We describe prototypic implementations and give some test results demonstrating time- and space-efficiency of the proposed algorithms. Recall that for each application we have to specify an underlying set M of elementary objects, a group G operating on this set, and a transversal $R \subset M$ of G-orbits to specify the inverted file index $\{G_\mathcal{D}(r) \mid r \in R\}$.

5.1   Content–Based Music Retrieval

Score-based polyphonic search has been used as a running example within this paper. In our PROMS-system ([Clausen et al., 2000]) we used the set $M := \mathbb{Z} \times [0 : 127]$ of notes consisting of onset-times and MIDI-pitches. The search is carried out w.r.t. the groups $G := (\mathbb{Z}, +)$ and $V := (16\mathbb{Z}, +)$ of time-shifts, where G shifts by metrical positions and V by whole measures each consisting of 16 metrical positions. As discussed above, the latter models prior knowledge about the metrical position of a query within a measure. The corresponding transversals R are defined as above.

Our database consists of 12,000 classical pieces of music given in the MIDI format. The pieces consist of a total of about 33 million notes. Response times for queries of various lengths are summarized in Table 1. The response times for each query length were averaged over 100 randomly generated queries. As the table demonstrates, our query processing is very fast. In addition, the index requires only a small amount of disk space and indexing of 330 MB of polyphonic music takes only 40 seconds. The uncompressed index requires 110 MB of disk space. Compressing the inverted lists using Golomb coding results in reducing the space requirement to 22 MB.

a	4	8	12	16	20	30	50	100
b	51	86	92	97	100	107	125	159
c	1	5	7	10	12	19	31	64

Table 1. Average total system response time (row b) in ms for different numbers of notes per query (row a). Row c: Disk access time for fetching inverted lists. (Pentium II, 333 MHz, 256 MB RAM).

5.2 Audio Identification

The task of audio identification may be described as follows. Given a short part q of an audio track and a database $x_1, \ldots, x_N$ of full-size audio tracks, locate all occurrences of q within the database tracks. In this, a pair (t, i) determines an occurrence of q in x_i iff $q = x_i[t : t + |q| - 1]$. Note that this problem may be stated in terms of our group-based approach since an audio signal $s : \mathbb{Z} \to \mathbb{R}$ may be interpreted via its graph as a subset $S \subset \mathbb{Z} \times \mathbb{R}$ where $G = (\mathbb{Z}, +)$ operates by addition in the $\mathbb{Z}$- (time) component. For robustness- and space-efficiency reasons, audio signals are preprocessed using a G-invariant feature extractor $F : \mathbb{R}^{\mathbb{Z}} \to \mathcal{P}(\mathbb{Z} \times X)$, where X denotes a set of feature classes (in this case G-invariance denotes the usual *time*-invariance). As an illustration we sketch a feature extractor which extracts significant local maxima from a smoothed version of a signal. For a detailed treatment of more robust feature extractors, we refer to [Ribbrock and Kurth, 2002]. F will be composed from several elementary *operators* which are each maps on the signal space, i.e., maps $\mathbb{R}^{\mathbb{Z}} \to \mathbb{R}^{\mathbb{Z}}$. First, an input signal s is smoothed by linear filtering. The corresponding operator is $C_f[s] : n \mapsto \sum_{k \in \mathbb{Z}} f(k)s(n - k)$, where f denotes a signal of finite support. Next,

K-significant local maxima are extracted by an operator

$$M_K[x] : n \mapsto \begin{cases} x(n) & \text{if } x(n-K) < \cdots < x(n) \land \\ & x(n) > \cdots > x(n+K), \\ 0 & \text{otherwise.} \end{cases}$$

The resulting signal is again processed by an operator $M'_{K'}$ extracting local maxima. $M'_{K'}$ is defined exactly as M_K, but regards only the support of the input signal. $M'_{K'}$ usually returns a very sparse output signal. The operator Δ assigns to each non-zero position of an input sequence the distance to the previous nonzero position (provided existence), and zero otherwise. Finally, let $\mathcal{Q}_{|X|}$ denote an, e.g., linear, quantizer which reduces a signal's amplitude to $|X|$ feature classes. Note that $\mathcal{Q}_{|X|}$ is an operator $\mathbb{R}^{\mathbb{Z}} \to \mathcal{P}(\mathbb{Z} \times X)$ which in addition to quantization discards zero-positions of a signal. Then our feature extractor may be written as $F := \mathcal{D}_c \circ \Delta \circ M'_{K'} \circ M_K \circ C_f$. In an example we could choose $K = 5$, $K' = 3$, and $X = [1 : 50]$. To construct our search index, we calculate $F[x_i] \subseteq \mathbb{Z} \times X$ for each signal x_i of our database. In this $\{[0, x] \mid x \in X\}$ serves as a transversal for index construction.

We briefly summarize some results of our extensive tests in the audio identification scenario. Our database consists of 4500 full-size audio tracks. This approximately amounts to 180 GB of original data or 13 days of high quality audio. Using the above significant maxima as features, we obtain an (uncompressed) index size of about 128 MB which is a compression ratio of about 1:1,400 as compared to the original data. Using different feature extraction methods, the index size may be further reduced to sizes of 1:5,000 or even lower. The query times range from only a few milliseconds (higher quality queries) to about one second. The required length of a query signal depends on the feature extractor and ranges from a few fractions of a second (significant maxima features) to 5-15 seconds (robust features and low quality queries) ([Clausen and Kurth, 2003]).

5.3   Content-Based Image Retrieval

In content-based 2D- or 3D-retrieval we are interested in finding possibly translated, rotated, or (uniformly) scaled versions of a query object in an underlying database. In this overview we shall consider translations and rotations only. Hence, the groups of interest are the group T_n of translations in $\mathbb{R}^n$, the orthogonal group O_n, and the group of *Euclidean motions* $\mathcal{E}_n := T_n \rtimes O_n$.

In content-based image retrieval, we are working with 2D images $D \subset \mathbb{R}^2 \times \mathbb{N} = P$ and are hence interested in the groups T_2 and $\mathcal{E}_2$. Assume a suitable feature extractor yielding a set of features $F(D) \subset \mathbb{R}^2 \times X$ for an image D. T_2 acts on P's first two components as described above. Hence, after feature extraction, we may create an index based on the transversal $\{[0, 0, x] \mid x \in X\} \subset P$. In our extensive tests described in [Röder, 2002] we investigated several kinds of feature extractors including corner detectors, gray value statistics and histograms.

Looking at retrieval under the group $\mathcal{E}_2$ acting on 2D points from P, we face the problem that each point $x \in \mathbb{R}^2$ is mapped to any other point $y \in \mathbb{R}^2$ by infinitely many elements from $\mathcal{E}_2$, resulting in inverted lists of infinite size. Hence we resort to indexing line segments, which are modeled by the set $\mathcal{P}_2(M) := \{L \subset M \mid |L| = 2\}$ of all two–element subsets of $\mathbb{R}^2$. Using $M' := \mathcal{P}_2(\mathbb{R}^2) \times X$ the line segments $\{[(0, a), (0, -a), x] \mid x \in X, a \geq 0\}$ serve as a transversal for indexing. In [Röder, 2002], several types of features were tested in the latter setting.

5.4 Searching 3D-Scenes

We investigated content-based search in 3D scenes for the case of a database of VRML (Virtual Reality Modeling Language) documents ([Mosig, 2001]). To obtain feasible inverted lists, the elementary objects were chosen to be all 3-sets in $\mathbb{R}^3$, i.e., $M = \mathcal{P}_3(\mathbb{R}^3)$, interpreted as the sets of all triangles in $\mathbb{R}^3$. Hence for indexing, all VRML documents were converted into documents consisting of triangles only. Indexing was performed for the groups T_3 of 3D translations and for the group $\mathcal{E}_3$ of Euclidean motions in $\mathbb{R}^3$. As a transversal of the $\mathcal{E}_3$-orbits we chose all sets of triangles with the origin as the center of gravity. Additionally, each representative is rotated such that one specific edge runs in parallel to the x-axis, this edge depending on the triangle having one, two, or three different side lengths. This way, one obtains a finite set of inverted lists for this application.

Fig. 6 illustrates the concept of 3D-retrieval using an underlying toy-database of 3D-objects (on the left). The top of the figure shows a part of an object which is used as a query to the database. An index-based search results in one object of the database matching the query (on the right). The position of that object matching the query is highlighted.

6. Related Work and Future Research

The techniques presented in this paper are related to work from across several communities. In this section we try to establish the most important relations to previous as well as ongoing research efforts.

From a classical database point of view, multimedia data may be modeled using a relational framework. In the usual approach, mul-

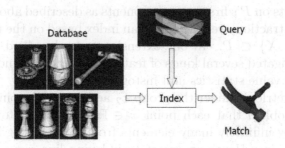

Figure 6.   Toy database of 3D-objects (left), query object (top), and matching database object (right). The matching position is highlighted.

timedia documents are preprocessed yielding certain feature vectors. The extracted features are then suitably stored in tables of a relational database ([Santini and Gupta, 2002]). Using relations, it is possible to model complex object dependencies like spatial constraints on regions in an image. Efficient retrieval methods have been proposed using approximate search like hill climbing ([Papadias, 2000]). Although our approach may be extended to a relational setting by introducing permutation groups, the methods proposed in this paper were primarily developed to exploit the structure of the underlying object set M and the group G acting on M (e.g., musical documents are structured by specific time- and pitch- intervals between single notes which are not changed by the group action). In this light, our model constitutes a special case of the general relational setting which, however, allows for very efficient query evaluation.

An important issue in multimedia indexing is the use of multidimensional access structures like k-d- or R*-Trees ([Lu, 2002]). Popular indexing approaches map multimedia documents such as time series to higher dimensional feature sequences and use multidimensional access structures for searching in those structures ([Faloutsos, 1996]). In our approach we, as far as possible, tried to avoid higher dimensional features in order to avoid problems resulting from the dimensionality curse. In indexing audio this became, e.g., possible by exploiting the fixed temporal relationships between the features. On the other hand, when dealing with more complex groups (e.g., the group of Euclidean motions in 3D leads to a 6-parameter representation for each element), our approach is also dependent on efficient algorithms for higher dimensional range and nearest neighbor search ([Agarwal, 1997]).

Specializing to time-series, there has been considerable recent interest in searching a query sequence in large sets of times–series w.r.t. various distance measures. Examples are Euclidean, ℓ^p, or dynamic time warping distances ([Keogh, 2002, Moon et al., 2002]). Whereas our approach has up to now only been applied to audio signals, its high performance suggests an extension to include general time-series search under the latter distance measures. For the particular case of audio identification, several methods suitable for large data collections have been proposed recently. Among those, the hashing algorithms proposed by ([Haitsma et al., 2001]) are the most similar to our approach, as hash tables are related to inverted files. Our approach has the advantage of needing significantly less memory for storing the index (about 100 MB as compared to 1–2 GB) with otherwise comparably (fast) performance data.

In music retrieval, most of the early work has concentrated on similarity based retrieval of melodies, see, e.g., [Uitdenbogerd and Zobel, 1999]. For a long time, retrieval in *polyphonic* music (see, e.g., [Lemström and Perttu, 2000]) already suffered from a lack of suitable data modeling. Our technique led to a breakthrough in allowing to model as well as efficiently search polyphonic music ([Clausen et al., 2000]). As our approach is up to now mainly focused on modeling and efficient retrieval, the use of music similarity measures is a natural challenge for future work.

An approach which is similar in spirit to our general technique is geometric hashing for object recognition proposed in Computer Vision, see [Wolfson and Rigoutsos, 1997]. This approach shares the modeling of shift operations which are considered for 2D/3D settings as well as the exploitation of the data's structural (geometric) properties. In our approach the data modeling is more general, which yields advantages in designing more efficient fault-tolerant retrieval algorithms.

An extension of our approach ([Clausen and Mosig, 2003]) including general distance measures between query and matching position leads to (shape-) matching problems which have been extensively treated in the area of computational geometry ([Veltkamp, 2001]). It will be a great challenge for future work to try to combine our approach, which is more tuned for use with larger datasets, with the sophisticated geometric matching techniques.

Many approaches to content-based image retrieval have been proposed in the last years, among which perhaps IBM's QBIC system (Query by Image Content) is the most popular. In comparison to those, a considerable advantage of our technique is the natural integration of partial matches and the ability to locate queries as subimages of database images at no additional cost. An interesting direction is the Blobworld approach by [Carson et al., 1999], of region-based image representation and retrieval, where image descriptors are created from a prior segmen-

54 COMPUTATIONAL NONCOMMUTATIVE ALGEBRA

tation into similarly textured regions. It would be very interesting to combine such texture descriptors with our approach to spatial indexing.

References

Agarwal, P. K. (1997). Geometric range searching. In *Handbook of Comp. Geometry*. CRC.

Carson, C., Thomas, M., Belongie, S., Hellerstein, J. M., and Malik, J. (1999). Blobworld: A system for region-based image indexing and retrieval. In *Third International Conference on Visual Information Systems*. Springer.

Clausen, M., Engelbrecht, R., Meyer, D., and Schmitz, J. (2000). PROMS: A Webbased Tool for Searching in Polyphonic Music. In *Proceedings Intl. Symp. on Music Information Retrieval 2000, Plymouth, M.A., USA*.

Clausen, M. and Kurth, F. (2003). A Unified Approach to Content-Based and Fault Tolerant Music Recognition. IEEE Transactions on Multimedia, Accepted for Publication.

Clausen, M. and Mosig, A. (2003). Approximately Matching Polygonal Curves under Translation, Rotation and Scaling with Respect to the Fréchet-Distance. In *19th European Workshop on Computational Geometry*.

Cormen, T. H., Leiserson, C. E., and Rivest, R. L. (1990). *Introduction to Algorithms*. MIT Press, Cambridge, MA.

Faloutsos, C. (1996). *Searching Multimedia Databases by Content*. Kluwer.

Haitsma, J., Kalker, T., and Oostven, J. (2001). Robust Audio Hashing for Content Identification. In *2nd Intl. Workshop on Content Based Multimedia and Indexing, Brescia, Italy*.

Keogh, E. (2002). Exact indexing of dynamic time warping. In *28th International Conference VLDB, Hong Kong*, pages 406–417.

Lemström, K. and Perttu, S. (2000). SEMEX - An Efficient Music Retrieval Prototype. In *Proceedings Intl. Symp. on Music Information Retrieval 2000, Plymouth, M.A., USA*.

Lu, G. (2002). Techniques and Data Structures for Efficient Multimedia Retrieval Based on Similarity. *IEEE Trans. on Multimedia*, 4(3):372–384.

Moon, Y.-S., Whang, K.-Y., and Han, W.-S. (2002). General Match: A Subsequence Matching Method in Time-Series Databases Based on Generalized Windows. In *SIGMOD Conference*, pages 382–393.

Mosig, A. (2001). Algorithmen und Datenstrukturen zur effizienten Konstellationssuche. Masters Thesis, Department of Computer Science III, University of Bonn, Germany.

Papadias, D. (2000). Hill Climbing Algorithms for Content-Based Retrieval of Similar Configurations. In *Proc. SIGIR, Greece*, pages 240–247.

Ribbrock, A. and Kurth, F. (2002). A Full-Text Retrieval Approach to Content-Based Audio Identification. In *Proc. 5. IEEE Workshop on MMSP, St. Thomas, Virgin Islands, USA*.

Röder, T. (2002). A Group Theoretical Approach to Content-Based Image Retrieval. Masters Thesis, Department of Computer Science III, University of Bonn, Germany.

Santini, S. and Gupta, A. (2002). Principles of Schema Design in Multimedia Data Bases. *IEEE Trans. on Multimedia*, 4(2).

Uitdenbogerd, A. L. and Zobel, J. (1999). Melodic Matching Techniques for Large Music Databases. In *Proc. ACM Multimedia.*

Veltkamp, R. C. (2001). Shape Matching: Similarity Measures and Algorithms. In *Shape Modelling International*, pages 188–199.

Wolfson, H. and Rigoutsos, I. (1997). Geometric Hashing: An Overview. *IEEE Computational Science and Engineering*, 4(4):10–21.

Yoshitaka, A. and Ichikawa, T. (1999). A Survey on Content-Based Retrieval for Multimedia Databases. *IEEE Transactions on Knowlegde and Data Engineering*, 11(1):81–93.

FOUR PROBLEMS
IN RADAR

Michael C. Wicks and Braham Himed
Air Force Research Laboratory
Sensors Directorate
26 Electronic Parkway
Rome, New York 13441-4514
Michael.Wicks@rl.af.mil
Braham.Himed@rl.af.mil

Keywords: AWACS, Hulsmeyer, PRF, RF, SBR, SNR, UHF, Watson-Watt, antenna, clutter, detection, frequency division multiplexing, microwave, modulator, phased array, pseudo-random, pulse compression, radar, radar range equation, sidelobe, spatial diversity, telemobiloskop, temporal diversity, tracker, transmitter, waveform.

1.   Introduction

Radar, short for RAdio Detection And Ranging, was invented almost a century ago by Christian Hulsmeyer in Düsseldorf, Germany. His patent for telemobiloskop, No. 165,546, issued 30 April 1904, was a collision prevention device for ships. However, Robert Watson-Watt is often given credit for inventing radar. Thirty years later, researchers were experimenting with radio transmission and reception at the Naval Research Laboratory in Washington DC. The experiment set up was such that the communications stations were on opposite sides of the Potomac River. Interference occurred each time a ship passed between these two communication stations, and proved to be a reliable indicator of an object present regardless of weather conditions. Since that time, great strides in radar have brought us air traffic control, airborne synthetic aperture radar for crop production assessment, and much more.

Radars operate via transmission through a directional antenna. Objects within the field of view scatter radio frequency (RF) energy in all directions, some towards the receiving antenna. Reflections from the clutter are undesired, and may mask signals from targets.

Throughout the twentieth century, great progress in and numerous applications of radar emerged as being commercially and militarily successful. Initially, investigations into electromagnetic theory supported

J. Byrnes (ed.) Computational Noncommutative Algebra and Applications, 57-73.

experimentation with radio wires. Then, in the 1930's, interest in the Ultra High Frequency (UHF) (300 MHz) band resulted in the first surveillance and fire control systems. By the 1940's, microwave radars operating in the 500 MHz to 10 GHz band were developed. This resulted in long range search and track radars by the late 1950's.

Advances in waveform and signal processing technology ushered in a new era starting in the 1960's. The first airborne radars were developed. Phased array technology made low sidelobe antennas available for the Airborne Warning And Control System (AWACS) wide area surveillance platform. Sidelobe cancellers emerged as the first adaptive signal processing technology designed to mitigate electromagnetic interference and jammers. Generalization of the sidelobe canceller concept, in conjunction with multi-channel phased array technology, led to space-time adaptive processing algorithms, architectures, and system concepts by the 1980's. During this same time, space-based radar (SBR) emerged as technically feasible for wide area surveillance for theater wide surveillance. As the 20th century came to a close, advances in digital technology, computing architectures and software, and solid state radio frequency devices offered some of the most exciting opportunities for fielding new radars with previously unheard of capabilities. All of this leads to the four challenge problems in radar discussed below.

2.   Radar Fundamentals

In Figure 1, a simplified block diagram presents the fundamental building blocks and inter-connectivity essential to the functioning of a radar. Here, a classical system using a reflector antenna is presented, while modern systems use phased arrays and multi-channel receivers.

In a radar system, the modulator generates a low power signal which drives the transmitter. In the transmitter, the signal may be converted in frequency from baseband, and is amplified, often to many kilowatts of peak power. The duplexer protects the receiver during transmit, and directs backscattered energy to the receiver. The antenna focuses transmit energy into narrow beams to localize in angle, as well as intercept returns from targets on receive. The receiver amplifies the return signal in order to overcome component noise, down converts the radar signal to a low intermediate frequency (1 MHz), match filters the radar returns, envelope detects the signal and digitizes it. The synchronizer is used for waveform generation, timing, and control of the transmit pulse, and measures range to the target on receive. The signal processor is designed to separate target returns from clutter and other interference, and estimate target parameters. The tracker further processes radar returns

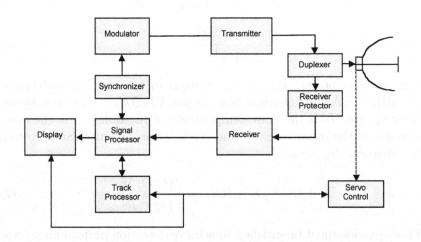

Figure 1. Basic Radar Architecture.

to present target history and predict future position for display to the operator.

The radar range equation is used to size a system to mission requirements. In its simplest form, the range equation starts with transmit power P_t, at a distance R from the radar with mainbeam transmit antenna gain G_t, the power density is

$$\frac{P_t G_t}{4\pi R^2}. \tag{1}$$

The power intercepted by a target of radar cross section σ and reradiated towards the radar is

$$\frac{P_t G_t \sigma}{4\pi R^2}. \tag{2}$$

The power density of the target return at the radar is

$$\frac{P_t G_t \sigma}{\left(4\pi R^2\right)^2}. \tag{3}$$

The received power is

$$\frac{P_t G_t \sigma A_r}{\left(4\pi R^2\right)^2}, \tag{4}$$

where

$$A_r = \frac{\lambda^2 G_r}{4\pi} \tag{5}$$

is the receive aperture at the radar. The result is

$$\frac{P_t G_t G_r \lambda^2 \sigma}{(4\pi)^3 R^4}. \tag{6}$$

The received power, P_r, can be written in terms of signal-to-noise ratio (SNR), S/N, and thermal noise power kT_0BN_F, where k is Boltzman's constant, T_0 is the noise temperature of the radar, B is the noise bandwidth of the radar receiver, and N_F is its noise figure. Substituting in to Equation (6), we get

$$P = (S/N)\, KT_0BN_F = \frac{P_t G_t G_r \lambda^2 \sigma}{(4\pi)^3 R^4}. \tag{7}$$

This equation must be satisfied to achieve detection performance commensurate with a given SNR. In addition to noise figure (N_F), numerous other losses, L, degrade the detection performance of a radar. They include hardware losses, atmospheric effects, and beam and filter shape losses, etc. The maximum range of a radar may then be computed using the formula

$$R^4_{\max} = \frac{P_t G_t G_r \lambda^2 \sigma}{(4\pi)^3 KT_0LBN_F\, (S/N)_{\text{req}}}. \tag{8}$$

In conventional radar, the output of the receiver is detection processed and compared to an adaptive threshold to determine target present (hypothesis H_1) or target absent (hypothesis H_0). From here, target declarations are handed off to the tracker for further analysis and handoff to the fighter/interceptor.

3.   Radar Waveforms

In airborne early warning radar for wide area surveillance, a number of tradeoffs must be considered before waveform parameters are selected. Among them are continuous wave or gated-wave modes of operation. In gated-wave operation, the pulse width and the pulse repetition rate must be selected to be comparable with the target characteristics as well as the radar hardware available to the systems engineer. Since Fourier analysis is a standard tool available for radar signal analysis, the gated waveform should be composed of a number of pulses regularly repeated at a high enough rate to meet the Nyquist sampling criterion. However, too high of a repetition rate causes other problems, most notably, ambiguities in range. With range ambiguities, distant target returns may be masked by very strong close-in clutter. As such, the pulse repetition rate may have to be lowered. The tradeoffs in pulse duration, pulse repetition rate, and pulse shape impact radar resolution, accuracy, and ambiguities not just in range, but in Doppler as well. Doppler is important in radar because it is proportional to target radial velocity. Through velocity measurements, target position and engagement procedures are established.

Considering the tradeoff between low, medium, and high pulse repetition frequency (PRF), several issues must be discussed. Low PRF waveforms offer precise unambiguous measurements of range, and permit simple elevation sidelobe clutter rejection. This must be balanced with difficulties in resolving Doppler ambiguities, and suffer from poor performance against ground moving targets.

In high PRF systems, range ambiguities compound the clutter rejection problem, but the clutter free Doppler may be quite large. An appropriate compromise may be medium PRF, if low elevation sidelobes are easily incorporated into the systems designed. Also complicating the waveform design are pulse compression and modulation characteristics. The purpose of pulse compression is to permit high resolution modes of operation with long duration large bandwidth pulses that meet energy on target requirements established by the radar range equation. Waveform diversity enabled by the plethora of possibilities in waveform generation, timing and control, is now commonly discussed in the literature. Waveform diversity can encompass many aspects of the signal design problem, including frequency division multiplexing, pseudo-random phase coding, and pulse compression chirp rate diversity. Similarly, this concept can easily be expanded to encompass both temporal and spatial diversity. One simple concept for waveform diversity, called spatial denial, prohibits non-cooperative receivers from interfering with mainlobe signals by altering the nature of the sidelobe structure. This can be accomplished through the application of individual waveform generation, timing and control electronics at each channel in a phased array, or by augmenting existing equipment with auxiliary antennas (at least two) designed to accomplish the same mission. This is illustrated in Figure 2.

Spatial denial is achieved as illustrated in Figure 3.

The effects of sidelobe modulation of the interferometric spatial denial antennas are to produce a broad null along the end fire axis of the antenna. If the two bracketing antennas are orthogonal to the orientation of the radar antenna, then mainlobe radar emissions are unaffected by the spatial denial radiation, while the radar sidelobes are completely obscured by this same energy. The resultant effect is to prevent the non-cooperative receiver from sampling the radar emissions, and cohering to them. In Figure 4, we illustrate the radiated waveforms that emerge from a waveform diverse transmit aperture.

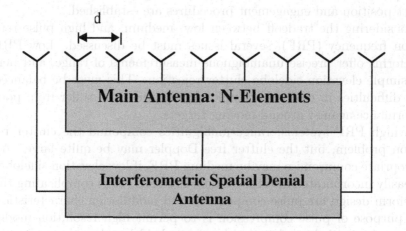

Figure 2. Interferometer Pattern.

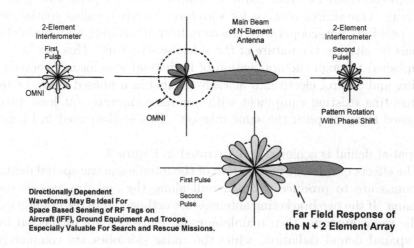

Figure 3. Spatial Denial Antenna Pattern.

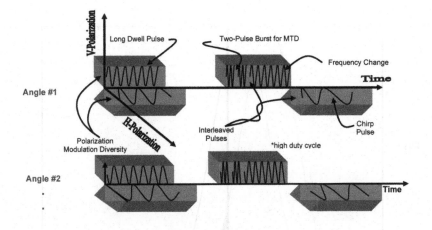

Figure 4. Transmit Waveform Diversity.

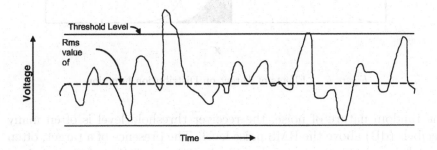

Figure 5. Typical Envelope of Radar Output.

Here, multi-mission, multi-mode waveforms are transmitted simultaneously in different directions, accomplishing different objectives, all without mutual interference.

4. Signal Processing

To achieve adequate detection in noise, the radar engineer must consider the minimum detectable signal given the radar receiver characteristics, the required SNR given the detection criteria and the radar range equation. The minimum detectable signal is determined by the ability of a receiver to detect a weak radar return as compared to the noise energy that occupies the same portion of the frequency band as the signal energy. This is illustrated in Figure 5.

The root mean-square (RMS) value of the in-band noise establishes the absolute limit to sensitivity exhibited by a radar receiver. Due to

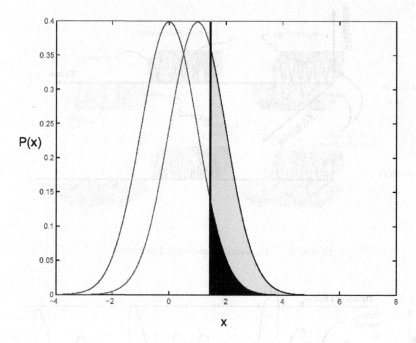

Figure 6.   Effects of Target on Distribution Function.

the random nature of noise, the receiver threshold level is often many decibels (dB) above the RMS noise level. The presence of a target, often modeled as a constant signal, will alter the statistics of the received signal by shifting the mean value of the baseband voltage waveform. This shift in mean value is what permits detection processing using the various criteria that have emerged including Neyman-Pearson, Bayes, etc. Figure 6 illustrates the effect of a target on the distribution of a received signal.

One problem in signal processing is false alarms due to thermal noise fluctuations. The ultimate tradeoff is between detection probability, which is a function of the range equation and the threshold level of the receiver, and the false alarm rate, which is a function of the noise statistics and the threshold level. The common element is the threshold setting. The probability of detection for a sine wave in noise as a function of the signal-to-noise (power) ratio and the probability of false alarm are presented in Figure 7.

The false alarm rate for wide area surveillance radars must be small because there are so many range, angle, Doppler biases with the potential for false alarms. For a 1 MHz bandwidth, there are on the order of 10^6 noise pulses per second. Hence, the false alarm probability of any pulse

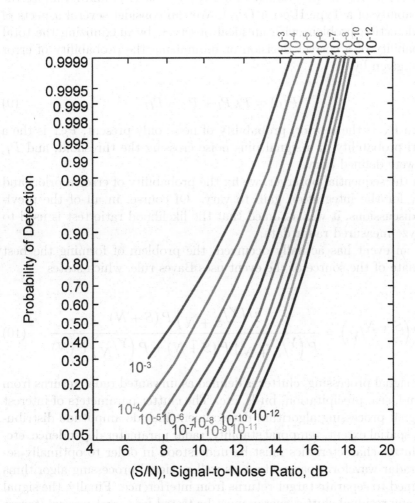

Figure 7. Probability of Detection vs. SNR.

must be small, less than one in a million. Non Gaussian clutter is more complicated and will further compound the false alarm control problem. In radar, we consider the classical hypothesis testing approach to be most appropriate. We have Type I errors, which are false alarms, and Type II errors, which are missed detections. In the Neyman-Pearson receiver, we fix the probability of a Type I error (P_I) and minimize the probability of a Type II error (P_{II}). We can consider several aspects of the detection problem with an ideal observer, by maximizing the total probability of a correct decision or minimizing the probability of error $P(e)$, given by

$$P(e) = P_N P_I + P_{S+N} P_{II} \tag{9}$$

where P_N is the a priori probability of noise only present, P_{S+} is the a priori probability of a signal plus noise crossing the threshold, and P_I, P_{II} were defined above.

In the sequential observer, we fix the probability of error a priori and allow for the integration gain to vary. Of course, in all of the previous discussions, it was assumed that the likelihood ratio test is used to analyze measured radar data.

If an event has actually occurred, the problem of forming the best estimate of the source of the event uses Bayes rule, which states

$$P\left(S + N/Y\right) = \frac{P\left(Y/S + N\right) P\left(S + N\right)}{P\left(Y/S + N\right) P\left(S + N\right) + P\left(Y/N\right) P\left(N\right)}. \tag{10}$$

In signal processing, clutter is defined as unwanted radar returns from ground, sea, precipitation, birds, etc. The clutter parameters of interest in signal processing algorithm development include amplitude distribution, spatial extent, temporal stability, radar parameter dependence, etc.

Clutter characteristics must be understood in order to optimally select radar waveform parameters, as well as signal processing algorithms designed to separate target returns from interference. Finally, the signal plus any residual clutter energy must be tested for target present/target absent. The basic assumption here is that adequate clutter rejection has been achieved such that statistical hypothesis testing against thermal noise is adequate. In realistic radar environments, that is seldom the case, and the detector circuit must be altered to adequately address the impact of clutter residue on detection performance. Clutter residue out of a simple two pulse canceller can be impacted by amplitude and phase errors in the radar system. Figure 8 illustrate these effects, where

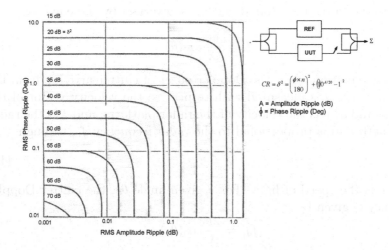

Figure 8. Cancellation Ratio

clutter residue (CR) is plotted as a function of RMS amplitude ripple (error) versus RMS phase ripple. As an example, an amplitude error of 0.1 dB limits the signal processor CR to less than 40 dB. If the interfering ground clutter is very strong, which it typically is in modern airborne radar, and exceeds the target return by more than 40 dB, then automatic detection processors will not be able to distinguish between threat targets and clutter residue, rendering the radar useless in this case.

5.   Space-Time Adaptive Processing

In radar signal processing research, modern Space-Time Adaptive Processing (STAP) developments incorporate numerous disciplines, including applied statistics, linear algebra, software engineering, advanced computing technology, transmit receive module technology, waveform generation, timing and control, antennas, and system engineering. Each of these topics requires years of study to become an expert in any one area. This section focuses on a systems engineering approach to understanding the need for STAP in modern radar.

Consider an airborne wide area surveillance radar with a transmit mainlobe pointing broadside (perpendicular to the velocity vector of the airborne radar). In this example, an airborne threat target is approaching the radar in the general direction of the broad transmit mainbeam. Our goal is to separate the target return from ground clutter returns.

The spectral spread of ground clutter is governed by the equation

$$f_d = \frac{2v}{\lambda} \cos(\theta) \,. \tag{11}$$

In this equation, f_d is the Doppler offset of clutter arriving from the angle q, where q is measured with respect to the velocity vector of the airborne radar, v is the velocity of the radar platform, and λ is the radar wavelength and is proportional to the radar frequency f_c, through

$$\lambda = c/f \,, \tag{12}$$

where c is the speed of light. For a given angle θ_1, the clutter Doppler frequency is given by

$$f_{d_1} = \frac{2v}{\lambda} \cos(\theta_1) \,.$$

If a mainlobe target is approaching the radar at radial velocity , then the target Doppler is

$$f_{d_2} = \frac{2v_2}{\lambda} \,.$$

If $f_{d_2} = f_{d_1}$, then sidelobe clutter from angle θ_1 is going to compete with mainlobe target traveling with radial velocity v_2. If the sidelobe clutter is effectively a much larger scattering center than the mainlobe airborne target, then the detection process could easily produce a type II error and the threat target will go unreported. One approach to rejecting clutter competing with mainlobe targets is to lower the sidelobes of the antenna to the point where unwanted interference is suppressed to below the thermal noise floor of the receiver. Then only mainlobe clutter will remain, and these unwanted returns can be rejected by Doppler processing. However, this is unrealistic.

In order to produce extremely low sidelobes, an antenna aperture would be very large. Additionally, the antenna would have to operate in the far field of any large scattering center. In modern airborne radar, this is unrealistic.

If the radial velocity of the threat target were known a priori, then a receive antenna pattern and a platform velocity vector could be selected to place a null on the clutter that would compete with the target in the detection process. However, this is unrealistic, even under the simplest conditions. The only way to accomplish the task of detecting targets and rejecting interference is to adaptively place nulls in the sidelobe pattern to optimize performance. However, since angle-Doppler coupling ties sidelobe clutter position to clutter Doppler frequency offset, the adaptive nulling problem is not one-dimensional (1-D) (angle) but two-dimensional (2-D) (angle-Doppler).

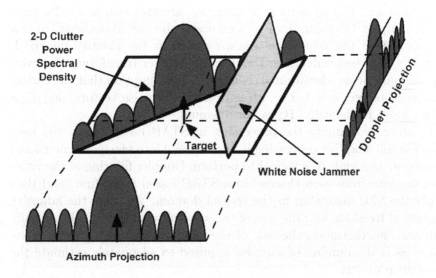

Figure 9. Angle-Doppler Profile.

In radar, angle and Doppler are derived variables. In a measurement system, data is collected in the space and time domains. It is important to note that space and angle are related by the Fourier transform, as are time and Doppler. This fact is exploited in STAP-based signal processing, and is illustrated in Figure 9.

As illustrated in Figure 9, mainlobe clutter and targets are widely separated in theory. In practice, the fundamental issue is leakage suppression among Doppler filters, or antenna beams. In space-time adaptive processing, the simplest approach to clutter rejection and target detection is through application of the sample matrix inversion (SMI) algorithm, which can be written as

$$lrt = \mathbf{s}^H \hat{\mathbf{R}}_k^{-1} \mathbf{x}_0 \,, \tag{13}$$

where s is the steering vector in angle, Doppler, or both, $\mathbf{x}_0$ is the measured data to be analyzed for target present $[H_1]$ or target absent $[H_0]$, and $\hat{\mathbf{R}}_k$ is the sample covariance matrix, where

$$\hat{\mathbf{R}}_k = \frac{1}{K} \sum_{k=1}^{K} \mathbf{x}_k \mathbf{x}_k^H \,. \tag{14}$$

The training data $\mathbf{x}_k$, $k = 1, 2, \ldots, K$, are selected from the immediate region where the data from the cell under test, $\mathbf{x}_0$, is collected. This helps to meet the independent, identically distributed assumption common in the literature. The measurement data, $\mathbf{x}_k$, are also assumed to be zero-mean complex Gaussian data vectors under the null hypothesis. In this simple test, if the sample covariance matrix is the identity matrix I, then the SMI test reduces to the discrete Fourier transform. However, $\hat{\mathbf{R}}_k$ will only be an identity matrix under the assumption that the clutter data vectors, $\mathbf{x}_k$, $k = 1, 2, \ldots, K$, are all white noise vectors, and K is large. More realistically, $\hat{\mathbf{R}}_k$ is not diagonal.

In order to simplify the processing in STAP based radar, the two-dimensional adaptive processing is decomposed into two one-dimensional processes. In radar, it is logical to perform Doppler filtering of the time-domain data from each channel in a STAP based radar first, and then apply the SMI algorithm in the spatial domain. As such, the adaptive degrees of freedom, i.e., the size of the steering vector s, is substantially reduced. Furthermore, the size of the sample covariance matrix is reduced as is the number of samples required to adequately estimate the covariance matrix.

In the 1980's, Kelly from MIT Lincoln Laboratory revisited the theoretical development that led to the SMI algorithm. The likelihood ratio test was developed under the Gaussian assumption where the data vectors were zero-mean under the null hypothesis. Additionally, the SMI algorithm was developed under the assumption of infinite iid training data, $K \to \infty$, for sample covariance matrix estimation, and that the sample covariance converges to the true covariance matrix as the number of samples tends towards infinity.

In Kelly's development, he applied one more condition to the likelihood ratio test, namely finite training data K. The generalization resulted in a new test statistic given by

$$lrt = \frac{\left| \mathbf{s}^H \hat{\mathbf{R}}^{-1} \mathbf{x}_0 \right|^2}{\left(\mathbf{s}^H \hat{\mathbf{R}}^{-1} \mathbf{s} \right) \left(1 + \mathbf{x}^H \hat{\mathbf{R}}^{-1} \mathbf{x} \right)}. \tag{15}$$

Kelly's generalized likelihood ratio test exhibits an embedded Constant False Alarm Rate (CFAR) characteristic, meaning that the probability of type I error is fixed. Additionally, it is important to note that the presence of a T2-test in the denominator prohibits the separation of the filter function from the false alarm control function in this formula. As such, the separation of filter function from false alarm control is not possible. A further modification to the SMI algorithm is to incorporate the effects of angle-Doppler coupling (structure in the covariance matrix) into its development. This could be accomplished by conditioning in angle-Doppler coupling in the probability $P_{x/H_i}(x/H_i)$, $i = 0, 1$, used in the formulation of the likelihood ratio test. Once this is accomplished, the effects of angle-Doppler coupling on other statistical tests are easily established.

6.   Four Problems in Radar

As analog hardware performance matures to a steady plateau, and Moore's Law provides for a predictable improvement in throughput and memory, it is only the advances in signal and data processing algorithms that offer potential for performance improvements in fielded sensor systems. However, it requires a revolution in system design and signal processing algorithms to dramatically alter the traditional architectures and concepts of operation. One important aspect of our current research emphasizes new and innovative sensors that are electrically small (on the order of 10 wavelengths or less), and operate in concert with a number of other electrically small sensor systems within a wide field of view (FOV). Our objective is to distribute the power and the aperture of the conventional wide area surveillance radar among a number of widely disbursed assets throughout the battlefield environment. Of course, we must have an algorithm for distributing those assets in real time as the dynamically changing surveillance demands. The mathematical challenge here relates to the traveling salesman problem. Classically, the traveling salesman must select his route judiciously in order to maximize potential sales. Recent analysis in the literature addresses multiples salesmen covering the same territory. This is analogous to our problem, where multiple unmanned aerial vehicle (UAV) based sensors are charged with the mission of detecting, tracking, and identifying all targets (friend or foe). Not only must these sensors detect and identify threat targets, they must also process data coherently across multiple platforms. Our mathematical challenge problem reduces to one in which the position and velocity all UAV-based sensors are selected to maximize detection performance and coverage area, and minimize revisit rate.

Enhancing one of the sensors described above, to be more like a classical radar with a large power-aperture product, leads to the second mathematical challenge problem to be addressed by this community. With a larger aperture and more precise estimates of target parameters (angle, Doppler), an opportunity to expand the hypothesis testing problem to include both detection and estimation emerges. Here, conventional wisdom dictates that we perform filtering and false alarm rate control as part of the detection process, yet perform track processing as a post-detection analysis, where the parameter estimation is focused upon target position and velocity history. Clearly, parameter estimation need not be accomplished as a post-detection process. Since this segmented approach to detection and track processing has been in effect for decades, it will require a dramatic demonstration of improvement before it will be embraced by the radar community.

A third challenge problem arises in the formulation of the Generalized Likelihood Ratio Test (GLRT). In Kelly's formulation of a GLRT, conditioning on finite sample support is incorporated into the basic test. As such, a statistical method developed under the assumption that only finite training data are available for sample covariance matrix formulation was made available. The next generalization to be made, in an extension of Kelly's GLRT, is to incorporate prior knowledge of the structure of the sample covariance matrix into the mathematical development of a statistical test. This mathematical structure arises due to the fact that the phase spectra of ground clutter as seen by an airborne radar is determined only by geometry, and remains independent of the underlying clutter statistics (except for initial phase). The effect of this geometric dependence is to localize the interference along a contour in the transform domain (Fourier analysis). Our objective is to formulate a single GLRT which incorporates the effects of finite training data as well as geometric dependence.

The fourth mathematical challenge facing the modern radar engineer is to incorporate adaptivity on transmit into the basic formulation of the signal processing algorithm. Since this is a new research topic, an opportunity exists to formulate the basic mathematical framework for fully adaptive radar on both transmit and receive. Further extensions arise by incorporating the above challenge problems into this analysis.

7. Conclusions

The rapid recent development in signal processing and waveform generation, timing, and control, has led to opportunities for fielding a vast array of new radar systems. Among these new sensor suites, we find that challenging problems remain unaddressed. Most notable is the incorporation of prior knowledge concerning the clutter environment into the likelihood ratio test. This work is expected to be a simple extension of Kelly's work in the 1980's.

Of course, extension of the thresholding process to a "post tracker" implementation offers further opportunities for performance enhancements and mathematical formulations. A further enhancement to radar is obtained through application of a wide variety of radar signals to a diverse number of radars, all operating in concert in a battlefield environment. This gives rise to a fourth opportunity in radar, that deals with the fielding and positioning of these radars, much like the traveling salesman problem.

All of this leads to new opportunities to exploit advances in applied mathematics, physics, electronics, and computer engineering to advance the state-of-the art in radar. While radar is nearing its 100-th year as a patented product, the prospects for advancement are as great as ever. Our defining moment in radar may have occurred in World War II; our impact on society is yet to occur, thanks to the mathematicians, scientists, and engineers who continue to advance the technology.

INTRODUCTION TO GENERALIZED CLASSICAL AND QUANTUM SIGNAL AND SYSTEM THEORIES ON GROUPS AND HYPERGROUPS

Valeriy Labunets
Urals State Tecnical University
Ekaterinburg, Russia
lab@rtf.ustu.ru

"Look," they say, "here is something new!" But no, it has all happened before, long before we have were born.

—*Good News Bible, Eccl.1:10*

Abstract   In this paper we develop two topics in parallel and show their inter- and crossrelation. The first centers on general notions of the classical signal/system theory on finite Abelian hypergroups. The second concerns the quantum hyperharmonic analysis of quantum signals (Hermitean operators associated with classical signals). We study classical and quantum generalized convolution hypergroup algebras of classical and quantum signals.

Keywords: classical and quantum signals/systems, classical and quantum Fourier transforms, Clifford algebra, hypergroups.

Introduction

The main F.Klein idea of the "Erlangen Program" lies in the correspondence of some group to a certain geometry. Thus, a group is the first (basic) notion of geometry and group can be interpreted as some

J. Byrnes (ed.) Computational Noncommutative Algebra and Applications, 75-100.

group of symmetries for the geometry. So, in general, every group of transformations (symmetries) determines its own geometry under the F.Klein correspondence **GEO** = f(**GROUP**). Quantum signal theory is a term referring to a collection of ideas and partial results, loosely held together, which assumes that there are deep connections between the worlds of quantum physics and classical signal/system theory, and that one should try to discover and develop these connections. The general topic of this paper is the following idea. If some algebraic structures arise together in quantum theory and classical signal/system theory in the same context, then one should try to make sense of this for more generalized algebraic structures. Here, the point is not to try to develop alternative theories as substitute models for quantum physics and signal/system theory, but rather to develop a "β-version" of a *unified scheme of general classical and quantum signal/system theory* based on the F.Klein "Erlangen Program". It is known that general building elements of the *Classical and Quantum Signal/System Theories* (**Cl-SST** and **Qu-SST**) are the following: 1) the Abelian group of real numbers **AR**, 2) the classical Fourier transform $\mathcal{F}$, and 3) the complex field **C**, i.e., these theories are associated with the triple $\langle\langle \mathbf{AR}, \mathcal{F}, \mathbf{C}\rangle\rangle$. Following F.Klein, we can write

$$\mathbf{Cl\text{–}SST} = f_{cl}\Big(\langle\langle \mathbf{AR}, \mathcal{F}, \mathbf{C}\rangle\rangle\Big), \quad \mathbf{Qu\text{–}SST} = f_{qu}\Big(\langle\langle \mathbf{AR}, \mathcal{F}, \mathbf{C}\rangle\rangle\Big)$$

for any F.Klein correspondences f_{cl} and f_{qu}, respectively. These correspondences mean that every triple $\langle\langle \mathbf{AR}, \mathcal{F}, \mathbf{C}\rangle\rangle$ determines certain theories **Cl–SST** and **Qu–SST**. In this paper we develop a new unified approach to the *Generalized Classical and Quantum Signal/System Theories* (**GCl-SST** and **GQu-SST**). They are based not on the triple $\langle\langle \mathbf{AR}, \mathcal{F}, \mathbf{C}\rangle\rangle$, but rather on other Abelian groups and hypergroups, on a large class of orthogonal and unitary transforms (instead of the classical Fourier transform), and involve other fields, rings and algebras (triplet color algebra, multiplet multicolor algebra, hypercomplex commutative algebras, Clifford algebras). In our approach, Generalized Classical and Quantum Signal/System Theories are two functions (correspondences) of a new triple:

$$\mathbf{GCl\text{–}SST} = f_{cl}\Big(\langle\langle \mathbf{HG}, \mathcal{F}, \mathcal{A}\rangle\rangle\Big), \quad \mathbf{GQu\text{–}SST} = f_{qu}\Big(\langle\langle \mathbf{HG}, \mathcal{F}, \mathcal{A}\rangle\rangle\Big),$$

where **HG** is a hypergroup, $\mathcal{F}$ is a unitary transform, and $\mathcal{A}$ is an algebra. When the triple $\langle\langle \mathbf{HG}, \mathcal{F}, \mathcal{A}\rangle\rangle$ is changed the theories **GCl–SST** and **GQu–SST** are changed too. For example, if $\mathcal{F}$ is the classical Fourier transform, **HG** is the group of real numbers **R** and $\mathcal{A}$ is the complex field **C**, then $\langle\langle \mathbf{R}, \mathcal{F}, \mathbf{C}\rangle\rangle$ describes free quantum particles. If

$\mathcal{F}$ is the classical Walsh transform (CWT), **HG** is an abelian dyadic group $\mathbf{Z}_2^n$ and $\mathcal{A}$ is the complex field **C**, then $\langle\langle\mathbf{Z}_2^n, \mathbf{CWT}, \mathbf{A}\rangle\rangle$ describes n-digital quantum registers. If $\mathcal{F}$ is the classical Vilenkin transform (CVT), **HG** is an abelian m-adic group $\mathbf{Z}_m^n$ and $\mathcal{A}$ is the complex field **C**, then $\langle\langle\mathbf{Z}_m^n, \mathbf{CWT}, \mathbf{C}\rangle\rangle$ describes n-digital quantum m-adic registers and so on. Every triple generates a wide class of classical and quantum signal processing methods. We develop these two topics in parallel and show their inter- and crossrelation. We study classical and quantum generalized convolution hypergroup algebras of signals and Hermitian operators. One of the main purposes of this paper is to demonstrate parallelism between the generalized classical hyperharmonic analysis and the generalized quantum hyperharmonic analysis.

1. Generalized classical signal/system theory on hypergroups

1.1 Generalized shift operators

The integral transforms and the signal representation associated with them are important concepts in applied mathematics and in signal theory. The Fourier transform is certainly the best known of the integral transforms and, with the Laplace transform, is also the most useful. Since its introduction by Fourier in the early 1800s, it has found use in innumerable applications. However, the Fourier transform is just one of many ways of signal representation, there are many other transforms of interest. An important aspect of many of these representations is the possibility to extract relevant information from a signal: the information that is actually present but hidden in its complex representation. But these transformations are not efficient analysis tools compared to the ordinary Fourier representation, since the latter is based on such useful and powerful tools of signal theory as linear and nonlinear convolutions, classical and higher-order correlations, invariance with respect to shift, ambiguity and Wigner distributions, etc. The other integral representations have no such tools. The ordinary group shift operators $(T_t^\tau x)(t) := x(t \oplus \tau)$ play the leading role in all the properties and tools of the Fourier transform mentioned above. In order to develop for each orthogonal transform a similar wide set of tools and properties as the Fourier transform has, we associate a family of generalized commutative shift operators with each orthogonal transform. Such families form *commutative hypergroups*. Only in particular cases are these hypergroups well-known Abelian groups. In 1934 F. MARTY [1, 2] and H.S. WALL [3, 4] independently introduced the notion of hypergroup.

Let $f(x) : \Omega \longrightarrow \mathcal{A}$ be an $\mathcal{A}$-valued signal, where $\mathcal{A}$ is an algebra. Usually, $\Omega = \mathbf{R}^n \times \mathbf{T}$, or $\Omega = \mathbf{Z}^n \times \mathbf{T}$, where $\mathbf{R}^n$, $\mathbf{Z}^n$ and $\mathbf{Z}_N^n$ are $n\mathrm{D}$ vector spaces over $\mathbf{R}$, $\mathbf{Z}$ and $\mathbf{Z}_N$, respectively, $\mathbf{T}$ is a compact (temporal) subset of $\mathbf{R}$, $\mathbf{Z}$, or $\mathbf{Z}_N$. Here, $\mathbf{R}$, $\mathbf{Z}$ and $\mathbf{Z}_N$ are the real field, the ring of integers, and the ring of integers modulo N, respectively. Let Ω^* be the space dual to Ω. The first one will be called the *spectral domain*, the second one is called the *signal domain* keeping the original notion of $x \in \Omega$ as "time" and $\omega \in \Omega^*$ as "frequency". Let

$$\mathbf{Sig}_0 = \mathbb{L}(\Omega, \mathcal{A}) := \{f(x)|\ f(x) : \Omega \longrightarrow \mathcal{A}\}\,,$$

$$\mathbf{Sp}_0 := \mathbb{L}(\Omega^*, \mathcal{A}) := \{F(\omega)|F(\omega) : \Omega^* \longrightarrow \mathcal{A}\}$$

be two vector spaces of $\mathcal{A}$-valued functions. In the following we assume that the functions satisfy certain general properties so that pathological cases where formulas would not hold are avoided. Let $\{\varphi_\omega(x)\}_{\omega \in \omega^*}$ be an orthonormal system of functions of $\mathbf{Sig}_0$. Then for any function $f(x) \in \mathbf{Sig}_0$ there exists a function $F(\omega) \in \mathbf{Sp}_0$ for which the following equations hold:

$$F(\omega) = \mathcal{CF}\{f\}(\omega) := \int_{x \in \Omega} f(x)\bar{\varphi}_\omega(x)d\mu(x), \tag{1}$$

$$f(x) = \mathcal{CF}^{-1}\{F\}(x) := \int_{\omega \in \Omega^*} F(\omega)\varphi_\omega(t)d\mu(\omega), \tag{2}$$

where $\mu(x), \mu(\omega)$ are certain suitable measures on the signal and spectral domains, respectively. The function $F(\omega)$ is called the $\mathcal{CF}$-spectrum of a signal $f(x)$ and expressions (1) and (2) are called the pair of *generalized classical Fourier transforms* (or $\mathcal{CF}$-transforms). In the following we will use the notation $f(x) \underset{\mathcal{CF}}{\longleftrightarrow} F(\omega)$ in order to indicate $\mathcal{CF}$-transform pairs. Along with the "time" and "frequency" domains we will work with "time-time" $\Omega \times \Omega$, "time-frequency" $\Omega \times \Omega^*$, "frequency-time" $\Omega^* \times \Omega$, and "frequency-frequency" $\Omega^* \times \Omega^*$ domains, and with four distributions, which are denoted by double letters $\mathrm{ff}(x, v) \in \mathbf{L}_2(\Omega \times \Omega, \mathcal{A})$, $\mathrm{Ff}(\omega, v) \in \mathbf{L}_2(\Omega^* \times \Omega, \mathcal{A})$, $\mathrm{fF}(x, \nu) \in \mathbf{L}_2(\Omega \times \Omega^*, \mathcal{A})$, and $\mathrm{FF}(\omega, \nu) \in \mathbf{L}_2(\Omega^* \times \Omega, \mathcal{A})$.

The classical shift operators in the "time" and "frequency" domains are defined as $(\widehat{T}_x^v f)(x) := f(x + v)$, $(\widehat{D}_\omega^\nu F)(\omega) := F(\omega + \nu)$. For $f(x) = e^{j\omega x}$ and $F(\omega) = e^{-j\omega x}$, we have $\widehat{T}_x^v e^{j\omega x} = e^{j\omega(x+v)} = e^{j\omega v}e^{j\omega x}$, and $\widehat{D}_\omega^\nu e^{-j\omega x} = e^{-j(\omega+\nu)x} = e^{-j\nu x}e^{-j\omega x}$, i.e., functions $e^{j\omega x}$, $e^{-j\omega x}$ are eigenfunctions of "time"-shift and "frequency"-shift operators $\widehat{T}_x^v$ and $\widehat{D}_\omega^\nu$ corresponding to eigenvalues $\lambda_v = e^{j\omega v}$ and $\lambda_\nu = e^{-j\nu x}$ respectively. We now generalize this result.

DEFINITION 1 *The operators*

$$(\widehat{T}_x^v \varphi_\omega)(x) = \varphi_\omega(x)\varphi_\omega(v), \quad (\widehat{T}_x^{\bar{v}}\varphi_\omega)(x) = \varphi_\omega(x)\bar{\varphi}_\omega(v), \tag{3}$$

$$(\widehat{D}_\omega^\nu \bar{\varphi}_\omega)(x) = \bar{\varphi}_\omega(x)\bar{\varphi}_\nu(x), \quad (\widehat{D}_\omega^{\bar{\nu}}\bar{\varphi}_\omega)(x) = \bar{\varphi}_\omega(x)\varphi_\nu(x). \tag{4}$$

are called the generalized commutative "time" and "frequency"-shift operators (GSOs) respectively.

It is known [5, 6] that two families of time GSOs $\{\widehat{T}_x^v\}_{v \in \Omega}$ and frequency GSOs $\{\widehat{D}_\omega^\nu\}_{\nu \in \Omega^*}$ form two commutative hypergroups. By definition, functions $\varphi_\omega(x)$ are eigenfunctions of GSOs: $\widehat{T}_x^v \varphi_\omega(x) = \varphi_\omega(v)\varphi_\omega(x)$, $\widehat{D}_\omega^\nu \bar{\varphi}_\omega(x) = \bar{\varphi}_\nu(x)\bar{\varphi}_\omega(x)$. For this reason, we can call them the *hypercharacters of the hypergroup*. The idea of a hypercharacter on a hypergroup encompasses characters of locally compact and finite Abelian groups and multiplication formulas for classical orthogonal polynomials. The theory of GSOs was initiated by LEVITAN [5, 6] and (in the terminology of hypergroup) by DUNCL [7] and JEWETT [8]. The class of commutative generalized translation hypergroups includes the class of locally compact and finite Abelian groups and semigroups. The theory for these hypergroups looks much like locally compact and finite Abelian group theory. We will show that many well-known harmonic analysis theorems extend to the commutative hypergroups associated with arbitrary Fourier transforms.

For a signal $f(x) \in \mathbf{Sig}_0$ we define its shifted copy by

$$\widehat{T}_x^v f(x) := f(x \boxplus v) = \widehat{T}_x^v \left(\int\limits_{\omega \in \Omega^*} F(\omega)\varphi_\omega(x)d\mu(\omega) \right) =$$

$$\int\limits_{\omega \in \Omega^*} F(\omega)\widehat{T}_x^v \left(\varphi_\omega \right)(x)d\mu(\omega) = \int\limits_{\omega \in \Omega^*} [F(\omega)\varphi_\omega(v)]\varphi_\omega(x)d\mu(\omega).$$

Analogously,

$$\widehat{T}_x^{\bar{v}} f(x) := f(x \boxminus v) = \int\limits_{\omega \in \Omega^*} [F(\omega)\bar{\varphi}_\omega(v)]\varphi_\omega(x)d\mu(\omega),$$

$$\widehat{D}_\omega^\nu F(\omega) := F(\omega \oplus \nu) = \int\limits_{x \in \Omega} [f(x)\bar{\varphi}_\nu(x)]\bar{\varphi}_\omega(x)d\mu(x),$$

$$\widehat{D}_\omega^{\bar{\nu}} F(\omega) := F(\omega \ominus \nu) = \int\limits_{x \in \Omega} [f(x)\varphi_\nu(x)]\bar{\varphi}_\omega(x)d\mu(x).$$

Here symbols $\boxplus, \oplus$ and $\boxminus, \ominus$ are the quasisums and quasidifferences, respectively. Obviously

$$\varphi_\omega(x \boxplus v) = \varphi_\omega(x)\varphi_\omega(v), \quad \varphi_\omega(x \boxminus v) = \varphi_\omega(x)\overline{\varphi}_\omega(v),$$

and

$$\varphi_{\omega \oplus \nu}(x) = \varphi_\omega(x)\varphi_\nu(x), \quad \varphi_{\omega \ominus \nu}(x) = \varphi_\omega(x)\overline{\varphi}_\nu(x).$$

We will need the following modulation operators:

$$(\widehat{M}_x^\nu f)(x) := \varphi_\nu(x)f(x), \quad (\widehat{M}_\omega^v F)\omega := \varphi_\omega(v)F(\omega).$$

$$(\widehat{M}_x^{\bar{\nu}} f)(x) := \bar{\varphi}_\nu(x)f(x), \quad (\widehat{M}_\omega^{\bar{v}} F)\omega := \bar{\varphi}_\omega(v)F(\omega).$$

From the GSOs definition we have:

THEOREM 1 *Shifts and modulations are connected as follows:*

$$\widehat{T}_x^v f(x) = f(x \boxplus v) \underset{C\mathcal{F}}{\longleftrightarrow} F(\omega)\varphi_\omega(v) = \widehat{M}_\omega^v F(\omega),$$

$$\widehat{T}_x^{\bar{v}} f(x) = f(x \boxminus v) \underset{C\mathcal{F}}{\longleftrightarrow} F(\omega)\bar{\varphi}_\omega(v) = \widehat{M}_\omega^{\bar{v}} F(\omega),$$

$$\widehat{M}_x^\nu f(x) = f(x)\bar{\varphi}_\nu(x) \underset{C\mathcal{F}}{\longleftrightarrow} F(\omega \oplus \nu) = \widehat{D}_\omega^\nu F(\omega),$$

$$\widehat{M}_x^\nu f(x) = f(x)\varphi_\nu(x) \underset{C\mathcal{F}}{\longleftrightarrow} F(\omega \ominus \nu) = \widehat{D}_\omega^{\bar{\nu}} F(\omega),$$

i.e.,

$$C\mathcal{F}\{\widehat{T}_x^v\}C\mathcal{F}^{-1} = \widehat{M}_\omega^v, \quad C\mathcal{F}\{\widehat{M}_x^\nu\}C\mathcal{F}^{-1} = \widehat{D}_\omega^{\bar{\nu}}, \qquad (5)$$

$$C\mathcal{F}\{\widehat{T}_x^{\bar{v}}\}C\mathcal{F}^{-1} = \widehat{M}_\omega^{\bar{v}}, \quad C\mathcal{F}\{\widehat{M}_x^{\bar{\nu}}\}C\mathcal{F}^{-1} = \widehat{D}_\omega^\nu, \qquad (6)$$

$$C\mathcal{F}^{-1}\{\widehat{D}_\omega^\nu\}C\mathcal{F} = \widehat{M}_x^{\bar{\nu}}, \quad C\mathcal{F}^{-1}\{\widehat{M}_\omega^v\}C\mathcal{F} = \widehat{T}_x^v, \qquad (7)$$

$$C\mathcal{F}^{-1}\{\widehat{D}_\omega^{\bar{\nu}}\}C\mathcal{F} = \widehat{M}_x^\nu, \quad C\mathcal{F}^{-1}\{\widehat{M}_\omega^{\bar{v}}\}C\mathcal{F}_0 = \widehat{T}_x^{\bar{v}}. \qquad (8)$$

The operators are noncommutative because

$$\widehat{M}_x^\nu \widehat{T}_x^v = \bar{\varphi}_\nu(v)\widehat{T}_x^v \widehat{M}_x^\nu, \quad \widehat{T}_x^v \widehat{M}_x^\nu = \varphi_\nu(v)\widehat{M}_x^\nu \widehat{T}_x^v,$$

$$\widehat{M}_\omega^v \widehat{D}_\omega^\nu = \bar{\varphi}_\nu(v)\widehat{D}_\omega^\nu \widehat{M}_\omega^v, \quad \widehat{D}_\omega^\nu \widehat{M}_\omega^v = \varphi_\nu(v)\widehat{M}_\omega^v \widehat{D}_\omega^\nu.$$

1.2   Some popular examples of GSOs

EXAMPLE 1 *In this example we consider GSOs on finite cyclic groups. Let $\Omega = \mathbb{Z}/N$ be an Abelian cyclic group. The ND vector Hilbert space of classical discrete $\mathcal{A}$-valued signals is* $\mathbf{Sig}_0 = \{f(x) | f(x) : \mathbb{Z}/N \longrightarrow \mathcal{A}\}$. *The characters of $\mathbb{Z}/N$ are discrete harmonic $\mathcal{A}$-valued signals $\chi_\omega(x) = \varepsilon^{\omega x}$, where $\omega \in (\mathbb{Z}/N)^* = \mathbb{Z}/N$, and ε is a primitive Nth root in an algebra $\mathcal{A}$. They form a unitary basis in* $\mathbf{Sig}_0$. *The Fourier transform in* $\mathbf{Sig}_0$ *is the discrete Fourier $\mathcal{A}$-valued transform*

$$f(x) = \mathcal{CF}_N^{-1}\{F\} = \sum_{\omega \in \mathbb{Z}/N} F(\omega)\varepsilon^{\omega x}$$

$$F(\omega) = \mathcal{CF}_N\{f\} = \sum_{x \in \mathbb{Z}/N} f(x)\varepsilon^{-\omega x}.$$

All Fourier spectra form the ND vector Hilbert spectral space $\mathbf{Sp}_0 = \{F(\omega) \mid F(\omega) : \mathbb{Z}/N \longrightarrow \mathcal{A}\}$. *The "time-frequency" and "frequency-time" domains are $\Omega \times \Omega^* = \Omega \times \Omega = \mathbb{Z}/N \times \mathbb{Z}/N$, i.e., the phase space is the 2D discrete torus $\mathbb{Z}/N \times \mathbb{Z}/N$. The "time" and "frequency"-shift operators $\widehat{T}_x^v$, $\widehat{D}_\omega^\nu$ are defined by $\widehat{T}_x^v f(x) := f(x \oplus v)$, $\widehat{D}_\omega^\nu F(\omega) := f(\omega \oplus \nu)$, where*

$$\widehat{T}_x^v := \begin{bmatrix} 0 & 1 & & & \\ & 0 & 1 & & \\ & & \ddots & & \\ & & & 0 & 1 \\ 1 & & & & 0 \end{bmatrix}^v, \quad \widehat{D}_\omega^\nu := \begin{bmatrix} 0 & 1 & & & \\ & 0 & 1 & & \\ & & \ddots & & \\ & & & 0 & 1 \\ 1 & & & & 0 \end{bmatrix}^\nu$$

and $\oplus$ is the symbol representing addition modulo N. It is obvious that $\mathcal{CF}_N\{\widehat{T}_x^v\}\mathcal{CF}_N^{-1} = \widehat{M}_\omega^v$, $\mathcal{CF}_N^{-1}\{\widehat{D}_\omega^\nu\}\mathcal{CF}_N = \widehat{M}_x^{-\nu}$. Here, modulation operators $\widehat{M}_x^\nu$ and $\widehat{M}_\omega^v$ are defined by $\widehat{M}_x^\nu f(x) := \varepsilon^{\nu x} f(x)$, $\widehat{M}_\omega^v F(\omega) := \varepsilon^{\omega v} F(\omega)$, where

$$\widehat{M}_x^\nu = \begin{bmatrix} 1 & & & & \\ & \varepsilon^1 & & & \\ & & \varepsilon^2 & & \\ & & & \ddots & \\ & & & & \varepsilon^{N-1} \end{bmatrix}^\nu, \quad \widehat{M}_\omega^v = \begin{bmatrix} 1 & & & & \\ & \varepsilon^1 & & & \\ & & \varepsilon^2 & & \\ & & & \ddots & \\ & & & & \varepsilon^{N-1} \end{bmatrix}^v.$$

The "time"-shift and "frequency"-shift operators induce the following pair of sets of noncommutative Heisenberg–Weyl operators:

$$\mathbb{HW}_x := \left\{ \widehat{\mathcal{E}}_x^{(\nu,v)} = \widehat{M}_x^\nu \widehat{T}_x^v \mid \nu \in \mathbb{Z}/N, \; v \in \mathbb{Z}/N \right\},$$

$$\mathrm{HW}_\omega := \left\{ \widehat{\mathcal{E}}_\omega^{(v,\nu)} = \widehat{M}_\omega^v \widehat{D}_\omega^\nu \mid v \in \mathbb{Z}/N, \ \nu \in \mathbb{Z}/N \right\}.$$

They act on $\mathbf{Sig}_0$ and $\mathbf{Sp}_0$ by the following rules:

$$\widehat{\mathcal{E}}_x^{(\nu,v)} f(x) := \widehat{M}_x^\nu \widehat{T}_x^v f(x) = \varepsilon^{\nu x} f(x \oplus v),$$

$$\widehat{\mathcal{E}}_\omega^{(v,\nu)} F(\omega) := \widehat{M}_\omega^v \widehat{D}_\omega^\nu F(\omega) = \varepsilon^{v\omega} F(\omega \oplus \nu).$$

$\square$

EXAMPLE 2 *Let* $\Omega_\mathbf{N}$ *and* $\Omega_\mathbf{N}^*$ *be two versions of a finite Abelian group of order* $\mathbf{N} := N_1 N_2 \cdots N_n$. *The fundamental structure theorem for finite Abelian groups implies that we may write* $\Omega_\mathbf{N}$ *and* $\Omega_\mathbf{N}^*$ *as the direct sums of cyclic groups, i.e.,* $\Omega_\mathbf{N} = \bigoplus_{l=1}^m \mathbb{Z}/N_l$, *and* $\Omega_\mathbf{N}^* = \bigoplus_{l=1}^m \mathbb{Z}^*/N_l$, *where both* $\mathbb{Z}/N_l$ *and* $\mathbb{Z}^*/N_l$ *are identified with the integers* $0, 1, \ldots, N_l - 1$ *under addition modulo* N_l. *Group elements* $x \in \Omega_\mathbf{N}$ *and* $\omega \in \Omega_\mathbf{N}^*$ *are identified with points* $x = (x_1, x_2, \ldots, x_m)$ *and* $\omega = (\omega_1, \omega_2, \ldots, \omega_m)$ *of the* mD *discrete torus, respectively. Let us embed finite groups* $\Omega_\mathbf{N}$ *and* $\Omega_\mathbf{N}^*$ *into two discrete segments* $\Omega_\mathbf{N} \longrightarrow \Omega := [0, \mathbf{N} - 1]$, $\Omega_\mathbf{N}^* \longrightarrow \Omega^* := [0, \mathbf{N} - 1]^*$ *using a mixed-radix number system*

$$x = \sum_i x_i \left(\prod_{j=0}^{i-1} N_j \right), \quad \omega = \sum_i \omega_i \left(\prod_{j=0}^{i-1} N_j \right).$$

The weights of x_1 *and* ω_1 *are unity* ($N_0 = 1$.) *The group addition induces "exotic" shifts in the segments* $\Omega := [0, \mathbf{N} - 1]$ *and* $\Omega^* := [0, \mathbf{N} - 1]^*$, *which we will denote as* $\underset{\mathbf{N}}{\oplus}$. *If* $x = (x_1, \ldots, x_m)$, $v = (v_1, \ldots, v_m)$ *and* $\omega = (\omega_1, \ldots, \omega_m)$, $\nu = (\nu_1, \ldots, \nu_m)$, *then*

$$x \underset{\mathbf{N}}{\oplus} v = (x_1, \ldots, x_m) \underset{\mathbf{N}}{\oplus} (v_1, \ldots, v_m) = (x_1 \underset{N_1}{\oplus} v_1, \ldots, x_m \underset{N_m}{\oplus} v_m)$$

and

$$\omega \underset{\mathbf{N}}{\oplus} \nu = (\omega_1, \ldots, \omega_m) \underset{\mathbf{N}}{\oplus} (\nu_1, \ldots, \nu_m) = (\omega_1 \underset{N_1}{\oplus} \nu_1, \ldots, \omega_m \underset{N_m}{\oplus} \nu_m).$$

The Fourier transforms in the space of all $\mathcal{A}$-*valued signals defined on the finite Abelian group* $\Omega_\mathbf{N} = \mathbb{Z}/N_1 \times \mathbb{Z}/N_2 \times \ldots \times \mathbb{Z}/N_m$ *in the form of* $\Omega = [0, \mathbf{N} - 1]$ *have a great interest for digital signal processing. Denote this space by* $\mathbf{Sig}_0 = \mathbb{L}(\Omega, \mathcal{A})$. *Let* ε_{N_l} *be a primitive* $\mathcal{A}$-*valued* N_l-*th root. The set of all characters of the group* $\Omega_\mathbf{N}$ *can be described by* $\chi_\omega(x) = \chi_{\omega_1}(x_1) \cdots \chi_{\omega_m}(x_m) = \varepsilon_1^{\omega_1 x_1} \cdots \varepsilon_m^{\omega_m x_m}$. *They form an orthogonal basis in the signal space* $\mathbb{L}(\Omega, \mathcal{A})$. *The Fourier transform of a signal* $f(x) \in \mathbb{L}(\Omega, \mathcal{A})$ *is defined as*

$$F(\omega) = \mathcal{CF}_\mathbf{N}\{f\}(\omega) = \sum_{t \in \Omega} f(x)\overline{\chi}_\omega(x), \quad \omega \in \Omega^*. \tag{9}$$

The inverse Fourier transform is

$$f(x) = \mathcal{CF}_\mathbf{N}^{-1}\{F\}(x) = \frac{1}{N} \sum_{\omega \in \Omega^*} F(\omega)\chi_\omega(x), \quad t \in \Omega. \tag{10}$$

The set of all functions $F(\omega)$ *forms the spectral space* $\mathbf{Sp}_0 = \mathbb{L}(\Omega^*, \mathcal{A})$.
The "time" and "frequency"-shift operators $\widehat{T}_x^v$, $\widehat{D}_\omega^\nu$ *are defined by*

$$\widehat{T}_x^v f(x) := f(x \underset{\mathbf{N}}{\oplus} v), \quad \widehat{D}_\omega^\nu F(\omega) := f(\omega \underset{\mathbf{N}}{\oplus} \nu),$$

where

$$\widehat{T}_x^v = \widehat{T}_{(x_1,x_2,...,x_m)}^{(v_1,v_2,...,v_m)} = \widehat{T}_{x_1}^{v_1} \otimes \widehat{T}_{x_2}^{v_2} \otimes \ldots \otimes \widehat{T}_{x_m}^{v_m},$$

and

$$\widehat{D}_\omega^\nu = \widehat{D}_{(\omega_1,\omega_2,...,\omega_m)}^{(\nu_1,\nu_2,...,\nu_m)} = \widehat{D}_{\omega_1}^{\nu_1} \otimes \widehat{D}_{\omega_2}^{\nu_2} \otimes \ldots \otimes \widehat{D}_{\omega_m}^{\nu_m}.$$

Obviously,

$$\mathcal{CF}_\mathbf{N}\{\widehat{T}_x^v\}\mathcal{CF}_\mathbf{N}^{-1} = \widehat{M}_\omega^v$$
$$\mathcal{CF}_\mathbf{N}^{-1}\{\widehat{D}_\omega^\nu\}\mathcal{CF}_\mathbf{N} = \widehat{M}_x^{-\nu}.$$

Here, modulation operators $\widehat{M}_x^\nu$ *and* $\widehat{M}_\omega^v$ *are defined by* $\widehat{M}_x^\nu f(x) := \chi_\omega(x)f(x)$ *and* $\widehat{M}_\omega^v F(\omega) := \chi_\omega(x)F(\omega)$, *where*

$$\widehat{M}_x^\nu = \widehat{M}_{(x_1,x_2,...,x_m)}^{(\nu_1,\nu_2,...,\nu_m)} = \widehat{M}_{x_1}^{\nu_1} \otimes \widehat{M}_{x_2}^{\nu_2} \otimes \ldots \otimes \widehat{M}_{x_m}^{\nu_m}$$

and

$$\widehat{M}_\omega^v = \widehat{M}_{(\omega_1,\omega_2,...,\omega_m)}^{(v_1,v_2,...,v_m)} = \widehat{M}_{\omega_1}^{v_1} \otimes \widehat{M}_{\omega_2}^{v_2} \otimes \ldots \otimes \widehat{M}_{\omega_m}^{v_m}.$$

$\square$

EXAMPLE 3 *Let* $\Omega = [a,b]$, $\Omega^* = \{0,1,2,...\} := \mathbf{N}_0$, *and let* $\varphi_k(t) \equiv p_k(t)$ *be a family of classical orthogonal polynomials. Then*

$$F(k) := \int_a^b f(t)p_k(t)\varrho(t)dt, \quad f(t) := \sum_{k=0}^\infty h_k^{-1}F(k)p_k(t) \tag{11}$$

is the pair of generalized Fourier transforms, where $k \in \mathbf{N}_0$, $t \in [a,b]$, $\varrho(t)dt = d\mu(t)$ *and* $d\mu(k) = h_k^{-1}$ *are measures on the signal and spectral domains respectively. We consider special cases of classical orthogonal polynomials. Case 1. Let* $\Omega = [-1, +1]$, $\varrho(t) = (1 - t)^\alpha (1 + t)^\beta$, $\alpha > \beta - 1$ *and* $Jac_k^{(\alpha,\beta)}(t)$ *be* (α, β)-*Jacobi polynomials. In this case, generalized Fourier transforms for each* α *and* β *are the Fourier–Jacobi transforms*

$$^{(\alpha,\beta)}F(k) = {}^{(\alpha,\beta)}\mathcal{CF}\{f\}(k) = \int_{-1}^{+1} f(t) Jac_k^{(\alpha,\beta)}(t)(1-t)^\alpha(1+t)^\beta dt, \quad (12)$$

$$f(t) := {}^{(\alpha,\beta)}\mathcal{CF}^{-1}\{F\}(k) = \sum_{k=0}^{\infty} h_k^{-1} \, {}^{(\alpha,\beta)}F(k) Jac_k^{(\alpha,\beta)}(t) \quad (13)$$

for special constants h_k. *If* $\alpha > \beta > -\frac{1}{2}$ *then the multiplication formula for* (α, β)-*Jacobi polynomials is*

$$Jac_k^{(\alpha,\beta)}(\tau) Jac_k^{(\alpha,\beta)}(t) = P_k^{(\alpha,\beta)}(t \boxplus \tau) = \int_0^1 \int_0^\pi Jac_k^{(\alpha,\beta)} \left[\frac{1}{2}(1+\tau)(1+t) + \right.$$

$$\left. \frac{1}{2}(1-\tau)(1-t)s^2 + \sqrt{(1-\tau^2)(1-t^2)}s\cos\theta - 1 \right] d\mu(s,\theta), \quad (14)$$

where $d\mu(s,\theta) = \frac{2\Gamma(\alpha+1)}{\sqrt{\pi}\Gamma(\alpha-\beta)\Gamma(\beta+\frac{1}{2})}(1-s^2)^{\alpha-\beta-1}s^{2\beta+1}(\sin\theta)^{2\beta}dsd\theta$. *There follows*

$$(T_t^\tau f)(t) = f(t \boxplus \tau) = \int_0^1 \int_0^\pi f \left[\frac{1}{2}(1+\tau)(1+t) + \right.$$

$$\left. +\frac{1}{2}(1-\tau)(1-t)s^2 + \sqrt{(1-\tau^2)(1-t^2)}s\cos\theta - 1 \right] d\mu(s,\theta). \quad (15)$$

Case 2. If $\alpha = \beta = 0$ *then* $\{Jac_k^{(0,0)}(t)\}_{k=0}^{\infty} = \{Leg_k(t)\}_{k=0}^{\infty}$ *is the Legendre basis. From (15) we obtain the Legendre GSOs*

$$(T_t^\tau f)(t) = f(t \boxplus \tau) = \frac{1}{2\pi} \int_{-1}^{1} f\left(\tau t + \sqrt{(1-\tau^2)(1-t^2)}s\right)(1-s^2)^{-1/2}ds, \quad (16)$$

associated with the Legendre transform. Case 3. If $\alpha = \beta = -0.5$, *then* $\{Jac_k^{(-0.5,-0.5)}(t)\}_{k=0}^{\infty} = \{Ch_k(t)\}_{k=0}^{\infty}$ *is the Legendre basis. In this case,* $\Omega = (-1, 1)$, $\varrho(t) = (1-t^2)^{-1/2}$, $h_0 = \frac{\pi}{2}$, $h_n = \pi$, $n \in \mathbf{N}_0 \equiv \Omega^*$. *For the Chebyshev polynomials the following multiplication formula is known:*

$$Ch_n(\tau)Ch_n(t) = Ch_n(t \boxplus \tau) =$$

$$\frac{1}{2}\left[Ch_n\left(\tau t+\sqrt{(1-\tau^2)(1-t^2)}\right)+Ch_n\left(\tau t-\sqrt{(1-t^2)(1-t^2)}\right)\right].$$
(17)

Therefore,

$$(T_t^\tau f)(t) = f(t \boxplus \tau) =$$

$$\frac{1}{2}\left[f\left(t\tau+\sqrt{(1-\tau^2)(1-t^2)}\right)+f\left(t\tau-\sqrt{(1-\tau^2)(1-t^2)}\right)\right].\quad (18)$$

□

EXAMPLE 4 *Finally, we consider the infinite interval* $\Omega = (-\infty, +\infty)$. *Let us introduce the signal and the spectrum spaces*

$$\mathbb{L}_2\left(\mathbb{R},\mathbb{C},w(t)\right)=\left\{f(t)\;\Big|\;\Big(f(t):\mathbb{R}\longrightarrow\mathbb{C}\Big)\&\Big(\int\limits_{-\infty}^{+\infty}|f(t)|^2w(t)dt<\infty\Big)\right\},$$

$$\mathbb{L}_2(\mathbb{N},\mathbb{C},\mu_n)=\left\{F(n)\;\Big|\;\Big(F(n):\mathbb{N}\longrightarrow\mathbb{C}\Big)\&\Big(\sum_{n\in\mathbb{N}}w_n|F(n)|^2<\infty\Big)\right\},$$

with the scalar products

$$(f,g):=\int\limits_{-\infty}^{+\infty}f(t)g(t)e^{-t^2/2}\,dt,\quad(F,G)=\sum_{n\in\mathbb{N}}\frac{1}{2^nn!\sqrt{\pi}}F(n)G(n),$$

where $\Omega^* = \mathbb{N} = \{0,1,2,\dots,\}$, $d\mu(t) = w(t)dt$, $w(t) = e^{-t^2/2}$, *and* $w_n = 1/2^nn!\sqrt{\pi}$. *In this case, the generalized classical Fourier transform of a signal* $f(t) \in \mathbb{L}_2\left(\mathbb{R},\mathbb{C},e^{-t^2/2}\right)$ *is the Fourier–Hermite transform*

$$F(n)=\mathcal{CF}\{f\}(n)=\int_{-\infty}^{+\infty}f(t)Her_n(t)e^{-t^2/2}\,dt,$$

where

$$f(t)=\mathcal{CF}^{-1}\{F\}(n)=\sum_{n=0}^{\infty}\frac{1}{2^nn!\sqrt{\pi}}F(n)Her_n(t),$$

where $Her_n(t)$ *are Hermite polynomials. Since*

$$Her_k(t)Her_k(\tau)=Her_k(t\boxplus\tau)=\frac{(-1)^k\Gamma(k+(3/2))^{2k+1}}{\sqrt{\pi}}\times$$

$$\times\int_0^\pi Her_k\Big[(t^2+\tau^2+2t\tau\cos\varphi)^{1/2}\exp(-t\tau\sin\varphi)\sin\varphi J_0(t\tau\sin\varphi)d\varphi\Big],$$
(19)

then

$$T_t^\tau f(t) = f(t \boxplus \tau) =$$

$$= \int_0^\pi f\left(\sqrt{t^2 + \tau^2 + 2t\tau \cos \varphi}\right) e^{-t\tau \cos \varphi} \sin \varphi J_0(t\tau \sin \varphi)d\varphi, \quad (20)$$

where $J_0(.)$ is the Bessel function. □

1.3 Generalized convolutions and correlations

It is well known that stationary linear dynamic systems (LDS) are described by convolution integrals. Using the GSO notion, we can formally generalize the notions of convolution and correlation [11]–[18].

DEFINITION 2 *The following functions*

$$y(x) := (h \Diamond f)(x) = \int_{v \in \Omega} h(v)f(x \boxminus v)d\mu(v), \quad (21)$$

$$Y(\omega) := (H \heartsuit F)(\omega) = \int_{\nu \in \Omega^*} H(\nu)F(\omega \ominus \nu)d\mu(\nu) \quad (22)$$

are called the $\Diamond$ and $\heartsuit$-convolutions respectively.

The spaces $\mathbf{Sig}_0$ and $\mathbf{Sp}_0$ equipped with multiplications $\Diamond$ and $\heartsuit$ form commutative signal and spectral convolution algebras $\langle\langle \mathbf{Sig}_0, \Diamond \rangle\rangle$ and $\langle\langle \mathbf{Sp}_0, \heartsuit \rangle\rangle$, respectively.

DEFINITION 3 *The expressions*

$$(f \clubsuit g)(v) := \int_{x \in \Omega} f(x)\overline{g}(x \boxminus v)d\mu(x), \quad (23)$$

$$(F \spadesuit G)(\nu) := \int_{\omega \in \Omega^*} F(\omega)\overline{G}(\omega \ominus \nu)d\mu(\omega) \quad (24)$$

are referred to as the cross $\clubsuit$ and $\spadesuit$-correlation functions of signals f, g and spectra F, G, respectively. If $f = g$ and $F = G$, then the crosscorrelation functions are called the $\clubsuit$ and $\spadesuit$-autocorrelation functions.

The measures indicating the similarity between fF-distributions and Ff-distributions and their time and frequency-shifted versions are their crosscorrelation functions.

DEFINITION 4 *The expressions*

$$(fF \clubsuit\spadesuit gG)(v,\nu) := \int_{t \in \Omega} \int_{\omega \in \Omega^*} fF(x,\omega)\overline{gG}(x \boxminus v, \omega \ominus \nu)d\mu(x)d\mu(\omega), \quad (25)$$

$$(Ff\spadesuit\clubsuit Gg)(\nu,v) := \int\limits_{\nu\in\Omega^*} \int\limits_{v\in\Omega} Ff(\omega,t)\overline{Gg}(\omega\ominus\nu,t\boxminus v)d\mu(x)d\mu(\omega). \quad (26)$$

are referred to as the ♣♠ *and* ♠♣*-crosscorrelation functions of the distributions respectively. If* $fF(x,\omega) = gG(x,\omega)$ *and* $Ff(\omega,t) = Gg(\omega,t)$, *then the crosscorrelation functions are called the autocorrelation functions.*

THEOREM 2 *Generalized classical Fourier transforms (1) and (2) map* $\diamondsuit$ *and* $\heartsuit$*-convolutions and* ♣ *and* ♠*-correlations into the products of spectra and signals, respectively,*

$$\mathcal{CF}\{h\diamondsuit f\} = \mathcal{CF}\{h\}\mathcal{CF}\{f\}, \quad \mathcal{CF}^{-1}\{H\heartsuit F\} = \mathcal{CF}^{-1}\{H\}\mathcal{CF}^{-1}\{F\}$$

$$\mathcal{CF}\{f\clubsuit g\} = \mathcal{CF}\{f\}\overline{\mathcal{CF}\{g\}}, \quad \mathcal{CF}^{-1}\{F\spadesuit G\} = \mathcal{CF}^{-1}\{F\}\overline{\mathcal{CF}^{-1}\{G\}}$$

Taking special forms of the GSOs, one can obtain known types of convolutions and crosscorrelations: arithmetic, cyclic, dyadic, m-adic, etc. Signal and spectral algebras have many of the properties associated with classical group convolution algebras. Many of them are catalogued in [9]–[19].

1.4 Generalized ambiguity functions and Wigner distributions

The Wigner distribution was introduced in 1932 by E. WIGNER [20] in the context of quantum mechanics. There he defined the probability distribution function of simultaneous values of the spatial coordinates and impulses. Wigner's idea was introduced in signal analysis in 1948 by J. VILLE [21], but it did not receive much attention there until 1953 when P. WOODWARD [22] reformulated it in the context of radar theory. Woodward proposed treating the question of radar signal ambiguity as part of the question of target resolution. For that, he introduced a function that described the correlation between a radar signal and its Doppler-shifted and time-translated version:

$$\mathrm{AW}^a[f](\nu,v) = \int\limits_{-\infty}^{+\infty} f(x)\bar{f}(x-v)e^{-j\nu x}dx = \underset{x\to\nu}{\mathcal{CF}}\{ff^a(x,v)\},$$

where $ff^a(x,v) := f(x)\bar{f}(x-v)$. The distribution $\mathrm{AW}^a[f](\nu,v)$ is called the *asymmetric Woodward ambiguity function*. It describes the local ambiguity of locating targets in range (time delay v) and in velocity (Doppler frequency ν). Its absolute value is called the *uncertainty function* since it is related to the *uncertainty principle* of radar signals.

The next time-frequency distribution is the so-called *symmetric Woodward ambiguity function*:

$$\text{AW}^s[f](\nu, v) := \underset{x \to \nu}{\mathcal{CF}} \left\{ f\left(x + \frac{v}{2}\right) \bar{f}\left(x - \frac{v}{2}\right) \right\} = \underset{x \to \nu}{\mathcal{CF}} \left\{ f f^s(x, v) \right\}, \quad (27)$$

where $f f^s(x, v) := f\left(x + \frac{v}{2}\right) \bar{f}\left(x - \frac{v}{2}\right)$. Analogously, we have expressions for computing $\text{AW}^s[F](v, \nu)$ in the frequency domain

$$\text{AW}^a[F](\nu, v) = \underset{v \leftarrow \omega}{\mathcal{CF}^{-1}} \{ F(\omega) \bar{F}(\omega - \nu) \} = \underset{v \leftarrow \omega}{\mathcal{CF}^{-1}} \{ F F^a(\nu, \omega) \},$$

$$\text{AW}^s[F](\nu, v) = \underset{v \leftarrow \omega}{\mathcal{CF}^{-1}} \{ F\left(\omega + \frac{\nu}{2}\right) \bar{F}\left(\omega - \frac{\nu}{2}\right) \} = \underset{v \leftarrow \omega}{\mathcal{CF}^{-1}} \{ F F^s(\nu, \omega) \}.$$

If $F = \mathcal{CF}\{f\}$, then from Parseval's relation we obtain

$$\text{AW}^a[f](\nu, v) = \text{AW}^a[F](\nu, v) \quad \text{and} \quad \text{AW}^s[f](\nu, v) = \text{AW}^s[F](\nu, v).$$

For this reason, we shall denote $\text{AW}^a[f](\nu, v)$, $\text{AW}^a[F](\nu, v)$ by $\text{AW}^a(\nu, v)$ and $\text{AW}^s[f](\nu, v)$, $\text{AW}^s[F](\nu, v)$ by $\text{AW}^s(\nu, v)$. Further, we use the symbol $\text{AW}(\nu, v)$ for both $\text{AW}^a(\nu, v)$ and $\text{AW}^s(\nu, v)$.

Important examples of time-frequency distributions are the so-called *asymmetrical and symmetrical Wigner–Ville distributions*. They can be defined as the 2D symplectic Fourier transform of $\text{AW}^a[f](\nu, v)$ and $\text{AW}^s[f](\nu, v)$, respectively,

$$\text{WV}^a[f](x, \omega) = \underset{x \leftarrow \nu}{\mathcal{CF}^{-1}} \underset{\omega \leftarrow v}{\mathcal{CF}} \{ \text{AW}^a[f](\nu, v) \} = f(x) \overline{F}(\omega) e^{-j\omega x}, \quad (28)$$

$$\text{WV}^s[f](x, \omega) = \underset{x \leftarrow \nu}{\mathcal{CF}^{-1}} \underset{\omega \leftarrow v}{\mathcal{CF}} \{ \text{AW}^s[f](\nu, v) \} = \int_{-\infty}^{\infty} f\left(x + \frac{v}{2}\right) \bar{f}\left(x - \frac{v}{2}\right) e^{-j\omega v} dv. \quad (29)$$

The 2D symplectic Fourier transform in (29) and (28) can be also viewed as two sequentially performed 1D transforms with respect to v and ν. The transform with respect to ν yields the *temporal autocorrelation functions*

$$\text{ff}^a(x, v) = \underset{x \leftarrow \nu}{\mathcal{CF}^{-1}} \{ \text{AW}^s[f](\nu, v) \} = f(x) f(x - v),$$

$$\text{ff}^s(x, v) = \underset{x \leftarrow \nu}{\mathcal{CF}^{-1}} \{ \text{AW}^s[f](\nu, v) \} = f\left(x + \frac{v}{2}\right) f\left(x - \frac{v}{2}\right).$$

The transform with respect to ν yields the *frequency autocorrelation functions*

$$\text{FF}^a(\nu, \omega) = \underset{\omega \leftarrow v}{\mathcal{CF}} \{ \text{AW}^s[F](\nu, v) \} = F(\omega) F(\omega - \nu),$$

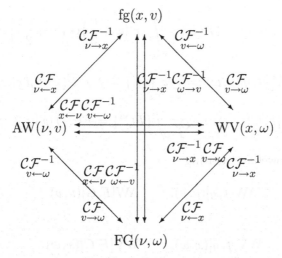

Figure 1. Diagram of relations between the different generalized 2D distributions

$$FF^s(\nu,\omega) = \underset{\omega \leftarrow v}{\mathcal{CF}}\{AW^s[F](\nu,v)\} = F\left(\omega + \frac{\nu}{2}\right) F\left(\omega - \frac{\nu}{2}\right).$$

We can formally generalize the notions of cross-ambiguity functions and Wigner-Ville distributions using the GSO notion.

DEFINITION 5 *The symmetric and asymmetric generalized Woodward distributions (cross-ambiguity functions) of two signals f, g and two spectra F, G are defined by*

$$AW^s[f,g](\nu,v) = \underset{\nu \leftarrow x}{\mathcal{CF}}\left\{fg^s\right\} = \int_{x \in \Omega} \left[f\left(x \boxplus \frac{v}{2}\right)\bar{g}\left(x \boxminus \frac{v}{2}\right)\right] \bar{\varphi}_\nu(x)d\mu(x),$$

$$AW^s[F,G](\nu,v) = \underset{v \leftarrow \omega}{\mathcal{CF}^{-1}}\left\{FG^s\right\} = \int_{\omega \in \Omega^*} \left[F\left(\omega \oplus \frac{v}{2}\right)\bar{G}\left(\omega \ominus \frac{v}{2}\right)\right] \varphi_\omega(v)d\mu(\omega),$$

$$AW^a[f,g](\nu,v) = \underset{\nu \leftarrow x}{\mathcal{CF}}\left\{fg^a\right\} = \int_{x \in \Omega} \left[f(x)\bar{g}(x \boxminus v)\right]\bar{\varphi}_\nu(x)d\mu(x),$$

$$AW^a[F,G](\nu,v) = \underset{v \leftarrow \omega}{\mathcal{CF}^{-1}}\left\{FG^a\right\} = \int_{\omega \in \Omega^*} \left[F(\omega)\bar{G}(\omega \ominus \nu)\right]\varphi_\omega(v)d\mu(\omega).$$

DEFINITION 6 *The generalized symmetric and asymmetric Wigner-Ville distributions of two signals f, g and two spectra F, G are defined by*

$$WV^s[f,g](x,\omega) = \underset{\omega \leftarrow v}{\mathcal{CF}}\{fg^s\} = \int_{v \in \Omega} \left[f\left(x \boxplus \frac{v}{2}\right)\bar{g}\left(x \boxminus \frac{v}{2}\right)\right]\bar{\varphi}_\omega(v)d\mu(v),$$

$$WV^s[F,G](x,\omega) = \underset{x \leftarrow \nu}{C\mathcal{F}}^{-1}\{FG^s\} = \int_{\nu \in \Omega^*} \left[F\left(\omega \oplus \frac{\nu}{2}\right) \bar{G}\left(\omega \oplus \frac{\nu}{2}\right) \right] \varphi_\nu(x) d\mu(\nu),$$

$$WV^a[f,g](x,\omega) := \underset{\omega \leftarrow v}{C\mathcal{F}}\{fg^a\} = f(x)\overline{F}(\omega)\bar{\varphi}_\omega(x),$$

$$WV^a[F,G](x,\omega) := \underset{x \leftarrow \nu}{C\mathcal{F}}^{-1}\{FG^a\} = F(\omega)\bar{f}(x)\varphi_\omega(x).$$

Figure 1 is a flowchart relating

$$\mathrm{AW}[f,g](\nu,v), \qquad \mathrm{AW}[F,G](\nu,v)$$

and

$$\mathrm{WV}[f,g](x,\omega), \qquad \mathrm{WV}[F,G](x,\omega)$$

and

$$fg(x,v), \qquad FG(\nu,\omega).$$

We can construct two vector Hilbert spaces of "time-frequency" and "frequency-time" distributions

$$\mathbf{WV} := \{\mathrm{WV}(x,\omega) \mid \mathrm{WV}(x,\omega) : \Omega \times \Omega^* \longrightarrow \mathcal{A}\},$$

$$\mathbf{AW} := \{\mathrm{AW}(\nu,v) \mid \mathrm{AW}(\nu,v) : \Omega^* \times \Omega \longrightarrow \mathcal{A}\}.$$

DEFINITION 7 *The generalized Woodward–Gabor ambiguity transforms (or short-time and short-frequency generalized Fourier transforms)* $\mathcal{WG}_g$ *and* $\mathcal{WG}_G$ *associated with functions* g *and* G *are defined as the following mappings:*

$$\mathcal{WG}_g : \mathbb{L}(\Omega, \mathcal{A}) \longrightarrow \mathbb{L}(\Omega^* \times \Omega, \mathcal{A}), \quad \mathcal{WG}_G : \mathbb{L}(\Omega^*, \mathcal{A}) \longrightarrow \mathbb{L}(\Omega^* \times \Omega, \mathcal{A})$$

given by

$$\mathcal{WG}_g\{f\}(\nu,v) := AW[f,g](\nu,v), \quad \mathcal{WG}_G\{F\}(\nu,v) := AW[F,G](v,\nu).$$

DEFINITION 8 *The generalized Wigner-Ville transforms* $\mathcal{WV}_g$ *and* $\mathcal{WV}_G$ *associated with functions* g *and* G *are defined as mappings*

$$\mathcal{WV}_g : \mathbb{L}(\Omega, \mathcal{A}) \longrightarrow \mathbb{L}(\Omega \times \Omega^*, \mathcal{A}), \quad \mathcal{WV}_G : \mathbb{L}(\Omega^*, \mathcal{A}) \longrightarrow \mathbb{L}(\Omega \times \Omega^*, \mathcal{A})$$

given by

$$\mathcal{WV}_g\{f\}(x,\omega) := WV[f,g](x,\omega), \quad \mathcal{WV}_G\{F\}(x,\omega) := WV[F,G](x,\omega).$$

2. Generalized quantum signal/system theory on hypergroups

2.1 Basic definitions

The *basic objects of quantum harmonic analysis* (QHA) are related not to classical signals and spectra f, F but to quantum signals and quantum spectra (Hermitian operators) $\widehat{f}$, $\widehat{F}$ associated with classical signals and spectra as follows:

$$f \to \text{AW}[f] \to \widehat{f}, \quad F \to \text{AW}[F] \to \widehat{F}.$$

These maps are called the *Weyl quantizations* of signals and spectra, respectively. There are also the *Schwinger quantizations* using Wigner–Ville distributions:

$$f \to \text{WV}[f] \to \widehat{f}, \quad F \to \text{WV}[F] \to \widehat{F}.$$

The functions $\text{AW}[f](\nu, v)$, $\text{AW}[F](\nu, v)$ (or $\text{WV}[f](x, \omega)$, $\text{WV}[F](x, \omega)$) are called the *symbols* (a symbol is not a kernel) of the quantum signal $\widehat{f}$ and the quantum spectra $\widehat{F}$, respectively, and are denoted by

$$\text{AW}[f](\nu, v) := sym\{\widehat{f}\}, \quad \text{AW}[F](v, \nu) := sym\{\widehat{F}\},$$

or

$$\text{WV}[f](x, \omega) := sym\{\widehat{f}\}, \quad \text{WV}[F](\omega, x) := sym\{\widehat{F}\}.$$

Vice versa, a quantum signal $\widehat{f}$ and quantum spectra $\widehat{F}$ are called the *operators associated with a classical signal f* and classical spectrum by symbols $\text{AW}[f]$, $\text{AW}[F]$ (or by $\text{WV}[f]$, $\text{WV}[F]$), respectively, and they are denoted by

$$\widehat{f} := Op\{\text{AW}[f]\}, \quad \widehat{F} := Op\{\text{AW}[F]\},$$

or

$$\widehat{f} := Op\{\text{WV}[f]\}, \quad \widehat{F} := Op\{\text{WV}[F]\}.$$

All quantum signals $\widehat{f}$ and quantum spectra $\widehat{F}$ form the following quantum spaces:

$$\mathbf{Sig}_1 := \{\widehat{f} \mid \widehat{f} \text{ are operators acting in } \mathbb{L}_2(\Omega, \mathcal{A})\},$$

$$\mathbf{Sp}_1 := \{\widehat{F} \mid \widehat{F} \text{ are operators acting in } \mathbb{L}_2(\Omega^*, \mathcal{A})\}.$$

Let $\mathbf{Sig}_1$ and $\mathbf{Sp}_1$ be the spaces of quantum signals and quantum spectra with the following scalar products and norms:

$$\langle \widehat{f_1} | \widehat{f_2} \rangle := \mathbf{Tr}(\widehat{f_1}\widehat{f_2}^\dagger), \quad \|\widehat{f}\| := \langle \widehat{f} | \widehat{f} \rangle = \mathbf{Tr}(\widehat{f}\widehat{f}^\dagger),$$

$$\langle \widehat{F}_1 | \widehat{F}_2^\dagger \rangle := \mathbf{Tr}(\widehat{F}_1 \widehat{F}_2^\dagger), \quad ||\widehat{F}|| := \langle \widehat{F} | \widehat{F} \rangle = \mathbf{Tr}(\widehat{F}\widehat{F}^\dagger),$$

where $\mathbf{Tr}(.)$ denotes the trace.

DEFINITION 9 *The spaces* $\mathbf{Sig}_1$, $\mathbf{Sp}_1$ *with the scalar products* $\langle \widehat{f}_1 | \widehat{f}_2 \rangle$, $\langle \widehat{F}_1 | \widehat{F}_2 \rangle$ *and norms* $||\widehat{f}||$, $||\widehat{F}||$ *are called the Hilbert–Liouville spaces.*

Let $\{\widehat{\varphi}_\lambda\}_{\lambda \in \Lambda}$ and $\{\widehat{\psi}_\lambda\}_{\lambda \in \Lambda}$ be two Λ-parametric families of operators, parametrized by the label $\lambda = (\lambda_1, \lambda_2, \ldots, \lambda_r) \in \Lambda \subset \mathbf{R}^r$ of a subset Λ of an rD space $\mathbf{R}^r$ endowed with a suitable measure $\mu(\lambda)$. These families are called the *quora* for any subalgebras $\mathbf{Alg}_1 \subset \mathbf{Sig}_1$ and $\mathbf{Alg}_1^* \subset \mathbf{Sp}_1$, if every quantum signal $\widehat{f} \in \mathbf{Alg}_1$ and quantum spectrum $\widehat{F} \in \mathbf{Alg}_1^*$ is determined by all scalar products

$$\mathrm{FT}(\lambda) = \langle \widehat{f} | \widehat{\varphi}_\lambda \rangle = \mathbf{Tr}(\widehat{f}\widehat{\varphi}_\lambda^\dagger), \ \ \mathrm{TF}(\lambda) = \langle \widehat{F} | \widehat{\psi}_\lambda \rangle = \mathbf{Tr}(\widehat{F}\widehat{\psi}_\lambda^\dagger)$$

for all $\widehat{\varphi}_\lambda$ and $\widehat{\psi}_\lambda$. The fundamental property of the quora is that any quantum signal and spectrum can be expressed as integral transforms

$$\widehat{f} = \mathcal{QF}^{-1}\{\mathrm{FT}(\lambda)\} = \int_{\lambda \in \Lambda} \mathrm{FT}(\lambda)\widehat{\varphi}_\lambda d\mu(\lambda) = Op\{\mathrm{FT}(\lambda)\}, \qquad (30)$$

$$\widehat{F} = \mathcal{QF}^{-1}\{\mathrm{TF}(\lambda)\} = \int_{\lambda \in \Lambda} \mathrm{TF}(\lambda)\widehat{\psi}_\lambda d\mu(\lambda) = Op\{\mathrm{TF}(\lambda)\}, \qquad (31)$$

where

$$\mathrm{FT}(\lambda) = \mathcal{QF}\{\widehat{f}\} = \langle \widehat{f} | \widehat{\varphi}_\lambda \rangle = \mathbf{Tr}(\widehat{f}\widehat{\varphi}_\lambda^\dagger) = sym\{\widehat{f}\}, \qquad (32)$$

$$\mathrm{TF}(\lambda) = \mathcal{QF}\{\widehat{f}\} := \langle \widehat{f} | \widehat{\psi}_\lambda^\dagger \rangle = \mathbf{Tr}(\widetilde{f}\widehat{\psi}_\lambda) = sym\{\widehat{F}\}. \qquad (33)$$

Usually, $\mathrm{FT}(\lambda)$ and $\mathrm{TF}(\lambda)$ are Wigner–Ville distribution and Woodward ambiguity functions, respectively. Further we shall use only Woodward ambiguity functions to design quantum signals and spectra.

DEFINITION 10 *Let* $\{\widehat{\varphi}_\lambda\}_{\lambda \in \Lambda}$ *and* $\{\widehat{\psi}_\lambda\}_{\lambda \in \Lambda}$ *be two r-parametric families of operators. Then transforms (30)–(33) are called the abstract quantum Fourier transforms for the algebras* $\mathbf{Alg}_1 \subset \mathbf{Sig}_1$ *and* $\mathbf{Alg}_1^* \subset \mathbf{Sp}_1$ *associated with two quora* $\{\widehat{\varphi}_\lambda\}_{\lambda \in \Lambda}$ *and* $\{\widehat{\psi}_\lambda\}_{\lambda \in \Lambda}$, *respectively.*

2.2   Classical Weyl quantization

It is well known that for the classical shift we have

$$\widehat{T}_x^v f(x) := f(x + v) = \sum_{k=0}^{\infty} \frac{v^k}{k!} \left(\frac{d}{dx}\right)^k f(x) = \left\{\sum_{k=0}^{\infty} \frac{v^k}{k!} \left(\frac{d}{dx}\right)^k\right\} f(x) =$$

$$\left\{ \sum_{k=0}^{\infty} \frac{(iv)^k}{k!} \left(-i\frac{d}{dx} \right)^k \right\} f(x) = \left\{ \sum_{k=0}^{\infty} \frac{(iv\widehat{\mathcal{D}}_x)^k}{k!} \right\} f(x) = e^{iv\widehat{\mathcal{D}}_x} f(x), \quad (34)$$

where $\widehat{\mathcal{D}}_x = -i\frac{d}{dx}$. This expression represents the decomposition of the ordinary finite shift into a series of powers of the differential operator $\frac{d}{dx}$ and is called the *infinitesimal representation* of translation shift. Analogously, we can obtain $\widehat{D}_\omega^\nu F(\omega) = F(\omega + \nu) = e^{iv\widehat{\mathcal{D}}_\omega} F(\omega)$, where $\widehat{\mathcal{D}}_\omega = -i\frac{d}{d\omega}$. Hence, $\widehat{T}_x^v = e^{iv\widehat{\mathcal{D}}_x}$, $\widehat{D}_\omega^\nu = e^{iv\widehat{\mathcal{D}}_\omega}$. In 1932 H. WEYL proposed [23] to modify the Fourier transform formula by changing its *complex-valued harmonics* into *operator-valued harmonics*. He used the following three quora for his quantization procedures of the signal space $\mathbf{Sig}_0$:

$$\left\{ \mathcal{E}_x^{[\nu,v]} = e^{i(\nu\widehat{\mathcal{M}}_x + v\widehat{\mathcal{D}}_x)} \right\}, \quad \left\{ \mathcal{E}_x^{(\nu,v)} = e^{iv\widehat{\mathcal{M}}_x} e^{iv\widehat{\mathcal{D}}_x} \right\}, \quad \left\{ \mathcal{E}_x^{(v,\nu)} = e^{iv\widehat{\mathcal{D}}_x} e^{iv\widehat{\mathcal{M}}_x} \right\}$$

associated with the classical Fourier transform, where multiplication $\widehat{\mathcal{M}}_x$ and differential $\widehat{\mathcal{D}}_x$ operators are given by

$$\widehat{\mathcal{M}}_x f(x) := x f(x), \quad \widehat{\mathcal{D}}_x f(x) := -i\frac{f(x)}{dx}.$$

Using the first quorum, H. Weyl wrote any quantum signal $\widehat{f} \in \mathbf{Sig}_1$ as

$$\widehat{f} := \mathcal{QF}_x \left\{ \mathrm{AW}[f] \right\} =$$

$$Op\left\{ \mathrm{AW}[f] \right\} = \int\limits_{\nu \in \Omega^*} \int\limits_{v \in \Omega} \mathrm{AW}[f](\nu, v) e^{i[\nu\widehat{\mathcal{M}}_x + v\widehat{\mathcal{D}}_x]} d\mu(\nu) d\mu(v), \quad (35)$$

where

$$\mathrm{AW}[f](\nu, v) = \mathcal{QF}_x^{-1}\{\mathrm{AW}[f]\} = Sym\{\widehat{f}\} = \mathbf{Tr}\left[\widehat{f} \, e^{-i[\nu\widehat{\mathcal{M}}_x + v\widehat{\mathcal{D}}_x]} \right].$$

$$(36)$$

Transformations (35) and (36) are called the direct and inverse *ordinary quantum Fourier transforms* in the quantum signal space. It is natural to view maps $\mathrm{AW}[f] \longrightarrow \widehat{f}$, $\widehat{f} \longrightarrow \mathrm{AW}[f]$ as operator-valued Fourier transforms. But we can write them in the explicit form of the integral kernels. For example, for the map $\mathrm{AW}[f] \longrightarrow \widehat{f}$ the kernel $f(x, y)$ of the operator $\widehat{f}$ has the form:

$$f(x, y) = \int\limits_{\nu \in \Omega^*} \mathrm{AW}[f](\nu, y - x) e^{i\nu x} e^{i\frac{\nu}{2}(y - x)} d\mu(\nu).$$

Of course, for quantization of the spectral space $\mathbf{Sp}_0$ one can use three dual quora

$$\left\{ \mathcal{E}_\omega^{[v,\nu]} = e^{i[v\hat{\mathcal{M}}_\omega + \nu\hat{\mathcal{D}}_\omega]} \right\}, \quad \left\{ \mathcal{E}_\omega^{(v,\nu)} = e^{iv\hat{\mathcal{M}}_\omega} e^{i\nu\hat{\mathcal{D}}_\omega} \right\}, \quad \left\{ \mathcal{E}_\omega^{(v,\nu)} = e^{i\nu\hat{\mathcal{D}}_\omega} e^{iv\hat{\mathcal{M}}_\omega} \right\},$$

where $\widehat{\mathcal{M}}_\omega F(\omega) := \omega F(\omega)$, $\widehat{\mathcal{D}}_\omega F(\omega) := -i\frac{F(\omega)}{d\omega}$. Using the first quorum, we can write any quantum spectrum $\widehat{F} \in \mathbf{Sp}_1$ as follows:

$$\widehat{F} := Q\mathcal{F}_\omega\left\{ \mathrm{AW}[F] \right\} =$$

$$Op\left\{ \mathrm{AW}[F] \right\} = \int\limits_{v\in\Omega} \int\limits_{\nu\in\Omega^*} \mathrm{AW}[F](v,\nu) e^{i[v\hat{\mathcal{M}}_\omega + \nu\hat{\mathcal{D}}_\omega]} d\mu(v) d\mu(\nu), \quad (37)$$

where

$$\mathrm{AW}[F](v,\nu) = Q\mathcal{F}_x^{-1}\{\mathrm{AW}[F]\} = Sym\{\widehat{F}\} = \mathbf{Tr}\left[\widehat{F}\, e^{-i[v\hat{\mathcal{M}}_\omega + \nu\hat{\mathcal{D}}_\omega]} \right],$$
$$(38)$$

Transformations (37) and (38) are called the direct and inverse *ordinary quantum Fourier transforms* in the quantum spectral space.

2.3   Generalized Heisenberg–Weyl operators

Let us construct generalized operator-valued hyperharmonics associated with an orthogonal basis $\{\varphi_\omega(x)\}_{\omega\in\Omega^*}$.

DEFINITION 11 *Operator $\widehat{\mathcal{D}}_x$, for which $\widehat{\mathcal{D}}_x\varphi_\omega(x) = \omega\varphi_\omega(x)$ is valid, is called the generalized differential operator.*

The generalized differential operator appears as an ordinary differential operator with variable coefficients, for example $\widehat{\mathcal{D}}_x = p_2(x)\frac{\widehat{D}^2}{dx^2} + p_1(x)\frac{d}{dx} + p_0(x)$, where $p_2(x)$, $p_1(x)$, $p_0(x)$ are some variable coefficients. Let us now find a connection between the GSOs $\widehat{T}_x^v$ and the generalized differential operator $\widehat{\mathcal{D}}_x$. It can be found using the Taylor expansion.

THEOREM 3 *Let $\{\varphi_\omega(x)\}_{\omega\in\Omega^*} \in \mathbf{Sig}_0$ be some Fourier basis, consisting of $\mathcal{A}$-valued basis functions. Then all GSOs associated with it have the infinitesimal representation: $\widehat{T}_x^v = \varphi_{\widehat{\mathcal{D}}_x}(v)$, $\widehat{D}_\omega^\nu = \varphi_\nu(\widehat{\mathcal{D}}_\omega)$, and are called the operator-valued hyperharmonics associated with an orthogonal basis $\{\varphi_\omega(x)\}_{\omega\in\Omega^*}$, where $\widehat{\mathcal{D}}_x\varphi_\omega(x) = \omega\varphi_\omega(x)$, $\widehat{\mathcal{D}}_\omega\varphi_\omega(x) = x\varphi_\omega(x)$.*

Proof: If the signals $\varphi_\omega(v)$ are decomposed into the following series of ω, $\varphi_\omega(v) = \sum_{k=0}^\infty X_k(v)(\omega)^k$, then we can construct the operators

$\varphi_{\widehat{\mathcal{D}}_x}(v) = \sum\limits_{k=0}^{\infty} X_k(v)\widehat{\mathcal{D}}_x^k$. For these operators we have

$$\left(\varphi_{\widehat{\mathcal{D}}_x}(v)\right)\varphi_\omega(x) = \left(\sum_{k=0}^{\infty} X_k(v)\widehat{\mathcal{D}}_x^k\right)\varphi_\omega(x) = \left(\sum_{k=0}^{\infty} X_k(v)(\omega)^k\right)\varphi_\omega(x) =$$

$$\varphi_\omega(v)\varphi_\omega(x) = \varphi_\omega(x \boxplus v) = \widehat{T}_x^v\varphi_\omega(x), \qquad (39)$$

i.e., $\widehat{T}_x^v = \varphi_{\widehat{\mathcal{D}}_x}(v)$. Analogously, $\widehat{D}_\omega^v = \varphi_v(\widehat{\mathcal{D}}_\omega)$. Obviously, $M_x^\nu = \varphi_\nu(\widehat{\mathcal{M}}_x)$ and $M_\omega^v = \varphi_{\widehat{\mathcal{M}}_\omega}(v)$. $\qquad\square$

Using the hyperharmonics $\widehat{T}_x^v = \varphi_{\widehat{\mathcal{D}}_x}(v)$ and $\widehat{D}_\omega^v = \varphi_v(\widehat{\mathcal{D}}_\omega)$ associated with the basis $\{\varphi_\omega(x)\}_{\omega\in\Omega^*}$, we can construct generalized Heisenberg–Weyl operators and quantum hyperharmonic analysis of quantum signals and spectra.

The "time"-shift and "frequency"-shift operators together acting on spaces $\mathbf{Sig}_0$ and $\mathbf{Sp}_0$ induce the following pair of sets of the *Heisenberg–Weyl operators*:

$$\mathbb{HW}_x := \left\{ \widehat{\mathcal{E}}_x^{(\nu,v)} = \widehat{M}_x^\nu \widehat{T}_x^v = \varphi_\nu(\widehat{\mathcal{M}}_x)\varphi_{\widehat{\mathcal{D}}_x}(v) \mid \nu \in \Omega^*,\ v \in \Omega\right\},$$

$$\mathbb{HW}_\omega := \left\{ \widehat{\mathcal{E}}_\omega^{(v,\nu)} = \widehat{M}_\omega^v \widehat{D}_\omega^\nu = \varphi_{\widehat{\mathcal{M}}_\omega}(v)\varphi_\nu(\widehat{\mathcal{D}}_\omega) \mid \nu \in \Omega^*,\ v \in \Omega\right\}.$$

They act on $\mathbf{Sig}_0$ and $\mathbf{Sp}_0$ by the following rules:

$$\widehat{\mathcal{E}}_x^{(\nu,v)} f(x) := \left(\widehat{M}_x^\nu \widehat{T}_x^v f\right)(x) = \varphi_\nu(x)f(x \boxplus v),$$

$$\widehat{\mathcal{E}}_\omega^{(v,\nu)} F(\omega) := \left(\widehat{M}_\omega^v \widehat{D}_\omega^\nu F\right)(\omega) = \varphi_\omega(v)F(\omega \oplus \nu).$$

Obviously,

$$\mathcal{F}_0 \left\{\widehat{\mathcal{E}}_x^{(\nu,v)} f(x)\right\} = \overline{\varphi}_\nu(v)\widehat{\mathcal{E}}_\omega^{(v,-\nu)} F(\omega)$$

and

$$\mathcal{F}_0^{-1} \left\{\widehat{\mathcal{E}}_\omega^{(v,\nu)} F(\omega)\right\} = \varphi_\nu(v)\widehat{\mathcal{E}}_x^{(\nu,v)} f(x).$$

Now we construct two sets of symmetric Heisenberg–Weyl operators:

$$\mathbb{SHW}_x = \left\{ \widehat{\mathcal{E}}_x^{[\nu,v]} = \varphi_\nu^{1/2}(v)\varphi_\nu(\widehat{\mathcal{M}}_x)\varphi_{\widehat{\mathcal{D}}_x}(v) \Big| \nu \in \Omega^*, v \in \Omega\right\},$$

$$\mathbb{SHW}_\omega = \left\{ \widehat{\mathcal{E}}_\omega^{[v,\nu]} = \overline{\varphi}_\nu^{1/2}(v)\varphi_{\widehat{\mathcal{M}}_\omega}(v)\varphi_\nu(\widehat{\mathcal{D}}_\omega) \Big| \nu \in \Omega^*, v \in \Omega\right\}.$$

These operators satisfy the following composition laws:

$$\widehat{\mathcal{E}}_x^{[\nu,v]} \widehat{\mathcal{E}}_x^{[\nu',v']} = \overline{\varphi}_\nu^{1/2}(v')\varphi_{\nu'}^{1/2}(v)\widehat{\mathcal{E}}_x^{[\nu+\nu',v+v']}$$

$$\widehat{\mathcal{E}}_{\omega}^{[v,\nu]}\widehat{\mathcal{E}}_{\omega}^{[v',\nu']} = \varphi_{\nu}^{1/2}(v')\overline{\varphi}_{\nu'}^{1/2}(v)\widehat{\mathcal{E}}_{\omega}^{[v\oplus v',\nu\oplus\nu']}$$

and the "commutation" relations

$$\widehat{\mathcal{E}}_{x}^{[\nu,v]}\widehat{\mathcal{E}}_{x}^{[\nu',v']} = \overline{\varphi}_{\nu}(v')\varphi_{\nu'}(v)\widehat{\mathcal{E}}_{x}^{[\nu',v']}\widehat{\mathcal{E}}_{x}^{[\nu,v]},$$

$$\widehat{\mathcal{E}}_{\omega}^{[v,\nu]}\widehat{\mathcal{E}}_{\omega}^{[v',\nu']} = \varphi_{\nu}(v')\overline{\varphi}_{\nu'}(v)\widehat{\mathcal{E}}_{\omega}^{[v',\nu']}\widehat{\mathcal{E}}_{\omega}^{[v,\nu]}.$$

2.4   Generalized Weyl quantizations

Let us consider the linear quantum spaces $\mathbf{Sig}_1$ and $\mathbf{Sp}_1$ of quantum signals $\widehat{f}$ and quantum spectra $\widehat{F}$, respectively. The inner product can be defined by $\langle\widehat{f}_1|\widehat{f}_2\rangle := \mathbf{Tr}(\widehat{f}_1\widehat{f}_2^\dagger)$, $\langle\widehat{F}_1|\widehat{F}_2\rangle := \mathbf{Tr}(\widehat{F}_1\widehat{F}_2^\dagger)$. It is easy to check that

$$\mathbf{Tr}\left[\widehat{\mathcal{E}}_{x}^{[\nu,v]}\left(\widehat{\mathcal{E}}_{x}^{[\nu',v']}\right)^\dagger\right] = \delta(\nu\boxminus\nu')\delta(v\boxminus v'), \tag{40}$$

$$\mathbf{Tr}\left[\widehat{\mathcal{E}}_{\omega}^{[v,\nu]}\left(\widehat{\mathcal{E}}_{\omega}^{[v',\nu']}\right)^\dagger\right] = \delta(v\ominus v')\delta(\nu\ominus\nu'). \tag{41}$$

The families $\left\{\widehat{\mathcal{E}}_{x}^{[\nu,v]}\right\}_{[\nu,v]\in\Omega^*\times\Omega}$ and $\left\{\widehat{\mathcal{E}}_{\omega}^{[v,\nu]}\right\}_{[v,\nu]\in\Omega\times\Omega^*}$ form two quora in quantum spaces. For this reason, any quantum signal $\widehat{f}\in\mathbf{Sig}_1$ and quantum spectra $\widehat{F}\in\mathbf{Sp}_1$ can be written as follows:

$$\widehat{f} = \mathcal{QF}_x\{\mathrm{AW}[f]\} = Op\{\mathrm{AW}[f]\} = \int\limits_{\nu\in\Omega^*}\int\limits_{v\in\Omega} \mathrm{AW}[f](\nu,v)\widehat{\mathcal{E}}_{x}^{[\nu,v]}d\mu(\nu)d\mu(v),$$

$$\tag{42}$$

$$\begin{aligned}\widehat{F} &= \mathcal{QF}_\omega\{\mathrm{AW}[F]\} \tag{43}\\ &= Op\{\mathrm{AW}[f]\}\\ &= \int\limits_{v\in\Omega}\int\limits_{\nu\in\Omega^*} \mathrm{AW}[F](v,\nu)\widehat{\mathcal{E}}_{\omega}^{[v,\nu]}\,d\mu(v)\,d\mu(\nu).\end{aligned}$$

Using (40) and (41), one can invert (42) and (43) as follows:

$$\mathrm{AW}[f](\nu,v) = \mathcal{QF}_x^{-1}\{\mathrm{AW}[f]\} = Sym\{\widehat{f}\} = \mathbf{Tr}\left[\widehat{f}\left(\widehat{\mathcal{E}}_{x}^{[\nu,v]}\right)^\dagger\right], \tag{44}$$

$$\mathrm{AW}[F](v,\nu) = \mathcal{QF}_x^{-1}\{\mathrm{AW}[F]\} = Sym\{\widehat{F}\} = \mathbf{Tr}\left[\widehat{F}\left(\widehat{\mathcal{E}}_{\omega}^{[v,\nu]}\right)^\dagger\right]. \tag{45}$$

The transformations (42) and (45) are called the *generalized quantum Fourier transforms*.

EXAMPLE 5 *In this example we consider the Weyl quantization on a finite cyclic group $\Omega = \Omega^* = \mathbb{Z}/p$, where p is a prime integer. In this case,*

$$\widehat{\mathcal{E}}_x^{[\nu,v]} = \varepsilon^{\frac{\nu v}{2}} \widehat{\mathcal{E}}_x^{(\nu,v)} = \varepsilon^{\frac{\nu v}{2}} \widehat{M}_x^\nu \widehat{T}_x^v =$$

$$= \varepsilon^{\frac{\nu v}{2}} \begin{bmatrix} 1 & & & & \\ & \varepsilon^1 & & & \\ & & \varepsilon^2 & & \\ & & & \ddots & \\ & & & & \varepsilon^{p-1} \end{bmatrix}^\nu \begin{bmatrix} 0 & 1 & & & \\ & 0 & 1 & & \\ & & & \ddots & \\ & & & 0 & 1 \\ 1 & & & & 0 \end{bmatrix}^v.$$

For this reason, the map

$$\widehat{f} = Q\mathcal{F}_x\left\{AW[f]\right\} = Op\left\{AW[f]\right\} = \sum_{\nu \in \mathbb{Z}/p} \sum_{v \in \mathbb{Z}/p} AW[f](\nu,v)\widehat{\mathcal{E}}_x^{[\nu,v]} =$$

$$= \sum_{\nu \in \mathbb{Z}/p} \sum_{v \in \mathbb{Z}/p} AW[f](\nu,v)\varepsilon^{\frac{\nu v}{2}} \begin{bmatrix} 1 & & & & \\ & \varepsilon^1 & & & \\ & & \varepsilon^2 & & \\ & & & \ddots & \\ & & & & \varepsilon^{p-1} \end{bmatrix}^\nu \begin{bmatrix} 0 & 1 & & & \\ & 0 & 1 & & \\ & & & \ddots & \\ & & & 0 & 1 \\ 1 & & & & 0 \end{bmatrix}^v$$

$$\tag{46}$$

is the discrete quantum Fourier transform associated with the cyclic group $\mathbb{Z}$.

2.5   Generalized quantum convolutions

For the product of two quantum signals $\widehat{f}$ and $\widehat{g}$ we have

$$\widehat{fg} = \int_{(\nu,v)} \int_{(\nu',v')} AW[f](\nu,v)\widehat{\mathcal{E}}_x^{[\nu,v]} AW[g](\nu',v')\widehat{\mathcal{E}}_x^{[\nu',v']} d\mu(\nu,v)d\mu(\nu',dv') =$$

$$\int_{(\omega,x)} \left(AW[f] \circledast AW[g]\right)(\omega,x)\widehat{\mathcal{E}}_x^{[\omega,x]} d\mu(\omega)dx =$$

$$Q\mathcal{F}_x\left\{AW[f] \circledast AW[g]\right\} = Op\left\{AW[f] \circledast AW[g]\right\},$$

where the expression

$$\left(AW[f] \circledast AW[g]\right)(\omega,x) = Q\mathcal{F}_x^{-1}\{\widehat{fg}\} = sym\{\widehat{fg}\} =$$

$$\int_{(\nu,v)} FT(\nu,v)GT(\omega \ominus \nu, x \boxminus v)\bar{\varphi}_\nu^{1/2}(v)\varphi_\nu^{1/2}(v')d\mu(\nu)dv \tag{47}$$

is called the *generalized twisted signal convolution*. Analogously,

$$\widehat{F}\widehat{G} = \int\limits_{(v,\nu)(v',\nu')} \int \mathrm{AW}[F](v,\nu)\widehat{\mathcal{E}}_\omega^{[v,\nu]}\mathrm{AW}[F](v',v')\widehat{\mathcal{E}}_\omega^{[v',\nu']}d\mu(\nu)dv\ d\mu(\nu')dv' =$$

$$\int\limits_{(x,\omega)} \left(\mathrm{AW}[F]\bigstar\mathrm{AW}[G]\right)(x,\omega)\widehat{\mathcal{E}}_\omega^{[x,\omega]}dx\ d\mu(\omega) =$$

$$\mathcal{QF}_\omega\left\{\mathrm{AW}[F]\bigstar\mathrm{AW}[G]\right\} = Op\left\{\mathrm{AW}[F]\bigstar\mathrm{AW}[G]\right\},$$

where

$$\left(\mathrm{AW}[F]\bigstar\mathrm{AW}[G]\right)(x,\omega) := \mathcal{QF}_\omega^{-1}\{\widehat{F}\widehat{G}\} = sym\left\{\widehat{F}\widehat{G}\right\} =$$

$$\int\limits_{(v,\nu)} \mathrm{AW}[F](v,\nu)\mathrm{AW}[G](x\boxminus v,\omega\ominus\nu)\varphi_{\nu'}^{1/2}(v)\bar{\varphi}_\nu^{1/2}(v')dv\ d\mu(\nu) \quad (48)$$

is called the *generalized twisted spectral convolution*.

According to the Pontryagin duality principle we can define the *generalized quantum convolution* of quantum signals by

$$\widehat{f}\circledast\widehat{g} := Op\{\mathrm{AW}[f]\mathrm{AW}[g]\} = \mathcal{QF}_x^{(s)}\{\mathrm{AW}[f]\mathrm{AW}[g]\},$$

where

$$\mathrm{AW}[f](\nu,v)\mathrm{AW}[g][\nu,v] =$$

$$sym\{\widehat{f}\circledast\widehat{g}\} = \mathcal{QF}_x^{-1}\{\widehat{f}\circledast\widehat{g}\} = \mathbf{Tr}\left[\left(\widehat{f}\circledast\widehat{g}\right)\left(\mathcal{E}_x^{[\nu,v]}\right)^\dagger\right],$$

and the *generalized quantum convolution* of quantum spectra by

$$\widehat{F}\bigstar\widehat{G} := Op\left\{\mathrm{AW}[F]\mathrm{AW}[G]\right\} = \mathcal{QF}_\omega\left\{\mathrm{AW}[F]\mathrm{AW}[G]\right\},$$

where

$$\mathrm{AW}[F](\nu,v)\mathrm{AW}[G][\nu,v] =$$

$$sym\left\{\widehat{F}\bigstar\widehat{G}\right\} = \mathcal{QF}_x^{-1}\left\{\widehat{F}\bigstar\widehat{G}\right\} = \mathbf{Tr}\left[\left(\widehat{F}\bigstar\widehat{G}\right)\left(\mathcal{E}_x^{[\nu,v]}\right)^\dagger\right].$$

THEOREM 4 *The quantum generalized convolutions and quantum generalized Fourier transforms are related by the expressions:*

$$\mathcal{QF}_x\left\{\mathrm{AW}[f]\circledast\mathrm{AW}[g]\right\} = \widehat{f}\widehat{g}, \quad \mathcal{QF}_x\left\{\mathrm{AW}[F]\bigstar\mathrm{AW}[G]\right\} = \widehat{F}\widehat{G},$$

and

$$\mathcal{QF}_x^{-1}\left\{\widehat{f}\circledast\widehat{g}\right\} = \mathrm{AW}[f](\nu,v)\mathrm{AW}[g][\nu,v],$$

$$\mathcal{QF}_x^{-1}\left\{\widehat{F}\bigstar\widehat{G}\right\} = \mathrm{AW}[F](\nu,v)\mathrm{AW}[G][\nu,v].$$

3. Conclusion

In this paper we have examined the idea of generalized shift operators associated with an arbitrary orthogonal transform and generalized linear and nonlinear convolutions based on these generalized shift operators. Such operators allow one to unify and generalize the majority of known methods and tools of signal processing based on the classical Fourier transform for generalized classical and quantum signal theories.

References

[1] Marty, F. (1934). Sur une generalization de la notion de groupe. *Sartryck ur Forhandlingar via Altonde Skandinavioka Matematiker kongressen i Stockholm,* pp. 45–49.

[2] Marty, F. (1935). Role de la notion d'hypergroupe dans l'etude des groupes non abelians. *Computes Rendus de l'Academie des Sciences,* **201**, pp. 636–638.

[3] Wall, H.S. (1934). Hypergroups. *Bulletin of the American Mathematical Socity,* **41**, 36–40 [Presented at the annual meeting of the American Mathematical Society, Pittsburgh, December 27–31, 1934].

[4] Wall, H.S. (1937). Hypergroups. *American Journal of Mathematics.* **59**, pp. 77–98.

[5] Levitan, B.M. (1949). The application of generalized displacement operators to linear differential equations of second order. *Uspechi Math. Nauk,* **4**, No 1(29), pp. 3–112 (English transl., *Amer. Mat. Soc. Transl.* (1), **10**, 1962, 408–541, MR 11, 116).

[6] Levitan, B.M. (1964). Generalized translation operators.*Israel Program for Scientific Translations,* Jerusalem, 120 p.

[7] Dunkl, C.F. (1966). Operators and harmonic analysis on the sphere. *Trans. Amer. Math. Soc.* **125**, 2, pp. 50–263.

[8] Jewett, R.I. (1975). Spaces with an abstract convolution of measures, *Advances in Math.,* **18**, pp. 1–101.

[9] Labunets, V.G., Sitnikov, O.P. (1976). Generalized harmonic analysis of **VP**-invariant systems and random processes. In: *Harmonic Analysis on Groups in Abstract Systems Theory.* (in Russian), Ural State Technical University, Sverdlovsk, pp. 44–67.

[10] Labunets, V.G. and Sitnikov, O.P. (1976). Generalized harmonic analysis of VP-invariant linear sequential circuits. In: *Harmonic Analysis on Groups in Abstract System Theory (in Russian).* Ural Polytechnical Institute Press: Sverdlovsk, pp. 67–83.

[11] Labunets-Rundblad, E.V., Labunets, V.G., and Astola, J. (2000). Algebraic frames for commutative hyperharmonic analysis of signals and images. *Algebraic Frames for the Perception-Action Cycle.* Second Inter. Workshop, AFPAC 2000, Kiel, Germany, September 2000. Lectures Notes in Computer Science, 1888, Berlin, 2000, pp. 294–308.

[12] Creutzburg, R., Labunets, E.V., and Labunets, V.G. (1998). Algebraic foundations of an abstract harmonic analysis of signals and systems. *Workshop on*

Transforms and Filter Banks, Tampere International Center for Signal Processing, pp. 30–68.

[13] Creutzburg, R., Labunets, E., and Labunets V. (1992) Towards an "Erlangen program" for general linear system theory. Part I. In: F. Pichler (Edt), *Lecture Notes in Computer Science*, Vol. 585, Springer: Berlin, pp. 32–51.

[14] Creutzburg, R., Labunets, E., and Labunets, V. (1994). Towards an "Erlangen program" for general linear systems theory. Part II. *Proceed. EUROCAST'93* (Las Palmas, Spain) F. Pichler, R. Moreno Diaz (Eds.) In: *Lecture Notes in Computer Science*, Vol. 763, Springer: Berlin, pp. 52–71.

[15] Labunets, V.G. (1993). Relativity of "space" and "time" notions in system theory. In: *Orthogonal Methods Application to Signal Processing and Systems Analysis* (in Russian). Ural Polytechnical Institute Press: Sverdlovsk, pp. 31–44.

[16] Labunets, V.G. (1982). Spectral analysis of linear dynamic systems invariant with respect to generalized shift operators. In: *Experimental Investigations in Automation* (in Russian). Institute of Technical Cybernetics of Belorussian Academy of Sciences Press: Minsk, pp. 33–45.

[17] Labunets, V.G. (1982). Algebraic approach to signals and systems theory: linear systems examples. In: *Radiioelectronics Devices and Computational Technics Means Design Automation* (in Russian). Ural Polytechnical Institute Press: Sverdlovsk, pp. 18–29.

[18] Labunets, V.G. (1980). Examples of linear dynamical systems invariant with respect to generalized shift operators. In: *Orthogonal Methods for the Application to Signal Processing and Systems Analysis* (in Russian): Ural Polytechnical Institute Press: Sverdlovsk, pp. 4–14.

[19] Labunets, V. G. (1980). Symmetry principles in the signal and system theories. In: *Synthesis of Control and Computation Systems* (in Russian). Ural Polytechnical Institute Press: Sverdlovsk, pp. 14–24.

[20] Wigner, E.R. (1932). On the quantum correction for thermo-dynamic equilibrium. *Physics Review,* **40**, pp. 749–759.

[21] Ville, J. (1948). Theorie et Applications de la Notion de Signal Analytique. *Gables et Transmission.* **2A**, pp. 61–74.

[22] Woodward, P.M. (1951). Information theory and design of radar receivers. *Proceedings of the Institute of Radio Engineers,* **39**, pp. 1521–1524.

[23] Weyl, H. (1932) *The Theory Group and Quantum Mechanics,* London: Methuen, 236 p.

LIE GROUPS
AND LIE ALGEBRAS
IN ROBOTICS

J.M. Selig
South Bank University
London SE1 0AA, U.K.
seligjm@sbu.ac.uk

Abstract

In this lecture the group of rigid body motions is introduced via its representation on standard three dimensional Euclidian space. The relevance for robotics is that the links of a robot are usually modelled as rigid bodies. Moreover the payload of a robot is also usually a rigid body and hence much of robotics is concerned with understanding rigid transformations and sequences of these transformations. Chasles's theorem is presented, that is: a general rigid body motion is a screw motion, a rotation about a line in space followed by a translation along the line.

The lower Reuleaux pairs are introduced. These are essentially surfaces which are invariant under some subgroup of rigid body motions. Such a surface can be the matting surface for a mechanical joint. In this way the basic mechanical joints used in robots can be classified. These surfaces turn out to have other applications in robotics. In robot gripping they are exactly the surfaces that cannot be immobilised using frictionless fingers. In robot vision the symmetries of these surfaces are motions which cannot be detected.

Next Lie algebras are introduced. The 1-parameter rigid motions about a joint are considered. The correspondence between elements of the Lie algebra and 1-degree-of-freedom joints is given. The exponential map from the Lie algebra to the group is defined and used to describe the forward kinematics of a serial robot using the product of exponentials formula. The Rodrigues formula for the exponential map is derived using a set of mutually annihilating idempotents.

The derivative of the exponential map is explored. In particular the Jacobian matrix for the forward kinematics is derived. More general results are also derived and these are used to show how the inverse kinematics problem can be cast as a set of ordinary differential equations.

J. Byrnes (ed.) Computational Noncommutative Algebra and Applications, 101-125.
© 2004 *Kluwer Academic Publishers. Printed in the Netherlands.*

Keywords: Clifford algebra, Lie algebras, Lie groups, rigid body motions, robotics.

1.   Introduction—Rigid Body Motions

A large amount of robotics is concerned with moving rigid bodies around in space. A robot here is usually an industrial robot arm as in figure 1, however most of the material presented here applies to parallel manipulators and closed loop mechanisms. There are also some applications to mobile robots and robot vision.

It is usual to think of the components that make up the links of the robot as rigid and also the payload or tool is often a rigid body. Hence, an important problem in robotics is to keep track of these rigid bodies.

It is well known that any rigid transformation can be composed from a rotation, a translation and a reflection. Physical machines cannot perform reflections and so we should really speak of proper rigid transformations, excluding the reflections. The proper rigid transformations can be represented using 4×4 matrices. These can be written in partitioned form as

$$A = \begin{pmatrix} R & \mathbf{t} \\ 0 & 1 \end{pmatrix}$$

where R is a 3×3 rotation matrix and $\mathbf{t}$ is a translation vector.

The action of these matrices on points in space is given by,

$$\begin{pmatrix} \mathbf{p}' \\ 1 \end{pmatrix} = \begin{pmatrix} R & \mathbf{t} \\ 0 & 1 \end{pmatrix} \begin{pmatrix} \mathbf{p} \\ 1 \end{pmatrix} = \begin{pmatrix} R\mathbf{p} + \mathbf{t} \\ 1 \end{pmatrix}$$

where $\mathbf{p}$ and $\mathbf{p}'$ are the original and transformed position vectors of a point.

There are two ways we can use these matrices to describe the position and orientation of a rigid body. If we use an active point of view we can agree on a standard 'home' position for the body, any subsequent position and orientation of the body is described by the rigid transformation which moves the home position of the body to its current position and orientation.

In the passive viewpoint a coordinate frame is fixed in the body, now the position and orientation of the body is given by the coordinate transform which expresses the coordinates of points in the current frame in terms of those in the home frame. For historical reasons, this passive viewpoint seems to be preferred in robotics even though the active viewpoint is often simpler.

These two viewpoints are related quite simply. The transformation given by the passive view is the inverse of the active transformation. To see this assume that the new frame has its origin at a point with position

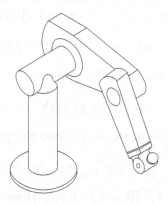

Figure 1. A six-joint industrial robot arm.

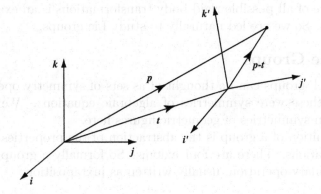

Figure 2. A vector referred to two coordinate frames.

vector $\mathbf{t}$ with respect to the original frame. Now an arbitrary point $\mathbf{p}$ in the original frame has position vector $\mathbf{p}' = \mathbf{p} - \mathbf{t}$ with respect to the new frame, see figure 2. In general the orientation of the new frame will be different to that of the original, assume it is given by a rotation R. That is, the basis vectors of the new frame $\mathbf{i}'$, $\mathbf{j}'$ and $\mathbf{k}'$, are given by,

$$\mathbf{i}' = R\mathbf{i}, \quad \mathbf{j}' = R\mathbf{j}, \quad \mathbf{k}' = R\mathbf{k}$$

If we write the position vector of the point as

$$\mathbf{p} = x\mathbf{i} + y\mathbf{j} + z\mathbf{k}$$

in the original frame, then in the new frame the point will have coordinates,

$$\mathbf{p}' = x'\mathbf{i}' + y'\mathbf{j}' + z'\mathbf{k}'$$

where $x' = \mathbf{p}' \cdot \mathbf{i}' = (\mathbf{p} - \mathbf{t})^T R\mathbf{i}$ and so forth. Hence the new coordinates can be written in terms of the old as

$$\begin{pmatrix} x' \\ y' \\ z' \end{pmatrix} = R^T \begin{pmatrix} x \\ y \\ z \end{pmatrix} - R^T \mathbf{t}$$

Compare this with the inverse of a general rigid transformation which is given by,

$$\begin{pmatrix} R & \mathbf{t} \\ 0 & 1 \end{pmatrix}^{-1} = \begin{pmatrix} R^T & -R^T \mathbf{t} \\ 0 & 1 \end{pmatrix}$$

The space of all possible rigid body transformations is an example of a Lie group. So we are led naturally to study Lie groups.

2. Lie Groups

In general, groups can be thought of as sets of symmetry operations. Originally these were symmetries of algebraic equations. We will be interested in symmetries of geometric figures here.

The definition of a group is the abstraction of the properties of symmetry operations. There are four axioms. So formally a group is a set G with a binary operation, usually written as juxtaposition.

- The group must be closed under its binary operation.

- The operation must be associative.

- The group must have a unique identity element.

- Every element of the group must have a unique inverse.

Notice that sets of square matrices can be groups quite easily, the group product is modelled by matrix multiplication, which is associative of course. The identity element of the group is represented by the identity matrix. The inverse of a group element is represented by the inverse matrix, so it is important that the matrices in the set are all non-singular.

Lie groups have to satisfy the following additional axioms.

- The set of group elements G must form a differentiable manifold.

- The group operation must be a differentiable map. As a map the group operation sends $G \times G \longrightarrow G$. On pairs of group elements $g_1, g_2 \in G$ the map has the effect $(g_1, g_2) \mapsto g_1 g_2$.

- The map from a group element to its inverse must be a differentiable mapping. On an element $g \in G$ this map has the effect, $g \mapsto g^{-1}$.

A differentiable manifold is essentially a space on which we can do calculus. Locally it looks like $\mathbb{R}^n$ but globally it could be quite different. It is usual to think of such a manifold as patched together from pieces of $\mathbb{R}^n$. The standard example is a sphere, which can be constructed from two patches, one covering the northern hemisphere and the other the southern hemisphere. The patches can also be thought of as local coordinate systems for the manifold, in general manifolds do not have global coordinate systems.

The set of non-singular $n \times n$ matrices is an example of such a Lie group. The underlying manifold of the group is simply an open set in $\mathbb{R}^{n^2}$, that is $\mathbb{R}^{n^2}$ with the closed set of determinant zero matrices deleted. The other axioms are clearly satisfied, matrix multiplication is clearly differentiable and so is the inversion map. This group is usually denoted $GL(n)$ for the general linear group of order n.

Subspaces of this group can also be groups. A good example for us is the rotation matrices. These are 3×3 matrices which also satisfy,

$$R^T R = I, \qquad \text{and} \qquad \det(R) = 1.$$

These equations define a non-singular algebraic variety in $GL(3)$. It is possible to show that such a subspace is a differentiable manifold. In this case the group manifold is 3-dimensional projective space $\mathbb{PR}^3$. This space can be thought of as the space of lines through the origin in $\mathbb{R}^4$ or alternatively as the 3-dimensional sphere with anti-podal points identified. The group is usually denoted $SO(3)$, the special orthogonal group of order three.

Next we look in detail at the Lie group that we are most concerned with; the group of rigid body transformations.

3.   Finite Screw Motions

In robotics the group of rigid transformations is called $SE(3)$, it is supposed to denote the special Euclidian group. It is the semi-direct product of $SO(3)$, the rotations about the origin, with $\mathbb{R}^3$, the translations.

$$SE(3) = SO(3) \rtimes \mathbb{R}^3$$

In a direct product the factors would not interact. The semi-direct product indicates that the rotations act on the translations. This can be easily seen from the 4×4 matrices, if we multiply two such matrices we get,

$$\begin{pmatrix} R_2 & \mathbf{t}_2 \\ 0 & 1 \end{pmatrix} \begin{pmatrix} R_1 & \mathbf{t}_1 \\ 0 & 1 \end{pmatrix} = \begin{pmatrix} R_2 R_1 & R_2 \mathbf{t}_1 + \mathbf{t}_2 \\ 0 & 1 \end{pmatrix}$$

Notice how the first translation vector is rotated before it is added to the second one.

The group manifold for a semi-direct product of groups is simply the Cartesian product of the manifolds of the factors. So the group manifold for $SE(3)$ is simply the Cartesian product of $\mathbb{PR}^3$ with $\mathbb{R}^3$. Notice that this is therefore also the configuration manifold for a rigid body, each position and orientation of the body corresponds to a group element and vice versa.

We end this section with a sketch of Chasles's theorem. This, originally geometric, theorem dates back to the 1830s, well before Lie's work on 'continuous groups'. The theorem states that every rigid transformation (with the exception of pure translations) is a finite screw motion. That is, a rotation about a line together with a translation along the line, see figure 3.

A finite screw motion about a line through the origin has the form

$$A(\theta) = \begin{pmatrix} R & \frac{\theta p}{2\pi} \hat{\mathbf{x}} \\ 0 & 1 \end{pmatrix}$$

where $\hat{\mathbf{x}}$ is a unit vector along the axis of the rotation, θ the angle of rotation, and p the pitch of the motion. Since the axis of rotation is $\hat{\mathbf{x}}$ we also expect that $R\hat{\mathbf{x}} = \hat{\mathbf{x}}$, in other words $\hat{\mathbf{x}}$ is an eigenvector of R with unit eigenvalue.

In general, if the line doesn't pass through the origin the transformation can be found by conjugation. Suppose $\mathbf{u}$ is a point on the line, then we can translate $\mathbf{u}$ back to the origin, perform the screw motion above

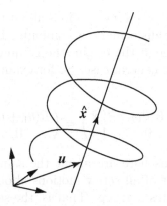

Figure 3. A finite screw motion.

and finally translate the origin back to **u**. This gives

$$\begin{pmatrix} I & \mathbf{u} \\ 0 & 1 \end{pmatrix} \begin{pmatrix} R & \frac{\theta p}{2\pi}\hat{\mathbf{x}} \\ 0 & 1 \end{pmatrix} \begin{pmatrix} I & \mathbf{u} \\ 0 & 1 \end{pmatrix} = \begin{pmatrix} R & \frac{\theta p}{2\pi}\hat{\mathbf{x}} + (I - R)\mathbf{u} \\ 0 & 1 \end{pmatrix}$$

Now Chasles's theorem amounts to the following: Given an arbitrary rigid transformation it can always be put in the above form. That is, we have to solve,

$$\begin{pmatrix} R & \mathbf{t} \\ 0 & 1 \end{pmatrix} = \begin{pmatrix} R & \frac{\theta p}{2\pi}\hat{\mathbf{x}} + (I - R)\mathbf{u} \\ 0 & 1 \end{pmatrix}$$

for p and **u** given **t** and R. Assuming we can find θ and $\hat{\mathbf{x}}$ from R then it is not too difficult to see that

$$\frac{\theta p}{2\pi} = \hat{\mathbf{x}} \cdot \mathbf{t}.$$

This gives the pitch p. Now we have a system of linear equations for **u**,

$$(I - R)\mathbf{u} = \frac{\theta p}{2\pi}\hat{\mathbf{x}} - \mathbf{t}.$$

These equations are singular, but the kernel of $(I - R)$ is clearly $\hat{\mathbf{x}}$ and so the equations are consistent and it will be possible to find a solution for **u** up to addition of an arbitrary multiple of $\hat{\mathbf{x}}$. In practice, it would be sensible to require that **u** be perpendicular to $\hat{\mathbf{x}}$. If the transformation is a pure translation then $R = I$ and the above fails, this is the only case where **u** cannot be found. Pure rotations correspond to pitch zero, $p = 0$. Screw motions and pure translations are often taken to be screw motions with infinite pitch.

Notice that if we have two screw motions about the same line with the same pitch then these transformations commute. It is only necessary to check this for lines through the origin, the conjugation can be used to extend this easily to the general case. So for example we have

$$\begin{pmatrix} R(\theta_2) & \frac{\theta_2 p}{2\pi}\hat{\mathbf{x}} \\ 0 & 1 \end{pmatrix} \begin{pmatrix} R(\theta_1) & \frac{\theta_1 p}{2\pi}\hat{\mathbf{x}} \\ 0 & 1 \end{pmatrix} = \begin{pmatrix} R(\theta_1 + \theta_2) & \frac{(\theta_1+\theta_2)p}{2\pi}\hat{\mathbf{x}} \\ 0 & 1 \end{pmatrix}$$

Remember, the rotations here are about the same axis. From this it can be seen that the set of all screw motions about the same line with the same pitch constitutes a group. That is, the set is closed under the group operation. These are the one-parameter subgroups of $SE(3)$.

4.   Mechanical Joints

At the end of the nineteenth century Franz Reuleaux described what he called "lower pairs" [7]. These were pairs of surfaces which can move relative to each other while remaining in surface contact. He took these to be idealisations for the most basic of mechanical joints. He found six possibilities, see figure 4.

It is possible to give a simple group theoretic proof that these are the only possibilities. The key observation is that these surfaces must be invariant under some subgroup of $SE(3)$. The subgroup represents the symmetries of the surface. For a joint, the subgroup will give the possible relative motions between the sides of the joint.

To find these surfaces, consider the surfaces invariant under one-parameter (1-dimensional) subgroups.

- The pitch zero subgroups correspond to rotations about a line. A surface invariant under such a subgroup is just a surface of rotation.

- Infinite pitch subgroups correspond to translations in a fixed direction, so any surface of translation is invariant under such a subgroup.

- The subgroups with finite, non-zero pitch have helicoidal surfaces as invariants. These give the first three lower Reuleaux pairs.

To find more Reuleaux pairs consider how the subgroups can be combined to form larger subgroups. Surfaces invariant with respect to these larger subgroups must have more than one of the above properties.

- So the cylinder is a surface of rotation and a surface of translation (and a helicoidal surface for any pitch).

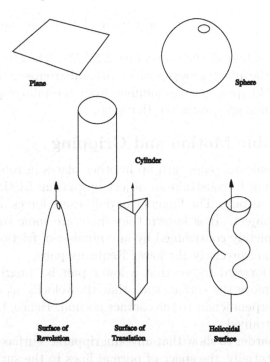

Figure 4. The six lower Reuleaux pairs.

- A sphere is invariant with respect to the subgroup of rotations in space $SO(3)$. That is, the sphere is a surface of rotation about any of its diameters.

- A plane with respect to $SE(2)$ the group of rigid transformations in the plane.

With a little care, it is possible to show that these are the only possibilities—Reuleaux found them all!

In mechanical engineering, the joints corresponding to lower Reuleaux pairs have special names. Simple hinge joints corresponding to the rotational pair are known as revolute joints. Sliding joints corresponding to the translational pair are called prismatic joints and the joints corresponding to the helicoidal pair are helical joints or sometimes screw joints. A ball-and-socket joint is more properly called a spherical joint and the last two pairs form cylindric and planar joints.

Practical robots may have any of these joints, but it is most common to actuate revolute or prismatic joints. Revolute joints can be accurately and easily controlled with electric motors. When larger forces are required hydraulic rams can be used as actuated prismatic joints.

However, for a hydraulic ram the length of travel is hard to control accurately.

It is possible to find all the subgroups of $SE(3)$. The zero-dimensional subgroups are the point groups, which are subgroups of the rotations only, and the 230 space groups familiar from crystallography. For the higher dimensional subgroups see Hervé [4].

5.   Invisible Motion and Gripping

The lower Reuleaux pairs turn up in other places in robotics.

Consider trying to constrain an object by placing frictionless fingers on the object's surface. The fingers can only exert forces normal to the surface of the object. It is known that there are some surfaces which cannot be completely constrained by any number of frictionless fingers. These surfaces are precisely the lower Reuleaux pairs.

It is straightforward to see that a lower pair is "ungrippable," the symmetry will move the surface such that the velocity at any point on the surface is perpendicular to the contact normal. Hence, these motions cannot be constrained.

It is a little harder to show that any 'ungrippable' surface must have a symmetry. Essentially, the space of normal lines to the surface are considered as six-dimensional force-torque vectors called wrenches. For an 'ungrippable' surface the normal do not span the six-dimension vector space of wrenches. Hence, there will be a dual six-dimensional vector which annihilates all normal lines, this vector will generate the symmetries. It is an element of the Lie algebra, see below.

In robot vision a common problem is to reconstruct the motion of objects given two or more successive images of the scene. Flow fields are constructed from differences in image intensity. However, it is well known that certain motions of objects cannot be detected in this way, Horn gives the example of a sphere rotating about any diameter, [5].

It is clear that motions of all Reuleaux pairs are undetectable in this way. It is also clear that these will be the only undetectable rigid body motions if we discount discrete symmetries of regular polyhedra exactly synchronised with the succession of images. For example, we would not expect to see the motion of a tetrahedron if it executes a symmetry between the images we capture. On the other hand we are not really limited to rigid motions here. There may be other motions which are invisible to such a system, but this would probably depend very much on the lighting conditions. For example, consider the motion of a sphere again, this time the sphere is moving away from the observer and at the same time is expanding at such a rate that the apparent contour in the image remains constant. This would be detectable under point lighting but there might be lighting conditions under which a move like this is undetectable.

6.   Forward Kinematics

From the above it is clear that motion about a 1-degree-of-freedom joint can be parameterised by a one-parameter subgroups.

The general case is a helical joint where the corresponding 1-parameter subgroup can be written

$$A(\theta) = \begin{pmatrix} R & \frac{\theta p}{2\pi}\hat{\mathbf{x}} + (I - R)\mathbf{u} \\ 0 & 1 \end{pmatrix}$$

where $\hat{\mathbf{x}}$ is the direction of the joint axis, p is the pitch of the joint and $\mathbf{u}$ the position vector of a point on the joint axis. Revolute joints are most common in practical machines, these correspond to subgroups of pitch zero,

$$A(\theta) = \begin{pmatrix} R & (I - R)\mathbf{u} \\ 0 & 1 \end{pmatrix}$$

Sometimes prismatic joints are used, the corresponding 1-parameter subgroup for such a joint has the form,

$$A(\theta) = \begin{pmatrix} I & \theta\hat{\mathbf{t}} \\ 0 & 1 \end{pmatrix}$$

where $\hat{\mathbf{t}}$ is the direction of the joint. Notice in each of the cases above, the identity element is given when the parameter is zero.

For a serial robot, such as the on illustrated in figure 1, it is important to find the transformation undergone by the end-effector. This problem is usually known as the forward kinematics of the robot. Let us fix a standard or home position for the robot. In the home configuration all the joint variables will be taken to be zero. Now, to move the robot to a position specified by a set of joint variables $\boldsymbol{\theta}^T = (\theta_1, \theta_2, \theta_3, \theta_4, \theta_5, \theta_6)$ we can perform the motions about the joints in turn beginning with the distal joint θ_6; the one furthest from the base of the robot. The motion of the end-effector about this joint will be given by the matrix $A_6(\theta_6)$. Next we move the fifth joint, the axis of this joint has not be affected by the motion about the last joint and so the effect on the tool is just $A_5(\theta_5)$. In this way we can work our way down the arm and deduce the overall transformation on the end-effector or tool to be,

$$K(\boldsymbol{\theta}) = A_1(\theta_1)A_2(\theta_2)A_3(\theta_3)A_4(\theta_4)A_5(\theta_5)A_6(\theta_6)$$

This expression represents the active transformation undergone by the robot's end-effector relative to the home configuration. It is straightforward to compute the 'A' matrices if we know the position of the robot's joint axes in the home configuration.

In the robotics literature it is also common to use tool-frame coordinates, and to give the forward kinematics in terms of this frame. Specifically, the tool-frame is a coordinate system fixed to the robot's end-effector. We seek the passive, coordinate transformation which converts coordinates in a frame fixed in the base link of the robot to coordinates in the tool frame. Suppose that in the robot's home position the active transformation from the base link frame to the tool frame is given by the matrix B. The overall transformation between the base link frame and the tool frame in an arbitrary configuration of the robot will be given by the transformation, B to the tool frame in the home position, followed by the transformation of the end-effector itself. That is $K(\theta)B$. This is still an active transformation, to convert it to a coordinate transformation, as we saw above in section 1, we simply invert the matrix. Hence, the kinematics in terms of tool-frame coordinates is given by,

$$\left(K(\boldsymbol{\theta})B\right)^{-1} = B^{-1}A_6^{-1}(\theta_6)A_5^{-1}(\theta_5)A_4^{-1}(\theta_4)A_3^{-1}(\theta_3)A_2^{-1}(\theta_2)A_1^{-1}(\theta_1)$$

7.   Lie Algebra

In 1900 Ball published his influential treatise on "the theory of screws" [1]. With the benefit of hindsight it is clear that Ball's screws were simply elements of the Lie algebra of $SE(3)$. More precisely, the twists correspond to the element of the Lie algebra, the screws are elements of the projective space formed from the Lie algebra. This distinctions will not be used in the following.

The theory of Lie groups and their Lie algebras was developed some years later by, Lie, Klein, Killing and Cartan among others. So now we are in a position to take advantage of the Lie theoretic viewpoint and update Ball's original screw theory.

The Lie algebra of a Lie group can be defined in several different ways. Historically, Lie algebra elements were thought of as infinitesimal group elements. Here we can think of the Lie algebra of a group as the tangent space at the identity element. To find the Lie algebra elements we take a curve in the group and find the derivative at the identity.

We can use use the 4×4 representation of $SE(3)$ to find its Lie algebra. As we saw above finite screw motion can be written,

$$A(\theta) = \begin{pmatrix} R & \frac{\theta p}{2\pi}\hat{\mathbf{x}} + (I - R)\mathbf{u} \\ 0 & 1 \end{pmatrix}$$

where, R is a rotation through angle θ about the axis $\hat{\mathbf{x}}$, and $\mathbf{u}$ is a point on the axis. This defines a sequence of rigid transformations parameterised by θ. Moreover, when $\theta = 0$ we get the identity transformation.

So taking the derivative at $\theta = 0$ gives a typical element of the Lie algebra,

$$S = \begin{pmatrix} \Omega & \frac{\omega p}{2\pi}\hat{\mathbf{x}} - \Omega\mathbf{u} \\ 0 & 0 \end{pmatrix}$$

Here $\Omega = dR(0)/d\theta$ is a 3×3 anti-symmetric matrix. This can be seen by differentiating the relation that ensures R is orthogonal

$$\frac{d}{d\theta}RR^T = \frac{dR}{d\theta}R^T + R\frac{dR^T}{d\theta} = \frac{dI}{d\theta} = 0$$

when $\theta = 0$ we have that $R = I$ and hence,

$$\frac{dR(0)}{d\theta} + \frac{dR(0)}{d\theta}^T = 0$$

Now a 3×3 anti-symmetric matrix corresponds to a 3-dimensional vector,

$$\Omega = \begin{pmatrix} 0 & -\omega_z & \omega_y \\ \omega_z & 0 & -\omega_x \\ -\omega_y & \omega_x & 0 \end{pmatrix}$$

where ω_i are the components of the vector, $\boldsymbol{\omega}$. With these definitions the product of the matrix Ω with any 3-dimensional vector exactly models the vector product with $\boldsymbol{\omega}$, that is, $\Omega\mathbf{p} = \boldsymbol{\omega} \times \mathbf{p}$ for any $\mathbf{p}$.

Hence, we can see that $\boldsymbol{\omega}$ is the angular velocity vector of the body and ω is its magnitude.

A general element of the Lie algebra $se(3)$, is a matrix with the partitioned form,

$$S = \begin{pmatrix} \Omega & \mathbf{v} \\ 0 & 0 \end{pmatrix}$$

These matrices form a 6-dimensional vector space and it is often useful to write the elements of the space as six-dimensional vectors,

$$\mathbf{s} = \begin{pmatrix} \boldsymbol{\omega} \\ \mathbf{v} \end{pmatrix}$$

8.   The Adjoint Representation

So far the group $SE(3)$ has been defined by it 4×4 matrix representation. Sometimes it is useful to think of the group as an abstract object and consider different representations of it. A representation here is a linear representation, that is, we represent group elements by matrices, or more genrerally linear transformations. The group product is modeled by matrix multiplication and inverses in the group by matrix inversion. For any group there are many such representations, here we look at a representation which can be defined for any group. Any Lie group acts linearly on its Lie algebra. This representation is known as the adjoint representation of the group.

For $SE(3)$ we can write this in terms of the 4×4 matrices as

$$S' = ASA^{-1}$$

where $A = A(g)$ is the 4×4 matrix corresponding to some group element g. In terms of the 6-dimensional vectors representing the Lie algebra, we get,

$$s' = \text{Ad}(g)s$$

Here $\text{Ad}(g)$ is a 6×6 matrix representing the group element g. The general form of the matrices in this adjoint representation is,

$$\text{Ad}(g) = \begin{pmatrix} R & 0 \\ TR & R \end{pmatrix}$$

again R is a 3×3 rotation matrix but T is a 3×3 anti-symmetric matrix corresponding to a translation vector $\mathbf{t}$.

This adjoint action of the group on its Lie algebra can be extended to an action of the Lie algebra on itself. In terms of the 4×4 representation of $SE(3)$ this can be explained as follows. Take S_2 to be an element of the Lie algebra, as a 4×4 matrix. Now suppose $A(\theta)$ is a one-parameter subgroups corresponding to a Lie algebra element S_1, that is $dA(0)/d\theta = S_1$. Now if we differentiate the adjoint action of the group, AS_2A^{-1} and set $\theta = 0$ we get the commutator,

$$S_1S_2 - S_2S_1 = [S_1, S_2]$$

In terms of 6-dimensional vectors the commutator can be written as

$$\text{ad}(\mathbf{s}_1)\mathbf{s}_2 = \begin{pmatrix} \Omega_1 & 0 \\ V_1 & \Omega_1 \end{pmatrix} \begin{pmatrix} \boldsymbol{\omega}_2 \\ \mathbf{v}_2 \end{pmatrix}$$

The notation ad, lower case, denotes the adjoint action of the Lie algebra on itself. The commutators of elements are a key feature of Lie algebras, in robotics these commutators turn up in the dynamics of robots and in many other applications. Notice that in the 6-dimensional, partitioned form of representation of the Lie algebra, commutators can be written,

$$[\mathbf{s}_1, \mathbf{s}_2] = \text{ad}(\mathbf{s}_1)\mathbf{s}_2 = \begin{pmatrix} \boldsymbol{\omega}_1 \times \boldsymbol{\omega}_2 \\ \mathbf{v}_1 \times \boldsymbol{\omega}_2 + \boldsymbol{\omega}_1 \times \mathbf{v}_2 \end{pmatrix}$$

9. The Exponential Map

The exponential map sends Lie algebra elements to the Lie group. In a region of the origin the map is guaranteed to be a homeomorphism, and hence we can use the Lie algebra as a coordinate system for the group but only in a patch containing the identity. For $SE(3)$, this patch is quite big, we just have to keep rotation angles to between $\pm\pi$. The exponential map provides the link between the Lie algebra and the group.

In general, the exponential of a matrix X is defined by the series,

$$e^X = I + X + \frac{1}{2!}X^2 + \frac{1}{3!}X^3 + \cdots$$

where X^2 is the matrix product $X^2 = XX$ etc. It is possible to show that the series converges and, if X is a matrix representing an element of a Lie algebra, then the exponential e^X will be a matrix representing an element of the corresponding Lie group. The mapping is essentially independent of the representation.

The product of a pair of matrix exponentials is generally not simply the exponential of the sum of the exponents. This is because the exponents may not commute. However, if S is a fixed screw, then it is clear that

$$e^{\theta_1 S} e^{\theta_2 S} = e^{(\theta_1+\theta_2)S}$$

So these are the 1-parameter subgroups again, and we can write,

$$A(\theta) = e^{\theta S}$$

This means that we can associate a screw S, to any 1-degree-of-freedom mechanical joint. We will see how to do this in detail in a moment. For now simply observe that the forward kinematics for a six-joint industrial robot, as outlined in section 6 above, can be written as

$$K(\boldsymbol{\theta}) = e^{\theta_1 S_1} e^{\theta_2 S_2} e^{\theta_3 S_3} e^{\theta_4 S_4} e^{\theta_5 S_5} e^{\theta_6 S_6}$$

The parameters θ_i, are usually referred to as the joint parameters, or joint angles if the joints are revolute. This form of forward kinematics of serial robots was introduced by Brockett [2].

As promised, we next look at how to associate a screw to a mechanical joint. To each joint we can associate a 1-parameter subgroup,

$$e^{\theta S} = \begin{pmatrix} R & \frac{\theta p}{2\pi}\hat{\mathbf{x}} + (I - R)\mathbf{u} \\ 0 & 1 \end{pmatrix}$$

where the screw S has the form,

$$S = \begin{pmatrix} \Omega & \mathbf{v} \\ 0 & 0 \end{pmatrix}$$

If we look at the rotation part first we expect that $\hat{\mathbf{x}}$ to be along the axis of rotation so that $R\hat{\mathbf{x}} = \hat{\mathbf{x}}$. Differentiating this with respect to θ and setting $\theta = 0$ we get,

$$\Omega\hat{\mathbf{x}} = \omega \times \hat{\mathbf{x}} = 0$$

From this we can conclude that ω is proportional to $\hat{\mathbf{x}}$. In fact, it makes sense to define $\omega = \hat{\mathbf{x}}$ so that the parameter θ, becomes the rotation angle. Next, we look at $\mathbf{u}$, a point on the axis of the screw motion. Here we expect,

$$A(\theta) \begin{pmatrix} \mathbf{u} \\ 1 \end{pmatrix} = \begin{pmatrix} \mathbf{u} + \frac{\theta p}{2\pi}\hat{\mathbf{x}} \\ 1 \end{pmatrix}$$

that is, a point on the axis of the screw is simply translated alog the axis. The derivative at $\theta = 0$ gives

$$S \begin{pmatrix} \mathbf{u} \\ 1 \end{pmatrix} = \begin{pmatrix} \frac{p}{2\pi}\hat{\mathbf{x}} \\ 0 \end{pmatrix}$$

That is,

$$\Omega\mathbf{u} + \mathbf{v} = \frac{\theta p}{2\pi}\hat{\mathbf{x}}$$

Replacing $\hat{\mathbf{x}}$ by ω and rearranging we get the result

$$\mathbf{v} = \mathbf{u} \times \omega + \frac{p}{2\pi}\omega$$

Notice that if the pitch of the joint vanishes $p = 0$, that is the joint is revolute, then the screw we associate with it,

$$\mathbf{s} = \begin{pmatrix} \omega \\ \mathbf{u} \times \omega \end{pmatrix}$$

is simply given by the Plücker coordinates for the line forming the axis of the joint. For helical joints we add the extra term $\frac{p}{2\pi}\omega$ to the $\mathbf{v}$ part of the screw, where p is the pitch of the joint. The above does not apply to prismatic joints, but they are easy to deal with. Using the same sort of argument as above, a prismatic joint which allows movement in the direction $\hat{\mathbf{v}}$ will be represented by a screw,

$$\mathbf{s} = \begin{pmatrix} \mathbf{0} \\ \hat{\mathbf{v}} \end{pmatrix}$$

In general it is quite difficult to compute matrix exponentials, but in the case of rigid body motions we are lucky, the problem is relatively straightforward.

Consider a typical element from the 4×4 representation,

$$S = \begin{pmatrix} \Omega & \mathbf{v} \\ 0 & 0 \end{pmatrix}$$

Any such matrix satisfies the polynomial equation,

$$S^4 + |\omega|^2 S^2 = 0$$

this can be checked by direct computation. The equation factorises as

$$S^4 + |\omega|^2 S^2 = S^2(S + i|\omega|I)(S - i|\omega|I)$$

Now we can find three matrix functions of S, labelled P_0, P_+ and P_-, which satisfy $P_i^2 = P_i$ and $P_i P_j = 0$ if $i \neq j$, such a system is called a system of mutually annihilating idempotents. These matrices are given by,

$$P_0 = \frac{1}{|\omega|^2}(S^2 + |\omega|^2 I),$$

$$P_+ = \frac{1}{2i|\omega|^3} S^2(S - i|\omega|I),$$

$$P_- = \frac{-1}{2i|\omega|^3} S^2(S + i|\omega|I).$$

here $i = \sqrt{-1}$ is the imaginary unit. We also require a nilpotent here, $N_0 = \frac{1}{|\omega|^2} S(S^2 + |\omega|^2 I)$. This nilpotent satisfies $N_0^2 = 0$, $N_0 P_+ = N_0 P_- = 0$ and $N_0 P_0 = N_0$.

The point of all this is that we can write our original matrix S as

$$S = N_0 - i|\omega|P_+ + i|\omega|P_-$$

and so when we take powers of S we only need to take powers of the coefficients of P_+ and P_- since there are no cross-terms and the powers of the matrices are simple. So exponentiating we get,

$$e^S = I + N_0 + (e^{-i|\omega|} - 1)P_+ + (e^{i|\omega|} - 1)P_-$$

It is also straightforward to see that $I = P_0 + P_+ + P_-$ so the exponential can be simplified to,

$$e^S = P_0 + N_0 + e^{-i|\omega|}P_+ + e^{i|\omega|}P_-$$

Finally, substituting back the Ss and replacing the complex exponentials with trigonometric functions we have the result

$$e^S = I + S + \frac{1}{|\omega|^2}(1 - \cos|\omega|)S^2 + \frac{1}{|\omega|^3}(|\omega| - \sin|\omega|)S^3$$

The corresponding formula for the group $SO(3)$ is known as the Rodrigues formula.

We can produce a similar result for the adjoint representation of $SE(3)$. A typical element of the Lie algebra in this representation is given by,

$$\text{ad}(s) = \begin{pmatrix} \Omega & 0 \\ V & \Omega \end{pmatrix}$$

In general we would expect a 6×6 matrix to satisfy a degree 6 polynomial equation, these matrices however, satisfy a degree 5 equation,

$$\text{ad}(s)^5 + 2|\omega|^2 \, \text{ad}(s)^3 + |\omega|^4 \, \text{ad}(s) = 0$$

This relation can be verified easily using the following relation between pairs of 3×3 anti-symmetric matrices,

$$\Omega^2 V \Omega^2 + |\omega|^2 (\Omega V + V \Omega) = -|\omega|^4 V$$

In turn this relation can be verified by using the connection between these matrices and vector products, $V\mathbf{x} = \mathbf{v} \times \mathbf{x}$ for any vector $\mathbf{x}$, and then using the familiar formula for vector triple products.

The degree 5 equation above factorises as

$$\text{ad}(s)^5 + 2|\omega|^2 \, \text{ad}(s)^3 + |\omega|^4 \, \text{ad}(s) = \text{ad}(s)(\text{ad}(s) + i|\omega|I)^2(\text{ad}(s) - i|\omega|I)^2$$

So this time we have two repeated factors and hence we seek three idempotents as before and two nilpotents, one associated with each repeated factor.

$$
\begin{aligned}
P_0 &= \frac{1}{|\omega|^4}(\text{ad}(s) + i|\omega|I)^2(\text{ad}(s) - i|\omega|I)^2 \\
&= (\text{ad}(s)^4 + 2|\omega|^2 \, \text{ad}(s)^2 + |\omega|^4 I)/|\omega|^4, \\
P_+ &= \frac{-1}{4|\omega|^4} \, \text{ad}(s)(\text{ad}(s) - i|\omega|I)^2(2\,\text{ad}(s) + 3i|\omega|I) \\
&= -(2\,\text{ad}(s)^4 - i|\omega| \, \text{ad}(s)^3 + 4|\omega|^2 \, \text{ad}(s)^2 - 3i|\omega|^3 \, \text{ad}(s))/4|\omega|^4, \\
N_+ &= \frac{1}{4|\omega|^4} \, \text{ad}(s)(\text{ad}(s) + i|\omega|I)(\text{ad}(s) - i|\omega|I)^2 \\
&= (\text{ad}(s)^4 - i|\omega| \, \text{ad}(s)^3 + |\omega|^2 \, \text{ad}(s)^2 - i|\omega|^3 \, \text{ad}(s))/4|\omega|^4, \\
P_- &= \frac{-1}{4|\omega|^4} \, \text{ad}(s)(\text{ad}(s) + i|\omega|I)^2(2\,\text{ad}(s) - 3i|\omega|I) \\
&= -(2\,\text{ad}(s)^4 + i|\omega| \, \text{ad}(s)^3 + 4|\omega|^2 \, \text{ad}(s)^2 + 3i|\omega|^3 \, \text{ad}(s))/4|\omega|^4, \\
N_- &= \frac{1}{4|\omega|^4} \, \text{ad}(s)(\text{ad}(s) + i|\omega|I)^2(\text{ad}(s) - i|\omega|I) \\
&= (\text{ad}(s)^4 + i|\omega| \, \text{ad}(s)^3 + |\omega|^2 \, \text{ad}(s)^2 + I|\omega|^3 \, \text{ad}(s))/4|\omega|^4.
\end{aligned}
$$

By inspection we have that

$$\mathrm{ad}(\mathbf{s}) = -i|\omega|P_+ - i|\omega|N_+ + i|\omega|P_- + i|\omega|N_-$$

and as usual we also have,

$$I = P_0 + P_+ + P_-$$

Notice that with matrices P and N such that $P^2 = P$, $N^2 = 0$ and $PN = N$ the kth power of their sum is simply, $(P + N)^k = P + kN$. Hence the exponential of a matrix $\mathrm{ad}(\mathbf{s})$ from the adjoint representation of $se(3)$ can be written as

$$e^{\mathrm{ad}(\mathbf{s})} = I + (e^{-i|\omega|} - 1)P_+ - i|\omega|e^{-i|\omega|}N_+ + (e^{i|\omega|} - 1)P_- + i|\omega|e^{i|\omega|}N_-$$

or

$$e^{\mathrm{ad}(\mathbf{s})} = P_0 + e^{-i|\omega|}P_+ - i|\omega|e^{-i|\omega|}N_+ + e^{i|\omega|}P_- + i|\omega|e^{i|\omega|}N_-$$

Expanding the idempotents and nilpotents in terms of powers of $\mathrm{ad}(\mathbf{s})$ we finally obtain the result

$$
\begin{aligned}
e^{\mathrm{ad}(\mathbf{s})} = I \ &+ \ \frac{1}{2|\omega|}(3\sin|\omega| - |\omega|\cos|\omega|)\,\mathrm{ad}(\mathbf{s}) \\
&+ \ \frac{1}{2|\omega|^2}(4 - 4\cos|\omega| - |\omega|\sin|\omega|)\,\mathrm{ad}(\mathbf{s})^2 \\
&+ \ \frac{1}{2|\omega|^3}(\sin|\omega| - |\omega|\cos|\omega|)\,\mathrm{ad}(\mathbf{s})^3 \\
&+ \ \frac{1}{2|\omega|^4}(2 - 2\cos|\omega| - |\omega|\sin|\omega|)\,\mathrm{ad}(\mathbf{s})^4
\end{aligned}
$$

10. Derivatives of Exponentials

Suppose we are given a path in a Lie group ($SE(3)$ here) as an exponential,

$$g(t) = e^{X(t)}.$$

How can we find the derivative of this formula and what does it represent?

In the previous section we met the simple case where the path in the Lie algebra $X(t)$ is just tX_0 where X_0 is a constant element in the algebra. The derivative of such a path is simply,

$$\frac{d}{dt}e^{tX_0} = X_0 e^{tX_0}$$

However, when $X(t)$ is more complicated this is no longer valid. The problem is that $dX/dt = \dot{X}$ does not necessarily commute with X so differentiating the expansion of the exponential gives

$$
\begin{aligned}
\frac{d}{dt}e^X = {} & \dot{X} + \frac{1}{2!}(\dot{X}X + X\dot{X}) + \frac{1}{3!}(\dot{X}X^2 + X\dot{X}X + X^2\dot{X}) + \cdots \\
& \cdots + \frac{1}{(k+1)!}(\dot{X}X^k + X\dot{X}X^{k-1} + \cdots + X^k\dot{X}) + \cdots
\end{aligned}
$$

In [3] Hausdorff showed that

$$
\left(\frac{d}{dt}e^X\right)e^{-X} = \dot{X} + \frac{1}{2!}[X, \dot{X}] + \frac{1}{3!}[X, [X, \dot{X}]] + \frac{1}{4!}[X, [X, [X, \dot{X}]]] + \cdots
$$

The right-hand side of this equation will be abbreviated to X_d. Notice that this matrix is composed of sums of iterated commutators and hence X_d is an element of the Lie algebra. The Hausdorff formula above implies that

$$
\frac{d}{dt}e^X = X_d e^X
$$

In the rotation group $SO(3)$ this means that X_d corresponds to the angular velocity of the motion. In $SE(3)$, the group of rigid body motions the corresponding vector is the velocity screw of the motion. The equation is important in several applications. In a typical dynamical simulation, for example, equations of motion would be solved numerically to find the velocity screw or, angular and linear velocities. From these quantities the transformation matrices usually have to be found. It is essentially the above equation that has to be solved to find the rigid transformation undergone by the rigid body under consideration. In most commercial applications a rudimentary approach to this problem is often taken. This is because the speed of computation is far more important than the accuracy.

Using vectors to represent the Lie algebra the Hausdorff formula can be written neatly as

$$
\mathbf{x}_d = \sum_{k=0}^{\infty} \frac{1}{(k+1)!}\,\mathrm{ad}^k(\mathbf{x})\dot{\mathbf{x}}
$$

with ad() the adjoint representation of the Lie algebra, as usual.

For $SE(3)$ we can evaluate this infinite sum using the idempotents and nilpotents found in section 9 above. This gives

$$\sum_{k=0}^{\infty} \frac{1}{(k+1)!} \text{ad}(\mathbf{s})^k = P_0 + \sum_{k=0}^{\infty} \frac{(-i|\omega|)^k}{(k+1)!} P_+ + \sum_{k=0}^{\infty} \frac{k(-i|\omega|)^k}{(k+1)!} N_+$$
$$+ \sum_{k=0}^{\infty} \frac{(i|\omega|)^k}{(k+1)!} P_- + \sum_{k=0}^{\infty} \frac{k(i|\omega|)^k}{(k+1)!} N_-$$

Evaluating the infinite sums gives

$$\sum_{k=0}^{\infty} \frac{1}{(k+1)!} \text{ad}(\mathbf{s})^k = P_0 + \frac{1}{-i|\omega|}(e^{-i|\omega|} - 1)P_+$$
$$+ \frac{1}{-i|\omega|}((-i|\omega| - 1)e^{-i|\omega|} + 1)N_+$$
$$+ \frac{1}{i|\omega|}(e^{i|\omega|} - 1)P_-$$
$$+ \frac{1}{i|\omega|}((i|\omega| - 1)e^{i|\omega|} + 1)N_-$$

Substituting for the idemponents and nilpotents gives

$$\sum_{k=0}^{\infty} \frac{1}{(k+1)!} \text{ad}(\mathbf{s})^k = I + \frac{1}{2|\omega|^2}(4 - |\omega|\sin|\omega| - 4\cos|\omega|)\,\text{ad}(\mathbf{s})$$
$$+ \frac{1}{2|\omega|^3}(4|\omega| - 5\sin|\omega| + |\omega|\cos|\omega|)\,\text{ad}(\mathbf{s})^2$$
$$+ \frac{1}{2|\omega|^4}(2 - |\omega|\sin|\omega| - 2\cos|\omega|)\,\text{ad}(\mathbf{s})^3$$
$$+ \frac{1}{2|\omega|^5}(2|\omega| - 3\sin|\omega| + |\omega|\cos|\omega|)\,\text{ad}(\mathbf{s})^4$$

This format is not particularly useful for dynamic simulation, it would be more useful to have $\dot{\mathbf{s}}$ as the subject of the equation. This would allow direct numerical methods to be used. Such an inversion is possible in general, see [3]. The result is another infinite series, however it is not necessary to follow this route since the above formula can be inverted more directly using the idempotents. Recall that $P_0 + P_+ + P_- = I$ so that for any constants a_i, b_i,

$$(a_0 P_0 + a_+ P_+ + b_+ N_+ + a_- P_- + b_- N_-)$$
$$* (\frac{1}{a_0} P_0 + \frac{1}{a_+} P_+ - \frac{b_+}{a_+^2} N_+ + \frac{1}{a_-} P_- - \frac{b_-}{a_-^2} N_-)$$
$$= (P_0 + P_+ + P_-) = I$$

The computations are a little more than can be comfortably done by hand, but are readily computed using a computer algebra package such as Maple or *Mathematica*.

Let, $\mathbf{s}_d$ be the Lie algebra element satisfying $\frac{d}{dt}e^{\mathrm{ad}(\mathbf{s})} = \mathrm{ad}(\mathbf{s}_d)e^{\mathrm{ad}(\mathbf{s})}$, then we have the result

$$\dot{\mathbf{s}} = \left(I - \frac{1}{2}\mathrm{ad}(\mathbf{s}) + \left(\frac{2}{|\omega|^2} \frac{|\omega| + 3\sin|\omega|}{4|\omega|(\cos|\omega| - 1)} \right) \mathrm{ad}(\mathbf{s})^2 + \right.$$
$$\left. + \left(\frac{1}{|\omega|^4} + \frac{|\omega| + \sin|\omega|}{4|\omega|^3(\cos|\omega| - 1)} \right) \mathrm{ad}(\mathbf{s})^4 \right) \mathbf{s}_d$$

Notice the absence of a term in $\mathrm{ad}(\mathbf{s})^3$ in the above.

Finally here, we look at the interpretation of these derivatives. Consider a general sequence of rigid motions parameterised by time t, then we can see from the above that

$$\frac{dA}{dt}A^{-1} = S_d$$

is an element of the Lie algebra and that it corresponds to the instantaneous velocity. The Ω or ω part is the angular velocity while the $\mathbf{v}$ part is the linear velocity of a point on the screw axis. We can find the velocity of a point on the body moving with this motion as follows. At time $t = 0$ the point will be assumed to have position vector $\mathbf{p}_0$, at any subsequent time its position will be given by,

$$\begin{pmatrix} \mathbf{p} \\ 1 \end{pmatrix} = A(t) \begin{pmatrix} \mathbf{p}_0 \\ 1 \end{pmatrix}$$

Differentiating with respect to time gives

$$\begin{pmatrix} \dot{\mathbf{p}} \\ 0 \end{pmatrix} = \dot{A}A^{-1} \begin{pmatrix} \mathbf{p} \\ 1 \end{pmatrix} = S_d \begin{pmatrix} \mathbf{p} \\ 1 \end{pmatrix}$$

Hence the velocity of the point is,

$$\dot{\mathbf{p}} = \omega \times \mathbf{p} + \mathbf{v}$$

The screw S_d or $\mathbf{s}_d$ will be called the velocity screw of the motion.

11.   Jacobians

The exponential form of robot kinematics is useful for deriving differential properties, for example Jacobians.

Consider a point attached to the end-effector of the robot, in the robot's home position the point has position vector $\mathbf{p}$. Hence, at any subsequent configuration of the robot the position of the point will be given by,

$$\begin{pmatrix} \mathbf{p}' \\ 1 \end{pmatrix} = K(\boldsymbol{\theta}) \begin{pmatrix} \mathbf{p} \\ 1 \end{pmatrix}$$

If the joints are moving with rates, $\dot{\theta}_1, \dot{\theta}_2, \ldots \dot{\theta}_6$ then the velocity of the point will be,

$$\begin{pmatrix} \dot{\mathbf{p}} \\ 0 \end{pmatrix} = \left(\dot{\theta}_1 \frac{\partial K}{\partial \theta_1} + \dot{\theta}_2 \frac{\partial K}{\partial \theta_2} + \cdots + \dot{\theta}_6 \frac{\partial K}{\partial \theta_6} \right) \begin{pmatrix} \mathbf{p} \\ 1 \end{pmatrix}$$

The partial derivatives are easy to evaluate at the home position,

$$\frac{\partial K}{\partial \theta_i} = e^{\theta_1 S_1} \cdots S_i e^{\theta_i S_i} \cdots e^{\theta_6 S_6} = S_i$$

So in the home position the velocity of the point is,

$$\begin{pmatrix} \dot{\mathbf{p}} \\ 0 \end{pmatrix} = (\dot{\theta}_1 S_1 + \dot{\theta}_2 S_2 + \cdots + \dot{\theta}_6 S_6) \begin{pmatrix} \mathbf{p} \\ 1 \end{pmatrix}$$

This is more general than it looks. If we choose the current configuration of the robot to be the home configuration then the above argument will apply. So, S_i here will be taken to mean the current position of the i-th joint. Alternatively, we could let $\mathbf{s}_i^0$ denote the home position of the robot's joint screws. Subsequent configurations for the joints will be given by,

$$\mathbf{s}_i = e^{\theta_1 \, \mathrm{ad}(\mathbf{s}_1^0)} e^{\theta_2 \, \mathrm{ad}(\mathbf{s}_2^0)} \cdots e^{\theta_{i-1} \, \mathrm{ad}(\mathbf{s}_{i-1}^0)} \mathbf{s}_i^0$$

Hence, we have that

$$\frac{\partial K}{\partial \theta_i} K^{-1} = S_i$$

So we can find the velocity screw of the robot's end-effector by simply substituting the adjoint representation of the Lie algebra. Hence,

$$\begin{pmatrix} \boldsymbol{\omega} \\ \mathbf{v} \end{pmatrix} = \mathbf{s}_1 \dot{\theta}_1 + \mathbf{s}_2 \dot{\theta}_2 + \cdots + \mathbf{s}_6 \dot{\theta}_6$$

If we compare this with the Jacobian relation for the velocity $\mathbf{s}_d = J\dot{\boldsymbol{\theta}}$, we can see that the columns of the 6×6 Jacobian matrix must be the current joint screws of the robot,

$$J = \left(\mathbf{s}_1 \,\middle|\, \mathbf{s}_2 \,\middle|\, \mathbf{s}_3 \,\middle|\, \mathbf{s}_4 \,\middle|\, \mathbf{s}_5 \,\middle|\, \mathbf{s}_6 \right)$$

A well known result.

An immediate application of the above is to robot control. Suppose we wanted the robots end-effector to follow a path given by a exponential, $e^{\mathbf{s}(t)}$. In a traditional approach to this problem, the rigid motions along the path would be computed and then a inverse kinematic routine would be used to find the corresponding joint angles. It is probably more efficient to find the joint angles by numerically integrating the equation,

$$J\dot{\boldsymbol{\theta}} = \mathbf{s}_d$$

For most current designs of industrial the Jacobian matrix J can be inverted symbolically, although this doesn't seem to be widely appreciated. Also $\mathbf{s}_d$ can be computed using the formulas given above.

12.   Concluding Remarks

In this brief lecture there has only been enough time to give an outline of how group theory can be used in robotics. Several important applications have been omitted in the interests of brevity. Although we have discussed the forward kinematics of serial manipulators the inverse kinematics is an key problem in the subject. Here we know where we want to place the robot's end-effector and we must compute the joint angles needed to achieve this. Although, an approach using differential equations has been sketched above, this clearly runs into difficulties when the robot's Jacobian becomes singular. The investigation of these robot singularities is another problem where group theory has been usefully applied, see [6]. For parallel mechanism, such as the Gough-Stewart platform, the inverse kinematics are straightforward, it is the forward kinematics which are hard.

Robot dynamics requires another couple of representation of the group $SE(3)$. First, the forces and torques acting on a rigid body can be combined into a single 6-dimensional vector called a wrench. The wrenches transform according to a representation dual to the adjoint representation, the coadjoint representation. An accidental property of $SE(3)$ means that the adjoint and coadjoint representations are similar. In Ball's original screw theory velocities and wrenches were treated as the same sort of object, however, to make progress it is simplest to separate them. The second representation we need is the symmetric product of the adjoint representation. Elements of this representation can then be interpreted as 6×6 inertia matrices.

There are many applications of group theory to robot vision. Now the group may be different. For example, there are formulations of the camera calibration problem that use the group $SL(4)$ to model the possible camera parameters. Certainly projective geometry will be important when discussing cameras and there are projective groups associated with this geometry. The group relevant to geometric optics is the symplectic group $Sp(6, \mathbb{R})$, this is the symmetry groups of a complex of lines in space.

Finally, the above presentation has used matrices and vectors exclusively. There is some merit to using Clifford algebras to represent the group of rigid body motions and the various spaces on which it acts. Relations and formulas in Clifford algebra tend to be neater than their matrix counterparts and this is useful for symbolic and hand computation. Another advantage of the Clifford algebra is that the group $SE(3)$ can be represented very compactly and this allows us to study the geometry of the group itself. See Joan Lasenby's lecture later in this meeting.

References

[1] R.S. Ball. *The Theory of Screws.* Cambridge University Press, Cambridge, 1900.

[2] R. Brockett. Robotic manipulators and the product of exponential formula. In P. Fuhrman ed. *Proc. Mathematical Theory of Networks and Systems* pp. 120–129, 1984.

[3] F. Hausdorff. Die Symbolische exponential formel in den gruppen theorie. Berichte de Sächicen Akademie de Wissenschaften (Math Phys Klasse) vol. 58, pp. 19–48, 1906.

[4] J.M. Hervé Analyse Structurelle de Mecanismes par Groupe des Displacements. *Mechanism and Machine Theory* **13**:437-450, 1978.

[5] B.K.P Horn. *Robot Vision* Cambridge, MA:MIT Press, 1986. 1987.

[6] A. Karger. Classification of serial robot-manipulators with non-removable singularities. *J. Mech. Design* **30**:202–208, 1996.

[7] F. Reuleaux *Theoretische Kinematic: Grunzüge einer Theorie des Maschinwesens* Braunschweig: Vieweg, 1875. Trans. A.B. W. Kennedy as *The Kinematics of Machinery*, London: Macmillan, 1876. Reprinted, New York: Dover, 1963.

QUANTUM/CLASSICAL INTERFACE:
A GEOMETRIC APPROACH
FROM THE CLASSICAL SIDE

William E. Baylis
Physics Dept., University of Windsor
Windsor, ON, Canada N9B 3P4
baylis@uwindsor.ca

Abstract   Classical relativistic physics in Clifford's geometric algebra has a spinorial formulation that is closely related to the standard quantum formalism. The algebraic use of spinors and projectors, together with the bilinear relations of spinors to observed currents, gives quantum-mechanical
form to many classical results, and the clear geometric content of the algebra makes it an illuminating probe of the quantum/classical interface.
The aim of this lecture is to close the conceptual gap between quantum
and classical phenomena while highlighting their essential differences.
The paravector representation of spacetime in APS is used in particular to provide insight into what many pioneering quantum physicists
have considered classically indescribable: spin-1/2 systems and their
measurement.

Keywords: algebra of physical space, Clifford algebra, entanglement, paravectors,
quantum/classical interface, quantum computing, quantum informa-
tion, relativity

1.   Introduction

This lecture continues the theme[1] that Clifford's geometric algebras, in particular the algebra of physical space (APS), provide the lu-
bricant for smooth paradigm shifts from Newtonian to post-Newtonian
paradigms. The focus here is on the shift to the quantum view of the
universe, and how APS can make the quantum/classical interface more
transparent.

After a brief review of paravector space and its ability to model spacetime, we start the lecture by relating the classical eigenspinor of a particle, that is the Lorentz rotor that takes it from rest to its motion in the

J. Byrnes (ed.) Computational Noncommutative Algebra and Applications, 127-154.

lab, to its relativistic wave function. We then show how the 2-valued measurement of the spin arises naturally from the algebraic properties of rotors, and how the g factor and magnetic moment of elementary fermions result. The spin itself arises classically as a gauge freedom of the eigenspinor, and as we will see, the choice of a spin rate proportional to the total energy of the system leads to de Broglie waves and momentum eigenstates of the Dirac equation. Finally, we consider work on entangled spin states which are of interest in quantum computation.

2.   Paravector Space as Spacetime

Recall from my previous lecture[1] that we use paravector space in APS to model spacetime. In addition to the three dimensions of physical space, paravector space also includes scalars. A typical paravector can be written

$$p = p^0 + \mathbf{p} \equiv p^\mu \mathbf{e}_\mu , \qquad (1)$$

where p^0 is a scalar and $\mathbf{p}$ a physical vector, $\{\mathbf{e}_1, \mathbf{e}_2, \mathbf{e}_3\}$ is an orthonormal basis of physical space, and $\mathbf{e}_0 = 1$. The *paravector basis* $\{\mathbf{e}_0, \mathbf{e}_1, \mathbf{e}_2, \mathbf{e}_3\}$ is the observer's *proper* basis (at rest). Paravector space in APS is thus a four-dimensional linear space. It has a metric determined by the quadratic form, which for the paravector p is proportional to the scalar $p\bar{p} = \bar{p}p = \left(p^0\right)^2 - \mathbf{p}^2$, where $\bar{p} = p^0 - \mathbf{p}$ is the *Clifford conjugate* of p. Since the quadratic form is a scalar expression, insertion of the basis expansion (1) gives

$$p\bar{p} = \langle p\bar{p} \rangle_S = p^\mu p^\nu \langle \mathbf{e}_\mu \bar{\mathbf{e}}_\nu \rangle_S \equiv p^\mu p^\nu \eta_{\mu\nu}, \qquad (2)$$

and the metric tensor $\eta_{\mu\nu} = \langle \mathbf{e}_\mu \bar{\mathbf{e}}_\nu \rangle_S$ automatically arising here is the *Minkowski metric* of spacetime.

Rotations in paravector space preserve the "square length" $p\bar{p}$ of paravectors and therefore also the scalar products $\langle p\bar{q} \rangle_S = \frac{1}{2}(p\bar{q} + q\bar{p})$ between paravectors. They have the form often called a *spin transformation*[2]

$$p \to LpL^\dagger, \qquad (3)$$

where L is a unimodular element $(L\bar{L} = 1)$ known as a *Lorentz rotor*. Lorentz rotations are the physical Lorentz transformations of relativity: boosts, spatial rotations, and their products. Several Lorentz rotations can be applied in succession, for example $p \to L_3 L_2 L_1 p L_1^\dagger L_2^\dagger L_3^\dagger$. All the information about the transformation is contained in the composition $L = L_3 L_2 L_1$, which of course is another Lorentz rotor. Indeed, the Lorentz rotors form a group isomorphic to $SL(2, \mathbb{C})$, the double covering group of $SO_+(1,3)$.

Lorentz rotors L are generated by biparavectors:

$$L = \pm \exp(\mathbf{W}/2) \in \$pin_+(1,3) \simeq SL(2,\mathbb{C}),$$

where $\mathbf{W}$ is a biparavector

$$\mathbf{W} = \frac{1}{2} W^{\mu\nu} \langle \mathbf{e}_\mu \bar{\mathbf{e}}_\nu \rangle_V.$$

If $\mathbf{W}$ is a vector (a timelike biparavector), then $L = B = B^\dagger$ is a boost, whereas if $\mathbf{W}$ is a bivector (a spacelike biparavector), then $L = R = \bar{R}^\dagger$ is a spatial rotation. Every rotor L can be uniquely factored into the product of a boost and a rotation[3]:

$$L = BR.$$

The spin transformations of paravector products are found by transforming each paravector factor (3).

3.   Eigenspinors

Recall[1, 4] that the *eigenspinor* Λ of a system is an *amplitude* for the motion and orientation of the system. More precisely, it is a Lorentz rotor of particular interest, relating the system reference frame to the observer (or lab). It transforms distinctly from paravectors and their products: a Lorentz transformation L applied in the lab transforms $\Lambda \to L\Lambda$. This characterizes Λ as a *spinor*. The *eigen* part of its name refers to its association with the system, which in the simplest case is a single particle. The eigenspinor is generally a reducible element of the spinor carrier space of Lorentz rotations. Any property of the system in the reference frame is easily transformed by Λ to the lab. For example, in units with $c = 1$ the proper velocity of a massive particle is taken to be $u = \mathbf{e}_0 = 1$ in the reference frame; that is, the reference frame at proper time τ is an inertial frame that moves with the particle at that instant. In the lab, the proper velocity is

$$u = \Lambda \mathbf{e}_0 \Lambda^\dagger = \Lambda\Lambda^\dagger, \tag{4}$$

which is also the timelike basis paravector of a frame moving with proper velocity u. The complete paravector basis of the moving frame (relative to the observer) is $\{\mathbf{u}_\mu = \Lambda \mathbf{e}_\mu \Lambda^\dagger\}$ with $u \equiv \mathbf{u}_0$. Thought of as a passive transformation, Λ transforms the observer from the reference frame to the lab.

3.1   Time evolution

The eigenspinor Λ of an accelerating or rotating system changes in time, and we let $\Lambda(\tau)$ be the Lorentz rotation at proper time τ from the reference frame to the commoving inertial frame of the system. Eigenspinors at different times are related by

$$\Lambda(\tau_2) = L(\tau_2, \tau_1) \Lambda(\tau_1),$$

where by the group property of rotations, the time-evolution operator

$$L(\tau_2, \tau_1) = \Lambda(\tau_2) \bar{\Lambda}(\tau_1)$$

is another Lorentz rotation. The time evolution of Λ is found by solving an equation of motion

$$\dot{\Lambda} = \frac{1}{2}\Omega\Lambda = \frac{1}{2}\Lambda\Omega_{\text{ref}}, \tag{5}$$

where the unimodularity of Λ ($\Lambda\bar{\Lambda} = 1$) implies that $\Omega \equiv 2\dot{\Lambda}\bar{\Lambda} = -2\Lambda\dot{\bar{\Lambda}}$ is a biparavector that can be identified as the instantaneous *spacetime rotation rate* of the system as seen in the lab, and $\Omega_{\text{ref}} = \bar{\Lambda}\Omega\Lambda$ is its value in the reference frame. Note generally that operators from the left on Λ act in the lab whereas those from the right act in the reference frame.

The equation of motion (5) allows us to compute time-rates of change of any covariant property known in the reference frame in terms of Ω or Ω_{ref}. For example, the acceleration in the lab can be expanded $\dot{u} = d\left(\Lambda\Lambda^\dagger\right)/d\tau = \dot{\Lambda}\Lambda^\dagger + \Lambda\dot{\Lambda}^\dagger = 2\left\langle \dot{\Lambda}\Lambda^\dagger \right\rangle_{\Re}$ and substitution of the equation of motion (5) gives

$$\dot{u} = \langle \Omega u \rangle_{\Re}, \tag{6}$$

Here, $\langle x \rangle_{\Re} = \frac{1}{2}\left(x + x^\dagger\right)$ indicates the *real* (grades $0 + 1$) part of x. In terms of Ω_{ref},

$$\dot{u} = \left\langle \Lambda\Omega_{\text{ref}}\Lambda^\dagger \right\rangle_{\Re} = \Lambda \left\langle \Omega_{\text{ref}} \right\rangle_{\Re} \Lambda^\dagger. \tag{7}$$

Note that relation (7) has the form of a Lorentz rotation of a paravector from the reference frame to the lab. Evidently the real (vector) part of the rotation rate Ω_{ref} in the reference frame is just the acceleration of the system there. Equation (6) is just the covariant Lorentz-force equation if we identify $\Omega = e\mathbf{F}/m$. This identification provides a convenient covariant operational definition of the electromagnetic field $\mathbf{F}$. However, this definition is more restrictive than is *required* by the Lorentz-force equation, and as we see below, a generalization is needed in the presence of spin.

The equation of motion (5) for eigenspinors, especially when supplemented by algebraic projectors, offers powerful new tools for finding classical trajectories of charges in electromagnetic fields. Advantages of the eigenspinor approach are nicely demonstrated in new analytic solutions for the autoresonant laser accelerator.[5]

3.2 Gauge Transformations

The eigenspinor equation (5) embodies more information than can ever be obtained from the Lorentz force equation alone: Λ gives not only the velocity $\Lambda\Lambda^\dagger$ of the particle, but also its *orientation*. In this section we look at the gauge freedom associated with orientation and see that it leads to the concept of spin.

The proper velocity $u = \Lambda\Lambda^\dagger$ of a massive particle is invariant under transformations of the form

$$\Lambda \to \Lambda R_0 , \tag{8}$$

where R_0 is any spatial rotation: $R_0 R_0^\dagger = R_0 \bar{R}_0 = 1$. Because R_0 is positioned on the right of Λ, it acts on the reference frame. Transformations that leave measured values unchanged are called *gauge transformations,* and if, as in Newtonian mechanics, we measure only the spacetime position (world line) of the particle and derivatives thereof, then the reorientation (8) of the reference frame is a gauge transformation. We therefore have the gauge freedom to choose the reference-frame orientation as we wish.

The rotation R_0 in (8) may be time dependent. Then, in order for the equation of motion (5) to remain valid, not only is the spacetime rotation rate in the reference frame transformed to the new reference frame, but it also adds an imaginary (bivector) part:

$$\Omega_{\text{ref}} \to \bar{R}_0 \left(\Omega_{\text{ref}} - i\boldsymbol{\omega}_0\right) R_0 , \tag{9}$$

where $-i\boldsymbol{\omega}_0 = 2\dot{R}_0 \bar{R}_0$ is the spatial rotation rate that generates R_0. We can confirm from Eq. (7) that the two transformations (8) and (9) together leave the acceleration (and by integration, the proper velocity and the world line) of the massive particle unchanged.

3.3 Free De Broglie Waves

The gauge freedom of the classical eigenspinor equations thus admits an arbitrary rotation of the reference frame. It is reasonable to assume that for a free particle the rotation rate $\boldsymbol{\omega}_0$ is constant, and since we are free to choose the initial orientation of the reference frame, we can take $\boldsymbol{\omega}_0 = \omega_0\mathbf{e}_3$. This rotation rate generates what we refer to as a "spin" rotation, with the spin rotor $R_0 = \exp\left(-i\mathbf{e}_3\omega_0\tau/2\right)$ acting in the reference frame. The eigenspinor of the free particle then has the form

$$\Lambda\left(\tau\right) = \Lambda\left(0\right) e^{-i\mathbf{e}_3\omega_0\tau/2} \tag{10}$$

and satisfies the equation of motion

$$\dot{\Lambda} = \frac{1}{2}\Lambda\Omega_{\text{ref}} \, , \quad \Omega_{\text{ref}} = -i\omega_0\mathbf{e}_3.$$

The linear form of the equation of motion (5) generally suggests the use of superpositions of solutions with different world lines, and hence with different proper times τ. To be able to combine these we need to express the proper time τ in particle coordinates x :

$$\tau = \langle x\bar{u}\rangle_S \; = \langle u\bar{x}\rangle_S \, , \tag{11}$$

where u is the proper velocity of the particle. When τ (11) is substituted in (10), the result has the form of *free de Broglie waves* in covariant formulation, except that the phase is now a real angle of the spin rotation. The wavelength is fixed at the de Broglie value by relating the rest energy m of the particle to ω_0 by

$$m = \frac{1}{2}\hbar\omega_0 \, .$$

In terms of the momentum paravector $p = mu$, the free eigenspinors (10) are then

$$\Lambda\left(x\right) = \Lambda\left(0\right)\exp\left[-i\mathbf{e}_3\left\langle p\bar{x}\right\rangle_S/\hbar\right]. \tag{12}$$

4. Spin

The experimentally verified existence of de Broglie waves suggests that the spin rotation in the free eigenspinor (12) is real and occurs at the high angular frequency associated with Zitterbewegung:

$$\omega_0 = 2m/\hbar = 1.55 \times 10^{21}\,\text{s}^{-1} \, .$$

As introduced above, the proper paravector basis $\{\mathbf{e}_\mu\}$ of the particle is transformed to the basis $\{\mathbf{u}_\mu = \Lambda\mathbf{e}_\mu\Lambda^\dagger\}$ of a *commoving particle frame* (relative to the observer) by the Lorentz transformation Λ. In particular the moving time axis $\mathbf{u}_0 = u = \Lambda\Lambda^\dagger$ is the proper velocity of that frame with respect to the observer. It is natural to ask whether any of the transformed *spatial* vectors, $\mathbf{u}_k = \Lambda\mathbf{e}_k\Lambda^\dagger$, carries any special physical meaning. By taking $\mathbf{e}_3$ to be the rotation axis in the reference frame, the corresponding unit paravector $\mathbf{u}_3$ is the only spatial axis of the commoving frame that is not spun into a blur in the lab. It is associated with the *spin*. We refer to the unit paravector $\mathbf{u}_3 = w$ as the (unit) *spin paravector*[1] and note that $w\bar{w} = -1$.

---

[1] The physical spin requires an extra scalar factor of $\hbar/2$ to ensure the correct magnitude and units.

Since $\mathbf{e}_0$ and $\mathbf{e}_3$ are orthogonal, so are $\mathbf{u}_0$ and $\mathbf{u}_3$ and hence u and w :

$$\langle \mathbf{u}_3 \bar{\mathbf{u}}_0 \rangle_S = \langle w\bar{u} \rangle_S = \langle \mathbf{e}_3 \bar{\mathbf{e}}_0 \rangle_S = 0,$$

and $\mathbf{u}_3 \bar{\mathbf{u}}_0 = w\bar{u}$ is a biparavector (a spacetime plane). Its dual is another biparavector

$$\mathbf{S} = \Lambda \mathbf{e}_1 \bar{\mathbf{e}}_2 \bar{\Lambda} = -i\Lambda \mathbf{e}_3 \bar{\Lambda} \tag{13}$$
$$= -iw\bar{u} ,$$

which in a frame commoving with the charge is the spin plane. The dual of the spin paravector $w = i\mathbf{S}u = w^\dagger$ is seen to be the *triparavector* $\mathbf{S}u$. This identifies w as the *Pauli-Lubański spin* paravector.

4.1 Factoring Λ

Before considering the equation of motion for the spin, recall that as with any Lorentz rotor, Λ can always be factored into the product of a real boost $B = B^\dagger$ and a unitary spatial rotation $R = \bar{R}^\dagger$: $\Lambda = BR$. The proper velocity of the particle is $u = \Lambda\Lambda^\dagger = B^2$. It follows that the boost part of the eigenspinor is

$$B = u^{1/2} = \frac{1+u}{\sqrt{2\langle 1+u \rangle_S}} = \frac{m+p}{\sqrt{2m(m+E)}}, \tag{14}$$

where $p = E + \mathbf{p} = mu$.

The second equality in (14) can be derived simply from the relations $u^2 = B$, $B\bar{B} = 1$ and $B + \bar{B} = \langle 2B \rangle_S$, since then $B(B + \bar{B}) = B\langle 2B \rangle_S = u + 1$, the scalar part of which determines $\langle B \rangle_S^2$. However, there is a more general geometrical derivation which makes the result easy to comprehend and remember. Recall from my previous lecture that any simple Lorentz rotation transforms a paravector v as

$$v \rightarrow u = LvL^\dagger = L^2 v^\triangle + v^\perp,$$

where $v^\triangle$ is the part of v coplanar with the paravector rotation plane of L. If this plane contains v, $v^\triangle = v$ and $u = L^2 v$. Let v be a timelike paravector of unit "length": $v\bar{v} = 1$. It is the time axis of a frame moving at proper velocity v (relative to the lab). Then Lv is the unit paravector in the plane of v and u, halfway between them; it is the result of rotating v with only half of the rotation parameter needed to get it to u. Thus, Lv should be parallel to $v + u$. Since it should have the same unit "length" as v, it must be

$$Lv = \frac{v+u}{\sqrt{(v+u)(\bar{v}+\bar{u})}} = \frac{v+u}{\sqrt{2(1+\langle u\bar{v} \rangle_S)}}$$

and therefore
$$L = (u\bar{v})^{1/2} = \frac{1 + u\bar{v}}{\sqrt{2\left(1 + \langle u\bar{v}\rangle_S\right)}}.$$

In the special case above (14), $v = 1$.

4.2   Spin Precession and the g-Factor

The charge-to-mass ratio of a classical charged particle is independent of its magnetic moment. The former determines the cyclotron frequency of the charge in a magnetic field, and the latter determines its precession rate. The magnetic moment of a spinning charge is proportional to its angular momentum, and the proportionality constant is given by the g-factor times the charge-to-mass ratio. Consequently, the g-factor determines the ratio of the precession rate to cyclotron frequency of a spin-1/2 fermion in a magnetic field.

The eigenspinor of a particle describes both its translational and rotational motion. Compound systems with several particles or other independent parts may require several eigenspinors to describe their motion, but a classical "elementary" particle can be defined as one whose motion is fully described by a *single* eigenspinor. As we show below, the equation of motion (5) for the eigenspinor remains linear for this eigenspinor only if $g = 2$.

The reference frame of a massive spinning particle at proper time τ, related to the lab by the eigenspinor $\Lambda(\tau)$, is taken to be a commoving inertial frame of the particle. It is a rest frame of the particle in that the proper velocity of the particle in its reference frame is 1. We assume that the eigenspinor Λ includes the spin rotor $R_0 = \exp\left(-i\mathbf{e}_3\omega_0\tau/2\right)$. Because of its spin rotation, its reference frames at slightly different τ are rotated by large angles with respect to one another. For many purposes, it is convenient to consider a *nonspinning rest frame,* which is intermediate between the reference and lab frames. Such a frame is defined to be related to the lab frame by a pure boost B. The eigenspinor Λ is the product BR of this boost with the spatial rotation R that rotates the reference frame into the nonspinning rest frame. Under most circumstances, R is dominated by the spin rotor R_0.

The equation of motion (5) can now be written in a third way

$$\dot{\Lambda} = \frac{1}{2}\Omega\Lambda = \frac{1}{2}\Lambda\Omega_{\text{ref}} = \frac{1}{2}B\Omega_{\text{rest}}R, \qquad (15)$$

where the spacetime rotation rate Ω_{rest} in the nonspinning rest frame is related to that in the lab (Ω) and in the reference frame (Ω_{ref}) by

$$\Omega_{\text{rest}} = \bar{B}\Omega B = R\Omega_{\text{ref}}\bar{R}. \qquad (16)$$

The resulting proper acceleration (7)

$$\dot{u} = \langle \Omega u \rangle_\Re = \Lambda \langle \Omega_{\text{ref}} \rangle_\Re \Lambda^\dagger = B \langle \Omega_{\text{rest}} \rangle_\Re B \tag{17}$$

is consistent with the Lorentz-force equation if and only if

$$\langle \Omega_{\text{rest}} \rangle_\Re = e \langle \mathbf{F}_{\text{rest}} \rangle_\Re / m, \tag{18}$$

where $\mathbf{E}_{\text{rest}} = \langle \mathbf{F}_{\text{rest}} \rangle_\Re$ is the electric field in the nonspinning rest frame. The minimal covariant extension of (18) is $\Omega = e\mathbf{F}/m$, but this is not acceptable in the presence of spin since the spin rotation rate remains even when the electromagnetic field $\mathbf{F}$ vanishes. Indeed, according to (9), because of the spin term, Ω_{rest} contains an extra bivector (imaginary) term $-i\omega_0 R\mathbf{e}_3 R^\dagger = -i\omega_0 \mathbf{s}$, where

$$\mathbf{s} = R\mathbf{e}_3 R^\dagger \tag{19}$$

is the unit spin paravector in the nonspinning rest frame. We therefore need to consider other covariant extensions of the relation (18).

The Lorentz-force equation is independent of the magnetic field in the rest frame of the charge. We need another equation of motion to relate Ω_{rest} to the magnetic field. An obvious one is spin precession. By differentiating $\mathbf{s}$ (19) with respect to proper time, we obtain the precession equation

$$\dot{\mathbf{s}} = \langle \Omega_R \mathbf{s} \rangle_\Re = \boldsymbol{\omega}_R \times \mathbf{s} \tag{20}$$

with $\Omega_R \equiv 2\dot{R}\bar{R} = -\Omega_R^\dagger = -i\boldsymbol{\omega}_R$. The spin paravector w in the lab is related to $\mathbf{s}$ by

$$w = \Lambda \mathbf{e}_3 \Lambda^\dagger = B\mathbf{s}B ,$$

and the equation of motion for w can be expressed using either the first or second equality:

$$\dot{w} = \frac{d}{d\tau}\left(\Lambda \mathbf{e}_3 \Lambda^\dagger\right) = \langle \Omega w \rangle_\Re = B \langle \Omega_{\text{rest}}\mathbf{s} \rangle_\Re B$$
$$= \langle \Omega_B w \rangle_\Re + B\dot{\mathbf{s}}B,$$

where $\Omega_B \equiv 2\dot{B}\bar{B}$. By now, we have defined a bunch of spacetime rotation rates, but fortunately, they are simply related. Note in particular

$$\Omega_{\text{rest}} = \bar{B}\Omega B = 2\bar{B}\Lambda\bar{\Lambda}B = 2\bar{B}\left(\dot{B}R + B\dot{R}\right)\bar{R}$$
$$= 2\left(\bar{B}\dot{B} + \dot{R}\bar{R}\right) = \Omega_B^\dagger + \Omega_R . \tag{21}$$

For simplicity, consider first the case that the electric field vanishes in the nonspinning rest frame. According to the Lorentz-force equation,

(17) with (18), in this case, the acceleration $\dot{u}$ vanishes, and consequently so do $\dot{B}$ and Ω_B. The precession rate is then proportional to $\mathbf{B}_{\text{rest}}$, and the g-factor is defined so that it is $-\frac{1}{2}eg\mathbf{B}_{\text{rest}}/m$. The full spacetime rotation rate Ω_{rest} has only a bivector part

$$\Omega_{\text{rest}} = i\frac{eg}{2m}\mathbf{B}_{\text{rest}} - i\omega_0\mathbf{s}, \qquad (22)$$

where the first term on the right gives the precession rate of the spin and the second term generates the spin rotation as mentioned above.

More generally, the presence of a rest-frame electric field adds an acceleration $e\mathbf{E}_{\text{rest}}/m$ to Ω_{rest} :

$$\Omega_{\text{rest}} = \frac{e}{m}\left(\mathbf{E}_{\text{rest}} + i\frac{g}{2}\mathbf{B}_{\text{rest}}\right) - i\omega_0\mathbf{s} .$$

Since

$$\mathbf{E}_{\text{rest}} = \frac{1}{2}\left(\mathbf{F}_{\text{rest}} + \mathbf{F}_{\text{rest}}^\dagger\right), \; i\mathbf{B}_{\text{rest}} = \frac{1}{2}\left(\mathbf{F}_{\text{rest}} - \mathbf{F}_{\text{rest}}^\dagger\right)$$

and $\mathbf{F}_{\text{rest}} = \bar{B}\mathbf{F}B$, transformation to the lab gives the rotation rate

$$\Omega = B\Omega_{\text{rest}}\bar{B} = \frac{e}{2m}\left[\mathbf{F} + u\mathbf{F}^\dagger\bar{u} + \frac{g}{2}\left(\mathbf{F} - u\mathbf{F}^\dagger\bar{u}\right)\right] + \omega_0\mathbf{S}$$

$$= \frac{e}{4m}\left[(2+g)\,\mathbf{F} + (2-g)\,u\mathbf{F}^\dagger\bar{u}\right] + \omega_0\mathbf{S} , \qquad (23)$$

where from (13) and (19) $\mathbf{S} = -iB\mathbf{s}\bar{B}$. The equation of motion for w is

$$\dot{w} = \langle\Omega w\rangle_\Re = \frac{e}{4m}\left[(2+g)\langle\mathbf{F}w\rangle_\Re + (2-g)\left\langle u\mathbf{F}^\dagger\bar{u}w\right\rangle_\Re\right] ,$$

which is exactly the algebraic form of the BMT equation.[6] The term involving $\mathbf{S}$ drops out since $\mathbf{S}w$ is a pure bivector and thus imaginary. If we demand that the eigenspinor evolution (5) is linear in Λ, then Ω should not depend on the proper velocity u.[7] Then, from (23), we must have $g = 2$ and the spacetime rotation rate Ω reduces to $e\mathbf{F}/m + \omega_0\mathbf{S}$.

Note that the imaginary part of relation (21) relates the precession rate in the nonspinning rest frame to that in the lab. The difference between the lab and rest-frame precession frequencies is known as the (proper) Thomas precession and is given by

$$\omega_{\text{Th}} = i\langle\Omega_R - \Omega_{\text{rest}}\rangle_\Im = i\langle\Omega_B\rangle_\Im = 2i\left\langle\dot{B}\bar{B}\right\rangle_\Im .$$

It is easy to evaluate ω_{Th} from the expression (14) for B, since the only imaginary contribution $\dot{B}\bar{B}$ arises from the term $\langle\dot{u}\bar{u}\rangle_\Im/(2\gamma + 2)$:

$$\omega_{\text{Th}} = \frac{2i}{2(\gamma+1)}\langle\dot{u}\bar{u}\rangle_\Im = \frac{\dot{\mathbf{u}} \times \mathbf{u}}{\gamma+1} . \qquad (24)$$

4.3   Magnetic Moment

The torque physically responsible for the precession of the spin in a magnetic field arises from the coupling of the magnetic dipole moment of the particle to the field. This coupling can be found independently from the spin rotation rate (22) in the nonspinning rest frame. If we assume a magnetic field $B_{\text{rest}} \ll m\omega_0/e$ ($\simeq 4.414 \times 10^9$ tesla when $\omega_0 = 2m/\hbar$), the rotation associated with the intrinsic spin dominates:

$$
\begin{aligned}
\boldsymbol{\Omega}_{\text{rest}} &= -i\omega_0 \mathbf{s}\left(1 - \frac{e}{m\omega_0}\mathbf{s}\mathbf{B}_{\text{rest}}\right) \\
&\simeq -i\omega_0 \mathbf{s}'\left(1 - \frac{e}{m\omega_0}\mathbf{s}\cdot\mathbf{B}_{\text{rest}}\right)
\end{aligned}
\tag{25}
$$

with

$$
\mathbf{s}' = R_1 \mathbf{s} R_1^\dagger \simeq \mathbf{s}\left(1 - \frac{e}{m\omega_0}\langle\mathbf{s}\mathbf{B}_{\text{rest}}\rangle_V\right),
$$

$$
R_1 = \exp\left(\frac{e}{2m\omega_0}\langle\mathbf{s}\mathbf{B}_{\text{rest}}\rangle_V\right).
$$

The shift in magnitude of the proper rotation frequency, together with the association $\hbar\omega_0 = 2m$ of that frequency with the mass of the charge, implies a potential energy that shifts the mass. From (25), this potential energy can be expressed

$$
-\frac{e\hbar}{2m}\mathbf{s}\cdot\mathbf{B}_{\text{rest}} = -\boldsymbol{\mu}\cdot B_{\text{rest}}
\tag{26}
$$

with a magnetic dipole moment that for the electron with $e < 0$ is

$$
\boldsymbol{\mu} = \frac{e\hbar}{2m}\mathbf{s} = -g\mu_0\frac{\mathbf{s}}{2} = -\mu_0\mathbf{s},
\tag{27}
$$

where $\hbar\mathbf{s}/2$ is the spin vector, $\mu_0 = |e|\,\hbar/(2m)$ is the Bohr magneton, and $g = 2$ is the g factor.

5.   Dirac Equation

A simple classical equation of motion[7, 8] follows from the Lorentz transformation $p = \Lambda m\Lambda^\dagger$ and the unimodularity of Λ :

$$
p\bar{\Lambda}^\dagger = m\Lambda .
\tag{28}
$$

This is the *classical Dirac equation,* which has the form of the quantum Dirac equation in momentum space. It is a real linear equation, in that any real linear combination of solutions Λ is another solution.

Elementary particles are often modeled classically as point charges. An obvious advantage of point charges is that they are simple and structureless. A disadvantage is that their electromagnetic energy is infinite, requires mass renormalization, and leads to preacceleration. An alternative approach is to assume no *a priori* distribution. Instead, a current density $j(x)$ is related by an eigenspinor *field* $\Lambda(x)$ to the reference-frame density ρ_{ref}:

$$j(x) = \Lambda(x) \rho_{\text{ref}}(x) \Lambda^\dagger(x) . \tag{29}$$

This form allows the velocity and orientation to be different at different spacetime positions x. The current density (29) can be written in terms of a density-normalized eigenspinor Ψ as

$$j = \Psi\Psi^\dagger, \ \Psi = \rho_{\text{ref}}^{1/2}\Lambda ,$$

where Ψ obeys the same classical Dirac equation as Λ:

$$p\bar{\Psi}^\dagger = m\Psi . \tag{30}$$

As with the eigenspinor equation of motion (5), this equation (30) is invariant under gauge rotations $\Psi \to \Psi R$, real linear combinations of solutions are also solutions, and free particle solutions have the form of de Broglie waves

$$\Psi(x) = \Psi(0) \exp\left[-i\mathbf{e}_3 \langle p\bar{x}\rangle_S /\hbar\right] . \tag{31}$$

Recall that we used our gauge freedom to choose the spin axis to be $\mathbf{e}_3$. This restricts further gauge rotations to also be about $\mathbf{e}_3$:

$$\Psi \to \Psi \exp\left(-i\mathbf{e}_3\phi\right), \tag{32}$$

where the scalar parameter ϕ effectively sets the angular position from which rotation angles about $\mathbf{e}_3$ are measured.

If Ψ is assumed to be a real linear superposition of classical de Broglie waves (31), the momentum p can be replaced by a differential operator:

$$p\bar{\Psi}^\dagger = i\hbar\partial\bar{\Psi}^\dagger\mathbf{e}_3 = m\Psi. \tag{33}$$

The differential form of p has implications for the remaining gauge rotations. If the rotation angle 2ϕ is fixed, the transformation (32) is said to be a *global gauge transformation* and there is no change in the differential form (33) of the Dirac equation. Indeed, we can multiply from the right by any constant element that commutes with $\mathbf{e}_3$. However, if ϕ is a function of position x, then (32) is a *local gauge transformation* and the Dirac equation (33) picks up an additional term:

$$i\hbar\partial\bar{\Psi}^\dagger\mathbf{e}_3 + \hbar\left(\partial\phi\right)\bar{\Psi}^\dagger = m\Psi.$$

In order to have an equation of motion that is invariant under local gauge transformations, we play the usual game of introducing the paravector gauge potential A as part of the "covariant derivative":

$$p\bar{\Psi}^{\dagger} = i\hbar\partial\bar{\Psi}^{\dagger}\mathbf{e}_3 - eA\bar{\Psi}^{\dagger} = m\Psi, \qquad (34)$$

and the gauge transformation (32) is accompanied by a gauge transformation of A :

$$eA \rightarrow eA + \hbar\left(\partial\phi\right).$$

The result (34) is a covariant algebraic form of the usual quantum Dirac equation in APS.

To cast the equation into a form that makes it easier to solve, we split it into two complementary minimal left ideals of APS with the projectors $\mathsf{P}_{\pm3} \equiv \frac{1}{2}\left(1 \pm \mathbf{e}_3\right)$:

$$p\bar{\Psi}^{\dagger}\mathsf{P}_{+3} = \left(i\hbar\partial - eA\right)\bar{\Psi}^{\dagger}\mathsf{P}_{+3} = m\Psi\mathsf{P}_{+3}$$
$$p\bar{\Psi}^{\dagger}\mathsf{P}_{-3} = \left(-i\hbar\partial - eA\right)\bar{\Psi}^{\dagger}\mathsf{P}_{-3} = m\Psi\mathsf{P}_{-3} \ ,$$

where we noted the "pacwoman" property $\mathbf{e}_3\mathsf{P}_{\pm3} = \pm\mathsf{P}_{\pm3}$. We can bar-dagger conjugate the second equation to project it into the same minimal left ideal of the algebra as the first:

$$\bar{p}\Psi\mathsf{P}_{+3} = \left(i\hbar\bar{\partial} - e\bar{A}\right)\Psi\mathsf{P}_{+3} = m\bar{\Psi}^{\dagger}\mathsf{P}_{+3} \ .$$

Then, defining

$$\psi^{(W)} = \frac{1}{\sqrt{2}}\left(\begin{array}{c}\Psi\mathsf{P}_{+3}\\ \bar{\Psi}^{\dagger}\mathsf{P}_{+3}\end{array}\right),$$

we obtain a pair of projected equations in the matrix form

$$\left(\begin{array}{cc}0 & p\\ \bar{p} & 0\end{array}\right)\psi^{(W)} = \left(\begin{array}{cc}0 & i\hbar\partial - eA\\ i\hbar\bar{\partial} - e\bar{A} & 0\end{array}\right)\psi^{(W)} = m\psi^{(W)}. \qquad (35)$$

This is fully equivalent to our single algebraic equation (34). The projector P_3 gobbles $\mathbf{e}_3$ and changes the spin rotor $\exp\left(-i\omega_0\tau\mathbf{e}_3/2\right)$ into a simple phase factor. If the standard matrix representation $\mathbf{e}_\mu \rightarrow \underline{\sigma}_\mu$ is used for APS in terms of the Pauli spin matrices $\underline{\sigma}_\mu$, Eq. (35) is the traditional form[9] of the Dirac equation $p^\mu\gamma_\mu\psi^{(W)} = m\psi^{(W)}$ with gamma matrices in the Weyl (or spinor) representation. The projection of the algebraic Ψ by $\mathsf{P}_{\pm3}$ picks out the upper and lower components of $\psi^{(W)}$ and is seen to be equivalent to multiplication of $\psi^{(W)}$ by the traditional chirality projectors $\frac{1}{2}\left(1 \pm \gamma_5\right)$ with $\gamma_5 = -i\gamma_0\gamma_1\gamma_2\gamma_3$:

$$\frac{1}{2}\left(1 \pm \gamma_5\right)\psi^{(W)} \Leftrightarrow \Psi\mathsf{P}_{\pm3} \ .$$

The Dirac bispinor in the Dirac-Pauli (or standard) representation is related by

$$\psi^{(DP)} = \frac{1}{\sqrt{2}} \begin{pmatrix} 1 & 1 \\ 1 & -1 \end{pmatrix} \psi^{(W)} = \begin{pmatrix} \langle \Psi \rangle_+ \, \mathsf{P}_{+3} \\ \langle \Psi \rangle_- \, \mathsf{P}_{+3} \end{pmatrix}, \tag{36}$$

where $\langle \Psi \rangle_\pm = \frac{1}{2} \left(\Psi \pm \bar{\Psi}^\dagger \right)$ are the even and odd parts of Ψ and correspond to the *large* and *small* components at low velocities:

$$\langle \Psi \rangle_+ = \rho_{\text{ref}}^{1/2} \langle B \rangle_+ R = \rho_{\text{ref}}^{1/2} \sqrt{\frac{m+E}{2m}} R \simeq \rho_{\text{ref}}^{1/2} R$$

$$\langle \Psi \rangle_- = \rho_{\text{ref}}^{1/2} \langle B \rangle_- R = \frac{\mathbf{p}}{m+E} \langle \Psi \rangle_+ \, .$$

The last expression for $\langle \Psi \rangle_+$ on the RHS is the low-velocity approximation. In the rest frame, the small component disappears and the eigenfunction is even. We say the particle has even *intrinsic parity*. More generally, the solutions $\psi^{(W)}$ (35) and $\psi^{(DP)}$ (36) are 4×2 matrices whose second columns are zero and whose first columns give the usual Dirac bispinors of quantum theory. If R is replaced by the de Broglie spin rotor $R = \exp\left[-ie_3 \langle p\bar{x} \rangle_S / \hbar\right]$, the solutions in the nonvanishing columns of $\psi^{(W)}$ and $\psi^{(DP)}$ are the usual momentum eigenstates of the Dirac equation.

5.1   Spin Distributions

To study spin distributions, the low-velocity limit

$$\Psi \simeq \langle \Psi \rangle_+ \simeq \rho_{\text{ref}}^{1/2} R \tag{37}$$

is sufficient. In terms of Euler angles ϕ, θ, χ, about space-fixed axes, the rotor R can be expressed by

$$R = \exp\left(-ie_3 \phi/2\right) \exp\left(-ie_2 \theta/2\right) \exp\left(-ie_3 \chi/2\right). \tag{38}$$

The classical spin direction in a static system is $\mathbf{s} = Re_3 R^\dagger$ [see (19) above], where R may be a function of position $\mathbf{r}$. A distribution of such spin directions is thus $\rho_{\text{ref}} Re_3 R^\dagger$, where the positive scalar $\rho_{\text{ref}} = \rho_{\text{ref}}(\mathbf{r})$ is the density of spins in the reference frame. As seen below, simple measurements of the spin direction give only one component at a time. The distribution of the component of the spin in the direction of an arbitrary unit vector $\mathbf{m}$ is

$$\rho_{\text{ref}} \, \mathbf{s} \cdot \mathbf{m} = \left\langle \rho_{\text{ref}} Re_3 R^\dagger \mathbf{m} \right\rangle_S .$$

In terms of the projector P_{+3}, since $e_3 = P_{+3} - P_{-3}$ and for any elements p, q, $\langle pq \rangle_S = \langle qp \rangle_S = \langle \overline{pq} \rangle_S$, the distribution is

$$2 \left\langle \rho_{\text{ref}} R P_{+3} R^\dagger \mathbf{m} \right\rangle_S = 2 \left\langle P_{+3} R^\dagger \rho_{\text{ref}}^{1/2} \mathbf{m} \rho_{\text{ref}}^{1/2} R P_{+3} \right\rangle_S$$

$$= \text{tr} \left\{ \psi^\dagger \mathbf{m} \psi \right\}, \tag{39}$$

where by ψ we mean the ideal spinor, whose standard matrix representation is

$$\psi \equiv \rho_{\text{ref}}^{1/2} R P_{+3} = e^{-i\chi/2} \rho_{\text{ref}}^{1/2} \begin{pmatrix} e^{-i\phi/2} \cos\theta/2 & 0 \\ e^{i\phi/2} \sin\theta/2 & 0 \end{pmatrix}.$$

If the column of zeros is dropped, the standard matrix representation of ψ is a two-component spinor, as familiar from the usual nonrelativistic Pauli theory. Such spinors carry an irreducible representation of the rotation group $SU(2)$.[2] It may be noted that the full rotor R can always be recovered from the ideal projection RP_3 as twice its even part:

$$R = 2 \langle RP_3 \rangle_+ .$$

The term $\psi^\dagger \mathbf{m} \psi$ is a scalar and tr can be omitted from (39). Although we derived the spin distribution as a classical expression, it has *precisely the quantum form* if we recognize that the matrix representation of the unit vector $\mathbf{m}$, namely $\mathbf{m} = m^j \mathbf{e}_j \rightarrow m^1 \sigma_x + m^2 \sigma_y + m^3 \sigma_z$, is traditionally (but misleadingly, since it represents a vector, not a scalar) written $\mathbf{m} \cdot \boldsymbol{\sigma}$. From the definition of ρ_{ref} the spinor ψ satisfies the usual normalization condition,

$$2 \int d^3 x \left\langle \psi^\dagger \psi \right\rangle_S \equiv \langle \psi | \psi \rangle = 1$$

and the average component of the spin in the direction $\mathbf{m}$ is

$$2 \int d^3 x \left\langle \psi^\dagger \mathbf{m} \psi \right\rangle_S \equiv \langle \psi | \mathbf{m} | \psi \rangle .$$

From expression (39) we see that the real paravector $P_{\mathbf{s}} = R P_{+3} R^\dagger = \frac{1}{2}(1 + \mathbf{s})$ embodies information about the classical spin state at a given point in space or in a homogeneous ensemble. It is equivalent to the quantum spin density operator $\varrho \sim \psi\psi^\dagger$ for the pure state of spin $\mathbf{s}$,

---

[2]More generally, once projected onto a minimal left ideal, any spinor $\Psi = \rho_{\text{ref}}^{1/2} BR$ is equivalent to a dilated spatial rotation, thereby reducing this representation of the noncompact Lorentz group to a scaling factor times the compact group $SU(2)$.

and it is also a projector that acts as a state filter. The component of the spin in the $\mathbf{m}$ direction is $2\langle \mathsf{P_s\,m}\rangle_S$, whose matrix representation is identical to the usual quantum expression, traditionally written $\frac{1}{2}\mathrm{tr}\{\varrho\,\mathbf{m}\cdot\sigma\}$. Note that the part of a rotation R around the spin axis becomes a phase factor of $\psi = \rho_{\mathrm{ref}}^{1/2}R\mathsf{P}_{+3}$, since the pacwoman property gives $e^{-i\chi\mathbf{e}_3/2}\mathsf{P}_{+3} = e^{-i\chi/2}\mathsf{P}_{+3}$. Thus, when phase factors of ψ are ignored, information about the axial rotation of R is lost. Good quantum calculations keep track of *relative* phase, and this appears to be the only aspect of the phase (rotation about $\mathbf{e}_3$) that can be determined experimentally.

One way of seeing whether the system is in a given state of spin $\mathbf{n}$ is to apply the state filter to the spin density operator ϱ and see what remains:

$$\mathsf{P_n}\varrho\mathsf{P_n} = \left(\mathsf{P_n}\varrho + \bar{\varrho}\bar{\mathsf{P}}_n\right)\mathsf{P_n} = 2\langle\mathsf{P_n}\varrho\rangle_S\,\mathsf{P_n}. \tag{40}$$

The scalar coefficient $2\langle\mathsf{P_n}\varrho\rangle_S = \langle(1+\mathbf{n})\,\varrho\rangle_S$ is the probability of finding the system described by ϱ in the state $\mathbf{n}$. For a system in the pure state $\varrho = \mathsf{P_s} = \frac{1}{2}(1+\mathbf{s})$, the probability is

$$2\langle\mathsf{P_n}\mathsf{P_s}\rangle_S = \frac{1}{2}\langle(1+\mathbf{n})(1+\mathbf{s})\rangle_S = \frac{1}{2}(1+\mathbf{n}\cdot\mathbf{s}). \tag{41}$$

This is unity if the system is definitely in the state $\mathbf{n}$, whereas it vanishes if the system is in a state *orthogonal* to $\mathbf{n}$. Thus, $\mathbf{s} = \mathbf{n}$ is required for the states to be the same and $\mathbf{s} = -\mathbf{n}$ for the states to be orthogonal. Note that the mathematics is the same as used in my first lecture to describe light polarization.[1]

5.2   Spin $\frac{1}{2}$ and State Expansions

The value of $\frac{1}{2}$ for the spin of elementary spinors considered here arises in several ways. It is the group-theoretical label for the irreducible spinor representation of the rotation group $SU(2)$ carried by ideal spinors. It is also required by the fact that any rotation can be expressed as a linear superposition of two independent orthogonal rotations defined for any direction in space. The Euler-angle form (38) of any rotor R can be rewritten

$$R = \exp\left(-i\mathbf{n}\theta/2\right)\exp\left[-i\mathbf{e}_3\left(\phi+\chi\right)/2\right] \tag{42}$$

$$= \left(\cos\frac{\theta}{2} - i\mathbf{n}\sin\frac{\theta}{2}\right)\exp\left[-i\mathbf{e}_3\left(\phi+\chi\right)/2\right],$$

where $\mathbf{n} = \exp\left(-i\mathbf{e}_3\phi/2\right)\mathbf{e}_2\exp\left(i\mathbf{e}_3\phi/2\right)$ is a unit vector in the $\mathbf{e}_1\mathbf{e}_2$ plane. Therefore, any rotor R is a real linear combination $\cos\frac{\theta}{2}R_\uparrow +$

$\sin\frac{\theta}{2}R_\downarrow$ of rotors $R_\uparrow = \exp\left[-i\mathbf{e}_3\left(\phi+\chi\right)/2\right]$ and $R_\downarrow = -i\mathbf{n}R_\uparrow$ that are mutually orthogonal: $\left\langle R_\uparrow R_\downarrow^\dagger\right\rangle_S = \langle -i\mathbf{n}\rangle_S = 0$.

By projecting the rotors with P_{+3} we obtain the equivalent relation of ideal spinors:

$$RP_{+3} = \cos\frac{\theta}{2}R_\uparrow P_{+3} + \sin\frac{\theta}{2}R_\downarrow P_{+3}$$

$$\psi \equiv \rho_{\text{ref}}^{1/2}RP_{+3} = \cos\frac{\theta}{2}\psi_\uparrow + \sin\frac{\theta}{2}\psi_\downarrow \tag{43a}$$

$$\psi_\uparrow = \rho_{\text{ref}}^{1/2}e^{-i(\phi+\chi)/2}P_{+3}, \; \psi_\downarrow = -i\mathbf{n}\psi_\uparrow \; .$$

Traditional orthonormality conditions hold:

$$2\left\langle\psi_\uparrow\psi_\downarrow^\dagger\right\rangle_S = 2\left\langle\psi_\uparrow\psi_\uparrow^\dagger i\mathbf{n}\right\rangle_S = 2\rho_{\text{ref}}\langle P_{+3}i\mathbf{n}\rangle_S = 0$$

$$2\left\langle\psi_\downarrow\psi_\downarrow^\dagger\right\rangle_S = 2\left\langle\psi_\uparrow\psi_\uparrow^\dagger\right\rangle_S = 2\rho_{\text{ref}}\langle P_{+3}\rangle_S = \rho_{\text{ref}} \; .$$

It follows that the amplitudes are

$$\langle\psi_\uparrow|\psi\rangle = 2\int d^3x \left\langle\psi\psi_\uparrow^\dagger\right\rangle_S = \cos\frac{\theta}{2}$$

$$\langle\psi_\downarrow|\psi\rangle = 2\int d^3x \left\langle\psi\psi_\downarrow^\dagger\right\rangle_S = \sin\frac{\theta}{2}$$

giving probabilities as found above in Eq. (41).

$$|\langle\psi_\uparrow|\psi\rangle|^2 = \cos^2\frac{\theta}{2} = \frac{1}{2}\left(1+\mathbf{s}\cdot\mathbf{e}_3\right) \tag{44}$$

$$|\langle\psi_\downarrow|\psi\rangle|^2 = \sin^2\frac{\theta}{2} = \frac{1}{2}\left(1-\mathbf{s}\cdot\mathbf{e}_3\right). \tag{45}$$

By tacking on an additional fixed rotation, the treatment can be extended to measurements along an arbitrary axis.

5.3 Stern-Gerlach Experiment

The basic measurement of spin is that of the Stern-Gerlach experiment [10], in which a beam of ground-state silver atoms is split by a magnetic-field gradient into distinct beams of opposite spin polarization. It is a building block of real and thought experiments in quantum measurement [11].

Consider a nonrelativistic beam of ground-state atoms that travels with velocity $\mathbf{v} = v_x\mathbf{e}_1$ through a static magnetic field $\mathbf{B}$ that vanishes everywhere except in the vicinity of the Stern-Gerlach magnet. The net effect of the magnet on the beam is a vertical force proportional to the z component μ_z of the magnetic dipole moment, which we take to be the magnetic dipole moment of an electron (27), corresponding to an atomic ground state with an unpaired electron in an S state:

$$\mu_z \frac{\partial B_z}{\partial z} = -\mu_0 \mathbf{s} \cdot \mathbf{e}_3 \frac{\partial B_z}{\partial z} .$$

We assume the beam is uniform along its length with $\rho_{\text{ref}} = 1$ and that $\partial B_z / \partial z < 0$. The eigenspinor for the initial beam in the nonrelativistic limit is then

$$\Psi = BR \simeq \left(1 + \frac{1}{2} \mathbf{v} \right) R .$$

The key point is that, as seen above (42), every rotor is a linear combination of "up" and "down" rotors with opposite spin directions. For a state in which $\mathbf{s}$ makes an angle θ with respect to $\mathbf{e}_3$, $R = \cos \frac{\theta}{2} R_\uparrow + \sin \frac{\theta}{2} R_\downarrow$ with

$$R_\uparrow \mathbf{e}_3 R_\uparrow^\dagger = \mathbf{e}_3, \ R_\downarrow \mathbf{e}_3 R_\downarrow^\dagger = -\mathbf{e}_3 . \tag{46}$$

The components of the spin unit vector

$$\mathbf{s} \cdot \mathbf{e}_3 = \left\langle R \mathbf{e}_3 R^\dagger \mathbf{e}_3 \right\rangle_S = \begin{cases} +1, & R = R_\uparrow \\ -1, & R = R_\downarrow \end{cases}$$

are also opposite for the rotors $R_\uparrow$ and $R_\downarrow$ so that the Stern-Gerlach magnet forces the eigenspinor components $R_\uparrow$ and $R_\downarrow$ of the beam eigenspinor upwards and downwards, respectively. This splits the incident beam into two isolated branches, analogous to the way a birefringent crystal splits a beam of light into two branches of orthogonal polarization. The fraction of the initial beam in the upper branch is $\cos^2 \theta/2 = \frac{1}{2} (1 + \cos \theta)$ whereas in the lower branch the fraction is $\sin^2 \theta/2 = \frac{1}{2} (1 - \cos \theta)$, just as found from the square amplitudes (44) and (45) above. The two-valued property of the measurement is a direct result of the decomposition of any rotor R into rotors for rotations about any two opposite directions, rotations that correspond to "spin-up" and "spin-down" components.

These relations can be re-expressed in the form of eigenspinor ideals, which is closer to the quantum formulation. Since the rotors $R_\uparrow, R_\downarrow$ are unitary, relation (46) can be cast in a form

$$\mathbf{e}_3 R_\uparrow = R_\uparrow \mathbf{e}_3, \ \mathbf{e}_3 R_\downarrow = -R_\downarrow \mathbf{e}_3$$

that becomes an eigenfunction equation when projected onto the P_{+3} ideals $\psi_{\uparrow\downarrow} = R_{\uparrow\downarrow} P_3$:

$$\mathbf{e}_3 \psi_\uparrow = \psi_\uparrow, \ \mathbf{e}_3 \psi_\downarrow = -\psi_\downarrow .$$

Evidently $e_3 = P_3 - \bar{P}_3$ is the spin operator corresponding to the z component $s \cdot e_3$ of the spin in units of $\hbar/2$ for the P_3 ideal representation of ψ. The projected rotor is split into $RP_3 = \psi = \cos\frac{\theta}{2}\psi_\uparrow + \sin\frac{\theta}{2}\psi_\downarrow$, the two parts of which experience opposite forces from the Stern-Gerlach magnet.

6. Bell's Theorem

Quantum systems are often tested against Bell's theorem. Violation of Bell's inequalities is generally held to be evidence of nonlocal quantum weirdness. A simple proof of the inequalities can be derived from polarization splitters acting on a single beam. The nice thing about this derivation is that the question of locality vs. nonlocality does not arise. It shows that Bell's inequality concerns directly the difference between classical probabilities for particles and quantum probability amplitudes in waves. Nonlocality arises only in EPR-type applications with entangled states, but the inequality itself is violated locally for a single beam of waves.

Consider a beam of silver atoms passing through a sequence of Stern-Gerlach magnets aligned to split the beam into spins polarized in opposite directions specified by unit vectors $\pm\mathbf{a}, \pm\mathbf{b}, \pm\mathbf{c}$. If one of the polarized beams leaving a given Stern-Gerlach magnet is blocked, that magnet serves as a polarization filter. Let $f(\mathbf{c}, \mathbf{b}, \mathbf{a})$ be the fraction of atoms passing through sequential polarization filters of types $\mathbf{a}, \mathbf{b}, \mathbf{c}$. Then, with "classical" atoms (the same math works for polarized photons), the fraction that pass through filters $\mathbf{a}$ and then $\mathbf{c}$ is just the sum of those that pass through $\mathbf{a}, \mathbf{b}, \mathbf{c}$ and $\mathbf{a}, -\mathbf{b}, \mathbf{c}$:

$$f(\mathbf{c}, \mathbf{a}) = f(\mathbf{c}, \mathbf{b}, \mathbf{a}) + f(\mathbf{c}, -\mathbf{b}, \mathbf{a}).$$

Similarly,

$$f(-\mathbf{c}, \mathbf{b}) = f(-\mathbf{c}, \mathbf{b}, \mathbf{a}) + f(-\mathbf{c}, \mathbf{b}, -\mathbf{a})$$
$$f(\mathbf{c}, \mathbf{b}, \mathbf{a}) + f(-\mathbf{c}, \mathbf{b}, \mathbf{a}) = f(\mathbf{b}, \mathbf{a}).$$

Adding these three relations we obtain

$$f(\mathbf{c}, \mathbf{a}) + f(-\mathbf{c}, \mathbf{b}) = f(\mathbf{b}, \mathbf{a}) + f(\mathbf{c}, -\mathbf{b}, \mathbf{a}) + f(-\mathbf{c}, \mathbf{b}, -\mathbf{a})$$
$$\geq f(\mathbf{b}, \mathbf{a}) \tag{47}$$

which is a form of Bell's inequality. We found above (41) that $f(\mathbf{c}, \mathbf{a}) = \frac{1}{2}(1 + \mathbf{c} \cdot \mathbf{a})$ so that Bell's inequality (47) can be written

$$1 + \mathbf{c} \cdot \mathbf{a} - \mathbf{c} \cdot \mathbf{b} \geq \mathbf{b} \cdot \mathbf{a}.$$

If we let $\mathbf{b}$ bisect $\mathbf{a}$ and $\mathbf{c}$, so that we can put $\mathbf{c} \cdot \mathbf{a} = \cos 2\theta$ and $\mathbf{c} \cdot \mathbf{b} = \mathbf{b} \cdot \mathbf{a} = \cos \theta$, we obtain

$$1 + \cos 2\theta \geq 2 \cos \theta.$$

However, this inequality is broken for $\theta = \pi/4$, for example. The error is in working with fractions or probabilities instead of amplitudes. Using our filter method (40), we find indeed that

$$f(\mathbf{c}, \mathbf{a}) = 2 \langle P_{\mathbf{c}} P_{\mathbf{a}} P_{\mathbf{c}} \rangle_S = \frac{1}{2}(1 + \mathbf{a} \cdot \mathbf{c})$$
$$= 2 \langle P_{\mathbf{c}} \left(P_{\mathbf{b}} + \bar{P}_{\mathbf{b}} \right) P_{\mathbf{a}} \left(P_{\mathbf{b}} + \bar{P}_{\mathbf{b}} \right) P_{\mathbf{c}} \rangle_S ,$$

but this is not the same as

$$f(\mathbf{c}, \mathbf{b}, \mathbf{a}) + f(\mathbf{c}, -\mathbf{b}, \mathbf{a}) = 2 \langle P_{\mathbf{c}} P_{\mathbf{b}} P_{\mathbf{a}} P_{\mathbf{b}} P_{\mathbf{c}} \rangle_S + 2 \langle P_{\mathbf{c}} \bar{P}_{\mathbf{b}} P_{\mathbf{a}} \bar{P}_{\mathbf{b}} P_{\mathbf{c}} \rangle_S$$
$$= \langle P_{\mathbf{c}} P_{\mathbf{b}} P_{\mathbf{c}} \rangle_S (1 + \mathbf{b} \cdot \mathbf{a})$$
$$+ \langle P_{\mathbf{c}} \bar{P}_{\mathbf{b}} P_{\mathbf{c}} \rangle_S (1 - \mathbf{b} \cdot \mathbf{a})$$
$$= \frac{1}{2}(1 + \mathbf{b} \cdot \mathbf{c} \, \mathbf{b} \cdot \mathbf{a})$$

unless $\mathbf{b} = \mathbf{a}$ or $\mathbf{b} = \mathbf{c}$.

7.   Qubits and Entanglement

The basis of quantum computation[12] lies in the replacement of the classical bit, with its discrete binary values $(1, 0)$ [for (on, off) or (true, false)], by a two-level quantum system, called a quantum bit or *qubit*, that can generally exist in a superposition of its two states. The unitary evolution of a coupled system of qubits in an arbitrary superposition of states simultaneously computes the evolution of all binary sequences in the superposition, and this massive quantum parallelism offers huge advantages for certain problems, such as factoring large numbers and some types of large searches.

The simplest example of a two-level quantum system is the spin-1/2 system considered above. The same formalism can be applied to qubits. Here, we mainly formulate the algebra in the minimum left ideal $\mathcal{C}\ell_3 P_3$ of APS and tensor products thereof, since these lie closest to the quantum formalism. We first consider operations on single spin systems in the language of qubits and quantum computing, and next present some extensions to systems of two or more qubits.

7.1   Qubit Operators

There is arbitrariness in the phase as well as in factors of $\mathbf{e}_3$ in $\psi_\uparrow$ and $\psi_\downarrow$. Let us therefore define a *standard ideal spinor basis*, which we denote in the ket notation

$$|\uparrow\rangle = P_3$$
$$|\downarrow\rangle = e_1 |\uparrow\rangle = e_1 P_3 |\uparrow\rangle = e_1 e_3 P_3 |\uparrow\rangle = -ie_2 P_3 |\uparrow\rangle \ .$$

The *unipotent* ($e_1^2 = 1$) element e_1 interchanges these: $|\downarrow\rangle = e_1 |\uparrow\rangle$ and $|\uparrow\rangle = e_1 |\downarrow\rangle$. We may think of the "up" state $|\uparrow\rangle$ as representing the bit value 1, and the "down" state $|\downarrow\rangle$ as the bit value 0. Evidently, e_1 acts as the NOT operator. It is also a *reflection operator*, since for any vector $\mathbf{v}$

$$\mathbf{v} \rightarrow e_1 \mathbf{v} e_1 = \mathbf{v}^\| - \mathbf{v}^\perp = 2\mathbf{v} \cdot e_1 \, e_1 - \mathbf{v}$$

keeps the component $\mathbf{v}^\|$ of $\mathbf{v}$ along e_1 constant and reverses components perpendicular to e_1. The dual to e_1 is $-ie_1 = \exp\left(-i\pi e_1/2\right)$, a rotor for the rotation of π about e_1. If it is applied twice, the result is a 2π rotation which changes the sign of any spinor for a spin-1/2 system: $(-ie_1)^2 = -1$. Note that any unit vector in the e_{12} plane can serve as the NOT operator. The resultant states differ only in their relative phase.

A state in which the spin is aligned with e_1 is an eigenstate of the NOT operator. Such a state is obtained by applying the rotor $\exp\left(-i\pi e_2/4\right)$ to P_3, which rotates the spin from e_3 by $\pi/2$ in the $e_3 e_1$ plane. It can be expressed in several ways:

$$e^{-i\pi e_2/4} P_3 = \left(\frac{1 - ie_2}{\sqrt{2}}\right) P_3 = \left(\frac{1 + e_1}{\sqrt{2}}\right) P_3 = \left(\frac{e_3 + e_1}{\sqrt{2}}\right) P_3 \ .$$

In the fourth form, we see the eigenstate of the NOT operator is also obtained by a reflection of the up basis state in the unit vector $\hat{\mathbf{h}} = (e_1 + e_3)/\sqrt{2} = e^{i\pi e_2/4} e_1 = e^{-i\pi e_2/4} e_3$. This reflection is called the *Hadamard transformation.*

Since any state can be expressed as a linear combination of the basis states P_3 and $e_1 P_3$, we can find the effect of an operation on any state if we know its effect of the basis states. In the Hadamard transformation,

$$|\uparrow\rangle \rightarrow \hat{\mathbf{h}} |\uparrow\rangle = \frac{|\uparrow\rangle + |\downarrow\rangle}{\sqrt{2}}$$

$$|\downarrow\rangle \rightarrow \hat{\mathbf{h}} |\downarrow\rangle = \hat{\mathbf{h}} e_1 |\uparrow\rangle = \frac{|\uparrow\rangle - |\downarrow\rangle}{\sqrt{2}} \ .$$

7.2   Exponential Forms

Rotations can be written in exponential form with a scalar parameter proportional to the angle of rotation. They are thus connected to unity (scalar parameter = 0) and can easily be interpolated. Reflections, on the other hand, are usually viewed as discrete transformations, performed entirely or not at all. However, unit vectors can also be extended to exponential form. Such exponential forms are useful for working out physical interactions in the Hamiltonian that induce the reflections through the time evolution operator.

Consider first the dual $-i\mathbf{n}$ of the real unit vector $\mathbf{n}$. It is a rotor for a rotation by angle π about the $\mathbf{n}$ axis and can be viewed as the continuous transformation $\exp\left(-i\mathbf{n}\alpha/2\right)$ when $\alpha = \pi$:

$$-i\mathbf{n} = \exp\left(-i\mathbf{n}\alpha/2\right), \ \alpha = \pi \ .$$

Thus the unit vector $\mathbf{n}$ can be expressed

$$\mathbf{n} = i\exp\left(-i\mathbf{n}\pi/2\right) = \exp\left[i\left(1 - \mathbf{n}\right)\pi/2\right]$$
$$= \exp\left[i\bar{\mathbf{P}}_{\mathbf{n}}\pi\right] = \exp\left[i\bar{\mathbf{P}}_{\mathbf{n}}\alpha\right], \ \alpha = \pi \ .$$

In other words, it is part of a continuous transformation generated by the projector $\bar{\mathbf{P}}_{\mathbf{n}}$. At intermediate angles α we note that

$$e^{i\alpha\bar{\mathbf{P}}} = e^{i\alpha\bar{\mathbf{P}}}\left(\mathbf{P} + \bar{\mathbf{P}}\right) = \mathbf{P} + e^{i\alpha\bar{\mathbf{P}}} \ .$$

7.3   Raising and Lowering Operators

We saw above that in the ideal-spinor representation, the unit vector $\mathbf{n}$ represents the spin component in the direction $\mathbf{n}$ in units of $\hbar/2$. One may verify that the commutation relations of unit vectors in space are just what one needs for the spin components, and of course that is why the Pauli spin matrices were introduced to represent electron spin. The operator for the squared spin is thus

$$S^2 = \left(\frac{\hbar}{2}\right)^2 \left(\mathbf{e}_1^2 + \mathbf{e}_2^2 + \mathbf{e}_3^2\right) = \frac{3}{4}\hbar^2 \ .$$

The *raising* operator s_+ is defined by $s_+\left|\downarrow\right\rangle = \left|\uparrow\right\rangle$, $s_+\left|\uparrow\right\rangle = 0$. Any non-null element that annihilates $\mathbf{P}_3 = \left|\uparrow\right\rangle$ must have the form $x\bar{\mathbf{P}}_3$, and

$$x\bar{\mathbf{P}}_3\left|\downarrow\right\rangle = x\bar{\mathbf{P}}_3\mathbf{e}_1\mathbf{P}_3 = x\mathbf{e}_1\left|\uparrow\right\rangle = \left|\uparrow\right\rangle$$

is satisfied by $x = \mathbf{e}_1$. Thus, we can put

$$s_+ = \mathbf{e}_1\bar{\mathbf{P}}_3 = \frac{1}{2}\left(\mathbf{e}_1 + i\mathbf{e}_2\right) = \mathbf{P}_3\mathbf{e}_1 \ . \tag{48}$$

Similarly, the lowering operator is

$$s_- = \mathbf{e}_1\mathbf{P}_3 = s_+^\dagger . \tag{49}$$

The operators $s_\pm$ are examples of nilpotent operators: $s_\pm^2 = 0$. Their products are

$$s_+ s_- = P_3, \ s_- s_+ = \bar{P}_3 \ ,$$

which gives $s_+ s_- + s_- s_+ = 1$ and $s_+ s_- - s_- s_+ = \mathbf{e}_3$.

7.4   Two-Qubit Systems

We represent multi-qubit systems very generally as tensor products of single qubit systems, that is, in tensor products of ideals of APS: $\mathcal{C}\ell_3 P_3 \otimes \mathcal{C}\ell_3 P_3$. This representation is sufficiently general to handle distinguishable quantum systems. If the systems are identical, additional symmetries apply. Tensor (or "Kronecker") products obey the fundamental relation

$$(a \otimes b)(c \otimes d) = ac \otimes bd$$

Label two qubits A and B. With two qubits or spin-1/2 systems, the projector is the tensor (or "Kronecker") product $P_3 \otimes P_3$ and the general pure state can be written

$$\psi_{AB} = \Psi_{AB}(P_3 \otimes P_3).$$

If Ψ_{AB} is a *product state*, it can be expressed as a single product term $\Psi_{AB} = \Psi_A \otimes \Psi_B$ so that

$$\psi_{AB} = \Psi_{AB}(P_3 \otimes P_3) = (\Psi_A \otimes \Psi_B)(P_3 \otimes P_3) = \Psi_A P_3 \otimes \Psi_B P_3$$

and the density operator is also a single product

$$\begin{aligned}
\varrho_{AB} = \psi_{AB}\psi_{AB}^\dagger &= (\Psi_A \otimes \Psi_B)(P_3 \otimes P_3)(\Psi_A \otimes \Psi_B)^\dagger \\
&= (\Psi_A \otimes \Psi_B)(P_3 \otimes P_3)\left(\Psi_A^\dagger \otimes \Psi_B^\dagger\right) \\
&= \left(\Psi_A P_3 \Psi_A^\dagger\right) \otimes \left(\Psi_B P_3 \Psi_B^\dagger\right) \\
&= \varrho_A \otimes \varrho_B \ .
\end{aligned}$$

Such systems, whose density operators can be expressed as a single tensor product of subsystem parts, are *decomposable*. In a decomposable system, measurements in one subsystem are independent of those in another. There is no entanglement and indeed no correlation between the subsystems. A system whose density operator is a mixture of decomposable density operators

$$\varrho_{AB} = \sum_j w_j \varrho_A^j \otimes \varrho_B^j$$

is called *disentangled* or *separable*. An entangled state is not separable.

A *reduced density operator* for one subsystem of a composite system is found by summing over all the states of the other subsystems; this is equivalent to tracing over the other subsystems. We can write

$$\varrho_{A\bar{B}} = \operatorname{tr}_B \{\varrho_{AB}\}$$

for the reduced density matrix of subsystem A after summing over all states of subsystem B. In the case of a product state $\varrho_{AB} = \varrho_A \otimes \varrho_B$,

$$\varrho_{A\bar{B}} = \operatorname{tr}_B \{\varrho_A \otimes \varrho_B\} = \varrho_A \otimes \operatorname{tr}_B \{\varrho_B\} = \varrho_A .$$

It is often stated that measurements in one subsystem *affect* the outcome of those in another, but such language is misleading and based on classical probability interpretations. It suggests a cause-and-effect relation that propagates faster than the speed of light, and such propagation is not physically possible by relativistic causality. In an entangled system, if measurements on subsystem A are only recorded when a given measurement result is obtained on subsystem B, then the results on A will depend on the measurement outcome selected on B. However, if results are recorded for A regardless of the measured value for B, then the A results cannot depend on the actual measurement performed on B. Mathematically, the density operator for A in that case is the subtrace of ϱ over B, and that subtrace is independent of basis and therefore independent of which property eigenstates are used.

Arbitrary states, including entangled (and hence inseparable) ones, can be expressed as linear combinations of product states:

$$\Psi_{AB}(P_3 \otimes P_3) = \sum_{j,k} c_{jk} [\Psi_A(\Omega_j) \otimes \Psi_B(\Omega_k)](P_3 \otimes P_3) .$$

The corresponding density operator is

$$\varrho_{AB} = \Psi_{AB}(P_3 \otimes P_3) \Psi_{AB}^\dagger .$$

The average value of spin A in the direction $\hat{\mathbf{a}}$ and, simultaneously, spin B in the direction $\hat{\mathbf{b}}$ is $2^2 \left\langle \varrho_{AB} \hat{\mathbf{a}} \otimes \hat{\mathbf{b}} \right\rangle_{S \oplus S}$ where the subscript $S \oplus S$ indicates that the scalar part is to be taken in the spaces of both spin A and spin B. Similarly, the probability that spin A, upon appropriate measurement, is found in state P_{s_A} and spin B is found in state P_{s_B} is $2^2 \langle \varrho_{AB} P_{s_A} \otimes P_{s_B} \rangle_{S \otimes S}$.

One way that linear superpositions of product states arise is in reduction of representations of the rotation group. If we rotate Ψ_{AB} by an additional rotation with rotor R we get

$$\Psi_{AB} \to R \otimes R \Psi_{AB}$$

and the rotation factor $R \otimes R$ is a product representation of $SU(2)$. For example, suppose we rotate each spin by π about the axis $\hat{a}$. Then $R = \exp(-i\hat{a}\pi/2) = -i\hat{a}$ and $R \otimes R = -\hat{a} \otimes \hat{a}$.

The rotation factor $R \otimes R$ can be reduced to a direct sum of spin-1 and spin-0 parts:

$$R \otimes R \simeq UR \otimes RU^{\dagger} = D^{(0)} \oplus D^{(1)},$$

where the elements of the 4×4 unitary matrix U are Clebsch-Gordan coefficients $\langle SM | s_1 m_1 s_2 m_2 \rangle$. The reduction entangles states that contribute to more than one irreducible representation. In particular, the $S, M = 1, 0$ and $S, M = 0, 0$ states are fully entangled. Consider the spin-0 part:

$$\psi_{AB}^{(0)} = \frac{1}{\sqrt{2}} (1 \otimes e_1 - e_1 \otimes 1)(P_3 \otimes P_3)$$

with a spin density

$$\varrho_{AB}^{(0)} = \psi_{AB}^{(0)} \psi_{AB}^{(0)\dagger} = \frac{1}{2} (1 - e_1 \otimes e_1)(\bar{P}_3 \otimes P_3 + P_3 \otimes \bar{P}_3).$$

Note that $\psi_{AB}^{(0)}$ for the singlet is antisymmetric under interchange, and $\varrho_{AB}^{(0)}$ is symmetric. One can also verify that $\psi_{AB}^{(0)}$ is invariant under rotations: $R \otimes R \psi_{AB}^{(0)} = \psi_{AB}^{(0)}$. It is easy to see that $\varrho_{AB}^{(0)}$ is idempotent and may be viewed as a 2-spin (or 2-qubit) projector. Thus, $\varrho_{AB}^{(0)}$ represents a pure state. Fully entangled states are pure. However, mixed states can also be partially entangled.

Consider the triplet state with $M_S = 0$. This is easily found within a normalization factor by starting with the $(S, M) = (1, 1)$ state

$$\psi^{(1,1)} = P_3 \otimes P_3$$

and applying the lowering operator [compare (49)] $S_- = e_1 P_3 \otimes 1 + 1 \otimes e_1 P_3$:

$$\psi_{AB}^{(1,0)} = \frac{1}{\sqrt{2}} (1 \otimes e_1 + e_1 \otimes 1)(P_3 \otimes P_3)$$

$$\varrho_{AB}^{(1,0)} = \psi_{AB}^{(1,0)} \psi_{AB}^{(1,0)\dagger} = \frac{1}{2} (1 + e_1 \otimes e_1)(\bar{P}_3 \otimes P_3 + P_3 \otimes \bar{P}_3).$$

Note $\frac{1}{2}(1 + e_1 \otimes e_1)$ and $\frac{1}{2}(1 - e_1 \otimes e_1)$ are complementary 2-spin projectors.

A mixture of equal amounts of fully entangled (and hence nonseparable) singlet and triplet states with $M = 0$ gives the state

$$\varrho' = \frac{1}{2} \left[\varrho_{AB}^{(0)} + \varrho_{AB}^{(1,0)} \right] = \frac{1}{2} (\bar{P}_3 \otimes P_3 + P_3 \otimes \bar{P}_3). \tag{50}$$

This is a *separable* state, that is a mixture of decomposable (uncorrelated) states. Through the mixture, the two spins do become correlated, but not entangled. The correlation is purely classical.

7.5   Total Spin and Nonlocal Projectors

However, we can project an entangled state out of the separable one (50) by applying a nonlocal projector. Consider the singlet-state projector for the two-spin system:

$$P_{S=0} = 1 - \frac{1}{2}S^2,$$

where

$$S^2 = \sum_j S_j^2 = \frac{1}{4}\sum_{j=1}^{3}(\mathbf{e}_j \otimes 1 + 1 \otimes \mathbf{e}_j)^2 = \frac{1}{2}\sum_j (1 + \mathbf{e}_j \otimes \mathbf{e}_j)$$

is the square of the total spin operator. Substitution of S^2 into $P_{S=0}$ gives directly the isotropic form

$$P_{S=0} = 1 - \frac{1}{4}[3 + \mathbf{e}_1 \otimes \mathbf{e}_1 + \mathbf{e}_2 \otimes \mathbf{e}_2 + \mathbf{e}_3 \otimes \mathbf{e}_3]$$

$$= \frac{1}{4}[1 - (\mathbf{e}_1 \otimes \mathbf{e}_1 + \mathbf{e}_2 \otimes \mathbf{e}_2 + \mathbf{e}_3 \otimes \mathbf{e}_3)] .$$

Applying $P_{S=0}$ to ϱ' (50) and noting that, with applications of the pacwoman property,

$$P_{S=0}(P_3 \otimes \bar{P}_3 + \bar{P}_3 \otimes P_3) = \frac{1}{2}(1 - \mathbf{e}_1 \otimes \mathbf{e}_1)(P_3 \otimes \bar{P}_3 + \bar{P}_3 \otimes P_3)$$

we find

$$P_{S=0}\varrho'P_{S=0} = \frac{1}{8}[1 - \mathbf{e}_1 \otimes \mathbf{e}_1](P_3 \otimes \bar{P}_3 + \bar{P}_3 \otimes P_3)[1 - \mathbf{e}_1 \otimes \mathbf{e}_1]$$

$$= \frac{1}{4}[1 - \mathbf{e}_1 \otimes \mathbf{e}_1](P_3 \otimes \bar{P}_3 + \bar{P}_3 \otimes P_3),$$

which is just $\frac{1}{2}\varrho_{AB}^{(0)}$, the singlet contribution to ϱ' (50). Note also that

$$P_{S=1} = \frac{1}{2}S^2 = 1 - P_{S=0}$$

is the projector onto the triplet state, and

$$P_{S=1}\varrho'P_{S=1} = \frac{1}{2}\varrho_{AB}^{(1,0)} .$$

The spin projectors $P_{S=0}$ and $P_{S=1}$ are examples of nonlocal projectors that can project unentangled states into entangled ones.

8. Conclusions

We have derived many quantum results for a classical system. The quantum/classical interface has become almost transparent in the spinor approach of Clifford's geometric algebra. Quantum effects demand *amplitudes*, and these are provided classically by rotors of the geometric algebra. In particular, the *eigenspinor* is an amplitude closely associated with quantum spinor wave functions. Linear equations for the classical eigenspinor suggest superposition and hence the possibility of quantum-like interference. Projectors onto minimal left ideals of the algebra are needed to reduce the spinor representation of Lorentz rotations in the algebra and to simplify the equation of motion, and by using them, many of the classical results take quantum form. The association helps clarify various quantum phenomena, and extensions to multiparticle systems promise to demystify such quantum phenomena as entanglement. The Q/C relation appears to be much deeper than superficial. The basic spinor representation gives spin 1/2 and changes in sign for rotations of 360°, and the linearity of the equation of motion gives $g = 2$.

While our results are concordant with most of quantum theory, there are some important differences in *interpretation*. In particular, fermion spin is given a classical foundation. Although quantum texts often give the magnitude of the spin as $\sqrt{3}\hbar/2$, there is no indication of any length greater than $\hbar/2$ in the classical approach. Since the projection of the spin in any direction is $\hbar/2$, a larger magnitude would precess about the quantization axis. Such a precession should have observable effects, but none has evidently been seen. The classical approach indicates that the prediction of a larger size is erroneous, arising from a misunderstanding of the spin component operators in an ideal spinor representation. A related statement from conventional quantum theory is that only one component of the spin can have a definite value at a time. This is usually justified by the uncertainty relation, which predicts uncertainties in the x and y components of the spin in any eigenstate of the z component. The classical approach suggests that in such an eigenstate, the x and y components vanish and are therefore definite. However, any such state is a linear combination of spin up and down along $\mathbf{e}_1$ and any *measurement* of the x component forces the system into one of these states. The uncertainty relation is therefore correct but its interpretation questionable. The classical interpretation is that a spin direction $\mathbf{s}$ exists for each pure state, and this is reinforced by the spin density operator approach in quantum theory. The spin density operator for a pure quantum state is $\varrho = \psi\psi^{\dagger} = \frac{1}{2}(1 + \mathbf{s})$, which indicates a spin direction $\mathbf{s}$ identical to $R\mathbf{e}_3 R^{\dagger}$ in the classical approach. Although ϱ cannot be determined by a single measurement, it can be found by a set of measurements on different samples of a pure-state ensemble, in complete analogy with the process of measuring the Stokes' parameters in a fully polarized beam of light. The mathematics is in fact identical to the treatment of polarized light and our interpretation should probably be analogous.

The quantum age has started and quantum technology is on its way. It's time to shift gears into the paradigm of relativistic quantum theory, on which the quantum age is based. APS formulates mechanics in a fashion that naturally incorporates relativity and much of the structure of quantum theory. It seems destined to play a central role in the revision of undergraduate curriculum needed to complete the paradigm shift.

Acknowledgement

Support from the Natural Sciences and Engineering Research Council of Canada is gratefully acknowledged.

References

[1] W. E. Baylis, "Geometry of Paravector Space with Applications to Relativistic Physics", an earlier chapter in this volume. See also W. E. Baylis, "Applications of Clifford Algebras in Physics", Lecture 4 in *Lectures on Clifford Geometric Algebras,* ed. by R. Abłamowicz and G. Sobczyk, Birkhäuser Boston, 2003.

[2] P. Lounesto, *Clifford Algebras and Spinors*, second edition, Cambridge University Press, Cambridge (UK) 2001.

[3] W. E. Baylis, *Electrodynamics: A Modern Geometric Approach*, Birkhäuser, Boston 1999.

[4] W. E. Baylis, editor, *Clifford (Geometric) Algebra with Applications to Physics, Mathematics, and Engineering*, Birkhäuser, Boston 1996.

[5] W. E. Baylis and Y. Yao, *Phys. Rev.* **A60**:785–795, 1999.

[6] V. Bargmann, L. Michel, and V. L. Telegdi, *Phys. Rev. Lett.* **2**:435–436, 1959.

[7] W. E. Baylis, *Phys. Rev.* **A45** (1992) 4293–4302.

[8] W. E. Baylis, *Adv. Appl. Clifford Algebras* **7(S)**:197–213, 1997.

[9] V. B. Berestetskii, E. M. Lifshitz, and L. P. Pitaevskii, *Quantum Electrodynamics* (Volume 4 of *Course of Theoretical Physics*), 2nd edn. (transl. from Russian by J. B. Sykes and J. S. Bell), Pergamon Press, Oxford, 1982.

[10] L. E. Ballentine, *Quantum Mechanics: a modern development*, World Scientific, Singapore 1998.

[11] D. Z. Albert, Sci. Am. **270**:(no. 5), 58–67, 1994.

[12] M. A. Nielsen, *Quantum Computation and Quantum Information*, Cambridge University Press, Cambridge (UK) 2000.

PONS, REED-MULLER CODES, AND GROUP ALGEBRAS

Myoung An, Jim Byrnes, William Moran, B. Saffari, Harold S. Shapiro, and Richard Tolimieri
Prometheus Inc.
21 Arnold Avenue
Newport, RI 02840
jim@prometheus-inc.com

Abstract   In this work we develop the family of Prometheus orthonormal sets (PONS) in the framework of certain abelian group algebras. Classical PONS, considered in 1991 by J. S. Byrnes, turned out to be a rediscovery of the 1960 construction by G. R. Welti [28], and of subsequent rediscoveries by other authors as well.

This construction highlights the fundamental role played by group characters in the theory of PONS. In particular, we will relate classical PONS to idempotent systems in group algebras and show that signal expansions over classical PONS correspond to multiplications in the group algebra.

The concept of a splitting sequence is critical to the construction of general PONS. We will characterize and derive closed form expressions for the collection of splitting sequences in terms of group algebra operations and group characters.

The group algebras in this work are taken over direct products of the cyclic group of order 2. PONS leads to idempotent systems and ideal decompositions of these group algebras. The relationship between these special systems and ideal decompositions, and the analytic properties of PONS, is an open research topic. A second open research topic is the extension of this theory to group algebras over cyclic groups of order greater than 2.

Keywords: character basis, companion row, crest factor, FE, Fejer dual, functional equation, generating function, generating polynomial, Golay, group algebra, Hadamard matrix, PONS, QMF, Reed-Muller Codes, Shapiro transform, Shapiro sequence, splitting property, splitting sequence, symmetric PONS, Thue-Morse sequence, Welti codes, Walsh-Hadamard matrices.

155

J. Byrnes (ed.) Computational Noncommutative Algebra and Applications, 155-196.
© 2004 *Kluwer Academic Publishers. Printed in the Netherlands.*

1.   Introduction

PONS, the *Prometheus Orthonormal Set*, has undergone considerable refinement, development, and extension since it was considered in [6]. First it turned out to be a rediscovery of Welti's 1960 construction [28] and of subsequent rediscoveries by others as well. Its application as an energy spreading transform for one-dimensional digital signals is discussed in [10], with further details of this aspect presented in the patent [9]. The construction of a smooth basis with PONS-like properties is given in [8], thereby answering in the affirmative a question posed by Ingrid Daubechies in 1991. A conceptually clear definition of the PONS sequences that comprise the original symmetric PONS matrix, via polynomial concatenations, is also given in [8]. An application to image processing of a preliminary version of multidimensional PONS can be found in [7]. An in-depth study of multidimensional PONS is currently being prepared. Proofs of the results given below will appear elsewhere.

2.   Analytic Theory of One-Dimensional PONS (Welti)

First we provide an account of some of the most basic mathematical concepts and results on *one-dimensional* PONS (Welti) sequences and the related PONS (Welti) matrices.

2.1   Shapiro Transforms of Unimodular Sequences

The whole mathematical theory of the PONS system, and also its applications to signal processing, turn out to derive from *one* fundamental idea, that of the *Shapiro transform of a unimodular sequence*. We begin by describing it.

Let $(\alpha_0, \alpha_1, \alpha_2, \ldots)$ be *any* infinite sequence of unimodular complex numbers. Then a sequence (P_m, Q_m) of polynomial pairs (with unimodular coefficients and common length 2^m) is inductively defined as follows: $P_0(x) = Q_0(x) = 1$, and:

$$\begin{cases} P_{m+1}(x) & = P_m(x) + \alpha_m x^{2^m} Q_m(x) \\ Q_{m+1}(x) & = P_m(x) - \alpha_m x^{2^m} Q_m(x) \end{cases} \quad \text{for all integers } m \geq 0. \quad (1)$$

Since $P_m(x)$ is a truncation of $P_{m+1}(x)$ for *every* $m \geq 0$, it follows that there is an *infinite* sequence $(\beta_0, \beta_1, \beta_2, \ldots)$ of unimodular complex numbers which *only* depends on the given sequence $(\alpha_0, \alpha_1, \alpha_2, \ldots)$ and such that for each $m \geq 0$ the *first* polynomial P_m of the pair (P_m, Q_m) is always the partial sum of length 2^m (i.e., of degree $2^m - 1$) of the *unique* power series $\sum_{k=0}^{\infty} \beta_k x^k$. Note that such a property does *not* hold for the polynomials Q_m, since Q_m is not a truncation of Q_{m+1}.

The explicit construction of the Shapiro transform $(\beta_k)_{k \geq 0}$ in terms of the original unimodular sequence $(\alpha_m)_{m \geq 0}$ is as follows: Let $k = \sum_{r \geq 0} \delta_r \cdot 2^r$ denote the expansion of an arbitrary integer $k \geq 0$ in base 2 (so that the "*binary digits*" δ_r take only the values 0 and 1, and $\delta_r = 0$ for all $r > \log k / \log 2$). Then we have

$$\beta_k = \epsilon_k \prod_{r \geq 0} \alpha_r^{\delta_r} \tag{2}$$

where

$$\epsilon_k = (-1)^{\sum_{r \geq 0} \delta_r \delta_{r+1}} \quad \text{(the classical Shapiro sequence [25]).} \tag{3}$$

We call $(\beta_k)_{k \geq 0}$ the *Shapiro transform* of the sequence $(\alpha_m)_{m \geq 0}$. In particular, when $\alpha_m = 1$ for all $m \geq 0$, we have $\beta_k = \epsilon_k$ (the classical Shapiro sequence).

The "*Shapiro transform power series*" $\sum_{k=0}^{\infty} \beta_k z^k$ and the related polynomial pairs (P_m, Q_m) have, respectively, all the remarkable properties of the classical Shapiro power series $\sum_{k=0}^{\infty} \epsilon_k z^k$ and those of the classical Shapiro polynomial pairs. (We will recall these properties in the following section, on the "original PONS matrix"). In addition, all the *unimodular complex* numbers $\alpha_0, \alpha_1, \alpha_2, \ldots$ are at our disposal, which can be useful in many situations. For our present purposes (PONS constructions) the parameters $\alpha_0, \alpha_1, \alpha_2, \ldots$ will only take the values ± 1. But the case when $\alpha_m = \pm 1$ or $\pm i$ can also be useful in signal processing (and leads to PONS-type Hadamard matrices with entries $\pm 1, \pm i$). Other choices of the unimodular parameters $\alpha_0, \alpha_1, \alpha_2, \ldots$ have other interesting applications that will be dealt with elsewhere.

2.2 The Original PONS (Welti) Matrix of Order 2^m

Before giving, in section 2.4, several (essentially equivalent) definitions of PONS matrices in full generality, and indicating structure theorems for such general PONS matrices, we start by recalling the *original PONS matrix* constructed in 1991 and published in 1994 [6]. Indeed, the proofs of structure results for *general* PONS matrices make use (in the inductive arguments) of properties of this original matrix of order 2^m, which we will denote by P_{2^m}. (There is no risk of confusion with the previously defined *polynomial P_m*).

The original way of defining the P_{2^m} uses an inductive method based on the *concatenation rule:*

$$\begin{pmatrix} A \\ B \end{pmatrix} \rightarrow \begin{pmatrix} A & B \\ A & -B \\ B & A \\ B & -A \end{pmatrix} \qquad (4)$$

where A and B are two consecutive matrix rows. More precisely, we start with the 2×2 matrix

$$P_2 := \begin{pmatrix} + & + \\ + & - \end{pmatrix}$$

(where, henceforth, we write $+$ instead of $+1$ and $-$ instead of -1, to ease notation). Thus the two rows of P_2 are $A = (++)$ and $B = (+-)$. Then the rule (4) means that the first row of the next matrix, P_4, is the concatenation of A and B, which here is $(+ + +-)$; the second row of P_4 is the concatenation of $A = (++)$ and $-B := (-+)$, which is therefore $(+ + -+)$; the third row of P_4 is the concatenation of B and A, i.e., $(+ - ++)$; and finally the fourth row of P_4 is the concatenation of $B = (+-)$ and $-A := (--)$. Thus

$$P_4 := \begin{pmatrix} + & + & + & - \\ + & + & - & + \\ + & - & + & + \\ + & - & - & - \end{pmatrix}.$$

To obtain the next matrix P_8, we first take the pair A, B to be the first two rows of P_4 (in that order) and use the concatenation rule (4) to obtain the first four rows of P_8. Then we take the pair A, B to be the next two rows of P_4 (in that order), and use the concatenation rule (4) to obtain the next four rows of P_8. Thus

$$P_8 := \begin{pmatrix} + & + & + & - & + & + & - & + \\ + & + & + & - & - & - & + & - \\ + & + & - & + & + & + & + & - \\ + & + & - & + & - & - & - & + \\ + & - & + & + & - & + & + & + \\ + & - & + & + & + & - & - & - \\ + & - & - & - & + & - & + & + \\ + & - & - & - & - & + & - & - \end{pmatrix}.$$

Similarly, the first two rows of P_8 and the concatenation rule (4) yield the first four rows of P_{16}, and so on. Already, from this definition, one can deduce many fundamental properties held by P_{2^m}, $(m = 1, 2, 3, \dots)$.

We list some of them and indicate later on that many of these properties also hold for the most general PONS (Welti) matrices that we shall define and characterize below. In addition, some of these properties (for example, the extremely important *bounded crest-factor property for all finite sections*) will also be seen to be valid for some very broad classes of Hadamard matrices generalizing the PONS (Welti) matrices.

We note that, in the context of codewords, a basically identical matrix to P_L was constructed by Welti [28]. Several additional authors [16, 18, 26, 29] have also reported on these *Welti codes* and their application to radar, and others [5, 3, 4, 21, 22, 19, 20] have discussed communications applications of similar constructions.

2.3 Some Properties of the Original PONS (Welti) matrix P_L, $(L = 2^m)$

As we said, most of the properties below will be seen to hold for the most general PONS matrices (that we shall define later, in section 2.4) and even for some very broad generalizations of PONS.

PROPERTY 1 P_L, with $L = 2^m$, is a Hadamard matrix of order 2^m.

PROPERTY 2 *Suppose the rows of P_L are ordered from 0 to $L - 1$ (i.e., the first row has rank 0 and the last row has rank $L-1$). Denote by $A_r(z)$ the polynomial "associated" to the r-th row (i.e., $A_r(z) = \sum_{k=0}^{L-1} a_k z^k$ if $(a_0, a_1, \ldots a_{L-1})$ denotes the r-th row, $r = 0, 1, \ldots L-1$). It is well known [2] that, with this notation, $A_1(z) = (-1)^{m+1} A_0^*(-z)$ where $A^*(z) = z^{\deg A} \overline{A}(1/z)$ denotes the "inverse" of the polynomial $A(z)$. This is a famous identity on the classical Shapiro pairs. Property 2 is that a similar identity $A_{2r+1}(z) = \lambda_{m,r} A_{2r}^*(-z)$ holds for all $r = 0, 1, \ldots L/2$, where $\lambda_{m,r}$ is an extremely interesting number (with values ± 1) expressible in terms of the "Morse sequence". The Morse sequence is the sequence of coefficients in the Taylor (or power series) expansion of the infinite product $\prod_{s=0}^{\infty} \left(1 - z^{2^s}\right)$.*

PROPERTY 3 *With the previous notation, for every $r = 0, 1, \ldots L/2$ the polynomials $A_{2r}(z)$ and $A_{2r+1}(z)$ are "Fejér-dual" (or "dual" for short), that is,*

$$|A_{2r}(z)|^2 + |A_{2r+1}(z)|^2 = \text{ constant } (= 2L, \text{ in this case}) \qquad (5)$$

for all $z \in \mathbb{C}$ with $|z| = 1$. Equivalently, the $(2r)$-th row and the $(2r+1)$-st row are always "Golay complementary pairs" [15].

PROPERTY 4 *[Much related to Properties 2 and 3] Every row-polynomial $A_r(z)$ is QMF, that is,*

$$|A_r(z)|^2 + |A_r(-z)|^2 = \text{ constant } (= 2L \text{ in this case}) \tag{6}$$

for all $z \in \mathbb{C}$ with $|z| = 1$.

PROPERTY 5 (THE "SPLITTING PROPERTY" OF ROWS) *For every $r = 0, 1, \ldots, L - 1$, the two "halves" of the row-polynomial $A_r(z)$ are dual, each of these two halves has dual halves, each of these halves (i.e., "quarters" of $A_r(z)$) has dual halves, and so on. This "splitting property" is, by far, the most important property, in view of its applications to "energy spreading". It extends to general PONS matrices and to a broader class of PONS-related Hadamard matrices that we will consider later.*

PROPERTY 6 (THE "QMF-SPLITTING PROPERTY")   *This is a finer form of the "QMF property" (Property 4) and an analog of the "splitting property" just described. We will postpone its definition until we come to the structure results for general PONS matrices (in section 2.4).*

PROPERTY 7 (THE "HYBRID SPLITTING PROPERTY")   *This is also a finer form of the "QMF property (above Property 4), and is as follows: Every row-polynomial $A_r(z)$ is QMF, i.e., (6) holds; and if we split $A_r(z)$ into two halves of equal length, then each of those halves is QMF. If we split these halves into halves of equal length, these in turn are QMF, and so on. We will return to this property in section 2.4 when we deal with the structure results for general PONS matrices.*

PROPERTY 8 (THE "CONSTANT ROW-SUMS PROPERTY") *If m is even, then each row-sum of P_L (with $L = 2^m$) has the constant value $\sqrt{L} = 2^{m/2}$. If m is odd, then the row sums are either zero or $\sqrt{2L} = 2^{(m+1)/2}$. This property is important but easy to check. However, this is a very special case of the deep (and still partly open) problem of the values of row-polynomials at various roots of unity.*

PROPERTY 9 ("BOUNDED CREST FACTOR PROPERTIES") *Not only is it true that every row-polynomial have crest factor $\leq \sqrt{2}$, but also every finite section of such a polynomial has crest factor not exceeding some absolute constant C. (Good values of C are known, but the optimal value of C is still an open problem.) We point out that these extremely important properties (for "energy spreading") are closely related to Property 5 (splitting properties of rows).*

PROPERTY 10 ("DUAL-COMPANION UNIQUENESS") *Every row has exactly one "companion row", that is, if $A_r(z)$ and $A_s(z)$ denote the associated row-polynomials, then $|A_r(e^{it})| = |A_s(e^{it})|$ for all $t \in \mathbb{R}$. These* two rows are "mirror images" of each other, except for a possible multiplication by -1. This possible sign change, and also the distribution of the location of s in terms of r, are quite surprising and can be expressed, here also, in terms of the Morse sequence. As a corollary, every row has exactly two "duals" (or, equivalently, two rows which are Golay-complements to it [15]).

2.4 General Definition of PONS (Welti) Matrices

In section 2.3 we described the original PONS matrix and some of its properties. Before considering any other special PONS matrices (such as, typically, the *symmetric* PONS matrices the very existence of which is really surprising), we will give a *general definition* of PONS matrices and also consider some of their useful generalizations. It is convenient to start by considering three very broad classes of finite sequences of length 2^m. These have an independent interest, with or without regard to Hadamard or PONS matrices.

So, let $S = (a_0, a_1, \ldots a_{L-1})$ denote any complex-valued sequence of length $L = 2^m$, $(m \geq 1)$. Its "*associated polynomial*" (or "*generating polynomial*") is

$$P(z) := \sum_{k=0}^{L-1} a_k z^k.$$

The two "halves" of $P(z)$ are:

$$A(z) = \sum_{k=0}^{L/2-1} a_k z^k \quad \text{and} \quad B(z) = \sum_{k=L/2}^{L-1} a_k z^k.$$

DEFINITION 11 *The sequence S is said to have the "splitting property" (or to be a "splitting sequence") if its generating polynomial $P(z)$ has "dual" halves, that is,*

$$|A(e^{it})|^2 + |B(e^{it})|^2 = \ constant \ (= \|P\|_2^2 = \sum_{k=0}^{L-1} |a_k|^2, \ necessarily),$$

and if each of the halves $A(z)$ and $B(z)$ has dual halves, and so on.

DEFINITION 12 *The sequence S is said to have the "QMF splitting property" (or to be a "QMF splitting sequence") if, first of all, it is QMF, which means that $P(z)$ and $P(-z)$ are dual, or, equivalently, that the even-index component $C(z) := \sum_{k=0}^{L/2-1} a_{2k} z^k$ and the odd-index component $D(z) := \sum_{k=0}^{L/2-1} a_{2k+1} z^k$ are dual; and if, in turn, both of $C(z)$ and $D(z)$ are QMF; and so on.*

DEFINITION 13 *The sequence S is said to have the "hybrid splitting property" (or to be a "hybrid splitting sequence") if it is QMF (i.e., $P(z)$ and $P(-z)$ are dual), and if it has QMF halves (i.e., $A(z)$ and $A(-z)$ are dual and also $B(z)$ and $B(-z)$ are dual), and if each of the halves has QMF halves, and so on.*

We point out that these are three (overlapping but) *pairwise distinct* classes, even if S is assumed to only take the values ± 1 (as long as $L \geq 16$). The smallest $L = 2^m$ for which one can find examples of ± 1 sequences of length L satisfying one of the above conditions and not the other two is precisely $L = 16$. However, we have the following two theorems:

THEOREM 14 *For any integer $m \geq 4$, there are Hadamard matrices of order $L = 2^m$ all of whose rows satisfy the requirements of any of the above three definitions, but not those of the other two.*

Thus we obtain three *pairwise distinct* (and very broad) classes of Hadamard matrices of order $L = 2^m$, $(L \geq 16)$, which we call respectively:

(A) the class of $2^m \times 2^m$ Hadamard matrices with *"splitting rows"*.

(B) the class of $2^m \times 2^m$ Hadamard matrices with *"QMF-splitting rows"*.

(C) the class of $2^m \times 2^m$ Hadamard matrices with *"hybrid splitting rows"*.

Our Theorem 15, stated below, says that the intersection of *any two* of the above three classes of Hadamard matrices is contained in the third class. (This will lead to the notion *and* complete identification of all *"general PONS matrices"*). We note that none of the classical Walsh-Hadamard matrices [1] lies in any of these three classes.

THEOREM 15 (UNIQUENESS AND STRUCTURE THEOREM) for General PONS Matrices *Suppose that all the rows of some $2^m \times 2^m$ Hadamard matrix P have any two of the above three properties (A), (B), (C).*

Then all the rows of P also have the third property, that is, all three of (A), (B) and (C) are satisfied by all the rows. In that case the rows of P constitute some permutation of the rows of the "original PONS matrix" P_L, with $L = 2^m$, after some of these rows have possibly been multiplied by -1.

REMARK 16 *The converse of Theorem 15 is obvious, in view of the properties of Shapiro transforms of ± 1 sequences.*

The proof of Theorem 15 is non-trivial. It is done by induction, and it uses (as a lemma) Property 10 of the *"original PONS matrix"* P_L and the fact that each row-polynomial $A_r(z)$ has *exactly two* duals $A_t(z)$ which are of the form $\pm A_r(-z)$ or $\pm A_r^*(-z)$. This lemma, in turn, rests on another lemma which involves the exact computation of the greatest common divisor between any two row-polynomials of P_L.

DEFINITION 17 *A general PONS (Welti) matrix of order 2^m is any $2^m \times 2^m$ Hadamard matrix satisfying the conditions of Theorem 15.*

COROLLARY 18 *All the rows of any $2^m \times 2^m$ general PONS matrix have all those properties of rows of the original PONS matrix P_L which are invariant by permutations of rows and sign changes.*

At this stage we mention the following result which also illustrates some structural difference between the PONS matrices and its generalizations.

PROPOSITION 19 *The largest length of runs of equal consecutive terms in any PONS-matrix row is 4. The largest length of runs of equal consecutive terms in any row of a "splitting Hadamard matrix" is 6.*

2.5 Symmetric PONS Matrices

If the "concatenation rule" (4) of section 2.2 is replaced by the new concatenation rule

$$\begin{pmatrix} A \\ B \end{pmatrix} \rightarrow \begin{pmatrix} A & B \\ A & -B \\ B & A \\ -B & A \end{pmatrix} \tag{7}$$

with the same starting matrix

$$P_2 := \begin{pmatrix} 1 & 1 \\ 1 & -1 \end{pmatrix},$$

then the result is a sequence of $2^m \times 2^m$ *symmetric* matrices which are, by construction and in view of the above Theorem 15, PONS matrices. A consequence of this extremely unexpected result is that, by standard operations on rows and suitable choice of the parameters α_j of section 2.2, we obtain *on the order of 2^m* symmetric $2^m \times 2^m$ PONS matrices.

3. Shapiro Sequences, Reed-Muller Codes, and Functional Equations

We begin our journey into the more algebraic aspects of PONS sequences by discussing their surprising relation to Reed-Muller codes.

Let $\mathbb{Z}_2^{2^m}$ be the set of binary 2^m-tuples, $m \geq 1$.

For each n, $1 \leq n \leq 2^m - 1$, and each j, $1 \leq j \leq m$, let $\delta_{j,n}$ be the coefficient of 2^{j-1} in the binary expansion of n and define $\delta_{0,n}$ to be $1, 0 \leq n \leq 2^m - 1$, so that

$$n = \sum_{j=1}^{m} 2^{j-1} \delta_{j,n}.$$

Define vectors $\vec{g}_j$ by

$$\vec{g}_j = \vec{g}_j(m) = \langle\, \delta_{j,0}\, \delta_{j,1}\, \delta_{j,2} \dots \delta_{j,2^m-1}\, \rangle,$$

$$\vec{g}_0 = \vec{g}_0(m) = \langle\, 111 \dots 1\, \rangle.$$

and the matrix $\mathbf{G}_m$ by

$$\mathbf{G}_m = \{\vec{g}_0, \vec{g}_1, \dots, \vec{g}_m\}$$

Example: $m = 3$

n	0	1	2	3	4	5	6	7
$\vec{g}_0$	1	1	1	1	1	1	1	1
$\vec{g}_1$	0	1	0	1	0	1	0	1
$\vec{g}_2$	0	0	1	1	0	0	1	1
$\vec{g}_3$	0	0	0	0	1	1	1	1

$$\mathbf{G}_3 = \{\ \langle 11111111 \rangle,\quad \langle 01010101 \rangle,$$
$$\langle 00110011 \rangle,\quad \langle 00001111 \rangle\ \}$$

The $\vec{g}_m$ are discretized versions of the Rademacher functions.

Claim. The elements of $\mathbf{G}_m$ are linearly independent.

The *Reed-Muller code* of rank m and order 0 is

$$RM(0,m) = \{\langle 00 \dots 0 \rangle, \langle 11 \dots 1 \rangle\},$$

where each vector (*codeword*) has 2^m entries. $RM(1,m)$ is the subgroup of $\mathbb{Z}_2^{2^m}$ generated by the codewords in $\mathbf{G}_m$, *i.e.*, the vector space over $\mathbb{Z}_2$ spanned by these codewords. $RM(1,m)$ contains 2^{m+1} codewords. Define *multiplication* $\cdot$ on $\mathbb{Z}_2^{2^m}$ by

$$\langle x_0\, x_1\ \dots\ x_{2^m-1} \rangle \cdot \langle y_0\, y_1\ \dots\ y_{2^m-1} \rangle = \langle x_0 y_0\ x_1 y_1\ \dots\ x_{2^m-1} y_{2^m-1} \rangle.$$

Augment $\mathbf{G}_m$ with all products $\vec{g}_i \cdot \vec{g}_j$, $1 \leq i < j \leq m$, to form $\mathbf{G}_m^{(2)}$.

Example: $m = 3$

n	0	1	2	3	4	5	6	7
$\vec{g}_0$	1	1	1	1	1	1	1	1
$\vec{g}_1$	0	1	0	1	0	1	0	1
$\vec{g}_2$	0	0	1	1	0	0	1	1
$\vec{g}_3$	0	0	0	0	1	1	1	1
$\vec{g}_1 \cdot \vec{g}_2$	0	0	0	1	0	0	0	1
$\vec{g}_1 \cdot \vec{g}_3$	0	0	0	0	0	1	0	1
$\vec{g}_2 \cdot \vec{g}_3$	0	0	0	0	0	0	1	1

$$\mathbf{G}_3^{(2)} = \mathbf{G}_3 \cup \{\langle 00010001 \rangle, \langle 00000101 \rangle, \langle 00000011 \rangle\}.$$

Claim. The $1 + m + \binom{m}{2}$ elements of $\mathbf{G}_m^{(2)}$ are linearly independent.
$RM(2, m)$ is the subgroup of $\mathbb{Z}_2^{2^m}$ generated by the codewords in $\mathbf{G}_m^{(2)}$.
$RM(2, m)$ contains $2^{1+m+\binom{m}{2}}$ codewords.
Augmenting $\mathbf{G}_m^{(2)}$ with all products of the form $\vec{g}_i \cdot \vec{g}_j \cdot \vec{g}_k$, $1 \leq i < j < k \leq m$, and continuing as above we get $\mathbf{G}_m^{(3)}$, $RM(3, m)$, etc.

Theorem. $RM(k, m)$ for $m \geq 1$, $0 \leq k \leq m$ is a subgroup of $\mathbb{Z}_2^{2^m}$ consisting of 2^N codewords, where $N = \sum_{i=0}^{k} \binom{m}{i}$. The minimum *Hamming weight* (*i.e.*, number of ones) of the nonzero codewords in $RM(k, m)$ is 2^{m-k}.

Proof. Exercise, or see Handbook of Coding Theory, V. Pless and W.C. Huffman, Editors, Vol. 1, pp. 122–126.
Let's examine a particular element $\vec{S}_m \in RM(2, m)$ given by

$$\vec{S}_m = \sum_{j=1}^{m-1} \vec{g}_j \cdot \vec{g}_{j+1} = \langle s_0 s_1 \ldots s_{2^m-1} \rangle.$$

Example. $m = 3$

n	0	1	2	3	4	5	6	7
$\vec{g}_1$	0	1	0	1	0	1	0	1
$\vec{g}_2$	0	0	1	1	0	0	1	1
$\vec{g}_3$	0	0	0	0	1	1	1	1
$\vec{S}_3$	0	0	0	1	0	0	1	0

Let $\phi(n)$ be the number of times that the *block* $B = [1\,1]$ occurs in the binary expansion of n, $0 \leq n \leq 2^m - 1$.

Claim.

$$s_n = \begin{cases} 0 & \text{if } \phi(n) \text{ is even} \\ 1 & \text{if } \phi(n) \text{ is odd.} \end{cases}$$

Let $\mathcal{G}_m = \{\vec{\gamma}_0, \vec{\gamma}_1, \vec{\gamma}_2, \ldots, \vec{\gamma}_{2^m-1}\}$ be the subgroup of $RM(1, m)$ generated by $\vec{g}_1, \vec{g}_2, \ldots, \vec{g}_m$.

Example. $m = 3$

	n	0	1	2	3	4	5	6	7
	$\vec{g}_1$	0	1	0	1	0	1	0	1
	$\vec{g}_2$	0	0	1	1	0	0	1	1
	$\vec{g}_3$	0	0	0	0	1	1	1	1
$\mathcal{G}_3$	$\vec{\gamma}_0 = 0 \cdot \vec{g}_1 + 0 \cdot \vec{g}_2 + 0 \cdot \vec{g}_3$	0	0	0	0	0	0	0	0
	$\vec{\gamma}_1 = 1 \cdot \vec{g}_1 + 0 \cdot \vec{g}_2 + 0 \cdot \vec{g}_3$	0	1	0	1	0	1	0	1
	$\vec{\gamma}_2 = 0 \cdot \vec{g}_1 + 1 \cdot \vec{g}_2 + 0 \cdot \vec{g}_3$	0	0	1	1	0	0	1	1
	$\vec{\gamma}_3 = 1 \cdot \vec{g}_1 + 1 \cdot \vec{g}_2 + 0 \cdot \vec{g}_3$	0	1	1	0	0	1	1	0
	$\vec{\gamma}_4 = 0 \cdot \vec{g}_1 + 0 \cdot \vec{g}_2 + 1 \cdot \vec{g}_3$	0	0	0	0	1	1	1	1
	$\vec{\gamma}_5 = 1 \cdot \vec{g}_1 + 0 \cdot \vec{g}_2 + 1 \cdot \vec{g}_3$	0	0	1	1	0	0	1	1
	$\vec{\gamma}_6 = 0 \cdot \vec{g}_1 + 1 \cdot \vec{g}_2 + 1 \cdot \vec{g}_3$	0	0	1	1	1	1	0	0
	$\vec{\gamma}_7 = 1 \cdot \vec{g}_1 + 1 \cdot \vec{g}_2 + 1 \cdot \vec{g}_3$	0	1	1	0	1	0	0	1

We now switch gears slightly, by rewriting all codewords in $RM(k, m)$ by mapping $0 \to 1$, $1 \to -1$. Since $\vec{g}_1, \vec{g}_2, \ldots, \vec{g}_m$ are discretized versions of the Rademacher functions, $\vec{\gamma}_0, \vec{\gamma}_1, \ldots, \vec{\gamma}_{2^m-1}$, are discretized versions of the Walsh functions. That is, $\mathcal{G}_m$ is the $2^m \times 2^m$ Sylvester Hadamard matrix, which we relabel H_m.

Example.

$$H_3 = \begin{array}{cccccccc} 1 & 1 & 1 & 1 & 1 & 1 & 1 & 1 \\ 1 & -1 & 1 & -1 & 1 & -1 & 1 & -1 \\ 1 & 1 & -1 & -1 & 1 & 1 & -1 & -1 \\ 1 & -1 & -1 & 1 & 1 & -1 & -1 & 1 \\ 1 & 1 & 1 & 1 & -1 & -1 & -1 & -1 \\ 1 & 1 & -1 & -1 & 1 & 1 & -1 & -1 \\ 1 & 1 & -1 & -1 & -1 & -1 & 1 & 1 \\ 1 & -1 & -1 & 1 & -1 & 1 & 1 & -1 \end{array}$$

Now $s_n = (-1)^{\phi(n)}$.

Since we have

$$s_{2n} = s_n, \quad s_{2n+1} = \begin{cases} s_n & \text{if } n \text{ is even} \\ -s_n & \text{if } n \text{ is odd} \end{cases},$$

the binary expansion of $2n$ is the binary expansion of n shifted one slot to the left with a 0 added on the right and the binary expansion of $2n + 1$ is the binary expansion of n shifted one slot to the left with a 1 added on the right. If n is even this does not change $\phi(n)$. If n is odd (*i.e.*, n ends in 1) then $\phi(2n+1) = \phi(n) + 1$. ■

Consider the *generating function* of $\{s_n\}$,

$$g(z) = \sum_{n=0}^{\infty} s_n z^n.$$

It can be shown that $g(z)$ satisfies the functional equation (FE) (Brillhart and Carlitz)

$$g(z) = g(z^2) + zg(-z^2).$$

Iterate this FE:

$$g(z^2) = g(z^4) + z^2 g(-z^4)$$
$$g(-z^2) = g(z^4) - z^2 g(-z^4), \quad \text{so}$$
$$g(z) = (1+z)g(z^4) + z^2(1-z)g(-z^4).$$

Repeat:

$$g(z^4) = g(z^8) + z^4 g(-z^8)$$
$$g(-z^4) = g(z^8) - z^4 g(-z^8), \quad \text{so}$$
$$g(z) = (1+z+z^2-z^3)g(z^8) + z^4(1+z-z^2+z^3)g(-z^8).$$

Continuing we see that, beginning with

$$g(z) = A(z)g(z^{2^m}) + z^{2^{m-1}} B(z)g(-z^{2^m})$$

and applying

$$g(z^{2^m}) = g(z^{2^{m+1}}) + z^{2^m} g(-z^{2^{m+1}})$$
$$g(-z^{2^m}) = g(z^{2^{m+1}}) - z^{2^m} g(-z^{2^{m+1}})$$

we get at the next step

$$g(z) = \left[A(z) + z^{2^{m-1}} B(z) \right] g(z^{2^{m+1}})$$
$$+ z^{2^m} \left[A(z) - z^{2^{m-1}} B(z) \right] g(-z^{2^{m+1}}) .$$

Renaming the initial $A(z)$ and $B(z)$ to $P_0(z)$ and $Q_0(z)$, respectively, and naming the (polynomial) coefficients of $g(z^{2^m})$ and $g(-z^{2^m})$ $P_{m-1}(z)$ and $Q_{m-1}(z)$, respectively, $m \geq 1$, the above yields

$$P_0(z) = Q_0(z) = 1$$
$$P_m(z) = P_{m-1}(z) + z^{2^{m-1}} Q_{m-1}(z)$$
$$Q_m(z) = P_{m-1}(z) - z^{2^{m-1}} Q_{m-1}(z) .$$

Thus, the $\{P_m(z)\}_{m=0}^{\infty}$ and $\{Q_m(z)\}_{m=0}^{\infty}$ are precisely the Shapiro Polynomials! $P_m(z)$ and $Q_m(z)$ are each polynomials of degree $2^m - 1$ with coefficients ± 1. For each m, the first 2^m coefficients of $g(z)$ are exactly the coefficients of $P_m(z)$. Therefore, for each m, the 2^m-truncation

$$\langle\, s_0\, s_1\, \dots\, s_{2^m-1}\,\rangle$$

of the *Shapiro sequence* $\{s_j\}_{j=0}^{\infty}$ is an element of $RM(2,m)$.
Why might that be important?

Recall the fundamental property of the Shapiro polynomials, namely that for each m P_m and Q_m are complementary:

$$|P_m(z)|^2 + |Q_m(z)|^2 = 2^{m+1} \quad \text{for all } |z| = 1.$$

Consequently P_m and Q_m each have *crest factor* (the ratio of the sup norm to the L^2 norm on the unit circle) bounded by $\sqrt{2}$ *independent of* m. *i.e.*, P_m and Q_m are *energy spreading*. So the coefficients of P_m are an energy spreading second order Reed-Muller codeword. Related results may be found in [11, 27].

Also, letting $\vec{h}_j$, $0 \le j \le 2^m - 1$, denote the rows of $\mathbf{H}_m$, the matrix $\mathbf{P}_m$ whose rows are $\vec{S}_m \cdot \vec{h}_j$, is a *PONS matrix*. Its 2^m rows can be split into 2^{m-1} pairs of complementary rows, with each row having crest factor (bounded by) $\sqrt{2}$.

Since each $\vec{h}_j \in RM(1,m)$ and $\vec{S}_m \in RM(2,m)$, the (rows of the) PONS matrix is a coset of the subgroup $RM(1,m)$ of $RM(2,m)$.

Thus we have constructed 2^m (really 2^{m+1} by considering $-\mathbf{H}_m$) energy spreading second order Reed-Muller codewords.

Note that blocks other than $B = [1\,1]$ appear in connection with higher-order Reed-Muller codes. For example, the block $[1\,1\,1]$ yields codewords in $RM(3,m)$. The generating functions of these blocks satisfy similar (although more complicated) FE's. An open question is whether these FE's yield corresponding crest factor bounds for subsets of $RM(k,m)$, $k \ge 3$, resulting in higher-order energy spreading Reed-Muller codes.

Blocks and FE's

Let $B = [\beta_1\, \beta_2\, \dots\, \beta_r]$, $\beta_j = 0$ or 1, $\beta_1 = 1$ be a *binary block* and $N = N(B) = \beta_r + 2\beta_{r-1} + \dots + 2^{r-1}\beta_1$ be the integer whose binary expansion is B. Let $\Psi_B(n)$ be the number of occurrences of B in the binary expansion of n and let $f_B(z)$ be the generating function of Ψ_B,

$$f_B(z) = \sum_{n=0}^{\infty} \Psi_B(n)z^n \quad .$$

Theorem. $f_B(z)$ satisfies the FE

$$f_B(z) = (1+z)f_B(z^2) + \frac{z^{N(B)}}{1-z^{2^r}} \ .$$

Now consider the *parity sequence* of $\Psi_B(n)$, $\delta_B(n) = (-1)^{\Psi_B(n)}$, and its generating function $g_B(z) = \sum_{n=0}^{\infty} \delta_B(n)z^n$. For the general case it will again be useful to split g_B into its even and odd parts,

$$E_B(z) = \sum_{n=0}^{\infty} \delta_B(2n)z^{2n}$$

$$O_B(z) = \sum_{n=0}^{\infty} \delta_B(2n+1)z^{2n+1}$$

Previous example: $B = [11]$, $\delta_B(n)$ is the Shapiro sequence, $g_B(z)$ satisfies the FE $g_B(z) = g_B(z^2) + zg_B(-z^2)$.

Example: $B = [1]$.
As before, $\Psi_B(2n) = \Psi_B(n)$ and $\Psi_B(2n+1) = \Psi_B(n)+1$ so that (writing δ_n for $\delta_B(n)$ to ease notation) $\delta_{2n} = \delta_n$, $\delta_{2n+1} = -\delta_n$. Hence $E_B(z) = g_B(z^2)$, $O_B(z) = -zg_B(z^2)$, and we have the FE $g_B(z) = (1-z)g_B(z^2)$. Iterating, $g_B(z) = (1-z)(1-z^2)(1-z^4)\ldots$ and δ_n is the Thue-Morse sequence $[1 - 1 - 1 1 - 1 1 1 - 1 \ldots]$. We drop the subscript B from now on.

Example: $\beta_r = 0$.
$\Psi(2n+1) = \Psi(n)$, so $\delta_{2n+1} = \delta_n$, so $O(z) = zg(z^2)$. Since $g(z)-g(-z) = 2O(z)$ we have the FE $g(z) = g(-z) + 2zg(z^2)$.

Example: $\beta_r = 1$.
As above, now $g(z) = -g(-z) + 2g(z^2)$.
Example (a typical case?): $B = [1 1 0 0 1 0], r = 6$.

$\Psi(2n+1) = \Psi(n)$. $\Psi(2n) = \Psi(n)$ unless the binary expansion of n ends in $[1 1 0 0 1]$, *i.e.*, unless $n \equiv K(\text{mod } 2^5)$, where $K = 2^4 + 2^3 + 2^0 = 25$, in which case $\Psi(2n) = \Psi(n) + 1$. So

$$\delta_{2n+1} = \delta_n, \quad \delta_{2n} = \begin{cases} -\delta_n & \text{if } n \equiv 25(\text{mod } 32) \\ \delta_n & \text{otherwise} \end{cases} .$$

So $O(z) = zg(z^2)$.

$$E(z) = \sum_{n=0}^{\infty} \delta_{2n} z^{2n} = \sum_{n=0}^{\infty} \delta_n z^{2n} - 2 \sum_{n \equiv 25 \pmod{32}} \delta_n z^{2n}$$

$$= g(z^2) - 2 \sum_{j=0}^{\infty} \delta_{32j+25} z^{64j+50} = g(z^2) - 2z^{50} F(z)$$

where $F(z) = \sum_{j=0}^{\infty} \delta_{32j+25} z^{64j}$.

But $\delta_{32j+25} = \delta_{2(16j+12)+1} = \delta_{16j+12} = \delta_{2(8j+6)} = \delta_{8j+6} = \delta_{2(4j+3)} = \delta_{4j+3} = \delta_{2(2j+1)+1} = \delta_{2j+1} = \delta_j$, where we have used the fact that neither $8j+6$ nor $4j+3$ can be congruent to $25 \pmod{32}$. So $F(z) = \sum_{j=0}^{\infty} \delta_j z^{64j} = g(z^{64})$, and we have the FE

$g(z) = (1+z)g(z^2) - 2z^{50} g(z^{64})$.

How *typical* is this example? Do we always get *Full Reduction* (FR) of the index of δ?

Consider the general case:

$$B = [\beta_1\, \beta_2\, \ldots\, \beta_r]$$
$$N = \beta_r + 2\beta_{r-1} + \ldots + 2^{r-1}\beta_1$$
$$K = \beta_{r-1} + 2\beta_{r-2} + \ldots + 2^{r-2}\beta_1 \quad .$$

Case I: $\beta_r = 0$. As above,

$$\delta_{2n+1} = \delta_n, \quad \delta_{2n} = \begin{cases} -\delta_n & \text{if } n \equiv K \pmod{2^{r-1}} \\ \delta_n & \text{otherwise} \end{cases} \quad .$$

$$O(z) = zg(z^2), \quad E(z) = g(z^2) - 2z^{2K} \sum_{j=0}^{\infty} \delta_{2^{r-1}j+K} z^{2^r j} \quad .$$

To get FR the index $I(1) = I_{j,K}(1) = 2^{r-1}j + K$ must reduce to j by repeated applications of the mapping $\mu(n)$:

$$\mu(2n+1) = n, \quad \mu(2n) = n \quad \text{unless } n \equiv K \pmod{2^{r-1}}.$$

Let $\{I(1), I(2), \ldots\}$ be the succession of indices that we get by repeating μ (assuming it works), and let I denote one of these indices. Whether $I = 2n+1$ or $I = 2n$, reduction to n occurs by dropping the last binary digit on the right of I and shifting what's left 1 slot to the right. For reduction to fail at the first step, $I(1)$ must be of the form $2n$ where $n \equiv K \pmod{2^{r-1}}$, or $n = 2^{r-1}m + K$ for some integer m, or $2n = 2^r m + 2K$.

The binary expansion (BE) of K is $(\beta_1\, \beta_2\, \ldots\, \beta_{r-1})$ so that of $2K$ is $(\beta_1\, \beta_2\, \ldots\, \beta_{r-1}\, 0)$.

So for the first reduction $I(1) \to I(2)$ to fail the BE of $I(1)$ must end in $(\beta_1 \beta_2 \ldots \beta_{r-1} 0)$. This is possible (*i.e.*, there are integers j which make it possible) iff the BE of $I(1)$ ends in $(\beta_2 \beta_3 \ldots \beta_{r-1} 0)$, or (since the BE of $I(1)$ ends in that of K)

$$(\beta_1 \beta_2 \ldots \beta_{r-1}) = (\beta_2 \beta_3 \ldots \beta_{r-1} 0) \quad .$$

Assuming this equation does not hold we get $I(2)$ whose BE ends in $(\beta_1 \beta_2 \ldots \beta_{r-2})$. As above, $I(2) \to I(3)$ fails iff the BE of $I(2)$ ends in $(\beta_1 \beta_2 \ldots \beta_{r-1} 0)$ which is possible (again, there are integers j which make it possible) iff $I(2)$ ends in $(\beta_3 \beta_4 \ldots \beta_{r-1} 0)$, or

$$(\beta_1 \beta_2 \ldots \beta_{r-2}) = (\beta_3 \beta_4 \ldots \beta_{r-1} 0) \quad .$$

Call the block $B = [\beta_1 \beta_2 \ldots \beta_r]$ *nonrepeatable* if

$$[\beta_1 \beta_2 \ldots \beta_\nu] \neq [\beta_{r-(\nu-1)} \beta_{r-(\nu-2)} \ldots \beta_r]$$

for each ν, $1 \leq \nu \leq r-1$.

Theorem. FR works iff B is nonrepeatable. When FR works we get the FE $g(z) = (1+z)g(z^2) - 2z^{2K}g(z^{2^r})$.

Case II: $\beta_r = 1$. The above argument works when B is nonrepeatable up to the last step, yielding:

Theorem. If $[\beta_1 \beta_2 \ldots \beta_\nu] \neq [\beta_{r-(\nu-1)} \beta_{r-(\nu-2)} \ldots \beta_r]$ for each ν, $2 \leq \nu \leq r-1$, and $\beta_1 = \beta_r = 1$, then reduction works up until the final step and we get the FE

$$g(z) = (1+z)g(z^2) - 2z^{2K+1-2^{r-1}} \left[g(z^{2^{r-1}}) - g(z^{2^r}) \right] \quad .$$

Other cases are not so neat.

Example. $B = [1\,1\,0\,1\,1\,1]$.

The FE is

$$g(z) = (1+z)g(z^2) - 2z^7 g(z^{16}) + 2z^7 g(z^{32}) + 2z^{23} g(z^{64}) \quad .$$

Example. $B = [1\,0\,1\,1\,0\,1]$.

The FE is

$$g(z) = (1+z)g(z^2) - 2z^5[g(z^8) - (1+z^8)g(z^{16})] - 2z^{13}[g(z^{32}) - g(z^{64})].$$

The general "1-1" case, $\beta_1 = \beta_r = 1$.

$$\delta_{2n} = \delta_n, \quad \delta_{2n+1} = \begin{cases} -\delta_n & \text{if } n \equiv K(\text{mod } 2^{r-1}) \\ \delta_n & \text{otherwise} \end{cases},$$

$$K = \beta_{r-1} + 2\beta_{r-2} + \ldots + 2^{r-2}\beta_1,$$

$$E(z) = g(z^2),$$

$$O(z) = zg(z^2) - 2 \sum_{\substack{n \equiv K \\ (\text{mod } 2^{r-1})}} \delta_n z^{2n+1} = zg(z^2) - 2G_B(z)$$

$$\text{where} \quad G_B(z) = \sum_{j=0}^{\infty} \delta_{2^{r-1}j+K} z^{2^r j + 2K + 1}.$$

Basic idea: Reduce the subscript of δ as much as possible, express $G_B(z)$ in terms of $G_B(z^{2^p})$ for some $p > 0$, replace $G_B(z^{2^p})$ by using $-2G_B(z^{2^p}) = O(z^{2^p}) - z^{2^p} g(z^{2^{p+1}}) = g(z^{2^p}) - g(z^{2^{p+1}}) - z^{2^p} g(z^{2^{p+1}})$ and then repeat to get the desired expression for $O(z) = g(z) - g(z^2)$. The result for the "fully repeatable" case, $\beta_j = 1$, $1 \le j \le r$, is:

$$g(z) = (1 - z)g(z^2) + 2z[g(z^4) + z^2 g(z^8) + z^6 g(z^{16})$$
$$+ \ldots + z^{2^{r-2}-2} g(z^{2^{r-1}}) + z^{2^{r-1}-2} g(z^{2^r})].$$

4.   Group Algebras

Consider a finite abelian group A with group composition written as multiplication. The *group algebra* $\mathbf{C}A$ is the vector space of formal sums

$$f = \sum_{u \in A} f(u)u, \quad f(u) \in \mathbf{C},$$

with algebra multiplication defined by

$$fg = \sum_{v \in A} \left(\sum_{u \in A} f(u)g(u^{-1}v) \right) v, \quad f, g \in \mathbf{C}A.$$

Identifying $u \in A$ with the formal sum u, we can view A as a subset of $\mathbf{C}A$. $\mathbf{C}A$ is a commutative algebra with identity. In this section we consider group algebras over direct products of the cyclic group of order 2.

Denote the direct product of N copies of the cyclic group of order 2 by

$$C_2(x_0, \ldots, x_{N-1})$$

and its group algebra by

$$A_2(x_0, \ldots, x_{N-1}).$$

$C_2(x_0, \ldots, x_{N-1})$ is the set of monomials

$$x_0^{j_0} \cdots x_{N-1}^{j_{N-1}}, \quad j_n = 0, 1, \quad 0 \le n < N,$$

with multiplication defined by

$$x_n^2 = 1, \quad 0 \le n < N, \tag{8}$$

$$x_m x_n = x_n x_m, \quad 0 \le m, n < N. \tag{9}$$

We call the factors x_n, for $0 \le n < N$, the *generators* of $C_2(x_0, \ldots, x_{N-1})$. For example,

$$C_2(x_0) = \{1, x_0\}$$

and

$$C_2(x_0, x_1) = \{1, x_0, x_1, x_0 x_1\}.$$

$A_2(x_0, \ldots, x_{N-1})$ is the algebra of polynomials

$$f = \sum_{j_0=0}^{1} \cdots \sum_{j_{N-1}=0}^{1} f(j_0, \ldots j_{N-1}) x_0^{j_0} \cdots x_{N-1}^{j_{N-1}},$$

with multiplication defined by (8) and (9). For example in $A_2(x_0)$ the algebra multiplication is given by

$$(a_0 + a_1 x_0)(b_0 + b_1 x_0) = c_0 + c_1 x_0,$$

where

$$\begin{bmatrix} c_0 \\ c_1 \end{bmatrix} = \begin{bmatrix} a_0 & a_1 \\ a_1 & a_0 \end{bmatrix} \begin{bmatrix} b_0 \\ b_1 \end{bmatrix}$$

and in $A_2(x_0, x_1)$ the algebra multiplication is given by

$$(a_0 + a_1 x_0 + a_2 x_1 + a_3 x_0 x_1)(b_0 + b_1 x_0 + b_2 x_1 + b_3 x_0 x_1) =$$
$$c_0 + c_1 x_0 + c_2 x_1 + c_3 x_0 x_1,$$

where

$$\begin{bmatrix} c_0 \\ c_1 \\ c_2 \\ c_3 \end{bmatrix} = \begin{bmatrix} a_0 & a_1 & a_2 & a_3 \\ a_1 & a_0 & a_3 & a_2 \\ a_2 & a_3 & a_0 & a_1 \\ a_3 & a_2 & a_1 & a_0 \end{bmatrix} \begin{bmatrix} b_0 \\ b_1 \\ b_2 \\ b_3 \end{bmatrix}.$$

In general multiplication in $A_2(x_0, \ldots, x_{N-1})$ can be described by the action of a $2^N \times 2^N$ block circulant matrix having 2×2 circulant blocks.

The monomials

$$x_0^{j_0} \cdots x_{N-1}^{j_{N-1}}, \quad j_n = 0, 1, \ 0 \le n < N,$$

define a basis, the *canonical basis* of the vector space $A_2(x_0, \ldots, x_{N-1})$. The basis is ordered by the lexicographic ordering on the exponents. For example, as listed,

$$1, x_0$$

is the canonical basis of $A_2(x_0)$ and

$$1, x_0, x_1, x_0 x_1$$

is the canonical basis of $A_2(x_0, x_1)$.

A nonzero element $\tau \in CA$ is called a character of A if $\tau(1) = 1$ and

$$v\tau = \tau(v^{-1})\tau, \quad v \in A.$$

Denote the collection of characters of A by A^*. The characters of

$$C_2(x_0, \ldots, x_{N-1})$$

are the set of products in $A_2(x_0, \ldots, x_{N-1})$

$$(1 + \epsilon_0 x_0) \cdots (1 + \epsilon_{N-1} x_{N-1}), \quad \epsilon_n = \pm 1, \ 0 \le n < N.$$

The characters of $C_2(x_0)$ are

$$1 + x_0, \ 1 - x_0.$$

The characters of $C_2(x_0, \ldots, x_{N-1})$ form a basis of the vector space $A_2(x_0, \ldots, x_{N-1})$, called the *character basis*. The characters are ordered such that the matrix H_{2^N} of the character basis relative to the canonical basis is the N-fold tensor product

$$H_{2^N} = H_2 \otimes \cdots \otimes H_2,$$

where

$$H_2 = \begin{bmatrix} 1 & 1 \\ 1 & -1 \end{bmatrix}$$

is the 2×2 Fourier transform matrix. The matrix of the character basis of $C_2(x_0)$

$$1 + x_0, \ 1 - x_0$$

is H_2 and the matrix of the character basis of $C_2(x_0, x_1)$

$$(1 + x_0)(1 + x_1), \ (1 - x_0)(1 + x_1), \ (1 + x_0)(1 - x_1), \ (1 - x_0)(1 - x_1)$$

is

$$H_4 = \begin{bmatrix} 1 & 1 & 1 & 1 \\ 1 & -1 & 1 & -1 \\ 1 & 1 & -1 & -1 \\ 1 & -1 & -1 & 1 \end{bmatrix}.$$

The set of group-translates of the characters of $C_2(x_0)$

$$(1 \pm x_0)x_1^{k_1} \cdots x_{N-1}^{k_{N-1}}, \quad k_n = 0, 1, \quad 0 \leq n < N,$$

forms a basis of $A_2(x_0, \dots, x_{N-1})$, called the *translate-character* basis. The translate-character basis is ordered by first forming the set of pairs

$$(1 + x_0)x_1^{k_1} \cdots x_{N-1}^{k_{N-1}}, \quad (1 - x_0)x_1^{k_1} \cdots x_{N-1}^{k_{N-1}} \tag{10}$$

and then lexicographically ordering the exponents in the set of pairs. The matrix of the translate-character basis relative to the canonical basis is the N-fold matrix direct sum

$$H_2 \oplus \cdots \oplus H_2.$$

The translate-character basis of $A_2(x_0, x_1)$ is

$$1 + x_0, \quad 1 - x_0, \quad (1 + x_0)x_1, \quad (1 - x_0)x_1.$$

The translate-character basis will be especially important in the development of PONS. The main reason is contained in the following discussion.

Consider $f \in A_2(x_0, \dots, x_{N-1})$ such that

$$f(j_0, \dots, j_{N-1}) = \pm 1, \quad j_n = 0, 1, \quad 0 \leq n < N.$$

The coefficients of the expansion of f over the translate-character basis are ± 1 or 0 and for each exponent set

$$k_1, \dots, k_{N-1}, \quad k_n = \pm 1, \quad 0 \leq n < N,$$

exactly one element in the pair (10) has nonzero coefficient. For example the element in $A_2(x_0, x_1)$

$$f = f_0 + f_1 x_0 + f_2 x_1 + f_3 x_0 x_1$$

can be written as

$$f = f_0(1 + f_0 f_1 x_0) + f_2(1 + f_2 f_3 x_0)x_1.$$

If $f_0, f_1, f_2, f_3 = \pm 1$, then $f_0 f_1, f_2 f_0 = \pm 1$.

5.   Reformulation of Classical PONS

We reformulate the classical PONS construction procedure as described in Section 2.2 using group algebra operations, and we distinguish these orthonormal bases within the group algebra framework. For the purpose of this work we will modify P_{2^N}, the $2^N \times 2^N$ PONS matrix, by row permutation and row multiplication by -1.

Specifically we explore certain relationships between PONS matrices and character basis matrices. In particular, we show that a PONS matrix is completely determined by its 0-th row and the equivalent size character basis matrix. For a classical PONS matrix, the 0-th row can be constructed arithmetically and provides an example of a splitting sequence. The concept of a splitting sequence will be developed in a group algebra framework in the next section.

By the original PONS construction as given in Section 2.2 the 4×4 PONS matrix is

$$\begin{bmatrix} 1 & 1 & 1 & -1 \\ 1 & 1 & -1 & 1 \\ 1 & -1 & 1 & 1 \\ 1 & -1 & -1 & -1 \end{bmatrix}.$$

Interchanging the 1st and 2nd row we have the matrix

$$P_4 = \begin{bmatrix} 1 & 1 & 1 & -1 \\ 1 & -1 & 1 & 1 \\ 1 & 1 & -1 & 1 \\ 1 & -1 & -1 & -1 \end{bmatrix}.$$

The component-wise product of any two rows of P_4 is a row of the character basis matrix H_4. In fact

$$P_4 = H_4 D_4,$$

where D_4 is the diagonal matrix formed by the 0-th row of P_4.

In general, by row permutation the classical $2^N \times 2^N$ PONS matrix can be transformed into the matrix

$$P_{2^N} = H_{2^N} D_{2^N},$$

where D_{2^N} is the diagonal matrix formed by the 0-th row of the classical PONS matrix.

The 0-th row of the original 4×4 PONS matrix can be defined arithmetically. From the binary representation of the integers $0 \leq n < 4$

$$00, \ 01, \ 10, \ 11$$

we see that -1 is placed in the position having two 1's. Once D_{2^N} is constructed independently of the original construction, we can define the $2^N \times 2^N$ PONS matrix as

$$P_{2^N} = H_{2^N} D_{2^N}.$$

In the following sections we will place this result in the group algebra setting.

Several other definitions are possible. For example

$$D_{2^N} H_{2^N} D_{2^N}$$

can be viewed as a *symmetrical form* of the $2^N \times 2^N$ PONS matrix.

6. Group Algebra of Classical PONS

For an integer $N > 0$, we will use group algebra operations to define a PONS sequence of size 2^N. The first 2^n, $n \leq N$, terms of this sequence are the same as the elements in the 0-th row of the classical $2^n \times 2^n$ PONS matrix. PONS sequences will be identified with elements in the group algebra having special properties. Below we will extend these results to give a group algebra characterization of general binary splitting sequences.

Denote the elements of the canonical basis of $A_2(x_0, \ldots, x_{n-1})$ by v_0, $v_1, \ldots, v_{2^n-1}$. For $\mathbf{a} \in \mathbf{C}^{2^n}$

$$\mathbf{a} = (a_0, a_1, \ldots, a_{2^n-1})$$

identify $\mathbf{a}$ with the element a in $A_2(x_0, \ldots, x_{n-1})$

$$a = \sum_{r=0}^{2^n-1} a_r v_r.$$

If $n = 4$,

$$a = a_0 1 + a_1 x_0 + a_2 x_1 + a_3 x_0 x_1.$$

Set

$$\alpha_2 = 1 + x_0, \quad \alpha_2^* = 1 - x_0$$

and

$$\alpha_4 = \alpha_2 + \alpha_2^* x_1, \quad \alpha_4^* = \alpha_2 - \alpha_2^* x_1.$$

α_4 corresponds to the sequence

$$1 \quad 1 \quad 1 \quad -1$$

which is the 0-th row of P_4.

Set
$$\alpha_{2^n} = \alpha_{2^{n-1}} + \alpha_{2^{n-1}}^* x_{n-1}$$
and
$$\alpha_{2^n}^* = \alpha_{2^{n-1}} - \alpha_{2^{n-1}}^* x_{n-1}.$$

The sequence corresponding to α_{2^n} is the 0-th row of P_{2^n}.

We will study the group algebra properties of the elements α_{2^n} and $\alpha_{2^n}^*$. The key to understanding the reason for expansions over the translate-character basis is contained in the character product formula

$$\alpha_2^2 = 2(1 + x_0), \quad (\alpha_2^*)^2 = 2(1 - x_0) \tag{11}$$

$$\alpha_2 \alpha_2^* = 0. \tag{12}$$

Implications of these formulas will be seen throughout this work.

Since
$$\alpha_4^2 = \alpha_2^2 + 2\alpha_2 \alpha_2^* x_1 + (\alpha_2^*)^2,$$

we have by (9) and (12)

$$\alpha_4^2 = 2(1 + x_0) + 2(1 - x_0) = 4. \tag{13}$$

In the same way
$$(\alpha_4^*)^2 = 4. \tag{14}$$

Since
$$\alpha_4 \alpha_4^* = \alpha_2^2 - (\alpha_2^*)^2,$$

by (9)

$$\alpha_4 \alpha_4^* = 2(1 + x_0) - 2(1 - x_0) = 4x_0. \tag{15}$$

We can write the important factorization

$$\alpha_4^* = \alpha_4 x_0. \tag{16}$$

By (16)
$$\alpha_8 = \alpha_4 + \alpha_4^* x_2 = \alpha_4(1 + x_0 x_2) \tag{17}$$

and
$$\alpha_8^* = \alpha_4(1 - x_0 x_2). \tag{18}$$

Since
$$(1 + x_0 x_2)^2 = 2(1 + x_0 x_2), \quad (1 - x_0 x_2)^2 = 2(1 - x_0 x_2),$$

and
$$(1 + x_0 x_2)(1 - x_0 x_2) = 0,$$

we have

$$\alpha_8^2 = 8(1 + x_0 x_2), \quad (\alpha_8^*)^2 = 8(1 - x_0 x_2),$$

and

$$\alpha_8 \alpha_8^* = 0.$$

α_4 and α_8 have very different group algebra properties reflecting the expansion of α_4

$$\alpha_4 = 1 + x_0 + (1 - x_0)x_1$$

in terms of the *conjugate* characters $1 + x_0$ and $1 - x_0$ and the expansion of α_8

$$\alpha_8 = \alpha_4(1 + x_0 x_2)$$

in which the character $1 + x_0 x_2$ is a factor. This result is general for α_{2^n} depending on whether n is even or odd.

Arguing as above

$$\alpha_{16} = \alpha_8 + \alpha_8^* x_3 = \alpha_4 \left[(1 + x_0 x_2) + (1 - x_0 x_2)x_3 \right],$$

$$\alpha_{16}^* = \alpha_8 - \alpha_8^* x_3 = \alpha_4 \left[(1 + x_0 x_2) - (1 - x_0 x_2)x_3 \right]$$

from which we have

$$\alpha_{16}^2 = (\alpha_{16}^*)^2 = 16,$$

$$\alpha_{16} \alpha_{16}^* = 16 x_0 x_2,$$

$$\alpha_{16}^* = \alpha_{16} x_0 x_2.$$

The same arguments show

$$\alpha_{32} = \alpha_{16} + \alpha_{16}^* x_4 = \alpha_{16}(1 + x_0 x_2 x_4)$$

$$\alpha_{32}^* = \alpha_{16}(1 - x_0 x_2 x_4).$$

In general if n is odd

$$\alpha_{2^n} = \alpha_{2^{n-1}}(1 + x_0 x_2 \cdots x_{n-1}),$$

$$\alpha_{2^n}^* = \alpha_{2^{n-1}}(1 - x_0 x_2 \cdots x_{n-1})$$

and

$$\alpha_{2^n}^2 = 2^n(1 + x_0 x_2 \cdots x_{n-1}), \quad (\alpha_{2^n}^*)^2 = 2^n(1 - x_0 x_2 \cdots x_{n-1})$$

$$\alpha_{2^n} \alpha_{2^n}^* = 0.$$

If n is even

$$\alpha_{2^n} = \alpha_{2^{n-2}} \left[(1 + x_0 x_2 \cdots x_{n-2}) + (1 - x_0 x_2 \cdots x_{n-2})x_{n-1} \right],$$

$$\alpha_{2n}^* = \alpha_{2n-2}\left[(1 + x_0 x_2 \cdots x_{n-2}) - (1 - x_0 x_2 \cdots x_{n-2})x_{n-1}\right]$$

and

$$\alpha_{2n}^2 = (\alpha_{2n}^*)^2 = 2^n,$$

$$\alpha_{2n}\alpha_{2n}^* = 2^n x_0 x_2 \cdots x_{n-2}$$

$$\alpha_{2n}^* = \alpha_{2n} x_0 x_2 \cdots x_{n-2}.$$

An important implication of these formulas is that if n is odd, α_{2n} is not invertible in the group algebra while if n is even, α_{2n} is invertible with inverse $2^{-n}\alpha_{2n}$.

7.   Group Algebra Convolution

In this section we relate *convolution* in $A_2(x_0, \ldots, x_{N-1})$ by the PONS element α_{2N} with the PONS matrix P_{2N}.

Consider

$$\alpha \in A_2(x_0, \ldots, x_{N-1}).$$

The mapping $C_{2N}(\alpha) : A_2(x_0, \ldots, x_{N-1}) \longrightarrow A_2(x_0, \ldots, x_{N-1})$ defined by

$$C_{2N}(\alpha)\beta = \alpha\beta, \quad \beta \in A_2(x_0, \ldots, x_{N-1})$$

is a linear mapping of $A_2(x_0, \ldots, x_{N-1})$ called *convolution* by α. The matrix of $C_{2N}(\alpha)$ relative to the canonical basis is

$$C_{2N}(\alpha) = \left[\alpha(y^{-1}x)\right]_{x,y \in C_2(x_0, \ldots, x_{N-1})}.$$

For

$$\alpha_4 = 1 + x_0 + x_1 - x_0 x_1$$

the matrix $C_4(\alpha_4)$ can be formed from the products

$$\begin{aligned}
x_0 \alpha_4 &= 1 + x_0 - x_1 + x_0 x_1 \\
x_1 \alpha_4 &= 1 - x_0 + x_1 + x_0 x_1 \\
x_0 x_1 \alpha_4 &= -1 + x_0 + x_1 + x_0 x_1.
\end{aligned}$$

$$C_4(\alpha_4) = \begin{bmatrix} 1 & 1 & 1 & -1 \\ 1 & 1 & -1 & 1 \\ 1 & -1 & 1 & 1 \\ -1 & 1 & 1 & 1 \end{bmatrix}.$$

Note that, as described in Section 2.5, the above is the 4×4 *symmetric* PONS matrix.

We have defined the 4×4 PONS matrix as

$$P_4 = \begin{bmatrix} 1 & 1 & 1 & -1 \\ 1 & -1 & 1 & 1 \\ 1 & 1 & -1 & 1 \\ 1 & -1 & -1 & -1 \end{bmatrix}.$$

Convolution by α_4 and P_4 are related by

$$C_4(\alpha_4) = D_4 P(4,2) P_4, \tag{19}$$

where D_4 is the diagonal matrix formed by the 0-th row of P_4 and $P(4,2)$ is the 4×4 stride by 2 permutation matrix

$$P(4,2) = \begin{bmatrix} 1 & 0 & 0 & 0 \\ 0 & 0 & 1 & 0 \\ 0 & 1 & 0 & 0 \\ 0 & 0 & 0 & 1 \end{bmatrix}.$$

Since $P_4 = H_4 D_4$,

$$C_4(\alpha_4) = D_4 P(4,2) H_4 D_4.$$

In group algebra terminology, $P(4,2)$ is the matrix relative to the canonical basis of the automorphism of $A_2(x_0, x_1)$ defined by the group automorphism of $C_2(x_0, x_1)$

$$x_0 \longrightarrow x_1, \quad x_1 \longrightarrow x_0.$$

By (19), up to row permutation and multiplication by -1, convolution by the classical PONS element α_4 in $A_2(x_0, x_1)$ coincides with the action of the 4×4 PONS matrix P_4. This will be the case whenever N is even and 2^N is the length of the classical PONS element. The length 16 case will be considered below.

For N odd, since

$$\alpha_8^2 = 8(1 + x_0 x_2)$$

and α_8 is not an invertible element in $A_2(x_0, x_1, x_2)$, convolution by α_8 in $A_2(x_0, x_1, x_2)$ cannot coincide with P_8 even after row permutation and multiplication by -1.

Set

$$J_2 = \begin{bmatrix} 0 & 1 \\ 1 & 0 \end{bmatrix}, \quad J_{2^N} = I_N \otimes J_2.$$

Since

$$\alpha_8 = \alpha_4(1 + x_0 x_2)$$

and

$$C_8(1 + x_0 x_2) = \begin{bmatrix} I_4 & J_4 \\ J_4 & I_4 \end{bmatrix},$$

we have

$$C_8(\alpha_8) = \begin{bmatrix} I_4 & J_4 \\ J_4 & I_4 \end{bmatrix} (I_2 \otimes C_4(\alpha_4)).$$

By (19)

$$C_8(\alpha_8) = \begin{bmatrix} D_4 & J_4 D_4 \\ J_4 D_4 & D_4 \end{bmatrix} (I_2 \otimes P(4,2)) (I_2 \otimes P_4). \tag{20}$$

$$P_8 = H_8 D_8,$$

where

$$H_8 = (H_2 \otimes I_4)(I_2 \otimes H_4)$$

and

$$D_8 = (D_4 \otimes I_2)(I_2 \otimes D_4).$$

By (19)

$$P_8 = (H_2 \otimes I_4) \left((I_2 \otimes H_4)(D_4 \otimes I_2)(I_2 \otimes H_4^{-1}) \right) (I_2 \otimes P_4).$$

Direct computation shows

$$(I_2 \otimes H_4)(D_4 \otimes I_2)(I_2 \otimes H_4^{-1}) = (I_2 \otimes P(4,2))(I_4 \oplus J_4)(I_2 \otimes P(4,2)),$$

where $\oplus$ is the matrix direct sum.

These results show that

$$C_8(\alpha_8) P_8^{-1} = \begin{bmatrix} D_4 & J_4 D_4 \\ J_4 D_4 & D_4 \end{bmatrix} (I_4 \oplus J_4)(H_2^{-1} \otimes I_4)(I_2 \otimes P(4,2)).$$

Since

$$J_4 D_4 J_4 = D_4^* = \begin{bmatrix} 1 & & & \\ & 1 & & \\ & & -1 & \\ & & & 1 \end{bmatrix},$$

we have the main result relating $C_8(\alpha_8)$ and P_8

$$C_8(\alpha_8) = \frac{1}{2} \begin{bmatrix} D_4 + D_4^* & D_4 - D_4^* \\ (D_4 + D_4^*)J_4 & (D_4^* - D_4)J_4 \end{bmatrix} (I_2 \otimes P(4,2)) P_8.$$

A direct computation shows

$$C_8(\alpha_8) = D_8 Q_8 (I_2 \otimes P(4,2)) P_8,$$

where $D_8 = D_4 \oplus D_4^*$ and

$$
Q_8 = \begin{bmatrix}
1 & 0 & 0 & 0 & 0 & 0 & 0 & 0 \\
0 & 1 & 0 & 0 & 0 & 0 & 0 & 0 \\
0 & 0 & 0 & 0 & 0 & 0 & 1 & 0 \\
0 & 0 & 0 & 0 & 0 & 0 & 0 & 1 \\
0 & 1 & 0 & 0 & 0 & 0 & 0 & 0 \\
1 & 0 & 0 & 0 & 0 & 0 & 0 & 0 \\
0 & 0 & 0 & 0 & 0 & 0 & 0 & 1 \\
0 & 0 & 0 & 0 & 0 & 0 & 1 & 0
\end{bmatrix}.
$$

Q_8 is a singular matrix. In fact if

$$
\mathbf{w} = P_8 \mathbf{v},
$$

then $\mathcal{C}_8(\alpha_8)\mathbf{v}$ computes only the components

$$
w_0, \ w_2, \ w_5, \ w_7.
$$

The computation of the remaining components is given by $\mathcal{C}_8(\alpha_8^*)\mathbf{v}$.
 Since

$$
\alpha_{16} = \alpha_4 \left[(1 + x_0 x_2) + (1 - x_0 x_2) x_3 \right],
$$

convolution by α_{16} in $A_2(x_0, x_1, x_2, x_3)$ can be written as

$$
\mathcal{C}_{16}(\alpha_{16}) = \begin{bmatrix}
\mathcal{C}_8(1 + x_0 x_2) & \mathcal{C}_8(1 - x_0 x_2) \\
\mathcal{C}_8(1 - x_0 x_2) & \mathcal{C}_8(1 + x_0 x_2)
\end{bmatrix} \mathcal{C}_{16}(\alpha_4).
$$

Arguing as before

$$
\mathcal{C}_{16}(\alpha_{16}) = \begin{bmatrix}
X_8 & X_8^* \\
X_8^* & X_8
\end{bmatrix} (I_4 \otimes D_4 P(4,2))(I_4 \otimes P_4),
$$

where

$$
X_8 = \begin{bmatrix} I_4 & J_4 \\ J_4 & I_4 \end{bmatrix}, \quad
X_8^* = \begin{bmatrix} I_4 & -J_4 \\ -J_4 & I_4 \end{bmatrix}.
$$

$$
P_{16} = H_{16} D_{16},
$$

where

$$
H_{16} = (H_4 \otimes I_4)(I_4 \otimes H_4)
$$

and

$$
D_{16} = (D_4 \otimes I_4)(I_2 \otimes D_4 \otimes I_2)(I_4 \otimes D_4).
$$

We can now write

$$
P_{16} = (P_4 \otimes I_4)(I_4 \otimes H_4)(I_2 \otimes D_4 \otimes I_2)(I_4 \otimes H_4^{-1})(I_4 \otimes P_4).
$$

By direct computation

$$(I_4 \otimes H_4)(I_2 \otimes D_4 \otimes I_2)(I_4 \otimes H_4^{-1})$$

$$= (I_4 \otimes P(4,2))(I_4 \oplus J_4 \oplus I_4 \oplus J_4)(I_4 \otimes P(4,2)).$$

Combining the preceding formulas

$$\mathcal{C}_{16}(\alpha_{16})P_{16}^{-1}$$

$$= \begin{bmatrix} X_8 & X_8^* \\ X_8^* & X_8 \end{bmatrix} (I_4 \otimes D_4)(I_4 \oplus J_4 \oplus I_4 \oplus J_4)(P_4^{-1} \otimes I_4)(I_4 \otimes P(4,2)).$$

Since

$$P_4^{-1} \otimes I_4 = \begin{bmatrix} I_4 & I_4 & I_4 & I_4 \\ I_4 & -I_4 & I_4 & -I_4 \\ I_4 & I_4 & -I_4 & -I_4 \\ -I_4 & I_4 & I_4 & -I_4 \end{bmatrix}$$

and

$$J_4 D_4 J_4 = D_4^*,$$

we can derive the main result relating $\mathcal{C}_{16}(\alpha_{16})$ and P_{16}.

$$\mathcal{C}_{16}(\alpha_{16}) = Y_{16}(I_4 \otimes P(4,2))P_{16},$$

where

$$Y_{16} = \frac{1}{2} \begin{bmatrix} D_4 + D_4^* & D_4 - D_4^* & 0 & 0 \\ 0 & 0 & (D_4 + D_4^*)J_4 & (D_4^* - D_4)J_4 \\ D_4 - D_4^* & D_4 + D_4^* & 0 & 0 \\ 0 & 0 & (D_4 - D_4^*)J_4 & -(D_4 + D_4^*)J_4 \end{bmatrix}.$$

Direct computation shows that

$$Y_{16} = D_{16}Q_{16},$$

where Q_{16} is the permutation matrix

$$Q_{16} = \begin{bmatrix}
1 & 0 & 0 & 0 & 0 & 0 & 0 & 0 & & & & & & & & \\
0 & 1 & 0 & 0 & 0 & 0 & 0 & 0 & & & & & & & & \\
0 & 0 & 0 & 0 & 0 & 0 & 1 & 0 & & & 0 & & & & 0 & \\
0 & 0 & 0 & 0 & 0 & 0 & 0 & 1 & & & & & & & & \\
 & & & & & & & & 0 & 1 & 0 & 0 & 0 & 0 & 0 & 0 \\
 & & 0 & & & & 0 & & 1 & 0 & 0 & 0 & 0 & 0 & 0 & 0 \\
 & & & & & & & & 0 & 0 & 0 & 0 & 0 & 0 & 0 & 1 \\
 & & & & & & & & 0 & 0 & 0 & 0 & 0 & 0 & 1 & 0 \\
0 & 0 & 0 & 0 & 1 & 0 & 0 & 0 & & & & & & & & \\
0 & 0 & 0 & 0 & 0 & 1 & 0 & 0 & & & 0 & & & & 0 & \\
0 & 0 & 1 & 0 & 0 & 0 & 0 & 0 & & & & & & & & \\
0 & 0 & 0 & 1 & 0 & 0 & 0 & 0 & & & & & & & & \\
 & & & & & & & & 0 & 0 & 0 & 0 & 0 & 1 & 0 & 0 \\
 & & 0 & & & & 0 & & 0 & 0 & 0 & 0 & 1 & 0 & 0 & 0 \\
 & & & & & & & & 0 & 0 & 0 & 1 & 0 & 0 & 0 & 0 \\
 & & & & & & & & 0 & 0 & 1 & 0 & 0 & 0 & 0 & 0
\end{bmatrix}.$$

From these formulas

$$\mathcal{C}_{16}(\alpha_{16}) = D_{16}Q_{16}(I_4 \otimes P(4,2))P_{16}.$$

The permutation matrix Q_{16} corresponds to the automorphism of the algebra $A_2(x_0, x_1, x_2, x_3)$ induced by the group automorphism defined by

$$x_0 \longrightarrow x_0$$
$$x_1 \longrightarrow x_1 x_3$$
$$x_2 \longrightarrow x_3$$
$$x_3 \longrightarrow x_0 x_2.$$

Combined with the automorphism defining $(I_4 \otimes P(4,2))$

$$x_0 \longrightarrow x_1$$
$$x_1 \longrightarrow x_0$$
$$x_2 \longrightarrow x_2$$
$$x_3 \longrightarrow x_3$$

the permutation matrix $Q_{16}(I_4 \otimes P(4,2))$ corresponds to the automorphism of the algebra defined by

$$x_0 \longrightarrow x_1 x_3$$
$$x_1 \longrightarrow x_0$$
$$x_2 \longrightarrow x_3$$
$$x_3 \longrightarrow x_0 x_2.$$

8.   Splitting Sequences

The concept of a splitting sequence was discussed in Sections 2.3 and 2.4. In this section we will describe the binary splitting sequences using group algebra operations. We begin by establishing notation.

The characters of $C_2(x_0)$ are

$$1 + x_0, \quad 1 - x_0. \tag{21}$$

The expressions

$$\pm(1 + x_0), \quad \pm(1 - x_0)$$

will be denoted by λ with or without subscripts. Typically

$$\lambda = a(1 + \epsilon x_0), \tag{22}$$

where $a = \pm 1$ and $\epsilon = \pm 1$. a is called the *coefficient* of λ and ϵ is called the *sign* of λ. In general

$$\lambda_1 \lambda_2 = 0$$

if and only if $sign(\lambda_1) = -sign(\lambda_2)$. λ is called a *directed character* of $C_2(x_0)$.

Identify sequences of length 2^N with linear combinations over the canonical basis of $A_2(x_0, \ldots, x_{N-1})$. A sequence of length 4

$$b_0, \, b_1, \, b_2, \, b_3$$

is identified with the element

$$\alpha = b_0 + b_1 x_0 + b_2 x_1 + b_3 x_0 x_1$$

in $A_2(x_0, x_1)$.

Only sequences of ± 1 will be considered. Every sequence of length 2^N of ± 1 uniquely determines a sequence of directed characters

$$\lambda_0, \, \lambda_1, \, \ldots, \, \lambda_{N-1}$$

and consequently a sequence of signs, the n-th sign equal to the sign of λ_n, and a sequence of coefficients, the n-th coefficient equal to the coefficient of λ_n. If

$$\alpha = -(1 + x_0) + (1 - x_0)x_1,$$

then the corresponding sequence of directed characters is

$$-(1 + x_0), \quad 1 - x_0,$$

the corresponding sequence of signs is

$$+ \; -$$

and the corresponding sequence of coefficients is

$$- \; + \, .$$

The splitting condition on sequences of ± 1 places conditions on the sign patterns of splitting sequences. The following tables describe the sign patterns for $N = 2$, 3 and 4.

Table 1. Sign patterns for splitting sequences

$N = 2$	$- +$	$+ -$		
$N = 3$	$- + - +$	$- + + -$	$+ - + -$	$+ - - +$
$N = 4$	$- + - + - + - +$	$- + - + + - + -$	$- + + - - + + -$	
	$+ - + - + - + -$	$+ - + - - + - +$	$+ - - + + - - +$	

Consider a length 4 splitting sequence α_4 and write

$$\alpha_4 = \lambda_0 + \lambda_1 x_1,$$

where λ_0, λ_1 are directed characters. The length 4 splitting condition implies

$$\lambda_0^2 + \lambda_1^2 = 1.$$

By the character product formula

$$\lambda_0 \lambda_1 = 0. \tag{23}$$

Condition (23) is also sufficient for splitting, proving the following result.

THEOREM 20 *α_4 is a splitting sequence of length 4 if and only if α_4 has the form*

$$\alpha_4 = \pm((1 + x_0) \pm (1 - x_0)x_1)$$

or

$$\alpha_4 = \pm((1 - x_0) \pm (1 + x_0)x_1).$$

The splitting sequences of length 4 having sign pattern

$$+ \; -$$

are given by

$$\alpha_4 = \pm((1 + x_0) \pm (1 - x_0)x_1)$$

while those having sign pattern

$$- +$$

are given by

$$\alpha_4 = \pm((1 - x_0) \pm (1 + x_0)x_1).$$

COROLLARY 21 *If α_4 is a splitting sequence of length 4, then $\alpha_4^2 = 4$.*

The condition $\alpha_4^2 = 4$ is also sufficient for a sequence of ± 1 of length 4 to be a splitting sequence.

Consider a splitting sequence α_8 of length 8 given by the directed characters

$$\lambda_0, \ \lambda_1, \ \lambda_2, \ \lambda_3. \tag{24}$$

α_8 can be written

$$\alpha_8 = \alpha_4 + \alpha_4^* x_2,$$

where

$$\alpha_4 = \lambda_0 + \lambda_1 x_1, \quad \alpha_4^* = \lambda_2 + \lambda_3 x_1,$$

are splitting sequences of length 4. The length 4 splitting condition implies

$$\alpha_4^2 + (\alpha_4^*)^2 = 8.$$

There are 4 cases to consider. Suppose α_4 and α_4^* have sign pattern

$$- + .$$

The length 8 splitting condition implies that the coefficients of the directed characters in (24) satisfy

$$-a_0 a_1 - a_2 a_3 = 0.$$

Since

$$\alpha_4 \alpha_4^* = 2 a_0 a_2 (1 - x_0) + 2 a_1 a_3 (1 + x_0),$$

we have

$$\alpha_4 \alpha_4^* = \pm 4 x_0$$

and

$$\alpha_8 = \alpha_4 (1 \pm x_0 x_2). \tag{25}$$

The same argument shows that if α_4 and α_4^* have the sign pattern

$$+ -,$$

then α_8 has the same form.

Suppose α_4 has the sign pattern

$$- +$$

and α_4^* has the sign pattern

$$+ - .$$

The splitting condition implies

$$-a_0 a_1 + a_2 a_3 = 0.$$

Since

$$\alpha_4 \alpha_4^* = (2a_0 a_3 (1 - x_0) + 2a_1 a_2 (1 + x_0)) x_1,$$

we have

$$\alpha_4 \alpha_4^* = \pm 4 x_1$$

and

$$\alpha_8 = \alpha_4 (1 \pm x_1 x_2). \tag{26}$$

The same result holds if we reverse the patterns of α_4 and α_4^*.

Since (25) and (26) define splitting sequences whenever α_4 is a splitting sequence of length 4, we have the following result.

THEOREM 22 *α_8 is a splitting sequence of length 8 if and only if α_8 has one of the two forms*

$$\alpha_8 = \alpha_4 (1 \pm x_0 x_2)$$

or

$$\alpha_8 = \alpha_4 (1 \pm x_1 x_2),$$

where α_4 is an arbitrary splitting sequence of length 4.

As the proof shows, the splitting sequences of length 8 having sign pattern

$$- + - +, \quad + - + -$$

are given by

$$\alpha_4 (1 \pm x_0 x_2)$$

while those having sign patterns

$$- + + -, \quad + - - +$$

are given by

$$\alpha_4 (1 \pm x_1 x_2).$$

COROLLARY 23 *If α_8 is a splitting sequence of length 8, then*

$$\alpha_8^2 = 8(1 \pm x_0 x_2)$$

or

$$\alpha_8^2 = 8(1 \pm x_1 x_3).$$

Consider a splitting sequence α_{16} of length 16 and write

$$\alpha_{16} = \alpha_8 + \alpha_8^* x_3,$$

where α_8 and α_8^* are length 8 splitting sequences. The splitting condition on α_{16} implies

$$\alpha_8^2 + (\alpha_8^*)^2 = 16.$$

By Theorem 22

$$\alpha_8 \alpha_8^* = 0. \tag{27}$$

Suppose α_8 has sign pattern

$$+ \; - \; + \; - \; .$$

By Theorem 22 α_8 has the form

$$\alpha_8 = \alpha_4(1 \pm x_0 x_2),$$

where α_4 is a length 4 splitting sequence having sign pattern

$$+ \; - \; .$$

Consider the case

$$\alpha_8 = \alpha_4(1 + x_0 x_2).$$

If α_4 is given by the directed characters

$$\lambda_0 \, \lambda_1,$$

then α_8 is given by the directed characters

$$\lambda_0 \, \lambda_1 \, \lambda_0 \; - \lambda_1. \tag{28}$$

By (27)

$$\alpha_8^* = \alpha_4^*(1 - x_0 x_2).$$

There are two cases. Suppose α_4^* has sign pattern

$$+ -$$

and is given by the directed characters

$$\lambda_2 \, \lambda_3.$$

α_8^* is given by the directed characters

$$\lambda_2 \, \lambda_3 \, - \lambda_2 \, \lambda_3. \tag{29}$$

The splitting condition on α_{16} applied to (28) and (29) implies

$$-a_0 a_1 + a_2 a_3 = 0.$$

Since

$$\alpha_4 \alpha_4^* = 2a_0 a_2 (1 + x_0) + 2a_1 a_3 (1 - x_0),$$

we have

$$\alpha_4 \alpha_4^* = \pm 4$$

and

$$\alpha_{16} = \alpha_4 [(1 + x_0 x_2) \pm (1 - x_0 x_2) x_3].$$

Suppose α_4^* has sign pattern

$$- \, + \, .$$

α_8^* is given by the directed characters

$$\lambda_2 \, \lambda_3 \, \lambda_2 \, - \lambda_3 \tag{30}$$

and has sign pattern

$$- \, + \, - \, + \, .$$

The splitting condition on α_{16} applied to (28) and (30) implies

$$-a_0 a_1 + a_2 a_3 = 0.$$

Since

$$\alpha_4 \alpha_4^* = 2a_0 a_3 (1 + x_0) x_1 + 2a_1 a_2 (1 - x_0) x_1,$$

we have

$$\alpha_4 \alpha_4^* = \pm 4 x_1$$

and

$$\alpha_{16} = \alpha_4 [(1 + x_0 x_2) \pm (1 - x_0 x_2) x_1 x_3].$$

Similar arguments prove the following result.

THEOREM 24 α_{16} *is a splitting sequence of length 16 if and only if* α_{16} *has one of the following forms.*

$$\text{Type 1.} \qquad \alpha_4[(1 + x_0x_2) \pm (1 - x_0x_2)x_3]$$
$$\alpha_4[(1 - x_0x_2) \pm (1 + x_0x_2)x_3]$$
$$\text{Type 2.} \qquad \alpha_4[(1 + x_0x_2) \pm (1 - x_0x_2)x_1x_3]$$
$$\alpha_4[(1 - x_0x_2) \pm (1 + x_0x_2)x_1x_3]$$
$$\text{Type 3.} \qquad \alpha_4[(1 + x_1x_2) \pm (1 - x_1x_2)x_3]$$
$$\alpha_4[(1 - x_1x_2) \pm (1 + x_1x_2)x_3]$$
$$\alpha_4[(1 + x_1x_2) + (1 - x_1x_2)x_0x_3]$$
$$\alpha_4[(1 - x_1x_2) + (1 + x_1x_2)x_0x_3],$$

where α_4 *is a length 4 splitting sequence.*

The length 16 splitting sequences in Theorem 24 have been organized according to their sign patterns.

$$
\begin{array}{lll}
\text{Type 1} & - + - + - + - + \\
& + - + - + - + - \\
\text{Type 2} & - + - + + - + - \\
& + - + - - + - + \\
\text{Type 3} & - + + - - + + - \\
& + - - + + - - +.
\end{array}
$$

COROLLARY 25 *If* α_{16} *is a length 16 splitting sequence, then* $\alpha_{16}^2 = 16$.

The condition
$$\alpha_{16}^2 = 16$$
is not sufficient for α_{16} to be a splitting sequence.

A similar argument proves the following result.

THEOREM 26 *The splitting sequences of length 32 have one of the following forms.*

Type 1 $\alpha_{16}(1 \pm x_0x_2x_4)$, *where* α_{16} *is Type 1.*

Type 2 $\alpha_{16}(1 \pm x_1x_3x_4)$, *where* α_{16} *is Type 1.*

Type 3 $\alpha_{16}(1 \pm x_0x_2x_4)$, $\alpha_{16}(1 \pm x_1x_3x_4)$, *where* α_{16} *is Type 2.*

Type 4 $\alpha_{16}(1 \pm x_1x_2x_4)$, $\alpha_{16}(1 \pm x_0x_3x_4)$, *where* α_{16} *is Type 3.*

The classification of length 32 splitting sequences into types corresponds to the sign patterns of the type.

Type 1	$- + - + - + - + - + - + - + - +$
	$+ - + - + - + - + - + - + - + -$
Type 2	$- + - + - + - + + - + - + - + -$
	$+ - + - + - + - - + - + - + - +$
Type 3	$- + - + + - + - - + - + + - + -$
	$+ - + - - + - + + - + - - + - +$
Type 4	$+ - - + + - - + + - - + + - - +$
	$- + + - - + + - - + + - - + + -$

9. Historical Appendix on PONS

What we have been calling PONS was actually first introduced by G. R. Welti [28], and subsequently rediscovered on several occasions by independent authors. However closely related (and most important) work had already been done, independently by Golay [13, 14] in 1949–1951 and by H. S. Shapiro [25] in 1951.

A remarkable feature of these various rediscoveries is that all of them emerged in *entirely different* contexts, radars, *etc.* In this short appendix, we briefly comment on some of these works, leaving out many aspects of this history which would have required a detailed study and considerable space to present it. For a much more complete study of this history, see [24].

In his 1949–1951 papers [13] and [14], M. J. E. Golay introduced the general concept of "complementary pairs" of finite sequences all of whose entries are ±1. This was motivated by a highly non-trivial application to *infra-red spectrometry*. Neither Golay nor any of his fellow engineers ever used the language and properties of generating polynomials, until the 1980's.

In 1951, H. S. Shapiro [25] introduced what became known, after 1963, as the "Rudin-Shapiro" polynomial pairs. Shapiro's work was entirely in pure mathematics (specifically, complex and Fourier analysis). The use of Rudin's name in "Rudin-Shapiro" polynomials is utterly unjustified, and seems due to an unfortunate "original mistake" in the 1963 book [17] by J.-P. Kahane and R. Salem.

In 1959 G. R. Welti wrote the paper [28] which appeared in print in 1960. In spite of its title, of particular interest for our present purpose is the *first half* of the paper (on binary sequences), in which Welti introduced *exactly* what J. S. Byrnes and others called "PONS matrices" in the 1990's. (However, the approach and motivation of Byrnes was completely different from those of Welti). Welti was unaware of the works of Golay and Shapiro. He obtained the first row of Welti's matrix (*i.e.*, the Shapiro sequence) by a method entirely different from that of Shapiro, and he obtained the remaining rows as Hadamard products of the first row with the rows of the Walsh matrix.

Also in 1959, W. Rudin (who had been a member of Shapiro's MIT thesis committee!) wrote a paper [23] in which he claimed to have "rediscovered" the Shapiro polynomial pairs. This was *not* a rediscovery. As we said earlier, full details of this matter will be given in [24].

In 1961, Golay [15], who in turn was unaware of the works of Shapiro and Welti (and even that of Rudin), obtained all the Welti rows (*i.e.*, the PONS rows), by a method quite close to that of Shapiro.

In 1981, Mendes France and Tenenbaum [12] (who had never seen Shapiro's 1951 work [25] but had heard of it only via Rudin's "rediscovery" [23], and also were unaware of the works of Golay and Welti), rediscovered all the Welti rows and named them "paper-folding sequences". This was a work in pure mathematics, related to fractal dimensions of plane curves.

In the early 1990's, J. S. Byrnes (who was aware of the works of Shapiro [25] and Rudin [23] but not of those of Golay, Welti, and Mendes France & Tenenbaum) rediscovered [6] the Welti matrices (which he later on named "PONS matrices"), in yet another context of pure mathematics: His motivation was to use them as a tool to prove an "uncertainty principle conjecture" of H. S. Shapiro. However, unlike the previous instances of discovery/rediscovery of these objects in a *pure mathematics* context, PONS soon became a tool for radar signal processing as well.

These various discoveries and re-discoveries have given rise to an enormous amount of further research and new open problems, many of which are deep. The ongoing research on such subjects is very active. We hope to return to these matters elsewhere.

We finish this appendix with two remarks on *this paper*. The first remark is that we left out the subject of *correlation properties* of PONS sequences. The reason is that these correlation properties are sometimes very good and sometimes very bad (and proving how bad they can be is itself a difficult task). We will return to these correlation matters elsewhere. The second remark is that the heterogeneous aspect of this paper is due to the fact that it was written over several years by several authors with different views. (According to some historians, the Old Testament was written by various authors who were sometimes several centuries apart!)

References

[1] K.G. Beauchamp. *Applications of Walsh and Related Functions.* Academic Press, London, 1984.

[2] J. Brillhart and L. Carlitz. Note on the shapiro polynomials. *Proc. AMS,* 25:114–118, 1970.

[3] S.Z. Budisin. New complementary pairs of sequences. *Electronics Letters,* 26(13):881–883, 21 June 1990.

[4] S.Z. Budisin. Efficient pulse compressor for Golay complementary sequences. *Electronics Letters,* 27(3):219–220, 31 January 1991.

[5] S.Z. Budisin, B.M. Popović, and L.M. Indjin. Designing radar signals using complementary sequences. *Proc. IEE Conf. RADAR 87,* pages 593–597, Oct. 1987.

[6] J.S. Byrnes. Quadrature mirror filters, low crest factor arrays, functions achieving optimal uncertainty principle bounds, and complete orthonormal sequences — a unified approach. *Applied and Computational Harmonic Analysis,* 1:261–266, 1994.

[7] J.S. Byrnes. A low complexity energy spreading transform coder. In Y. Zeevi and R. Coifman, editors, *Signal and Image Representation in Combined Spaces,* Haifa, 1997.

[8] J.S. Byrnes, W. Moran, and B. Saffari. Smooth PONS. *The Journal of Fourier Analysis and Applications,* 6(6):663–674, 2000.

[9] J.S. Byrnes, M.A. Ramalho, G.K. Ostheimer, and I. Gertner. Discrete one dimensional signal processing method and apparatus using energy spreading coding. U.S. Patent number 5,913,186, 1999.

[10] J.S. Byrnes, B. Saffari, and H.S. Shapiro. Energy spreading and data compression using the Prometheus orthonormal set. In *Proc. 1996 IEEE Signal Processing Conf.,* Loen, Norway, 1996.

[11] J.A. Davis and J. Jedwab. Peak-to-mean power control in OFDM, Golay complementary sequences, and Reed-Muller codes. *IEEE Trans. Information Theory,* 45(7):2397–2417, November 1999.

[12] M. Mendes France and G. Tenenbaum. Dimensions des courbes planes, papiers pliés et suites de Rudin-Shapiro. *Bull. Soc. Math. France,* 109:207–215, 1981.

[13] M.J.E. Golay. Multislit spectrometry. *J. Optical Society Am.,* 39:437, 1949.

[14] M.J.E. Golay. Static multislit spectrometry and its application to the panoramic display of infrared spectra. *J. Optical Society Am.,* 41:468, 1951.

[15] M.J.E. Golay. Complementary series. *IEEE Trans. Information Theory,* 7:82–87, April 1961.

[16] John E. Gray and Soon H. Leong. On a subclass of Welti codes and Hadamard matrices. *IEEE Trans. Electromagnetic Compatibility,* 12(2):167–170, 1990.

[17] J.-P. Kahane and G. Tenenbaum. *Ensembles parfaits et séries trigonométriques.* Hermann, Paris, 1963.

[18] Hans Dieter Lüke. Sets of one and higher dimensional Welti codes and complementary codes. *IEEE Trans. Aerospace and Electronic Systems,* 21(2):170–178, March 1985.

[19] B.M. Popović. Power efficient multitone signals with flat amplitude spectrum. *IEEE Trans. Communications*, 39(7):1031–1033, July 1991.

[20] B.M. Popović. New RACH preambles with low autocorrelation sidelobes and reduced detector complexity. In *Proc. of the 4th CDMA International Conference, Vol. 2*, pages 157–161, September 1999.

[21] B.M. Popović. Spreading sequences for multicarrier CDMA systems. *IEEE Trans. Communications*, 47(6):918–926, June 1999.

[22] B.M. Popović, N. Suehiro, and P.Z. Fan. Orthogonal sets of quadriphase sequences with good correlation properties. *IEEE Trans. Information Theory*, 48(4):956–959, April 2002.

[23] W. Rudin. Some theorems on Fourier coefficients. *Proc. Amer. Math. Soc.*, 10:855–859, 1959.

[24] B. Saffari. History of Shapiro polynomials and Golay complementary pairs. (In preparation).

[25] H.S. Shapiro. Extremal problems for polynomials and power series. Sc.M. thesis, Massachusetts Institute of Technology, 1951.

[26] R. J. Turyn. Hadamard matrices, Baumert-Hall units, four symbol sequences, pulse compression and surface wave-encoding. *Journal of Combinatorial Theory*, 16:313–333, 1974.

[27] S. R. Weller, W. Moran, and J. S. Byrnes. On the use of the PONS sequences for peak-to-mean power control in OFDM. In *Proc. of the Workshop on Defense Applications in Signal Processing*, pages 203–209, LaSalle, Illinois, USA, August 1999.

[28] G. R. Welti. Quaternary codes for pulsed radar. *IRE Trans. Inf. Theory*, 6:400–408, 1960.

[29] Roland Wilson and John Richter. Generation and performance of quadraphase Welti codes for radar and synchronization of coherent and differentially coherent PSK. *IEEE Trans. Communications*, 27(9):1296–1301, September 1979.

CLIFFORD ALGEBRAS AS UNIFIED LANGUAGE FOR IMAGE PROCESSING AND PATTERN RECOGNITION

Valeriy Labunets
Urals State Tecnical University
Ekaterinburg, Russia
lab@rtf.ustu.ru

> *How wonderful are the things the Lord does! All who are delighted with them want to understand them.*
>
> —*Good News Bible, Psalm 111:2*

Abstract   The main goal of the paper is to show that commutative hypercomplex algebras and Clifford algebras can be used to solve problems of multicolor image processing and pattern recognition in a natural and effective manner.

Keywords: Clifford algebra, color images, hyperspectral images, invariants, multicolor images, pattern recognition.

1.   Introduction

The concept of color and multispectral image recognition connects the topics we consider in this work. The term "multispectral (multicolor, multicomponent) image" is defined for an image with more than one component. An RGB image is an example of a color image featuring three separate image components R(red), G(green), and B(blue). We know that primates and animals with different evolutionary histories

J. Byrnes (ed.) Computational Noncommutative Algebra and Applications, 197-225.
© 2004 *Kluwer Academic Publishers. Printed in the Netherlands.*

have color visual systems of different dimensionality. For example, the human brain uses three channel (RGB) images to recognize color (RGB)-images. Primates have dichromacy and trichromacy visual systems, and they can use various 2D and 3D channels for the same purpose. Non-primates have monochromacy and dichromacy visual systems. Images of such systems are real—or complex—valued functions. Reptiles have multichromacy visual systems. For example, the tortoise visual system has five types of color photoreceptors (R,G,B, DC,UV). Shrimps have the biggest known dimension of the visual system. They use ten spectral types of photoreceptors in their eyes to recognize fine spectral details. Our approach to multicolor image processing is to use so-called multiplet (multicolor or hypercomplex) numbers [1]–[7] to describe multicolor images and to operate directly on multi-channel images as on single-channel multiplet-valued images. In the classical approach every multicolor pixel (in particular, color pixel) is associated to a point of a kD multicolor vector space (to a point of a 3D RGB vector space for color images). In our approach, each image multicolor pixel is considered not as a kD vector, but as a multiplet (triplet) number, and multicolor (color) space is identified with the so-called multiplet (triplet) algebra. Note that both these assumptions (vector and hypercomplex natures of multicolor images) are only hypotheses. We have no biological evidence in the form of experiments that would verify that the brain actually uses any of the algebraic properties arising from structures of the vector spaces or the multiplet (triplet) algebra. It is our aim to show that the use of Clifford algebras fits more naturally to the tasks of multicolor image processing and recognition of multicolor patterns than does the use of color vector spaces. We give algebraic models of animals' visual systems using different hypercomplex and Clifford algebras. Our hypotheses are

1. Brains of primates operate with hypercomplex numbers during image processing and recognition.

2. Brains use different algebras on two levels (retina and Visual Cortex) for two general goals: image processing and pattern recognition, respectively. Multicolor images appear on the retina as functions with values in a multiplet kD algebra (k-cycle algebra) where k is the number of image spectral channels. But multicolor images in an animals' Visual Cortex (VC) are functions with values in a 2^k-D Clifford algebra.

3. Visual systems of animals with different evolutionary histories use different hypercomplex algebras for color and multicolor image processing.

One of the main and interesting problems of information science is clarification of how animals' eyes and brain recognize objects in the real world. Practice shows that they successfully cope with this problem and recognize objects at different locations, of different views and illumination, and with different degrees of blurring. But how is it done by the brain? How do we see? How do we recognize moving and changing objects of the surrounding world? A moving object is fixed in the retina as a sequence of different images. As in the famous aphorism of HERACLITUS, who pointed out that one cannot step into the same river twice, we literally never see the same object twice. No individual image allows reaching a conclusion about the true shape of the object. This means that a set of sequential images appearing in the retina must contain a constant "something," thanks to which we see and recognize the object as a whole. This constant "something" is called *invariant*. For example, all letters 'F' in Fig. 1 we interpret as the same for different geometric distortions.   This fact means that all geometrically

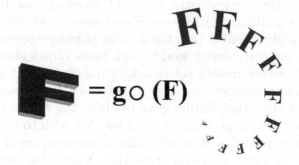

Figure 1.   Geometrical distorted versa of letter 'F'

distorted letters 'F' contain invariant features, which are not changed, when the shape of 'F' is changed. Our brain can extract these invariant features. In Fig. 2 we see hyperbolic (non-Euclidean) motions of grey-level mice and color fish. All transformed figures are interpreted as being the same. This fact means that all figures contain invariant features with respect to hyperbolic motions (and color transformations), and our brain can extract these invariant features from images, too. So, we see, we live in 3D Euclidean space but our brain can calculate invariants of images with respect to non-Euclidean transformations. In order for an artificial pattern recognition system to perform in the same way as any biological visual system, the recognition result should be invariant with respect to various *transformation groups* of the patterns such as translation, rotation, size variation, and change in illumination and

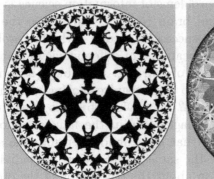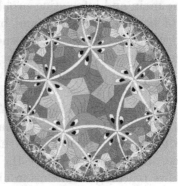

Figure 2. Non-Euclidean motions

color. The present work describes new methods of image recognition based on an algebraic-geometric theory of invariants. In this approach, each color or multicolor pixel is considered not as a kD vector, but as a kD hypercomplex number (k is the number of image spectral channels). Changes in the surrounding world which cause object shape and color transformations are treated not as matrix transforms, but as the action of some Clifford numbers in physical and perceptual spaces. We shall present some clues that Nature gives us about the role and importance of computing with hypercomplex numbers. We wish to review some of the reasons why such a state of affairs is necessary from a computational point of view. One can argue that Nature has also learned to utilize (through evolution) properties of hypercomplex numbers. Thus, the Visual Cortex might have the ability to operate as a Clifford algebra device. We don't agree with KRONECKER that *"The Lord created the integers and the rest is the work of man."* We think that the Lord knew Clifford algebras, and he was the first engineer who used these algebras to design animals' visual systems.

2. Clifford algebras as models of physical spaces

2.1 Algebras of physical spaces

We suppose that a brain calculates some hypercomplex-valued invariants of an image when recognizing it. Hypercomplex algebras generalize the algebras of complex numbers, quaternions and octonions. Of course, the algebraic nature of hypercomplex numbers must correspond to the spaces with respect to geometrically perceivable properties. For recognition of 2D, 3D and nD images we turn the spaces $\mathbf{R}^2$, $\mathbf{R}^3$ and $\mathbf{R}^n$ into corresponding algebras of hypercomplex numbers. Let "small" nD space $\mathbf{R}^n$ be spanned by the orthonormal basis of n *space hyperimaginary units* I_i, $i = 1, 2, \ldots n$. We assume

$$I_i^2 = \begin{cases} +1 & \text{for } i = 1, 2, \ldots, p, \\ -1 & \text{for } i = p+1, p+2, \ldots, p+q, \\ 0 & \text{for } i = p+q+1, p+q+2, \ldots, p+q+r = n, \end{cases}$$

and $I_i I_j = -I_j I_i$. Now, we construct the "big" 2^nD hypercomplex space $\mathbf{R}^{2^n}$. Let $\mathbf{b} = (b_1, b_2, \ldots, b_n) \in \mathbf{B}_2^n$ be an arbitrary n-bit vector, where $b_i \in \mathbf{B}_2 = \{0, 1\}$ and $\mathbf{B}_2^n$ is the nD Boolean algebra. Let us introduce $\mathbf{I^b} := I_1^{b_1} I_2^{b_2} \cdots I_n^{b_n}$. Then 2^n elements $\mathbf{I^b}$ form a basis of 2^nD space, i.e., for all $\mathcal{C} \in \mathbf{R}^{2^n}$ we have $\mathcal{C} := \sum_{\mathbf{b} \in \mathbf{B}_2^n} c_{\mathbf{b}} \mathbf{I_b}$. If $\mathcal{C}_1, \mathcal{C}_2 \in \mathbf{R}^{2^n}$, then we can define their product $\mathcal{C}_1 \mathcal{C}_2$. There are 3^n possibilities for $I_i^2 = +1, 0, -1$, $\forall i = 1, 2, \ldots, n$. Every possibility generates one algebra. Therefore, the space $\mathbf{R}^{2^n}$ with 3^n rules of the multiplication forms 3^n different 2^nD algebras, which are called the *space Clifford algebras* [8]. We denote these algebras by $\mathcal{A}_{2^n}^{Sp(p,q,r)}(\mathbf{R}|I_1, \ldots, I_n)$, $\mathcal{A}_{2^n}^{Sp(p,q,r)}$ or $\mathcal{A}_{2^n}^{Sp}$, if $I_1, \ldots, I_n$, p, q, r are fixed.

EXAMPLE 1 *We start with the space* $\mathbf{R}^2$ *and provide it with the algebraic frame of algebras of generalized complex numbers:* $\mathbf{R}^2 \rightarrow \mathcal{A}_2(\mathbf{R}) := \mathbf{R} + \mathbf{R}I = \{\mathbf{z} = x_1 + Ix_2 \mid x_1, x_2 \in \mathbf{R}\}$, *where* I *is a generalized imaginary unit.*

- *If* $I^2 = -1$, *i.e.,* $I = i$, *then* $\mathcal{A}_2(\mathbf{R}|i) = \mathbf{COM} = \{x_1 + ix_2 \mid x_1, x_2 \in \mathbf{R}; i^2 = -1\}$ *is the field of complex numbers.*

- *If* $I^2 = +1$, *i.e.,* $I = e$, *then* $\mathcal{A}_2(\mathbf{R}|e) = \mathbf{DOU} = \{x_1 + ex_2 \mid x_1, x_2 \in \mathbf{R}; e^2 = 1\}$ *is the ring of double numbers.*

- *If* $I^2 = 0$, *i.e.,* $I = \varepsilon$, *then* $\mathcal{A}_2(\mathbf{R}|\varepsilon) = \mathbf{DUA} = \{x_1 + \varepsilon x_2 \mid x_1, x_2 \in \mathbf{R}; \varepsilon^2 = 0\}$ *is the ring of dual numbers.*

EXAMPLE 2 *Quaternions, as constructed by Hamilton, form the 4D algebra* $\mathcal{A}_4 = \mathcal{A}_4(\mathbf{R}) = \mathcal{A}_4(\mathbf{R}|1, i, j, k) = \mathbf{R} + \mathbf{R}i + \mathbf{R}j + \mathbf{R}k$ *spanned by 4 hyperimaginary units* $1, i, j, k$. *The identities* $i^2 = j^2 = k^2 = -1$, $ij = -ji = k$ *are valid for these units.* $i^2 = j^2 = k^2 = \delta \in \{-1, 0, 1\}$ *can be set. Here, the two latter values (0 and 1) result in non-classical quaternions that were proposed by Clifford [8]. Introducing notations* I, J, K *for new hyperimaginary units, we get nine spatial algebras of generalized quaternions,* $\mathcal{A}_4(\mathbf{R}|1, I, J, K) := \mathcal{A}_4 = \mathbf{R} + \mathbf{R}I + \mathbf{R}J + \mathbf{R}K$ *depending on which of nine possibilities resulting from* $I^2 \in \{1, 0, -1\}$, $J^2 \in \{1, 0, -1\}$ *is valid for the two independent hyperimaginary units.*

Every generalized quaternion $\mathbf{q}$ *has the unique representation of the form* $\mathbf{q} = q_0 + q_1 I + q_2 J + q_3 K = Sc(\mathbf{q}) + Vec(\mathbf{q})$, *where* q_0, q_1, q_2, q_3 *are real numbers and* $Sc(\mathbf{q}) := q_0$, $Vec(\mathbf{q}) := q_1 I + q_2 J + q_3 K$ *are scalar and vector parts of the quaternion* $\mathbf{q}$, *respectively.*

We can make $\mathcal{A}_{2^n}^{Sp}$, be a *ranked* and $\mathbf{Z}/2\mathbf{Z}$-*graded algebra*. Let $r(\mathbf{b})$ be the *Hamming weight* (= *rank*) of $\mathbf{b}$, i.e., a functional $r : \mathbf{B}_2^n \longrightarrow [0, n-1]$ defined by $r(\mathbf{b}) := \sum_{i=1}^n b_i$, and let $\partial(\mathbf{b}) = r(\mathbf{b})$ (mod 2) be the *grad* of $\mathbf{b}$. Then $\mathcal{A}_{2^n}^{Sp}$ can be represented as the ranked and $\mathbf{Z}/2\mathbf{Z}$-graded sums $\mathcal{A}_{2^n}^{Sp} = \bigoplus_{r=0}^n \mathcal{A}_{2^n}^{[r]}$ and $\mathbf{R}^{2^n} = \bigoplus_{\partial=0}^1 \mathcal{A}_{2^n}^{\{\partial\}}$, where the dimension of the vector space $\mathcal{A}_{2^n}^{[k]}$ equals the binomial coefficient C_n^k and $\sum_{k=0}^n C_n^k = 2^n$. The dimensions of $\mathcal{A}_{2^n}^{\{0\}}$ and $\mathcal{A}_{2^n}^{\{1\}}$ are equal to 2^{n-1}. The subspaces $\mathcal{A}_{2^n}^{[k]}$ are spanned by the k-products of units $I_{i_1} I_{i_2} \ldots I_{i_k}$ ($i_1 < i_2 < \ldots < i_k$), i.e., by all basis vectors $\mathbf{I_b}$ with $r(\mathbf{b}) = k$. Every element $\mathcal{C} := \sum_{\mathbf{b} \in \mathbf{B}_2^n} c_\mathbf{b} \mathbf{I_b}$ of $\mathcal{A}_{2^n}^{Sp}$ has the representations: $\mathcal{C} = \mathcal{C}^{[0]} + \mathcal{C}^{[1]} + \ldots + \mathcal{C}^{[n]}$ and $\mathcal{C} = \mathcal{C}^{\{0\}} + \mathcal{C}^{\{1\}}$, where $\mathcal{C}^{[0]} \in \mathcal{A}_{2^n}^{[0]}$ is the scalar part of the Clifford numbers, $\mathcal{C}^{[1]} \in \mathcal{A}_{2^n}^{[1]}$ is its vector part, $\mathcal{C}^{[2]} \in \mathcal{A}_{2^n}^{[2]}$ is its bivector part, $\ldots$, $\mathcal{C}^{[n]} \in \mathcal{A}_{2^n}^{[n]}$ is its n-vector part, and, finally, $\mathcal{C}^{\{0\}}$ and $\mathcal{C}^{\{1\}}$ are even and odd parts of the Clifford number $\mathcal{C}$. If $\mathcal{C} \in \mathcal{A}_{2^n}^{\{l\}}$, we put $\partial(\mathcal{C}) = l$ and say that l is the *degree of* $\mathcal{C}$. Multiplication of two Clifford numbers of ranks k and s gives the sum of Clifford numbers from $|k - s|$ to $p = \min(k + s, 2n - k - s)$ with increment 2, i.e., $\mathcal{A}^{[k]} \mathcal{B}^{[s]} = \mathcal{C}^{[|k-s|]} + \mathcal{C}^{[|k-s|+2]} + \ldots + \mathcal{C}^{[p]}$.

2.2 Geometries of physical spaces

The conjugation operation in $\mathcal{A}_{2^n}^{Sp(p,q,r)}$ maps every Clifford number $\mathcal{C} := c_0 I_0 + \sum_{\mathbf{b} \neq 0} c_\mathbf{b} \mathbf{I_b}$ to the number $\overline{\mathcal{C}} = c_0 I_0 - \sum_{\mathbf{b} \neq 0} c_\mathbf{b} \mathbf{I_b}$. The algebras $\mathcal{A}_{2^n}^{Sp(p,q,r)}$ are transformed into 2^nD pseudometric spaces designated by $\mathcal{CL}_{2^n}^{Sp(p,q,r)}$ or $\mathcal{CL}_{2^n}^{p,q,r}$, if the pseudodistance between two Clifford numbers $\mathcal{A}$ and $\mathcal{B}$ is defined by $\rho(\mathcal{A}, \mathcal{B}) = \sqrt{(\mathcal{A} - \mathcal{B})\overline{(\mathcal{A} - \mathcal{B})}}$. Subspaces of pure vector Clifford numbers $x_1 I_1 + \ldots + x_n I_n \in Vec^1\left(\mathcal{A}_{2^n}^{Sp}\right)$ are nD spaces $\mathbf{R}^n := \mathcal{GR}_n^{Sp(p,q,r)}$. The pseudometrics constructed in $\mathcal{CL}_{2^n}^{Sp(p,q,r)}$ induce corresponding pseudometrics in $\mathcal{GR}_n^{Sp(p,q,r)}$.

EXAMPLE 3 *In* $\mathcal{A}_2(\mathbf{R})$ *we introduce a conjugation operation which maps every element* $\mathbf{z} = x_1 + I x_2$ *to the element* $\bar{\mathbf{z}} = x_1 - I x_2$. *Now, the generalized complex plane is turned into the pseudometric space* $\mathcal{A}_2(\mathbf{R}) \longrightarrow$

$\mathcal{GC}_2^{Sp(p,q,r)}$ *if one defines the pseudodistance as:*

$$\rho(\mathbf{z}_1, \mathbf{z}_2) = \begin{cases} \sqrt{(x_2 - x_1)^2 + (y_2 - y_1)^2}, & \text{if } \mathbf{z} \in \mathcal{A}_2(\mathbf{R}|i), \\ \sqrt{(x_2 - x_1)^2 - (y_2 - y_1)^2}, & \text{if } \mathbf{z} \in \mathcal{A}_2(\mathbf{R}|e), \\ |x_2 - x_1|, & \text{if } \mathbf{z} \in \mathcal{A}_2(\mathbf{R}|\varepsilon), \end{cases}$$

where $\mathbf{z}_1 := (x_1, x_2) = x_1 + Ix_2$, $\mathbf{z}_2 := (y_1, y_2) = y_1 + Iy_2$. *So, the plane of the classical complex numbers is the 2D Euclidean space* $\mathcal{GC}_2^{Sp(2,0,0)}$, *the double numbers plane is the 2D Minkowskian space* $\mathcal{GC}_2^{Sp(1,1,0)}$, *and the dual numbers plane is the 2D Galilean space* $\mathcal{GC}_2^{Sp(1,0,1)}$. *When one speaks about all three algebras (or geometries) simultaneously, then the corresponding algebra (or geometry) is that of* generalized complex numbers, *denoted by* $\mathcal{A}_2^{Sp(p,q,r)}$ *(or* $\mathcal{GC}_2^{Sp(p,q,r)}$*).*

EXAMPLE 4 *In* $\mathcal{A}_4(\mathbf{R})$ *we introduce a conjugation operation which maps every quaternion* $\mathbf{q} = q_0 + Iq_1 + Jq_2 + Kq_3$ *to the element* $\overline{q} = q_0 - Iq_1 - Jq_2 - Kq_3$. *If the pseudodistance* $\rho(\mathbf{p}, \mathbf{q})$ *between two generalized quaternions* $\mathbf{p}$ *and* $\mathbf{q}$ *is defined as the modulus of their difference*

$$\mathbf{u} = \mathbf{p} - \mathbf{q} = t + xI + yJ + zK : \rho(\mathbf{p}, \mathbf{q}) = \sqrt{(\mathbf{p} - \mathbf{q})\overline{(\mathbf{p} - \mathbf{q})}} = $$

$\sqrt{\mathbf{u}\overline{\mathbf{u}}}$, *then nine spatial algebras* $\mathcal{A}_4(\mathbf{R})$ *are transformed into nine 4D pseudometric spaces designed as* $\mathcal{GH}_4^{Sp(p,q,r)}$, *where* p, q *and* r *stand for the number of basis vectors with squares* $1, -1$ *and* 0, *respectively, and* $p + q + r = n$. *Thus, the pseudodistance can take positive, negative and pure imaginary values. There are only 5 different geometries* $\mathcal{GH}_4^{Sp(p,q,r)}$: $\mathcal{GH}_4^{Sp(4,0,0)}, \mathcal{GH}_4^{Sp(2,2,0)}, \mathcal{GH}_4^{Sp(2,0,2)}, \mathcal{GH}_4^{Sp(1,3,0)}, \mathcal{GH}_4^{Sp(1,2,1)}$. *The subspaces of pure vector-valued generalized quaternions* $xI + yJ + zK$ *are 3D spaces* $\mathcal{GR}_3^{S(p,q,r)} := \mathbf{Vec}\{\mathcal{GH}_4^{Sp(p,q,r)}\}$. *The pseudometrics introduced in* $\mathcal{GH}_4^{Sp(p,q,r)}$ *induce only three different pseudometrics in* $\mathcal{GR}_3^{Sp(p,q,r)}$:

$$\rho(\mathbf{Vec}\{\mathbf{p}\}, \mathbf{Vec}\{\mathbf{q}\}) = |\mathbf{Vec}\{\mathbf{p} - \mathbf{q}\}| = |\mathbf{Vec}\{\mathbf{u}\}| = \begin{cases} \sqrt{x^2 + y^2 + z^2}, \\ \sqrt{x^2 - y^2 - z^2}, \\ \sqrt{x^2} = |x|. \end{cases}$$

They will be denoted by $\mathcal{GR}_3^{Sp(3,0,0)}$ $\mathcal{GR}_3^{Sp(1,0,2)}$, $\mathcal{GR}_3^{Sp(1,2,0)}$. *They form Euclidean, Minkowskian, and Galilean 3D pseudometric spaces, respectively.*

All even Clifford numbers $\mathcal{E}_0 \in \mathcal{A}_{2^n}^{\{0\}}$ *of unit modulus represent the rotation group of the corresponding space* $\mathcal{GR}_n^{Sp(p,q,r)}$, *which is called the spinor group and is denoted by* $\mathbf{Spin}\left(\mathcal{A}_{2^n}^{Sp(p,q,r)}\right)$. *Generalized complex* numbers and quaternions of unit modulus have the forms: $\mathbf{e}_0 = e^{I\varphi} = \cos\varphi + I\sin\varphi$, $\mathcal{Q}_0 = e^{\mathbf{u}_0\varphi} = \cos\varphi + \mathbf{u}_0\sin\varphi$, where $\cos\varphi$ and $\sin\varphi$ are trigonometric functions in the corresponding nD $\mathcal{GC}_n^{Sp(p,q,r)}$-geometries, φ is a rotation angle around the vector-valued quaternion $\mathbf{u}_0$ of unit modulus ($|\mathbf{u}_0| = 1$, $\mathbf{u}_0 = -\overline{\mathbf{u}_0}$). Clifford numbers $\mathcal{E}_0 \in \mathrm{Spin}\left(\mathcal{A}_{2n}^{Sp(p,q,r)}\right)$ with unit modulus have the analogous form $\mathcal{E}_0 = e^{\mathbf{u}_0\varphi} = \cos\varphi + \mathbf{u}_0\sin\varphi \in \mathrm{Spin}\left(\mathcal{A}_{2n}^{Sp(p,q,r)}\right)$ for the appropriate bivector $\mathbf{u}_0$.

THEOREM 1 *[8]*. *All motions in 2D, 3D and nD spaces $\mathcal{GR}_2^{Sp(p,q,r)}$, $\mathcal{GR}_3^{Sp(p,q,r)}$, $\mathcal{GR}_n^{Sp(p,q,r)}$ are represented in the forms:*

$$\mathbf{z}' = \mathbf{e}_0\mathbf{z}\mathbf{e}_0 + \mathbf{w}, \quad \mathbf{x}' = \mathcal{Q}_0\mathbf{x}\mathcal{Q}_0^{-1} + \mathbf{w}, \quad \mathbf{x}' = \mathcal{E}_0\mathbf{x}\mathcal{E}_0^{-1} + \mathbf{w},$$

where $\mathbf{e}_0 = e^{I\varphi/2}$, $\mathcal{Q}_0 = e^{\mathbf{u}_0\varphi/2}$, $\mathcal{E}_0 = e^{\mathbf{u}_0\varphi/2}$, and $|\mathbf{e}_0| = |\mathcal{Q}_0| = |\mathcal{E}_0| = 1$. If $|\mathbf{e}_0|, |\mathcal{Q}_0|, |\mathcal{E}_0| \neq 1$, then the latter transformations form the "small" affine groups $\mathbf{Aff}\left(\mathcal{GR}_2^{Sp(p,q,r)}\right)$, $\mathbf{Aff}\left(\mathcal{GR}_3^{Sp(p,q,r)}\right)$, $\mathbf{Aff}\left(\mathcal{GR}_n^{Sp(p,q,r)}\right)$, *respectively.*

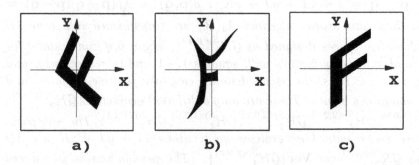

a)   b)   c)

Figure 3.   Rotations in a) 2D Euclidean space $\mathcal{GC}_2^{Sp(2,0,0)}$, b) 2D Minkowskian space $\mathcal{GC}_2^{Sp(1,1,0)}$ and c) 2D Galilean space $\mathcal{GC}_2^{Sp(1,0,1)}$.

Using this theorem, we can describe geometric distortions of images in the language of Clifford algebras. These distortions will be caused by: 1) nD translations $\mathbf{x} \longrightarrow \mathbf{x} + \mathbf{w}$; 2) nD rotations $\mathbf{x} \longrightarrow \mathcal{E}_0(\mathbf{x} + \mathbf{w})\mathcal{E}_0^{-1}$; 3) dilatation: $\mathbf{x} \longrightarrow \lambda\mathbf{x}$, where $\lambda \in \mathbf{R}^+$. If $f(\mathbf{x})$ is an initial image and $_{\lambda\mathcal{E}_0\mathbf{w}}\mathbf{f}(\mathbf{q})$ is its distorted version, then $_{\lambda\mathcal{E}_0\mathbf{w}}\mathbf{f}(\mathbf{x}) := \mathbf{f}\left(\lambda\mathcal{E}_0(\mathbf{x} + \mathbf{w})\mathcal{E}_0^{-1}\right)$, where λ is a scale factor, $\mathbf{x}, \mathbf{w} \in \mathcal{GR}_n^{Sp(p,q,r)}$. We suppose that the human brain can use the spinor and "small" affine groups for mental rotations (see Fig. 3) and motions of images (for example, in a dream), which are contained in the brain memory on the so-called "screen of mind."

3. Clifford Algebras as Models of Perceptual Multicolor Spaces

Early in the 19th century YOUNG (1802) proposed that the human visual system contains three color mechanisms. This theory was later supported by HELMHOLTZ (1852) and became known as the Young–Helmholtz theory of color vision. Later, HERING (1878) proposed that color vision is mediated by red-green and blue-yellow opponent mechanisms. Thus, for a time it appeared there were two conflicting theories of color vision. As a result of experimental work, it has since become recognized that the two theories describe different levels in the visual system. In agreement with the Young–Helmholtz theory, there are three photoreceptor mechanisms, i.e., cone types, in the human retina, and, in accordance with Hering's theory, there are color-opponent neurons at higher levels in the VC. The multicomponent color image appears on the retina as the kD vector

$$\mathbf{f}_{Mcol}^{Ret}(\mathbf{x}) = \begin{bmatrix} f_1(\mathbf{x}) \\ f_2(\mathbf{x}) \\ \dots \\ f_k(\mathbf{x}) \end{bmatrix} = \begin{bmatrix} \int_\lambda S^{obj}(\lambda, \mathbf{x}) H_1(\lambda) d\lambda \\ \int_\lambda S^{obj}(\lambda, \mathbf{x}) H_2(\lambda) d\lambda \\ \dots \\ \int_\lambda S^{obj}(\lambda, \mathbf{x}) H_k(\lambda) d\lambda \end{bmatrix}, \tag{1}$$

where $S^{obj}(\lambda, \mathbf{x})$ is the color spectrum received from the point $\mathbf{x}$ of an object and $H_1(\lambda), H_2(\lambda), \dots, H_k(\lambda)$ are sensor sensitivity functions. We give algebraic models for two levels (retina and VC) of visual systems using different hypercomplex $\mathbf{Z}/k\mathbf{Z}$-graded Clifford algebras.

3.1 Algebraization of the Young–Helmholtz model

3.1.1 The Young–Helmholtz model of color images.
We shall represent RGB-color images that appear on the human retina as triplet-valued functions: $\mathbf{f}_{col}^{Ret}(\mathbf{x}) = f_R(\mathbf{x})1_{col} + f_G(\mathbf{x})\varepsilon_{col} + f_B(\mathbf{x})\varepsilon_{col}^2$, where $1_{col}, \varepsilon_{col}^1, \varepsilon_{col}^2$ are hyperimaginary units, $\varepsilon_{col}^3 = 1$. Numbers of the form $\mathcal{C} = x1 + y\varepsilon_{col} + z\varepsilon_{col}^2$ ($\varepsilon_{col}^3 = 1$) were considered by GREAVES [10]. According to Greaves, these numbers are called the *triplet* or *3-cycle numbers*. We shall call them the *color numbers*. The product of two color numbers $\mathcal{C}_1 = a_0 + a_1\varepsilon_{col} + a_2\varepsilon_{col}^2$ and $\mathcal{C}_2 = b_0 + b_1\varepsilon_{col} + b_2\varepsilon_{col}^2$ is given by

$$\mathcal{C}_1\mathcal{C}_2 = (a_0 + a_1\varepsilon_{col} + a_2\varepsilon_{col}^2)(b_0 + b_1\varepsilon_{col} + b_2\varepsilon_{col}^2) =$$

$$(a_0b_0 + a_1b_2 + a_2b_1) + (a_1b_0 + a_2b_2 + a_0b_1)\varepsilon_{col} + (a_2b_0 + a_1b_1 + a_2b_0)\varepsilon_{col}^2.$$

Thus, the color product is isomorphic to the 3-point cyclic convolution. The *color conjugate* $\overline{\mathcal{C}}$ of a color number $\mathcal{C} = x + y\varepsilon_{col} + z\varepsilon_{col}^2$ is defined by $\overline{C} = \overline{x + y\varepsilon_{col} + z\varepsilon_{col}^2} = x + y\overline{\varepsilon}_{col} + z\overline{\varepsilon}_{col}^2 = x + z\varepsilon_{col} + y\varepsilon_{col}^2$. The norm $||\mathcal{C}|| = \overline{\mathcal{C}\mathcal{C}}$ is given by $||\mathcal{C}||^2 = \mathcal{C}\overline{\mathcal{C}} = (x^2 + y^2 + z^2) - (xy + xz + yz)$. Each color number has three modules

$$||\mathcal{C}||_1 = |x + y + z|, \quad ||\mathcal{C}||_2 = \sqrt{x^2 + y^2 + z^2 - xy - xz - yz},$$

$$||\mathcal{C}||_3 = \sqrt[3]{x^3 + y^3 + z^3 - 3xyz}$$

possessing the properties: $||\mathcal{C}_1\mathcal{C}_2||_i = ||\mathcal{C}_1||_i||\mathcal{C}_2||_i$, $i = 1, 2, 3$ and $||\mathcal{C}||_3^3 = ||\mathcal{C}||_2^2||\mathcal{C}||_1$. Triplets, strictly speaking, do not form a 3D field, but form an associative so-called *triplet (color) algebra*

$$\mathcal{A}_3^{col} = \mathcal{A}_3^{col}(\mathbf{R}) = \mathcal{A}_3(\mathbf{R}|1, \varepsilon_{col}, \varepsilon_{col}^2) := \mathbf{R}1 + \mathbf{R}\varepsilon_{col} + \mathbf{R}\varepsilon_{col}^2.$$

Greaves [10] considered a color number $x + y\varepsilon + z\varepsilon^2$ (here, $\varepsilon \equiv \varepsilon_{col}$)

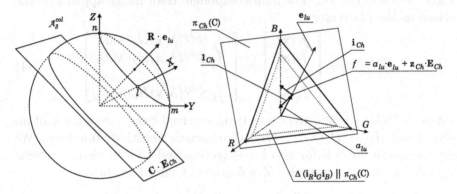

Figure 4. Geometric interpretation of triplet algebra

as a point of 3D space and gave the following geometric interpretation of triplet operations. He constructed a sphere (see Fig. 4) with the center at the origin and denoted intersection points of the sphere with the coordinate axes X, Y, Z in the positive direction by l, m, n, respectively. He drew a circle via points l, m, n; the "symmetrical axis" $\mathbf{e}_{lu}$ was drawn via the center of this circle and the origin; the "symmetrical plane" $\mathbf{E}_{ch}$ is perpendicular to the axis $\mathbf{e}_{lu}$. Then Greaves considered rectangular projections of color numbers $\mathcal{C}$ on the axis $\mathbf{e}_{lu}$ and plane $\mathbf{E}_{ch}$. Furthermore, he proved that every triplet $\mathcal{C}$ can be represented as a sum of a real number being depicted by a rectangular projection of the triplet on the axis $\mathbf{e}_{lu}$ and a complex number being depicted by a rectangular projection of the triplet on the plane $\mathbf{E}_{ch}$. Therefore, the color algebra is the direct sum of the real $\mathbf{R}$ and complex $\mathbf{C}$ fields: $\mathcal{A}_3(\mathbf{R}, \mathbf{C}) = \mathbf{R} \cdot \mathbf{e}_{lu} + \mathbf{C} \cdot \mathbf{E}_{ch} = \mathbf{R} \oplus \mathbf{C}$, where $\mathbf{e}_{lu} := (1 + \varepsilon + \varepsilon^2)/3$,

$\mathbf{E}_{ch} := (1 + \omega_3\varepsilon^2 + \omega_3^2\varepsilon)/3$ are orthogonal "real" and "complex" idempotents ($\mathbf{e}_{lu}^2 = \mathbf{e}_{lu}$, $\mathbf{E}_{ch}^2 = \mathbf{E}_{ch}$, $\mathbf{e}_{lu}\mathbf{E}_{ch} = \mathbf{E}_{ch}\mathbf{e}_{lu} = 0$), respectively, and $\omega_3 := e^{\frac{2\pi i}{3}}$. Therefore, every color number $\mathcal{C} = x + y\varepsilon^1 + z\varepsilon^2$ is a linear combination $\mathcal{C} = x + y\varepsilon^1 + z\varepsilon^2 = a_{lu} \cdot \mathbf{e}_{lu} + \mathbf{z}_{ch} \cdot \mathbf{E}_{ch} = (a_{lu}, \mathbf{z}_{ch})$ of the "scalar" $a_{lu}\mathbf{e}_{lu}$ and "complex" parts $\mathbf{z}_{ch}\mathbf{E}_{ch}$ in the idempotent basis $\{\mathbf{e}_{lu}, \mathbf{E}_{ch}\}$. We will call the real numbers $a_{lu} \in \mathbf{R}$ the *luminance (intensity) numbers*, and the complex numbers $\mathbf{z}_{ch} = b + jc \in \mathbf{C}$ the *chromaticity numbers*. Obviously, $a_{lu}\mathbf{e}_{lu} := \mathcal{C}\mathbf{e}_{lu} = (x + y\varepsilon^1 + z\varepsilon^2)\mathbf{e}_{lu} = (x+y+z)\mathbf{e}_{lu}$, $\mathbf{z}_{ch}\mathbf{E}_{ch} := \mathcal{C}\mathbf{E}_{ch} = (x+y\varepsilon^1+z\varepsilon^2)\mathbf{E}_{ch} = (x+\omega_3^1 y+z\omega_3^2)\mathbf{E}_{ch}$, where $a_{lu} = x+y+z$, $\mathbf{z}_{ch} = x+\omega_3^2 y+z\omega_3^1$. In the new representation two main arithmetic operations have very simple forms:

$$\mathcal{C}_1 \pm \mathcal{C}_2 = (a_1\mathbf{e}_{lu} + \mathbf{z}_1\mathbf{E}_{ch}) \pm (a_2\mathbf{e}_{lu} + \mathbf{z}_2\mathbf{E}_{ch}) = (a_1 \pm a_2)\mathbf{e}_{lu} + (\mathbf{z}_1 \pm \mathbf{z}_2)\mathbf{E}_{ch},$$

$$\mathcal{C}_1\mathcal{C}_2 = (a_1\mathbf{e}_{lu} + \mathbf{z}_1\mathbf{E}_{ch})(a_2\mathbf{e}_{lu} + \mathbf{z}_2\mathbf{E}_{ch}) = (a_1 a_2)\mathbf{e}_{lu} + (\mathbf{z}_1\mathbf{z}_2)\mathbf{E}_{ch},$$

or, briefly, $\mathcal{C}_1 \pm \mathcal{C}_2 = (a_1, \mathbf{z}_1) \pm (a_2, \mathbf{z}_2) = (a_1 \pm a_2, \mathbf{z}_1 \pm \mathbf{z}_2)$, $\mathcal{C}_1\mathcal{C}_2 = (a_1, \mathbf{z}_1) \cdot (a_2, \mathbf{z}_2) = (a_1 a_2, \mathbf{z}_1\mathbf{z}_2)$. For $\overline{\mathcal{C}}$ we have $\overline{\mathcal{C}} = x + y\varepsilon^1 + z\varepsilon^2 = a_{lu} \cdot \mathbf{e}_{lu} + \overline{\mathbf{z}}_{ch} \cdot \mathbf{E}_{ch} = (a_{lu}, \overline{\mathbf{z}}_{ch})$. Therefore,

$$||\mathcal{C}||_1 = |x + y + z| = |a_{lu}|, \quad ||\mathcal{C}||_2^2 = |x^2 + y^2 + z^2 - xy - xz - yz| = ||\mathbf{z}_{ch}||^2,$$

$$||\mathcal{C}||_3^3 = |x^3 + y^3 + z^3 - 3xyz| = |a_{lu}|||\mathbf{z}_{ch}||^2.$$

The norms $||.||_1$, $||.||_2$ are called the *luminance* and *chromaticity norms*, respectively. We can consider a color image in the two forms:

$$\mathbf{f}_{col}^{Ret}(x, y) = f_R(x, y)\mathbf{1}_{col} + f_G(x, y)\varepsilon_{col} + f_B(x, y)\varepsilon_{col}^2 =$$

$$f_{lu}(x, y)\mathbf{e}_{lu} + \mathbf{f}_{ch}(x, y)\mathbf{E}_{ch}.$$

In the second form we have separated the color image into two terms: the *luminance* (intensity) term $f_{lu}(x, y)$ and the *chromacity term* $\mathbf{f}_{ch}(x, y)$ (color information). This color transformation is a linear projection (see Fig. 4) of the color vector-valued pixel in the color space on the diagonal vector $\mathbf{e}_{lu} := (1, 1, 1) = \mathbf{i}_R + \mathbf{i}_G + \mathbf{i}_B$ and on the 2D plane $\pi_{ch}(\mathbf{C})$ which is orthogonal to the diagonal vector $\mathbf{e}_{lu} : \pi_{ch}(\mathbf{C}) \perp \mathbf{e}_{lu}$. The vector $\mathbf{e}_{lu}$ is called the *luminance (white) vector* and the 2D plane $\pi_{ch}(\mathbf{C})$ is called the *chromacy plane* of RGB-space. The triangle $\triangle(\mathbf{i}_R\mathbf{i}_G\mathbf{i}_B) \parallel \pi_{ch}(\mathbf{C})$ drawn between three primaries $\mathbf{i}_R\mathbf{i}_G\mathbf{i}_B$ is called the *Maxwell triangle*. The plain $\pi_{ch}(\mathbf{C})$ is equipped with the structure of the complex field $\mathbf{C}$. Therefore, we can consider an RGB image as a pair of images $\mathbf{f}_{col}^{Ret}(\mathbf{x}) = (f_{lu}(\mathbf{x}), \mathbf{f}_{ch}(\mathbf{x}))$, where $f_{lu}(\mathbf{x}) := \langle \mathbf{e}_{lu}|\mathbf{f}_{col}^{Ret}(\mathbf{x})\rangle = f_R(\mathbf{x}) + f_G(\mathbf{x}) + f_B(\mathbf{x})$, $\mathbf{f}_{ch}(\mathbf{x}) := \mathbf{f}_{col}(\mathbf{x}) - f_{lu}(\mathbf{x})\mathbf{e}_{lu}$, and $\langle f_{lu}(\mathbf{x})|\mathbf{f}_{ch}(\mathbf{x})\rangle = 0$. Here, $f_{lu}(\mathbf{x}) :$ $\mathbf{R}^2 \longrightarrow \mathbf{R}^+$ is a real-valued image (grey-level image) and $\mathbf{f}_{ch}(\mathbf{x}) : \mathbf{R}^2 \longrightarrow$

$\pi_{ch}(\mathbf{C})$ is a complex-valued image. Hence, each pixel is represented both by a real number $f_{lu}(\mathbf{x})$ for the luminance and a complex number $\mathbf{f}_{ch}(\mathbf{x})$ for the chromatic information. The argument of the complex number $\mathbf{f}_{ch}(\mathbf{x})$ is directly equivalent to the traditional definition of *hue*, and the modulus of the complex number is similar to *saturation*. Changes in both the surrounding world and mental space of reality such as intensity, color or illumination can be treated in the triplet algebra language as the action of some transformation groups in the perceptual color space $\mathcal{A}_3^{col}$. Let us consider some of them.

1. Let $\mathcal{A} = (1, e^{i\varphi})$. Then transformation of color images $\mathbf{f}_{col}^{Ret}(\mathbf{x}) \longrightarrow \mathcal{A}\mathbf{f}_{col}(\mathbf{x}) = (1, e^{i\varphi}) \cdot (f_{lu}(\mathbf{x}), \mathbf{f}_{ch}(\mathbf{x})) = (f_{lu}(\mathbf{x}), e^{i\varphi}\mathbf{f}_{ch}(\mathbf{x}))$ changes only the hue of the initial image. The set of all such transformations forms the *hue orthogonal group* $\mathbb{HOG}(2) := \{(1, e^{i\varphi}) | \ e^{i\varphi} \in \mathbf{C}\}$.

2. Now let $\mathcal{A} = (1, \lambda)$, $\lambda > 0$. Then transformation of color images $\mathbf{f}_{col}^{Ret}(\mathbf{x}) \longrightarrow \mathcal{A}\mathbf{f}_{col}(\mathbf{x}) = (1, \lambda) \cdot (f_{lu}(\mathbf{x}), \mathbf{f}_{ch}(\mathbf{x})) = (f_{lu}(\mathbf{x}), \lambda\mathbf{f}_{ch}(\mathbf{x}))$ changes only the saturation of the initial image. The set of all such transformations forms the *saturation dilation group* $\mathbb{SDG}(2) := \{(1, a) | \ a \in \mathbf{R}^+\}$.

3. If $\mathcal{A} = (1, \mathbf{z}_{ch}) = (1, |\mathbf{z}_{ch}|e^{i\varphi})$, then transformations $\mathbf{f}_{col}^{Ret}(\mathbf{x}) \longrightarrow \mathcal{A}\mathbf{f}_{col}(\mathbf{x}) = (1, |\mathbf{z}_{ch}|e^{i\varphi}) \cdot (f_{lu}(\mathbf{x}), \mathbf{f}_{ch}(\mathbf{x})) = (f_{lu}(\mathbf{x}), |\mathbf{z}_{ch}|e^{i\varphi}\mathbf{f}_{ch}(\mathbf{x}))$ change both hue and saturation of the initial image. The set of all such transformations forms the *chromatic group* $\mathbb{CG}(2) := \{(1, \mathbf{z}_{ch}) = (1, |\mathbf{z}|e^{i\varphi}) \ | \ |\mathbf{z}_{ch}|e^{i\varphi} \in \mathbf{C}\}$.

4. If $\mathcal{A} = (a_{lu}, \mathbf{z}_{ch}) = (a_{lu}, |\mathbf{z}_{ch}|e^{i\varphi})$, where $a > 0$, then transformations $\mathbf{f}_{col}^{Ret}(\mathbf{x}) \longrightarrow \mathcal{A}\mathbf{f}_{col}(\mathbf{x}) = (a_{lu}, |\mathbf{z}_{ch}|e^{i\varphi}) \cdot (f_{lu}(\mathbf{x}), \mathbf{f}_{ch}(\mathbf{x})) = (a_{lu}f_{lu}(\mathbf{x}), |\mathbf{z}_{ch}|e^{i\varphi}\mathbf{f}_{ch}(\mathbf{x}))$ change luminance, hue and saturation of the initial image. The set of all such transformations forms the *luminance-chromatic group* $\mathbb{LCG}(2) = \left\{(a_{lu}, \mathbf{z}_{ch}) \ | \ (a_{lu} \in \mathbf{R}^+) \& \ (\mathbf{z}_{ch} \in \mathbf{C})\right\}$.

3.1.2 The Young–Helmholtz k-cycle model of multicolor images.

We will interpret multicolor images (1) as multiplet-valued signals

$$\mathbf{f}_{Mcol}(\mathbf{x}) = f_0(\mathbf{x})1_{Mcol} + f_1(\mathbf{x})\varepsilon_{Mcol}^1 + \ldots + f_{k-1}(\mathbf{x})\varepsilon_{Mcol}^{k-1}, \quad \mathbf{x} \in \mathbf{R}^n,$$

which take values in the multiplet k-cycle algebra

$$A_k^{Mcol} := \mathbf{R}1_{Mcol} + \mathbf{R}\varepsilon_{Mcol}^1 + \ldots + \mathbf{R}\varepsilon_{Mcol}^{k-1},$$

where $1, \varepsilon^1_{Mcol}, \dots, \varepsilon^{k-1}_{Mcol}$ ($\varepsilon^k_{Mcol} = 1$) are *multicolor hyperimaginary units*. We will denote them by $1, \varepsilon^1, \dots, \varepsilon^{k-1}$. This algebra is called the *multiplet (multicolor) algebra*. One can show that this algebra is the direct sum of the real and complex fields:

$$A^{Mcol}_k = \sum_{i=1}^{k_{lu}} [\mathbf{R} \cdot \mathbf{e}^i_{lu}] + \sum_{j=1}^{k_{ch}} [\mathbf{C} \cdot \mathbf{E}^j_{ch}] = \mathbf{R}^{k_{lu}} \oplus \mathbf{C}^{k_{ch}},$$

where $k_{lu} = 1, 2$ and $k_{ch} = \frac{k}{2}, \frac{k-1}{2}$ if k is odd or even, respectively, and $\mathbf{e}^i_{lu}$ and $\mathbf{E}^j_{ch}$ are orthogonal idempotent units such that $(\mathbf{e}^i_{lu})^2 = \mathbf{e}^i_{lu}$, $\mathbf{e}_i \mathbf{e}_j = \mathbf{e}_j \mathbf{e}_i$, $(\mathbf{E}^j_{ch})^2 = \mathbf{E}^j_{ch}$, $\mathbf{E}^i_{ch} \mathbf{E}^j_{ch} = \mathbf{E}^j_{ch} \mathbf{E}^i_{ch}$, and $\mathbf{e}^i_{lu} \mathbf{E}^j_{ch} = \mathbf{E}^j_{ch} \mathbf{e}^i_{lu} = 0$, for all i, j. Every multiplet $\mathcal{C}$ can be represented as a linear combination of k_{lu} "scalar" parts and k_{ch} "complex" parts:

$$\mathcal{C} = \sum_{i=1}^{k_{lu}} (a_i \cdot \mathbf{e}^i_{lu}) + \sum_{j=1}^{k_{ch}} (\mathbf{z}_j \cdot \mathbf{E}^j_{ch}).$$

The real numbers $a_i \in \mathbf{R}$ are called the *multi-intensity numbers* and complex numbers $\mathbf{z}_j = b + ic \in \mathbf{C}$ are called the *multi-chromacity numbers*. Now we will interpret the multicolor nD image appearing on the nD retina as a multiplet-valued nD signal of the form:

$$\mathbf{f}^{Ret}_{Mcol}(\mathbf{x}) = \sum_{i=1}^{k_{lu}} [f^i_{lu}(\mathbf{x}) \cdot \mathbf{e}^i_{lu}] + \sum_{j=1}^{k_{ch}} [f^j_{ch}(\mathbf{x}) \cdot \mathbf{E}^j_{ch}] =$$

$$\left(f^1_{lu}(\mathbf{x}), \dots, f^{k_{lu}}_{lu}(\mathbf{x}); \ f^1_{ch}(\mathbf{x}), \dots, f^{k_{ch}}_{ch}(\mathbf{x}) \right).$$

Here, the argument $\mathbf{x}$ belongs to the nD vector part $\mathcal{GR}^{p,q,r}_n$ of the space algebra $\mathcal{A}^{Sp}_{2^n}$. Changes in both the surrounding world and mental spaces of reality, such as multi-intensity and multi-color, can be treated in the language of multiplet algebra as the action of the following transformation group in the perceptual multicolor space A^{Mcol}_k. Letting

$$\mathcal{A} = (a^1_{lu}, \dots, a^{k_{lu}}_{lu}, \ \mathbf{z}^1_{ch}, \dots, \mathbf{z}^{k_{ch}}_{ch}) =$$

$$(a^1, \dots, a^{k_{lu}}; \ |\mathbf{z}^1_{ch}| e^{i\varphi_1}, |\mathbf{z}^2_{ch}| e^{i\varphi_2}, \dots, |\mathbf{z}^{k_{ch}}_{ch}| e^{i\varphi_{k_{ch}}}),$$

the following transformations of a multicolor image

$$\mathbf{f}_{Mcol}(\mathbf{x}) \longrightarrow \mathcal{A}\mathbf{f}_{Mcol}(\mathbf{x}) =$$

$$\left(a^1 f^1_{lu}(\mathbf{x}), \dots, a^{k_{lu}} f^{k_{lu}}_{lu}(\mathbf{x}); \ |\mathbf{z}^1_{ch}| e^{i\varphi_1} f^1_{ch}(\mathbf{x}), \dots, |\mathbf{z}^{k_{ch}}_{ch}| e^{i\varphi_{k_{ch}}} f^{k_{ch}}_{ch}(\mathbf{x}) \right)$$

change multi-luminancies, multi-hues and multi-saturations of the initial multicolor image. The set of all such transformations forms the *multi-luminance/ multi-chromatic group* $\mathbb{MLCG}(k) := \{(a_{lu}^1, \ldots, a_{lu}^{k_{lu}};\ \mathbf{z}_{ch}^1, \ldots,$ $\mathbf{z}_{ch}^{k_{ch}})\ |\ a_{lu}^i \in \mathbf{R}^+,\ \mathbf{z}_{ch}^j \in \mathbf{C}\}.$

3.1.3 Multiorthounitary transforms of multicolor images.

A 2D discrete multicolor image can be defined as a 2D array $\mathbf{f}_{Mcol}^{Ret} = \left[\mathbf{f}_{Mcol}^{Ret}(i,j)\right]_{i,j=1}^N$, i.e., as a 2D discrete $\mathcal{A}_k^{Mcol}$-valued function of one of the following forms:

$$\mathbf{f}_{col}^{Ret}(i,j) : \mathbf{Z}_N^2 \longrightarrow \mathcal{A}_k^{Mcol}, \quad \mathbf{f}_{col}^{Ret}(i,j) : \mathbf{Z}_N^2 \longrightarrow \mathbf{R}^{k_{lu}} \oplus \mathbf{C}^{k_{ch}}.$$

Here, every color pixel $\mathbf{f}_{Mcol}^{Ret}(i,j)$ at position (i,j) is a multicolor number of the type $\mathbf{f}_{Mcol}^{Ret}(i,j) = f_0(i,j)\mathbf{1}_{Mcol} + f_1(i,j)\varepsilon_{Mcol}^1 + \ldots + f_{k-1}(i,j)\varepsilon_{Mcol}^{k-1}$ or of the type $\mathbf{f}_{Mcol}^{Ret}(i,j)) = \sum_{i=1}^{k_{lu}}[f_{lu}^i(i,j) \cdot \mathbf{e}_{lu}^i] + \sum_{j=1}^{k_{ch}}[f_{ch}^j(i,j) \cdot \mathbf{E}_{ch}^j] = \left(f_{lu}^1(i,j), \ldots, f_{lu}^{k_{lu}}(i,j);\ f_{ch}^1(i,j), \ldots, f_{ch}^{k_{ch}}(i,j)\right)$. In particular, for color images, we have $\mathbf{f}_{col}^{Ret}(i,j) = f_r(i,j) + f_g(i,j)\varepsilon^1 + f_b(i,j)\varepsilon^2$ and $\mathbf{f}_{col}^{Ret}(i,j) = f_{lu}(i,j)\mathbf{e}_{lu} + \mathbf{f}_{ch}(i,j)\mathbf{E}_{ch}$. The set of all such images forms N^2D Greaves–Hilbert space $\left(\mathcal{A}_k^{Mcol}\right)^{N^2}$. The vector space structure of this space is defined with multiplication by triplet-valued scalars $(\mathcal{C}f)(i,j) := \mathcal{C}f(i,j)$.

Our N^2D Greaves–Hilbert space $\left(\mathcal{A}_k^{Mcol}\right)^{N^2}$ over $\mathcal{A}_k^{Mcol}$ can be interpreted as kN^2D Hilbert space over $\mathbf{R}$, i.e., as $\left(\mathcal{A}_k^{Mcol}\right)^{N^2} = \mathbf{R}^{N^2}\mathbf{1} + \mathbf{R}^{N^2}\varepsilon^1 + \ldots + \mathbf{R}^{N^2}\varepsilon^{k-1}$, or as a direct sum of N^2D real and complex Hilbert spaces

$$\left(\mathcal{A}_k^{Mcol}\right)^{N^2} = \left[\bigoplus_{i=1}^{k_{lu}} \mathbf{R}^{N^2}\mathbf{e}_{lu}^i\right] + \left[\bigoplus_{j=1}^{k_{ch}} \mathbf{C}^{N^2}\mathbf{E}_{ch}^j\right] = \left[\bigoplus_{i=1}^{k_{lu}} \mathbf{R}^{N^2}\right] \oplus \left[\bigoplus_{j=1}^{k_{ch}} \mathbf{C}^{N^2}\right].$$

In particular, for color images $\left(\mathcal{A}_3^{col}\right)^{N^2} = \mathbf{R}^{N^2} + \mathbf{R}^{N^2}\varepsilon^1 + \mathbf{R}^{N^2}\varepsilon^2$, or $\left(\mathcal{A}_3^{col}\right)^{N^2} = \mathbf{R}^{N^2} \oplus \mathbf{C}^{N^2}$, where $\mathbf{R}^{N^2}$, $\mathbf{R}^{N^2}\varepsilon^1$, $\mathbb{R}^{N^2}\varepsilon^2$ are real N^2 Hilbert spaces of red, green, and blue images, respectively, $\mathbf{R}^{N^2}$ is N^2D real space of gray-level images, and $\mathbf{C}^{N^2}$ is N^2D complex space of chromatic images. Let $\left(\mathcal{A}_k^{Mcol}\right)^{N^2}$ be a N^2 Greaves–Hilbert space over $\mathcal{A}_k^{Mcol}$. We say that the operator $\mathcal{L}_{2D} : \left(\mathcal{A}_k^{Mcol}\right)^{N^2} \to \left(\mathcal{A}_k^{Mcol}\right)^{N^2}$, $\mathcal{L}_{2D}[\mathbf{f}_{Mcol}^{Ret}] = \mathbf{F}_{Mcol}^{Ret}$ is *multicolor linear* if and only if for all $\mathbf{f}_{Mcol}^{Ret}, \mathbf{g}_{Mcol}^{Ret} \in \left(\mathcal{A}_k^{Mcol}\right)^{N^2}$ and for all $\mathcal{C} \in \mathcal{A}_k^{Mcol}$ $\mathcal{L}_{2D}[\mathbf{f}_{Mcol}^{Ret} + \mathbf{g}_{Mcol}^{Ret}] = \mathcal{L}_{2D}[\mathbf{f}_{Mcol}^{Ret}] + \mathcal{L}_{2D}[\mathbf{g}_{Mcol}^{Ret}]$, $\mathcal{L}_{2D}[\mathcal{C}\mathbf{f}_{Mcol}^{Ret}] = \mathcal{C}\mathcal{L}_{2D}[\mathbf{f}_{Mcol}^{Ret}]$. Otherwise, we call $\mathcal{L}_{2D}$ *multicolor nonlinear*. If $\mathcal{L}_{2D}$ is a multicolor linear operator and $||\det(\mathcal{L}_{2D})||_2^2$ does not vanish, then we call operator $\mathfrak{L}_{2D}$ *nonsingular*. Otherwise, we call $\mathfrak{L}_{2D}$ *singular*. The inverse $\mathfrak{L}_{2D}^{-1}$ of an operator $\mathfrak{L}_{2D}$ exists if and only if $\mathfrak{L}_{2D}$ is nonsingular. $\mathfrak{L}_{2D}^{-1}$ is calculated in the same way as an ordinary inverse matrix. The set of nonsingular multicolor linear operators form the *general multicolor linear groups over multiplet algebra* A_k^{Mcol} and is denoted by $\mathbf{GL}(N, A_k^{Mcol})$. It is then easy to define the adjoint operator $\mathfrak{L}_{2D}^*$ for the color linear operator $\mathfrak{L}_{2D}$ whose essential properties are $\langle \mathbf{f}_{Mcol}^{Ret} | \mathfrak{L}_{2D} \mathbf{g}_{Mcol}^{Ret} \rangle = \langle \mathfrak{L}_{2D}^* \mathbf{f}_{Mcol}^{Ret} | \mathbf{g}_{Mcol}^{Ret} \rangle$, and for $_1\mathfrak{L}_{2D}\, _2\mathfrak{L}_{2D}$ the property $\left(_1\mathfrak{L}_{2D}\, _2\mathfrak{L}_{2D} \right)^* = _2\mathfrak{L}_{2D}^*\, _1\mathfrak{L}_{2D}^*$ is true.

DEFINITION 1 *A multicolor linear operator* $\mathfrak{L}_{2D}$ *on* $\left(A_k^{Mcol} \right)^{N^2}$ *is said to be* multi-orthounitary *if* $\mathfrak{L}_{2D}^{-1} = \mathfrak{L}_{2D}^*$.

Multi-orthounitary operators form the multi-orthounitary group $\mathbb{MOU}$ $= \mathbb{MOU}(A_k^{Mcol})$. This group is isomorphic to the direct sum of k_{lu} orthogonal and k_{ch} unitary groups, $\mathbb{MOU}(A_k^{Mcol}) =$

$$\left[\bigoplus_{i=1}^{k_{lu}} \mathbb{O}(\mathbf{R}) \mathbf{e}_{lu}^i \right] + \left[\bigoplus_{j=1}^{k_{ch}} \mathbb{U}(\mathbf{C}) \mathbf{E}_{ch}^j \right] = \left[\bigoplus_{i=1}^{k_{lu}} \mathbb{O}(\mathbf{R}) \right] \oplus \left[\bigoplus_{j=1}^{k_{ch}} \mathbb{U}(\mathbf{C}) \right].$$

In particular, the orthounitary group of transformations for color images has the decomposition: $\mathbb{OU}(A_3^{col}) = \mathbb{O}(\mathbf{R})\mathbf{e}_{lu} + \mathbb{U}(\mathbf{C})\mathbf{E}_{ch}$. Every element of $\mathbb{MOU}(A_k^{Mcol})$ and $\mathbb{OU}(A_3^{col})$ has the representation:

$$\mathfrak{L}_{2D} = \left[\bigoplus_{i=1}^{k_{lu}} \mathbf{O}_{2D}^i \mathbf{e}_{lu}^i \right] + \left[\bigoplus_{j=1}^{k_{ch}} \mathcal{U}_{2D}^j \mathbf{E}_{ch}^j \right], \quad \mathfrak{L}_{2D} = \mathbf{O}_{2D}\mathbf{e}_{lu} + \mathcal{U}_{2D}\mathbf{E}_{ch}, \quad (2)$$

where $\mathbf{O}_{2D}^i \in \mathbb{O}^i(\mathbf{R})$, $\mathbf{O}_{2D} \in \mathbb{O}(\mathbf{R})$, and $\mathcal{U}_{2D}^j \in \mathbb{U}^j(\mathbf{C})$ $\mathcal{U}_{2D} \in \mathbb{U}(\mathbf{C})$ are orthogonal and unitary transforms, respectively. For multicolor image processing we shall use separable $2D$ transforms.

DEFINITION 2 *We call the multi-orthounitary transform* $\mathfrak{L}_{2D}[\mathbf{f}_{Mcol}^{Ret}]$ *separable if it can be represented by* $\mathfrak{L}_{2D}[\mathbf{f}_{Mcol}^{Ret}] = \mathfrak{L}_{1D}[\mathbf{f}_{Mcol}^{Ret}]\mathfrak{M}_{1D}$, *i.e.*, $\mathfrak{L}_{2D} = \mathfrak{L}_{1D} \otimes \mathfrak{M}_{1D}$ *is the tensor product of two* $1D$ *multi-orthounitary transforms of the form:*

$$\mathfrak{L}_{2D} = \mathfrak{L}_{1D} \otimes \mathfrak{L}_{1D} = \left[\bigoplus_{i=1}^{k_{lu}} \left(\mathbf{O}_1^i \otimes \mathbf{O}_2^i \right) \mathbf{e}_{lu}^i \right] + \left[\bigoplus_{j=1}^{k_{ch}} \left(\mathcal{U}_1^j \otimes \mathcal{U}_2^j \right) \mathbf{E}_{ch}^j \right] \quad (3)$$

for multicolor images, and the tensor product of two $1D$ *orthounitary transforms* $\mathfrak{L}_{2D} = \mathfrak{L}_{1D} \otimes \mathfrak{L}_{1D} = (\mathbf{O}_1 \otimes \mathbf{O}_2)\mathbf{e}_{lu} + (\mathcal{U}_1 \otimes \mathcal{U}_2)\mathbf{E}_{ch}$ *for color images, where* $\mathbf{O}_1^i, \mathbf{O}_2^i$ *and* $\mathcal{U}_1^j, \mathcal{U}_2^j$ *are* $1D$ *orthogonal and unitary transforms, respectively.*

Figure 5.   Color Walsh–Fourier $\mathfrak{WF} = \mathbf{We}_{lu} + \mathcal{F}\mathbf{E}_{ch}$, Hartley–Fourier $\mathfrak{HIF} = \mathbf{Hte}_{lu} + \mathcal{F}\mathbf{E}_{ch}$ and Haar–Fourier $\mathfrak{HRF} = \mathbf{Hre}_{lu} + \mathcal{F}\mathbf{E}_{ch}$ transforms, respectively

In particular, we can obtain any orthounitary transform, using any two pairs of orthogonal $\mathbf{O}_1, \mathbf{O}_2$ and unitary transforms $\mathcal{U}_1, \mathcal{U}_2$. It is possible to use one pair of orthogonal and unitary transforms, when $\mathbf{O}_1 = \mathbf{O}_2 = \mathbf{O}$ and $\mathcal{U}_1 = \mathcal{U}_2 = \mathcal{U}$. In this case we obtain a wide family of orthounitary transforms of the form: $\mathfrak{L}_{2D} = (\mathbf{O} \otimes \mathbf{O})\mathbf{e}_{lu} + (\mathcal{U} \otimes \mathcal{U})\mathbf{E}_{ch}$, using different 1D orthogonal transforms. The following table shows some of the possibilities:

	$\mathcal{F}$	$\mathcal{CW}$	$\mathcal{CFP}$	$\mathcal{CH}$
W	$\mathbf{W}e_{lu} + \mathcal{F}\mathbf{E}_{ch}$	$\mathbf{W}e_{lu} + \mathcal{CW}\mathbf{E}_{ch}$	$\mathbf{W}e_{lu} + \mathcal{CFP}\mathbf{E}_{ch}$	$\mathbf{W}e_{lu} + \mathcal{CH}\mathbf{E}_{ch}$
Hd	$\mathbf{Hd}e_{lu} + \mathcal{F}\mathbf{E}_{ch}$	$\mathbf{Hd}e_{lu} + \mathcal{CW}\mathbf{E}_{ch}$	$\mathbf{Hd}e_{lu} + \mathcal{CFP}\mathbf{E}_{ch}$	$\mathbf{Hd}e_{lu} + \mathcal{CH}\mathbf{E}_{ch}$
Ht	$\mathbf{Ht}e_{lu} + \mathcal{F}\mathbf{E}_{ch}$	$\mathbf{Ht}e_{lu} + \mathcal{CW}\mathbf{E}_{ch}$	$\mathbf{Ht}e_{lu} + \mathcal{CFP}\mathbf{E}_{ch}$	$\mathbf{Ht}e_{lu} + \mathcal{CH}\mathbf{E}_{ch}$
Hr	$\mathbf{Hr}e_{lu} + \mathcal{F}\mathbf{E}_{ch}$	$\mathbf{Hr}e_{lu} + \mathcal{CW}\mathbf{E}_{ch}$	$\mathbf{Hr}e_{lu} + \mathcal{CFP}\mathbf{E}_{ch}$	$\mathbf{Hr}e_{lu} + \mathcal{CH}\mathbf{E}_{ch}$
Wv	$\mathbf{Wv}e_{lu} + \mathcal{F}\mathbf{E}_{ch}$	$\mathbf{Wv}e_{lu} + \mathcal{CW}\mathbf{E}_{ch}$	$\mathbf{Wv}e_{lu} + \mathcal{CFP}\mathbf{E}_{ch}$	$\mathbf{Wv}e_{lu} + \mathcal{CH}\mathbf{E}_{ch}$

where **W**, **Hd**, **Ht**, **Hr**, **Wv** are Walsh, Hadamard, Hartley, Haar, and Wavelet orthogonal transforms, and $\mathcal{F}$, $\mathcal{CW}$, $\mathcal{CFP}$, and $\mathcal{CH}$ are Fourier, complex Walsh, complex Fourier–Prometheus, complex Haar transforms. Every pair $(\mathbf{O}, \mathcal{U})$ of an orthogonal $\mathbf{O}$ and a unitary $\mathcal{U}$ transform generates an orthounitary (color) transform $\mathfrak{L} = \mathbf{O}e_{lu} + \mathcal{U}\mathbf{E}_{ch}$. Some examples of basis color functions of color transforms $\mathfrak{W}\mathfrak{F} = \mathbf{W}e_{lu} + \mathcal{F}\mathbf{E}_{ch}$, $\mathfrak{H}\mathfrak{T}\mathfrak{F} = \mathbf{Ht}e_{lu} + \mathcal{F}\mathbf{E}_{ch}$, $\mathfrak{H}\mathfrak{R}\mathfrak{F} = \mathbf{Hr}e_{lu} + \mathcal{F}\mathbf{E}_{ch}$ are shown in Fig. 5.

3.2 Algebraization of the Hering model

3.2.1 The Hering Z/2Z-graded model of color images.

Let us consider 3D color space $\mathbf{R}^3_{col} = \mathbf{R}J_R + \mathbf{R}J_G + \mathbf{R}J_B$ spanned by three new hyperimaginary units J_i $(i = 1, 2, 3$ or $i = R, G, B)$: $J_1 = J_R$ (red unit), $J_2 = J_G$ (green unit), $J_3 = J_B$ (blue unit). We assume $J_R^2 = a_r$, $J_G^2 = a_g$, $J_B^2 = a_b$, where $a_r, a_g, a_b \in \{+1, 0, -1\}$. We assume $J_i^2 = +1$ for $i = 1, \dots, \alpha$, $J_i^2 = -1$ for $i = \alpha + 1, \dots, \alpha + \beta$, $J_i^2 = 0$ for $i = \alpha + \beta + 1, \dots, \alpha + \beta + \gamma$, where $\alpha + \beta + \gamma = 3$. Further, we construct a new color Clifford algebra $\mathcal{A}_{2^3}^{col(\alpha,\beta,\gamma)}(J_R, J_G, J_B) = \mathbf{R}J_{Bl} + (\mathbf{R}J_R + \mathbf{R}J_G + \mathbf{R}J_B) + (\mathbf{R}J_{RG} + \mathbf{R}J_{RB} + \mathbf{R}J_{GB}) + \mathbf{R}J_{Wh}$, where $J_{Bl} = 1$, $J_{Wh} = J_R J_G J_B$ are black and white units, $I_{RG} := J_R J_G$, $J_{RB} := J_R J_B$, $J_{GB} := J_G J_B$. This is an algebra of generalized color octonions with signature (α, β, γ).

DEFINITION 3 *The functions* $\mathbf{f}_{col}^{VC} : \mathcal{GR}_n^{p,q,r} \to \mathcal{A}_8^{col(\alpha,\beta,\gamma)}(J_R, J_G, J_B)$ *of the form:*

$$\mathbf{f}_{col}^{VC}(\mathbf{x}) = f_{Bl}(\mathbf{x})J_{\emptyset} + \left(f_R(\mathbf{x})J_R + f_G(\mathbf{x})J_G + f_B(\mathbf{x})J_B\right) +$$

$$+ \left(f_{RG}(\mathbf{x})J_{RG} + f_{RB}(\mathbf{x})J_{RB} + f_{GB}(\mathbf{x})J_{GB}\right) + f_{Wh}(\mathbf{x})J_{Wh}$$

are called the $\mathcal{A}_{2^3}^{col(\alpha,\beta,\gamma)}$*-valued color nD images appearing in the human Visual Cortex.*

The second opponent cells map R,G,B components on the 4D unit sphere $f_{Bl}^2 + f_{RG}^2 + f_{RB}^2 + f_{GB}^2 = 1$, where $f_{RG}, f_{RB}, f_{GB}, f_{Bl}$ are black, red-green, red-blue and green-blue components, respectively. Therefore, resulting from the capacity of this algebraic model of color image, we can formulate the *spin-valued function*

$$\mathbf{f}_{col}^{VC}(\mathbf{x}) := f_{Bl}(\mathbf{x})J_{Bl} + \Big(f_{RG}(\mathbf{x})J_{RG} + f_{RB}(\mathbf{x})J_{RB} + f_{GB}(\mathbf{x})J_{GB}\Big)$$

which has values in the spin part $\mathrm{Spin}\Big(\mathcal{A}_8^{col}(J_R, J_G, J_B)\Big)$ of the color Clifford algebra.

DEFINITION 4 *The functions* $\mathbf{f}_{col}^{VC} : \mathbf{R}^2 \longrightarrow \mathrm{Spin}\Big(\mathcal{A}_8^{col}(J_R, J_G, J_B)\Big)$, *and* $\mathbf{f}_{col}^{VC} : \mathbf{R}^n \longrightarrow \mathrm{Spin}\Big(\mathcal{A}_8^{col}(J_R, J_G, J_B)\Big)$ *are called the* spin-valued color 2D *and* nD *images,* respectively.

For this model, color changes in both the surrounding world and mental spaces can be described as the action of the color spinor transformation group in the perceptual color spaces $\mathcal{A}_8^{col}$ and $\mathrm{Spin}\Big(\mathcal{A}_8^{col}\Big)$.

If $\mathcal{C}_0 \in \mathrm{Spin}\Big(\mathcal{A}_8^{col}\Big)$, then the transformations $\mathbf{f}_{col}^{VC}(\mathbf{x}) \longrightarrow \mathcal{C}_0\mathbf{f}_{col}^{VC}(\mathbf{x})$, $\mathbf{f}_{col}^{VC}(\mathbf{x}) \longrightarrow \mathbf{f}_{col}^{VC}(\mathbf{x})\mathcal{C}_0^{-1}$ and $\mathbf{f}_{col}^{VC}(\mathbf{x}) \longrightarrow \mathcal{C}_0\mathbf{f}_{col}^{VC}(\mathbf{x})\mathcal{C}_0^{-1}$ are called the left, right and two-sided spinor-color transformations of $\mathbf{f}_{col}^{VC}(\mathbf{x})$, respectively.

3.2.2   The Hering $\mathbf{Z}/k\mathbf{Z}$-graded model of multicolor images.

To form the algebraic model of multicolor images in the animals' VC, we consider a k^mD *multicolor quantum Clifford algebra*

$$\mathcal{QCA}_{k^m}^{Mcol(\alpha_0,\alpha_1,\dots,\alpha_{k-1})}(J_1,\dots,J_m)$$

(i.e., Clifford algebra with signature $(\alpha_0, \alpha_1, \dots, \alpha_{k-1})$ and deformed by a k primitive root of unity ω_k) generated by multicolor hypercomplex units $J_1, \dots, J_m$ with relations $J_i^k = \omega^0$ for α_0 hypercomplex units, $J_i^k = \omega^1$ for α_1 hypercomplex units, $\dots$, $J_i^k = \omega^{k-1}$ for α_{k-1} hypercomplex units, and $J_j J_i = \omega_k J_i J_j$, if $i < j$, where $\alpha_0 + \alpha_1 + \dots + \alpha_{k-1} = m$, $\omega := \sqrt[k]{1}$ [9]. The elements $\mathbf{J^s} = J_1^{s_1} \cdots J_m^{s_m}$ form a basis of $\mathcal{QCA}_{k^m}^{Mcol(\alpha_0,\alpha_1,\dots,\alpha_{k-1})}(J_1,\dots,J_m)$, where $\mathbf{s} = (s_1,\dots,s_m) \in \mathbf{B}_k^m$, $s_i \in \mathbf{B}_k = \{0,\dots,k-1\}$. We shall denote this algebra by $\mathcal{QCA}_{k^m}^{Mcol}$, if $J_1,\dots,J_m$, and $(\alpha_0,\alpha_1,\dots,\alpha_{k-1})$ are fixed. In the particular case when $k = 2$ ($\omega_k = -1$), $m = 3$ and $\alpha_0 = \alpha$, $\alpha_1 = \beta$, $\alpha_2 = \gamma$, the k^mD quantum Clifford algebra $\mathcal{QCA}_{k^m}^{Mcol(\alpha_0,\alpha_1,\dots,\alpha_{k-1})}(J_1,\dots,J_m)$ is the generalized color octonion algebra $\mathcal{A}_8^{col(\alpha,\beta,\gamma)}$. We can make $\mathcal{QCA}_{k^m}^{Mcol}$ be a *ranked* and $\mathbf{Z}/k\mathbf{Z}$-*graded algebra*. Let $r(\mathbf{s})$ be the *Hamming weight* of $\mathbf{s}$, i.e., a functional $r : \mathbf{B}_k^m \longrightarrow [0, m-1]$ defined by $r(\mathbf{b}) := \sum_{i=1}^m \left[1 - \delta_{0,s_i}\right]$, and let $\partial(\mathbf{s}) = \sum_{i=1}^{m} s_i \pmod{k}$ be the *grad* of $\mathbf{s}$. We set $\mathcal{QCA}_{km}^{[r]} = \mathbf{SPAN}\{\mathbf{J^s}| \ r(\mathbf{s}) = r\}$, obtaining the ranked sum

$$\mathcal{QCA}_{km}^{Mcol} = \bigoplus_{r=0}^{m} \mathcal{QCA}_{km}^{[r]}.$$

By setting $\mathcal{QCA}_{km}^{\{l\}} = \mathbf{SPAN}\{\mathbf{J^s}| \ \partial(\mathbf{s}) \equiv l \bmod k\}$, we get the $\mathbf{Z}/k\mathbf{Z}$-graded $\mathbf{R}$-algebra $\mathcal{QCA}_{km}^{Mcol} = \bigoplus_{i=0}^{k-1} \mathcal{QCA}_{km}^{\{l\}}$. We put $\partial(\mathcal{C}) = l$ and say that l is the *degree of* $\mathcal{C}$. . We say that Φ_g and Φ_r are graded and ranked automorphisms if $\Phi_g\left(\mathcal{QCA}_{km}^{\{l\}}\right) = \mathcal{QCA}_{km}^{\{l\}}$ and $\Phi_r\left(\mathcal{QCA}_{km}^{[r]}\right) = \mathcal{QCA}_{km}^{[r]}$ for all $l = 0, 1, \ldots, k-1$, and $r = 0, 1, \ldots, m$, respectively. Let $\mathcal{QCA}_{km}^{[1]} = \mathbf{SPAN}\{\mathbf{J^s} \ | \ r(\mathbf{s}) = 1\}$ be the vector part of the quantum Clifford algebra $\mathcal{QCA}_{km}^{Mcol}$. Then the set of automorphisms of unit modulus $\Phi_r\left(\mathcal{QCA}_{km}^{[1]}\right) = \mathcal{QCA}_{km}^{[1]}$ forms a transformation group of $\mathcal{QCA}_{km}^{[1]}$. We shall call it the *quantum spinor group* and denote it by

$$\mathrm{QSpin}\left(\mathcal{QCA}_{km}^{Mcol}\right).$$

DEFINITION 5 *The functions*

$$\mathbf{f}_{Mcol}^{VC}(\mathbf{x}): \mathcal{GR}_n^{Sp(p,q,r)} \longrightarrow \mathcal{QCA}_{km}^{Mcol(\alpha_0, \alpha_1, \ldots, \alpha_{k-1})}(J_1, \ldots, J_m),$$

$$\mathbf{f}_{Mcol}^{VC}(\mathbf{x}): \mathcal{GR}_n^{Sp(p,q,r)} \longrightarrow \mathrm{QSpin}\left(\mathcal{QCA}_{km}^{Mcol(\alpha_0, \alpha_1, \ldots, \alpha_{k-1})}(J_1, \ldots, J_m)\right)$$

are called the nD Cliffordean-valued *and* quantum-spin-valued *images appearing in the animals' VC.*

For this model, multicolor changes in both the surrounding world and mental spaces can be described as the action of the quantum spinor transformation group in the perceptual color spaces $\mathrm{QSpin}\left(\mathcal{QCA}_{km}^{Mcol}\right)$ and $\mathcal{QCA}_{km}^{Mcol}$. If $\mathcal{C}_0 \in \mathrm{QSpin}\left(\mathcal{QCA}_{km}^{Mcol}\right)$, then the transformations

$$\mathbf{f}_{Mcol}^{VC}(\mathbf{x}) \longrightarrow \mathcal{C}_0 \mathbf{f}_{Mcol}^{VC}(\mathbf{x}),$$

$$\mathbf{f}_{Mcol}^{VC}(\mathbf{x}) \longrightarrow \mathbf{f}_{Mcol}^{VC}(\mathbf{x})\mathcal{C}_0^{-1} \quad \text{and} \quad \mathbf{f}_{Mcol}^{VC}(\mathbf{x}) \longrightarrow \mathcal{C}_0 \mathbf{f}_{Mcol}^{VC}(\mathbf{x})\mathcal{C}_0^{-1}$$

are called the left, right and two-sided spinor-color transformations of $\mathbf{f}_{Mcol}^{VC}(\mathbf{x})$, respectively. Further, we interpret an image as an embedding of a manifold in a spatial-multicolor Clifford algebra of higher dimension. The embedding manifold is a "hybrid" (spatial-multicolor) space that includes spatial coordinates as well as color coordinates. For example, a 2D color image is considered as a 3D manifold in the 5D spatial-color space $\mathbf{R}^5_{SpCol}(I_1, I_2; J_R, J_G, J_B) = (\mathbf{R}I_1 + \mathbf{R}I_2) \oplus (\mathbf{R}J_R + \mathbf{R}J_G + \mathbf{R}J_B) = \mathbf{R}^2_{Sp} \oplus \mathbf{R}^3_{col}$, whose coordinates are (x, y, f_R, f_G, f_B), where $x \in \mathbf{R}I_1$, $y \in \mathbf{R}I_2$ are spatial coordinates and $f_R \in \mathbf{R}J_R$, $f_G \in \mathbf{R}J_G$, $f_B \in \mathbf{R}J_B$, are color coordinates. It is clear that the geometrical, color and spatial-multicolor spaces $\mathbf{R}^n_{Sp}$, $\mathbf{R}^k_{Mcol}$, $\mathbf{R}^{n+m(k-1)}_{SpMcol}$ generate *spatial, color and quantum spatial-multicolor Clifford algebras* $\mathcal{A}^{Sp(p,q,r)}_{2^n}$, $QC\mathcal{A}^{Mcol(\alpha_0,\alpha_1,...,\alpha_{k-1})}_{k^m}$, $QC\mathcal{A}^{SpMcol(p,q,r;\alpha_0,\alpha_1,...,\alpha_{k-1})}_{2^n k^m}$, respectively. Here, all spatial hyperimaginary units commute with all multicolor units.

3.2.3   Clifford-unitary transforms of multicolor images.

2D discrete Cliffordean-valued images appearing in the animals' VC can be defined as a 2D array $\mathbf{f}^{VC}_{Mcol} := [\mathbf{f}^{VC}_{Mcol}(i,j)]^N_{i,j=1}$. Here, every pixel $\mathbf{f}^{VC}_{Mcol}(i,j)$ at position (i,j) is a quantum Clifford number of the type $\mathbf{f}^{VC}_{Mcol}(i,j) = \sum\limits_{\mathbf{s} \in \mathbf{B}^m_k} f_{\mathbf{s}}(i,j) \mathbf{J}^{\mathbf{s}}$. The set of all such images forms the N^2D

Clifford–Hilbert space $\mathbb{L}\left(\mathbf{Z}^2_N, QC\mathcal{A}^{Mcol}_{k^m}\right) = \left(QC\mathcal{A}^{Mcol}_{k^m}\right)^{N^2}$. The vector structure of this space is defined with multiplication by Clifford-valued scalars $(\mathcal{C}f)(i,j) := \mathcal{C}f(i,j)$. We say that the operator

$$\mathfrak{L}_{2D} : \left(QC\mathcal{A}^{Mcol}_{k^m}\right)^{N^2} \longrightarrow \left(QC\mathcal{A}^{Mcol}_{k^m}\right)^{N^2}, \quad \text{i.e.,} \quad \mathfrak{L}_{2D}[\mathbf{f}^{VC}_{Mcol}] = \mathbf{F}^{VC}_{Mcol}$$

is *Clifford linear* if and only if for all $\mathbf{f}^{VC}_{Mcol}, \mathbf{g}^{VC}_{Mcol} \in \left(QC\mathcal{A}^{Mcol}_{k^m}\right)^{N^2}$ and for all $\mathcal{C}_1, \mathcal{C}_2, \mathcal{B}_1, \mathcal{B}_2 \in QC\mathcal{A}^{Mcol}_{k^m}$

$$\mathfrak{L}_{2D}\left[\mathcal{C}_1 \mathbf{f}^{VC}_{Mcol} \mathcal{B}_1 + \mathcal{C}_2 \mathbf{g}^{VC}_{Mcol} \mathcal{B}_2\right] = \mathcal{C}_1 \mathfrak{L}_{2D}[\mathbf{f}^{VC}_{Mcol}] \mathcal{B}_1 + \mathcal{C}_2 \mathfrak{L}_{2D}[\mathbf{g}^{VC}_{Mcol}] \mathcal{B}_2.$$

Otherwise we call $\mathfrak{L}_{2D}$ *Clifford nonlinear*. If $\mathfrak{L}_{2D}$ is a Clifford linear operator and $\|\det(\mathfrak{L}_{2D})\|^2_2$ does not vanish, then we call the operator $\mathfrak{L}_{2D}$ *nonsingular*. Otherwise $\mathfrak{L}_{2D}$ is *singular*. The inverse $\mathfrak{L}^{-1}_{2D}$ of the operator $\mathfrak{L}_{2D}$ exists if and only if $\mathfrak{L}_{2D}$ is nonsingular. The set of nonsingular Clifford linear operators forms the *general Clifford linear groups* over the quantum Clifford algebra $QC\mathcal{A}^{Mcol}_{k^m}$ and is denoted by $\mathbb{GCL}\left(N, QC\mathcal{A}^{Mcol}_{k^m}\right)$. It is then easy to define the adjoint operator $\mathfrak{L}^*_{2D}$ for the Clifford linear operator $\mathfrak{L}_{2D}$, whose essential properties are

$$\langle \mathbf{f}^{VC}_{Mcol} | \mathfrak{L}_{2D} \mathbf{g}^{VC}_{Mcol}\rangle = \langle \mathfrak{L}^*_{2D} \mathbf{f}^{VC}_{Mcol} | \mathbf{g}^{VC}_{Mcol}\rangle,$$

and for the product $_1\mathfrak{L}_{2D} {}_2\mathfrak{L}_{2D}$ we have

$$\left(_1\mathfrak{L}_{2D} {}_2\mathfrak{L}_{2D}\right)^* = {}_2\mathfrak{L}^*_{2D} {}_1\mathfrak{L}^*_{2D}.$$

The Clifford linear operator $\mathfrak{L}_{2D}$ is said to be *Clifford-unitary* if

$$\mathfrak{L}_{2D}^{-1} = \mathfrak{L}_{2D}^*.$$

Clifford-unitary operators form the Clifford-unitary multicolor group $\mathbb{CU}\left(N, QCA_{km}^{Mcol}\right)$. For multicolor image processing we shall use separable $2D$ transforms. The Clifford-unitary transform $\mathfrak{L}_{2D}[\mathbf{f}_{Mcol}^{VC}] = \mathbf{F}_{Mcol}^{VC}$ is called separable if it can be represented by $\mathbf{F}_{Mcol}^{VC} = \mathfrak{L}_{2D}\left[\mathbf{f}_{Mcol}^{VC}\right] = \mathfrak{L}_{1D}\left[\mathbf{f}_{Mcol}^{VC}\right]\mathfrak{M}_{1D}$, i.e., $\mathfrak{L}_{2D} = \mathfrak{L}_{1D} \otimes \mathfrak{M}_{1D}$ is the tensor product of two $1D$ Clifford-unitary transforms.

4. Hypercomplex-valued invariants of nD multicolor images

4.1 Clifford-valued invariants

Let us assume that $\mathbf{f}_{Mcol}(\mathbf{x}) : \mathbf{R}_{Sp}^n \longrightarrow A_k^{Mcol}$ is an image of a multicolor nD object. It appears on the nD eye retina as a function $\mathbf{f}_{Mcol}^{Ret}(\mathbf{x})$ of space variables $\mathbf{x} \in \text{Vec}^1(A_{2n}^{Sp}) = GR_n^{Sp(p,q,r)}$ with values in the multicolor algebra $A_k^{Mcol} : \mathbf{f}_{Mcol}^{Ret}(\mathbf{x}) : GR_n^{Sp(p,q,r)} \longrightarrow A_k^{Mcol}$. This image can be considered in the VC as a function $\mathbf{f}_{Mcol}^{VC}(\mathbf{x})$ of the same space variables $\mathbf{x}$, but with values in the multicolor quantum Clifford algebra QCA_{kn}^{Mcol}, i.e., as $\mathbf{f}_{Mcol}^{VC}(\mathbf{x}) : GR_n^{Sp(p,q,r)} \longrightarrow QCA_{kn}^{Mcol}$. We shall denote both algebras A_k^{Mcol} and QCA_{kn}^{Mcol} by A^{Mcol}. Changes in the surrounding world can be treated in the language of the spatial-multicolor algebra as an action of two groups: the space affine group $\mathbf{Aff}(GR_n^{p,q,r})$ acting on the physical space $\text{Vec}^1(A_{2n}^{Sp}) = GR_n^{p,q,r}$ and the multi-color group $\mathbb{MLCG}(k)$ acting on A_k^{Mcol} (if $A^{Mcol} = A_k^{Mcol}$) or the quantum spin group $\text{Spin}(QCA_{km}^{Mcol})$ on QCA_{kn}^{Mcol} (if $A^{Mcol} = QCA_{kn}^{Mcol}$). We shall denote both groups $\text{Spin}(QCA_{km}^{Mcol})$ and $\mathbb{MLCG}(k)$ by $\mathcal{MCGR}^{Mcol}$. Let $\mathbf{G}^{SpMcol} = \mathbf{Aff}(GR_n^{Sp(p,q,r)}) \times \mathcal{MCGR}^{Mcol}$ be a spatial-multicolor group, and $(\mathbf{g}^{Sp}, \mathbf{g}^{Mcol}) \in \mathbf{G}^{SpMcol}$, where $\mathbf{g}^{Sp} \in \mathbf{Aff}(GR_n^{p,q,r})$, $\mathbf{g}^{Mcol} \in \mathcal{MCGR}^{Mcol}$. If $\mathbf{x} \in GR_n^{Sp(p,q,r)}$ is a generalized space Clifford number and $\mathcal{C} \in A^{Mcol}$ is a multicolor quantum Clifford number, then all products of the form $\mathbf{x}\mathcal{C}$ are called *spatial-color numbers*. They form a *space-color algebra* $A^{SpMcol} := A_{2n}^{Sp} \otimes A^{Mcol}$. Here, we assume that all spatial hyperimaginary units commute with all color units.

DEFINITION 6 *The A^{SpMcol}-valued functional $\mathcal{J} = \mathcal{M}[\mathbf{f}_{Mcol}(\mathbf{x})]$ of the image $\mathbf{f}_{Mcol}(\mathbf{x})$ is called the relative $\mathbf{G}^{SpMcol}$-invariant if*

$$\mathcal{J} = \mathcal{M}\left\{\mathbf{g}^{Mcol} \circ \mathbf{f}_{Mcol}(\mathbf{g}^{Sp} \circ \mathbf{x})\right\} = \mathcal{C} \cdot \mathcal{M}\{\mathbf{f}_{Mcol}(\mathbf{x})\} \cdot \mathcal{C}^{-1},$$

$\forall \mathbf{g} \in \mathbf{G}^{SpMcol}$, *where* $\mathcal{C}, \mathcal{C}^{-1}$ *are left and right* $\mathcal{A}^{SpMcol}$*-valued multipliers. If* $\mathcal{C} = 1$ *then* $\mathcal{J}$ *is called the* absolute invariant *and denoted by* $\mathcal{I}$.

Let $\mathbf{c}$ be the centroid of the image $\mathbf{f}_{Mcol}(\mathbf{x})$.

DEFINITION 7   *The functionals*

$$\mathcal{M}_p := \mathcal{M}\{\mathbf{f}_{Mcol}\} = \int_{\mathbf{x} \in \mathcal{GR}_n^{Sp}} (\mathbf{x} - \mathbf{c})^p \mathbf{f}_{Mcol}(\mathbf{x}) \mathbf{dx}$$

are called the central $\mathcal{A}^{SpMcol}$*-valued moments of the* n*D image* $\mathbf{f}_{Mcol}(\mathbf{x})$, *where* $p \in \mathbf{Q}$ *are rational numbers.*

Let us clarify the rules of moment transformations with respect to distortions of color and geometry of the initial images. If $\mathbf{f}_{Mcol}(\mathbf{x})$ is the initial image, then $\mathbf{f}_{\lambda, \mathcal{E}_0, \mathbf{w}}^{\mathcal{C}_0}(\mathbf{x}^*) = \mathcal{C}_0 \Big\{ \mathbf{f}_{Mcol} \left(\lambda \mathcal{E}_0 (\mathbf{x} + \mathbf{w}) \mathcal{E}_0^{-1} \right) \Big\} \mathcal{C}_0^{-1}$ denotes its $\mathcal{A}^{Mcol}$-multicolor and $\mathcal{GR}_n^{p,q,r}$-geometrical distorted copy. Here $\mathbf{v}, \mathbf{w}$ are nD vectors. Summing $\mathbf{v}$ with $\mathbf{w}$ brings us to image translation by the vector $\mathbf{w}$, two-sided multiplications $\lambda \mathcal{E}_0(\mathbf{x} + \mathbf{w}) \mathcal{E}_0^{-1}$ by $\lambda \mathcal{E}_0$ and $\mathcal{E}_0^{-1}$ equivalent to both an nD rotation of the vector $\mathbf{z} + \mathbf{w}$ and a dilatation given by the factor λ. Here, $\mathbf{f}_{Mcol} \longrightarrow \mathcal{C}_0 \mathbf{f}_{Mcol} \mathcal{C}_0^{-1}$ is a multicolor transformation of the initial image.

THEOREM 2   *The central moments* $\mathcal{M}_p$ *of the multicolor images* $\mathbf{f}_{Mcol}(\mathbf{x})$ *are relative* $\mathcal{A}^{SpMcol}$*-valued invariants*

$$\mathcal{J}_p\{\lambda \mathcal{E}_0 \mathbf{w} \mathbf{f}_{Mcol}^{\mathcal{C}_0}\} = \mathcal{M}_p\{\lambda \mathcal{E}_0 \mathbf{w} \mathbf{f}_{Mcol}^{\mathcal{C}_0}\} = \left(\lambda^{p+n} \mathcal{C}_0 \mathcal{E}_0^p \right) \mathcal{M}_p\{\mathbf{f}_{Mcol}\} \left(\mathcal{E}_0^{-p} \mathcal{C}_0^{-1} \right)$$

$$(4)$$

with respect to the spatial-multicolor group $\mathbf{G}^{SpMcol}$ *with both* $\mathcal{A}^{SpMcol}$*-valued left multipliers* $\mathcal{C}_0 \lambda^{p+3} \mathcal{E}_0^p$ *and* $\mathcal{A}_{2^n}^{Sp}$*-valued right multipliers* $\mathcal{E}_0^{-p}$, *respectively (see Fig. 6), and the normalized central moments*

$$|\mathcal{N}_p\{\lambda \mathcal{E}_0 \mathbf{w} \mathbf{f}_{Mcol}^{\mathcal{C}_0}\}| = |\mathcal{M}_p\{\lambda \mathcal{E}_0 \mathbf{w} \mathbf{f}_{Mcol}^{\mathcal{C}_0}\} \mathcal{M}_0^{p-1}\{\lambda \mathcal{E}_0 \mathbf{w} \mathbf{f}_{Mcol}^{\mathcal{C}_0}\}| \Big/ |\mathcal{M}_1^p\{\lambda \mathcal{E}_0 \mathbf{w} \mathbf{f}_{Mcol}^{\mathcal{C}_0}\}|$$

are absolute scalar-valued invariants, with respect to the same group, i.e., $\mathcal{I}_p = |\mathcal{N}_p\{\lambda \mathcal{E}_0 \mathbf{w} \mathbf{f}_{Mcol}^{\mathcal{C}_0}\}| = |\mathcal{N}_p\{\mathbf{f}_{Mcol}\}|$.

Now we consider the most important cases for invariants of grey-level, color and multicolor 2D and 3D images.

4.2   Complex and quaternion invariants of 2D and 3D grey-level images

Let $\mathcal{GC}_2^{Sp} = \{\mathbf{z} = x + I y \mid x, y \in \mathbf{R}; \ I^2 = -1, 0, +1\}$ be a generalized spatial complex plane. Then a grey-level 2D image $f(x, y)$ can be considered as a function of a generalized complex variable, i.e., $f(x, y) = f(\mathbf{z})$, where $\mathbf{z} = x + I y \in \mathcal{GC}_2^{Sp}$. Let $\mathbf{c}$ be the centroid of the image $f(\mathbf{z})$.

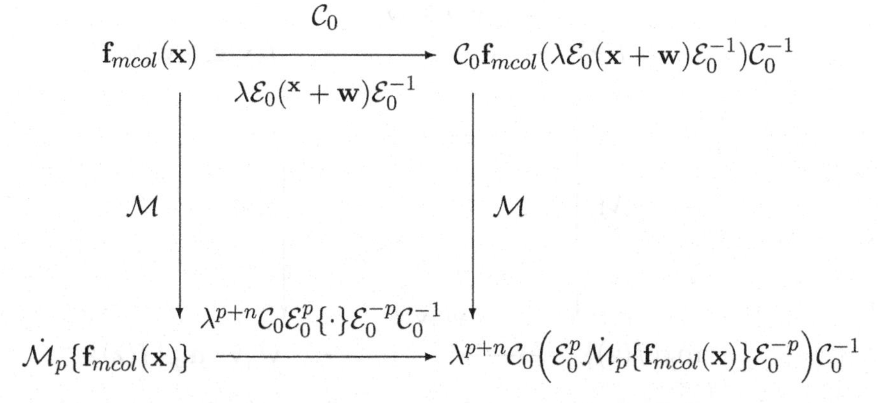

$$\mathbf{f}_{mcol}(\mathbf{x}) \xrightarrow[\lambda \mathcal{E}_0(^{\mathbf{x}} + \mathbf{w})\mathcal{E}_0^{-1}]{\mathcal{C}_0} \mathcal{C}_0 \mathbf{f}_{mcol}(\lambda \mathcal{E}_0(\mathbf{x} + \mathbf{w})\mathcal{E}_0^{-1})\mathcal{C}_0^{-1}$$

$$\downarrow \mathcal{M} \qquad\qquad\qquad \downarrow \mathcal{M}$$

$$\dot{\mathcal{M}}_p\{\mathbf{f}_{mcol}(\mathbf{x})\} \xrightarrow{\lambda^{p+n}\mathcal{C}_0\mathcal{E}_0^p\{\cdot\}\mathcal{E}_0^{-p}\mathcal{C}_0^{-1}} \lambda^{p+n}\mathcal{C}_0\left(\mathcal{E}_0^p\dot{\mathcal{M}}_p\{\mathbf{f}_{mcol}(\mathbf{x})\}\mathcal{E}_0^{-p}\right)\mathcal{C}_0^{-1}$$

Figure 6. Transformations of $\mathcal{A}^{SpMcol}$-valued moments with respect to the spatial-multicolor group $\mathbf{G}_{n,k}^{SpMcol}$

DEFINITION 8 *Functionals of the form*

$$\dot{\mathbf{m}}_p\{f\} = \int_{\mathbf{z} \in \mathcal{GC}_2^{Sp}} (\mathbf{z} - \mathbf{c})^p f(\mathbf{z}) d\mathbf{z}$$

are called the one-index central $\mathcal{A}_2^{Sp}$-valued moments *of the image* $f(\mathbf{z})$, *where* $\mathbf{dz} := dxdy$, *and* $p \in \mathbf{Q}$ *are rational numbers.*

Let us clarify the rules of $\mathcal{A}_2^{Sp}$-valued moment transformations under geometrical distortions of the initial 2D images. We will consider translation, rotation and scaling transformations. If $f(\mathbf{z})$ is the initial image, then $f_{\mathbf{v},\mathbf{w}}(\mathbf{z}) = f(\mathbf{v}(\mathbf{z} + \mathbf{w}))$ denotes its geometrical distorted copy.

THEOREM 3 *The central moments* $\dot{\mathbf{m}}_p\{f\}$ *of the image* $f(\mathbf{z})$ *are relative* $\mathcal{A}_2^{Sp}$-valued invariants

$$\mathcal{J}_p\{f_{\mathbf{v},\mathbf{w}}\} := \dot{\mathbf{m}}_p\{f_{\mathbf{v},\mathbf{w}}\} = \mathbf{v}^p|\mathbf{v}|^2\dot{\mathbf{m}}_p\{f\}$$

with respect to the affine group $\mathbf{Aff}(\mathcal{GC}_2^{Sp})$ *with* $\mathcal{A}_2^{Sp}$-valued *multipliers* $\mathbf{v}^p|\mathbf{v}|^2 = e^{Ip\varphi}|\mathbf{v}|^{p+2}$ *(see Fig. 7), and the normalized central* $\mathcal{A}_2^{Sp}$-valued *moments* $\mathcal{N}_p\{f_{\mathbf{v},\mathbf{w}}\} := \dot{\mathbf{m}}_p\{f_{\mathbf{v},\mathbf{w}}\}\dot{\mathbf{m}}_0^{p-1}\{f_{\mathbf{v},\mathbf{w}}\}/\dot{\mathbf{m}}_1^p\{f_{\mathbf{v},\mathbf{w}}\}$ *are absolute* $\mathcal{A}_2^{Sp}$-valued *invariants with respect to the same group, i.e.,* $\mathcal{I}_p = \mathcal{N}_p\{f_{\mathbf{v},\mathbf{w}}\} = \mathcal{N}_p\{f\}$.

Let us consider a 3D grey-level image $f(x, y, z)$. This image can be considered as a function of a generalized quaternion variable $\mathbf{q} = (xI + yJ +$

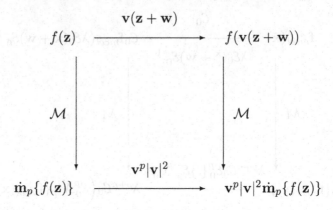

Figure 7.   Transformations of $\mathcal{A}_2^{Sp}$-valued moments with respect to the affine group $\mathbf{Aff}(\mathcal{GC}_2^{Sp})$

zK), i.e., $f(x, y, z) = f(\mathbf{q})$, where $\mathbf{q} \in \mathbf{Vec}\{\mathcal{A}_4^{Sp}\} = \mathcal{GR}^3$. Let $\mathbf{c}$ be the centroid of the image $f(\mathbf{q})$.

DEFINITION 9 *Functionals of the form*

$$\mathcal{M}_p\{f\} := \int_{\mathbf{q} \in \mathcal{GR}^3} (\mathbf{q} - \mathbf{c}) f(\mathbf{q}) d\mathbf{q}$$

are called the one-index central $\mathcal{A}_4^{Sp}$-valued (quaternion-valued) *moments of the grey-level 3D image* $f(\mathbf{q})$, *where* $p \in \mathbf{Q}$, *and* $d\mathbf{q} := dxdydz$.

Let us clarify the rules of moment transformations with respect to geometrical distortions of 3D images. These distortions will be caused by 1) 3D translations $\mathbf{q} \longrightarrow \mathbf{q} + \mathbf{w}$, 2) 3D rotations $\mathbf{q} \longrightarrow \mathcal{Q}_0(\mathbf{q} + \mathbf{w})\mathcal{Q}^{-1}$, where $\mathcal{Q}_0 = e^{\mathbf{u}_0 \phi / 2}$, 3) dilatation: $\mathbf{q} \longrightarrow \lambda \mathbf{q}$, where $\lambda \in \mathbf{R}^+$. If $f(\mathbf{q})$ is an initial image and $f_{\lambda \mathcal{Q}_0 \mathbf{w}}(\mathbf{q})$ its distorted version, then $f_{\lambda \mathcal{Q}_0 \mathbf{w}}(\mathbf{q}) := f(\lambda \mathcal{Q}_0(\mathbf{q} + \mathbf{w})\mathcal{Q}_0^{-1})$.

THEOREM 4 *The central moments* $\mathcal{M}_p\{f\}$ *of the 3D grey-level image* $f(\mathbf{q})$ *are relative* $\mathcal{A}_4^{Sp}$-valued invariants

$$\mathcal{J}_p\{f_{\lambda \mathcal{Q}_0 \mathbf{w}}\} := \mathcal{M}_p\{f_{\lambda \mathcal{Q}_0 \mathbf{w}}\} = \left(\lambda^{p+3} \mathcal{Q}_0^p\right) \mathcal{M}_p\{f\} \left(\mathcal{Q}_0^{-p}\right)$$

with respect to the group $\mathbf{Aff}(\mathcal{GC}_3^{Sp})$ *with left* $\mathcal{A}_4^{Sp}$-valued multipliers $\lambda^{p+3} \mathcal{Q}_0^p$ *and with right* $\mathcal{A}_4^{Sp}$-valued multipliers $\mathcal{Q}_0^{-p}$, *respectively (see Fig. 8), and the absolute values of normalized central moments*

$$|\mathcal{N}_p| = \left| \mathcal{M}_p\{f_{\lambda \mathcal{Q}_0 \mathbf{w}}(\mathbf{q})\} \mathcal{M}_0^{p-1}\{f_{\lambda \mathcal{Q}_0 \mathbf{w}}(\mathbf{q})\} \right| \Big/ \left| \mathcal{M}_1^{p-1}\{f_{\lambda \mathcal{Q}_0 \mathbf{w}}(\mathbf{q})\} \right|$$

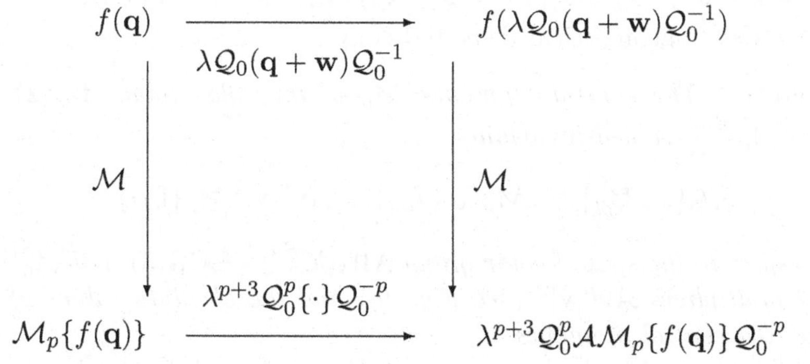

$$f(\mathbf{q}) \xrightarrow{\quad \lambda\mathcal{Q}_0(\mathbf{q}+\mathbf{w})\mathcal{Q}_0^{-1} \quad} f(\lambda\mathcal{Q}_0(\mathbf{q}+\mathbf{w})\mathcal{Q}_0^{-1})$$

$$\mathcal{M} \Big\downarrow \qquad\qquad\qquad \mathcal{M} \Big\downarrow$$

$$\mathcal{M}_p\{f(\mathbf{q})\} \xrightarrow{\quad \lambda^{p+3}\mathcal{Q}_0^p\{\cdot\}\mathcal{Q}_0^{-p} \quad} \lambda^{p+3}\mathcal{Q}_0^p \mathcal{A}\mathcal{M}_p\{f(\mathbf{q})\}\mathcal{Q}_0^{-p}$$

Figure 8. Transformations of $\mathcal{A}_4^{Sp}$-valued moments with respect to the group $\mathbf{Aff}(\mathcal{G}\mathcal{C}_3^{Sp})$

are absolute scalar-valued invariants with respect to the same group, i.e.,
$$\mathcal{I}_p = |\mathcal{N}_p\{f_{\lambda\mathcal{Q}_0\mathbf{w}}\}| = |\mathcal{N}_p\{f\}|.$$

4.3 Moments and invariants of color 2D and 3D images

Let $\mathcal{G}\mathcal{C}_2^{Sp} := \{\mathbf{z} = x_1 + Ix_2 \mid x_1, x_2 \in \mathbf{R}; I^2 = -1, 0, +1\}$ be a generalized spatial complex plane. Then the color image $\mathbf{f}_{col}(x, y)$ can be considered as a triplet-valued function of the generalized complex variable $\mathbf{z} = x_1 + Ix_2$, i.e., as $\mathbf{f}_{col}(x, y) = \mathbf{f}_{col}(\mathbf{z})$, where $\mathbf{z} \in \mathcal{G}\mathcal{C}_2^{Sp}$. Let $\mathbf{z} \in \mathcal{G}\mathcal{C}_2^{Sp}$ be spatial and $\mathcal{A} \in \mathcal{A}_3^{Col}$ be color triplet numbers. Then all products $\mathbf{z}\mathcal{A}$ will be called *spatial-color numbers* (or *Hurwitz numbers*). They form the 6D *space-color Hurwitz algebra* $\mathcal{A}_6^{SpCol} := \mathcal{A}_2^{Sp} \otimes \mathcal{A}_3^{Col}$. We assume that all *spatial hyperimaginary units* commute with all *color units*. Therefore, $\mathcal{A}_6^{SpCol} = \mathcal{A}_2^{Sp} \otimes \mathcal{A}_3^{Col} = \mathcal{A}_3^{Col} \otimes \mathcal{A}_2^{Sp}$. Let $\mathbf{c}$ be the centroid of the image $\mathbf{f}_{col}(\mathbf{z})$.

DEFINITION 10 *Functionals of the form*

$$\dot{\mathcal{M}}_p = \int_{\mathbf{z}\in\mathcal{A}_2^{Sp}} (\mathbf{z} - \mathbf{c})^p \mathbf{f}_{col}(\mathbf{z}) d\mathbf{z}$$

are called the central $\mathcal{A}_6^{SpCol}$-valued moments of the color image $\mathbf{f}_{col}(\mathbf{z})$.

Let $\mathcal{A} = (a_{lu}, \mathbf{z}_{ch}) \in \mathcal{A}_3^{Col}$. Let us clarify the rules of moment transformations with respect to distortions of color and geometry of the initial images. If $\mathbf{f}_{col}(\mathbf{z})$ is the initial image, then $_{\mathbf{v},\mathbf{w}}\mathbf{f}_{col}^{\mathcal{A}}(z) = \mathcal{A}\mathbf{f}_{col}(\mathbf{v}(\mathbf{z}+\mathbf{w})) =$

$a_{lu}f_{lu}(\mathbf{v}(\mathbf{z} + \mathbf{w})) \cdot \mathbf{e}_{lu} + \mathbf{z}_{ch}f_{ch}(\mathbf{v}(\mathbf{z} + \mathbf{w})) \cdot \mathbf{E}_{ch}$ denotes its $\mathbb{LCG}(3)$-color and $\mathbf{Aff}(\mathcal{GC}_2^{Sp})$-geometrical distorted copy.

THEOREM 5 *The central moments* $\dot{\mathcal{M}}_p$ *of the color image* $\mathbf{f}_{col}(\mathbf{z})$ *are relative* $\mathcal{A}_6^{SpCol}$*-valued invariants*

$$\mathcal{J}_p\{_{\mathbf{v},\mathbf{w}}\mathbf{f}_{col}^{\mathcal{A}}\} := \dot{\mathcal{M}}_p\{_{\mathbf{v},\mathbf{w}}\mathbf{f}_{col}^{\mathcal{A}}\} = \mathcal{A}\mathbf{v}^p|\mathbf{v}|^2\dot{\mathcal{M}}_p\{\mathbf{f}_{col}\} \qquad (5)$$

with respect to the spatial-color group $\mathbf{Aff}(\mathcal{GC}_2^{Sp}) \times \mathbb{LCG}(3)$ *with* $\mathcal{A}_6^{SpCol}$*-valued multipliers* $\mathcal{A}\mathbf{v}^p|\mathbf{v}|^2$ *(see Fig. 9), and the absolute values of the normalized central moments*

$$|\dot{\mathcal{N}}_p| = \left|\dot{\mathcal{M}}_p\{_{\mathbf{v},\mathbf{w}}\mathbf{f}_{col}^{\mathcal{A}}\}\dot{\mathcal{M}}_0^{p-1}\{_{\mathbf{v},\mathbf{w}}\mathbf{f}_{col}^{\mathcal{A}}\}\right|\Big/\left|\dot{\mathcal{M}}_1^p\{_{\mathbf{v},\mathbf{w}}\mathbf{f}_{col}^{\mathcal{A}}\}\right|$$

are absolute scalar-valued invariants, with respect to the same group, i.e., $\mathcal{I}_p = \left|\dot{\mathcal{N}}_p\{_{\mathbf{v},\mathbf{w}}\mathbf{f}_{col}^{\mathcal{A}}\}\right| = \left|\dot{\mathcal{M}}_p\{\mathbf{f}_{col}\}\right|.$

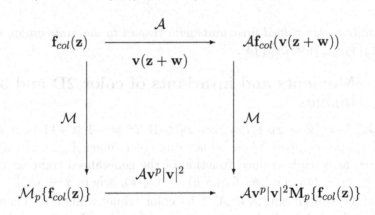

Figure 9. Transformations of relative $\mathcal{A}_6^{SpCol}$-valued moments with respect to the spatial-color group $\mathbf{Aff}(\mathcal{GC}_2^{Sp}) \times \mathbb{LCG}(3)$

Let $\mathbf{f}_{col}(\mathbf{q})$ be a color 3D image depending on the pure vector generalized quaternion variable $\mathbf{q} \in \mathbf{Vec}\{\mathcal{A}_4^{Sp}\} = \mathcal{GR}^3$. If $\mathbf{q} \in \mathcal{CR}_3 \subset \mathcal{A}_4^{Sp}$ is a generalized quaternion and $\mathcal{A} \in \mathcal{A}^{Col}$ is a color number, then all products of the form $\mathbf{q}\mathcal{A}$ are called *spatial-color quaternions*. They form a *space-color algebra* $\mathcal{A}^{SpCol} := \mathcal{A}_4^{Sp} \otimes \mathcal{A}^{Col} = \mathcal{A}^{Col} \otimes \mathcal{A}_4^{Sp}.$

DEFINITION 11 *Functionals of the form*

$$\dot{\mathcal{M}}_p\{\mathbf{f}_{col}\} := \int_{\mathbf{q} \in \mathcal{GR}^3} (\mathbf{q} - \mathbf{c})\mathbf{f}_{col}(\mathbf{q})d\mathbf{q}$$

are called the one-index central $\mathcal{A}^{SpCol}$-valued moments *of the color 3D image* $\mathbf{f}_{col}(\mathbf{q})$, *where* $p \in \mathbf{Q}$, *and* $\mathbf{dq} := dxdydz$.

Let us clarify the rules of moment transformations with respect to geometrical and color distortions of 3D color images. If $f_{col}(\mathbf{q})$ is an initial image and $_{\lambda \mathcal{Q} \mathbf{w}} \mathbf{f}_{col}^{\mathcal{A}}(\mathbf{q})$ its distorted version, then $_{\lambda \mathcal{Q} \mathbf{w}} \mathbf{f}_{col}^{\mathcal{A}}(\mathbf{q}) := \mathcal{A} \mathbf{f}_{col}(\lambda \mathcal{Q}(\mathbf{q} + \mathbf{w}) \mathcal{Q}^{-1})$.

$$
\begin{array}{ccc}
& \mathcal{A} & \\
\mathbf{f}_{col}(\mathbf{q}) & \xrightarrow{\hspace{3cm}} & \mathcal{A}\mathbf{f}_{col}(\lambda \mathcal{Q}(\mathbf{q} + \mathbf{w})\mathcal{Q}^{-1}) \\
& \lambda \mathcal{Q}_0(\mathbf{q} + \mathbf{w})\mathcal{Q}_0^{-1} & \\
\Big\downarrow \mathcal{M} & & \Big\downarrow \mathcal{M} \\
& \lambda^{p+3} \mathcal{Q}_0^p \mathcal{A}\{\cdot\}\mathcal{Q}_0^{-p} & \\
\dot{\mathcal{M}}_p\{\mathbf{f}_{col}(\mathbf{q})\} & \xrightarrow{\hspace{3cm}} & \lambda^{p+3} \mathcal{Q}_0^p \mathcal{A}\dot{\mathcal{M}}_p\{\mathbf{f}_{col}(\mathbf{q})\}\mathcal{Q}_0^{-p}
\end{array}
$$

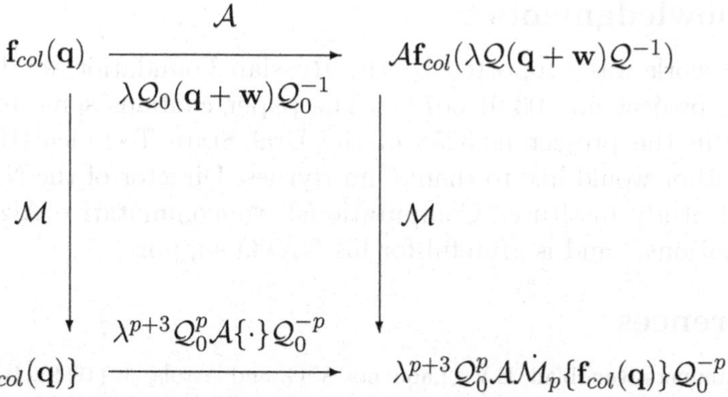

Figure 10. Transformations of $\mathcal{A}^{SpCol}$-valued moments with respect to the spatial-color group $\mathbf{Aff}(\mathcal{GC}_3^{Sp}) \times \mathbb{LCG}(3)$

THEOREM 6 *The central moments* $\dot{\mathcal{M}}_p\{\mathbf{f}_{col}\}$ *of the 3-dimensional color image* $f_{col}(\mathbf{q})$ *are relative* $\mathcal{A}^{SpCol}$-valued invariants

$$
\mathcal{J}_p\left\{\mathbf{f}_{\lambda \mathcal{Q}_0 \mathbf{w}}^{\mathcal{A}}\right\} := \mathcal{M}_p\left\{\mathbf{f}_{\lambda \mathcal{Q}_0 \mathbf{w}}^{\mathcal{A}}\right\} = \left(\mathcal{A}\lambda^{p+3}\mathcal{Q}_0^p\right)\mathcal{M}_p\{\mathbf{f}_{col}\}\left(\mathcal{Q}_0^{-p}\right)
$$

with respect to the group $\mathbf{Aff}(\mathcal{GC}_3^{Sp}) \times \mathbb{LCG}(3)$ *with left* $\mathcal{A}^{SpCol}$-valued *multipliers* $\mathcal{A}\lambda^{p+3}\mathcal{Q}_0^p$ *and with right* $Spin(\mathcal{A}_4^{Sp})$-valued *multipliers* $\mathcal{Q}_0^{-p}$, *respectively (see Fig. 10), and the absolute values of the normalized central moments*

$$
|\mathcal{N}_p\{_{\lambda \mathcal{E}_0 \mathbf{w}}\mathbf{f}_{col}^{\mathcal{C}_0}\}| = \left|\mathcal{M}_p\{_{\lambda \mathcal{E}_0 \mathbf{w}}\mathbf{f}_{col}^{\mathcal{C}_0}\}\mathcal{M}_0^{p-1}\{_{\lambda \mathcal{E}_0 \mathbf{w}}\mathbf{f}_{col}^{\mathcal{C}_0}\}\right|\Big/\left|\mathcal{M}_1^p\{_{\lambda \mathcal{E}_0 \mathbf{w}}\mathbf{f}_{col}^{\mathcal{C}_0}\}\right|
$$

are relative scalar-valued invariants, with respect to the same group, i.e., $\mathcal{I}_p = |\mathcal{N}_p\{_{\lambda \mathcal{E}_0 \mathbf{w}}\mathbf{f}_{col}^{\mathcal{C}_0}\}| = |\mathcal{N}_p\{\mathbf{f}_{col}\}|.$

5. Conclusions

We have shown how Clifford algebras can be used in the formation and computation of invariants of 2D, 3D and nD color and multicolor objects of different Euclidean and non-Euclidean geometries. The theorems stated show how simple and efficient the methods of calculation of invariants are that use spatial and color Clifford algebras. But how fast can the invariants be calculated? The answer to this question the interested reader can find in [2], where Fourier–Clifford and Fourier–Hamilton number theoretic transforms are used for this purpose.

Acknowledgments

The work was supported by the Russian Foundation for Basic Research, project no. 03-01-00735. The paper contains some results obtained in the project no.3258 of the Ural State Technical University. The author would like to thank Jim Byrnes, Director of the NATO Advanced Study Institute "Computational Noncommutative Algebra and Applications," and is grateful for his NATO support.

References

[1] Labunets-Rundblad, E.V., Labunets, V.G. and Astola, J. (1999). Is the Visual Cortex a Fast Clifford Algebra Quantum Compiler? *A NATO Advanced Research Workshop Clifford Analysis and Its Applications*, pp. 173–183.

[2] Labunets-Rundblad, E.V. (2000). *Fast Fourier–Clifford Transforms Design and Application in Invariant Recognition*. PhD thesis, Tampere University Technology, Tampere, Finland, 262 p.

[3] Labunets, V.G., Rundblad, E.V. and Astola, J., (2001). Is the brain "Clifford algebra quantum computer? *Proc. of SPIE "Materials and Devices for Photonic Circuits"*, 2001 Vol. 4453, pp. 134–145

[4] Labunets, V.G., Rundblad, E.V. and Astola, J.,(2001). Fast invariant recognition of color 3D images based on spinor–valued moments and invariants. *Proc. SPIE "Vision Geometry X"*, 2001, Vol. 4476, pp. 22–33

[5] Labunets, V.G., Maidan, A., Rundblad–Labunets, E.V. and Astola J., (2001). Colour triplet–valued wavelets and splines, *Image and Signal Processing and Analysis ISPA'01, June 19–21, Pula, Croatia*, 2001, pp. 535–541

[6] Labunets, V.G., Maidan, A., Rundblad–Labunets, E.V. and Astola J., (2001) Colour triplet–valued wavelets, splines and median filters, *Spectral Methods and Multirate Signal Processing, SMMSP'2001, June 16–18, Pula, Croatia*, 2001, pp. 61–70

[7] Labunets-Rundblad, E.V. and Labunets, V.G. (2001). Chapter 7. Spatial-Colour Clifford Algebra for Invariant Image Recognition, pp. 155–185. (In: *Geometric Computing with Clifford Algebra)*, Ed. G. Sommer. Springer, Berlin Heideberg, 2001, 452 p.

[8] Lasenby, A.N., Doran, C.J.L. and Gull, S.F. (1996). Lectures in Geometric Algebra, In: W. E. Baylis, Ed., *Clifford (Geometric) Algebras with Applications to Physics, Mathematics and Engineering*, Birkhouser, Boston, 256 p.

[9] Vela, M. (2000). Explicit solutions of Galois Embedding problems by means of generalized Clifford algebras, *J. Symbolic Computation*, **30**, pp. 811–842

[10] Ch. Greaves, (1847). On algebraic triplets, *Proc. Irish Acad.*, **3**, pp. 51–54, 57–64, 80–84, 105–108

RECENT PROGRESS
AND APPLICATIONS
IN GROUP FFTS

Daniel N. Rockmore
Dartmouth College
Hanover, NH 03755
rockmore@cs.dartmouth.edu

Abstract

The Cooley-Tukey FFT can be interpreted as an algorithm for the efficient computation of the Fourier transform for finite cyclic groups, a compact group (the circle), or the non-compact group of the real line. These are all commutative instances of a "Group FFT." We give a brief survey of some recent progress made in the direction of noncommutative generalizations and their applications.

Keywords: Fast Fourier transform, discrete Fourier transform, sampling, Gel'fand-Tsetlin bases.

1.   Introduction

The *Fast Fourier Transform* or *FFT* is an efficient algorithm to compute the *Discrete Fourier Transform* (*DFT*). This is a linear transformation, specifically realized in terms of the the $(n \times n)$ *DFT matrix*:

$$\widehat{f} = \left(e^{2\pi i j k / n} \right)_{jk} f, \tag{1}$$

which takes a vector of samples realized as a function $f \in \mathbb{C}^n$, and returns a collection of Fourier coefficients $\widehat{f} \in \mathbb{C}^n$.

The DFT plays a crucial role in a wide range of applied activities, principally in the analysis of *time series* data. These natural quantifications of temporal phenomena presumably owe their origins to the observations of the first priestly mathematicians who made it their work to chart the course of heavenly bodies in the construction of the first calendars. Other early examples would include the analysis of the time course of temperatures, rainfall, and various meteorological data, pos-

227

J. Byrnes (ed.) Computational Noncommutative Algebra and Applications, 227-254.
© 2004 *Kluwer Academic Publishers. Printed in the Netherlands.*

sibly in the service of the study of the time series of agriculture (crop yields, etc.). Closer to our hearts and heads are the time series of health that are our EKGs and EEGs, which are perhaps, in turn, influenced by the seemingly random walk that are the time series which reflect the health of the financial markets.

One approach to time series analysis is to view these phenomena as well-explained as the superposition of basic periodic phenomena: weather as a combination of diurnal and annual effects, blood pressure or hormone levels tracking some invisible and evolved pacemaker. Herein the DFT is that transformation of the data that teases out the periodicities, taking the discrete or discretized signal and transforming it to a representation in terms of weighted frequencies.

Direct calculation of a DFT of length n is effected by the multiplication of the $n \times n$ DFT matrix with a vector of length n and so requires n^2 operations. When n is large even this quadratic calculation is too large. The great success of the FFT is the reduction in complexity to $O(n \log_2 n)$ operations (with implicit small constant).

1.1   Brief History of the Classical FFT

The original FFT was indeed due to Gauss, and has the astronomical (although not religious) origins indicated above. Its story provides another proof that necessity is indeed often the mother of invention. Gauss was confronted with the computationally daunting problem of interpolating—by hand!—the periodic orbit of the asteroid Ceres, which had suddenly gone missing. Gauss determined a means of building the interpolation on n points from two interpolations of $n/2$ points, and in so doing discovered the basic step in what is now the standard divide-and-conquer efficient algorithm. His discovery languished for centuries (cloaked in Latin and hidden away in little known writings) while a few centuries later it was rediscovered by (among others) Danielson and Lanczos in the service of crystallography but, surely most famously, by Cooley and Tukey [12] in the mid 1960s, this time not in the service of the discovery of missing heavenly bodies, but instead for the detection of hidden nuclear tests in the Soviet Union, as well as stealthy Soviet nuclear submarines. For full histories see [22].

The technical motivations for Cooley and Tukey's rediscovery were

(1) The efficient computation of the power spectrum of time series (especially sampled time series of very long length, so equivalently, the calculation of high frequency contributions) and

(2) Efficient filtering (smoothing).

Item (2) is equivalent to the efficient computation of (cyclic) *convolution*. For vectors f and g of length n this is defined as

$$f \star g(x) = \sum_{y=0}^{n-1} f(x - y)g(y)$$

where the arguments are interpreted as integers mod n. (*Linear convolution* is that which corresponds to the generation of a vector of length $2n$ from f by interpreting the samples $f(i), g(k)$ as coefficients in a polynomial of degree $n - 1$, and then asking for the coefficient of x^k in the product. Note that this could be obtained by computing cyclic convolution of f and g *zero-padded* to vectors of length $2n$.)

Note that direct computation of the convolution requires n^2 operations. The identity

$$\widehat{f \star g}(k) = \widehat{f}(k)\widehat{g}(k) \tag{2}$$

shows how application of the DFT permits the filtering of f to be performed directly in the frequency domain via the assignation of a particular frequency profile for g. When $\widehat{g}$ takes only values zero and one, it has the form of a *bandpass filter*, and if the ones are restricted to a subsequence of indices, this nonzero interval is the *passband*. *Lowpass* filters restrict the passband to an initial segment and a terminal segment for *highpass* filters.

The FFT enables fast convolution via the algorithm

$$f, g \longrightarrow \widehat{f}, \widehat{g} \longrightarrow \widehat{f} \bigodot \widehat{g} \longrightarrow f \star g$$

where $\bigodot$ is meant to indicate pointwise multiplication of the two vectors it separates. Note that the last step is accomplished via an **inverse** FFT, so that in total, the algorithm requires three FFTs and a single n point pointwise multiplication for a total of $O(n \log n)$ operations.

1.2   Group Theoretic Interpretations

"The FFT" is actually a family of algorithms, all designed to compute efficiently the DFT (1). This linear transformation can be cast as a particular instance of any of a variety of mathematical operations, but the focus in this chapter is a group theoretic, indeed, representation theoretic point of view. Within this, there are at least three different interpretations, corresponding to either the case of finite, compact, or non-compact groups. We summarize these below, for each of them presents its own challenges for generalization.

(1) **Finite Groups—** In this setting we view f as a function on the cyclic group of order n, C_n (isomorphic to $\mathbb{Z}/n\mathbb{Z}$). The FFT is the

efficient change of basis algorithm that takes a function written in terms of the basis of delta functions and re-expresses it in terms of the basis of sampled exponentials,

$$\left\{ \sum_{x \in C_n} f(x)\delta_x \right\} \longrightarrow \left\{ \sum \widehat{f}(k)e_k \right\}$$

where $e_k(j) = e^{2\pi ijk/n}$.

(2) **Compact Groups**— In this case, the vector f is viewed as samples of a function on the circle S^1 (i.e., samples of a periodic function). Any such function has a Fourier expansion, $f(t) = \sum_{\ell \in \mathbb{Z}} \widehat{f}(\ell)e^{-2\pi i\ell t}$ where the Fourier coefficients are computed by an integral

$$\widehat{f}(\ell) = \int_0^1 f(e^{2\pi it})e^{2\pi i\ell t}dt.$$

In general, the FFT can be used to compute efficiently an approximation to these Fourier coefficients, but in the interesting *bandlimited* case, in which the function's Fourier expansion is finite (i.e., there exists $B \geq 0$ such that $\widehat{f}(\ell) = 0$ for $\ell \geq B$), there is an exact *quadrature* or *sampling* rule that provides an exact formula for the (potentially) nonzero Fourier coefficients in terms of a DFT of length $2B + 1$.

(3) **Non-compact Groups** – In this last case, we view our discrete set of samples as arising from a complex-valued function f defined on the real line $\mathbb{R}$. Once again, the Fourier transform is a linear transformation from time (or space) to frequency, this time given as the integral operator (for each x),

$$\widehat{f}(x) = \int_{\mathbb{R}} f(y)e^{-2\pi iyx}dy.$$

As in the compact case the DFT might be used to approximate this integral, and once again there is a bandlimited theory (i.e., the case in which the Fourier transform only has finite support). In this case, the function f is determined by its equispaced samples along the entire real line (i.e., so-called "Shannon sampling"). Consequently, the FFT provides a means for an efficient and quantifiable approximation to the computation of f's frequency content.

In summary, the FFT makes possible the efficient analysis of: (1) discrete periodic data viewed as a function on the discrete circle, that is, a cyclic group of finite order (C_n); (2) continuous periodic data, viewed as a function on the circle; and (3) continuous data, viewed as a function on the line.

1.3   Noncommutative generalizations

The groups C_n, S^1, and $\mathbb{R}$ are all *commutative* groups, i.e., the law we use to combine them obeys a commutative rule: $x + y = y + x$. Each of the above commutative group theoretic interpretations has, over the past generation, found generalization to the *noncommutative* setting, and the purpose of this chapter is to provide a window into this work.

Abstractly, a group is simply a set closed under some associative multiplication rule such that there is an identity element, and to each element there is an inverse. Classically, these arose as the symmetries of roots of polynomials, i.e., those arithmetic transformations that leave invariant a given polynomial, and from this they grew to encompass the notion of symmetry throughout mathematics and physics. They are in general noncommutative, i.e., usually $xy \neq yx$ (think matrices!). As indicated above, they come in at least three general flavors - the three in which we are interested: *Finite, Compact* and *Non-compact*.

(1) **Finite groups**—The most familiar commutative examples are the aforementioned cyclic groups, C_n, while of the noncommutative examples, the symmetric groups, S_n, the group of permutations of n elements, commonly realized as the group of all card shuffles of a deck of size n is perhaps the most familiar.

(2) **Compact groups**—Standard examples come from the matrix groups whose entries are bounded in size. The orthogonal groups $O(n)$ and the special orthogonal groups $SO(n)$ (also called *rotation groups*—symmetries of the $n - 1$-dimensional sphere), and their complex analogues—unitary groups $U(n)$ and special unitary groups $SU(n)$, are examples. With their length-preserving properties they are effectively the symmetries of space.

(3) **Non-compact groups**—The invertible complex or real matrices, $GL_n(\mathbb{C})$ or $GL_n(\mathbb{R})$ are well-known examples, and within these, the *Euclidean motion groups* are particularly useful. These are (for any n) the matrices of the form $\begin{pmatrix} a & b \\ 0 & 1 \end{pmatrix}$ where $a \in SO(n)$ and $b \in \mathbb{R}^n$. These occur naturally as symmetries of n-dimensional affine space.

1.3.1   Noncommutative DFTs.   Given a complex-valued function defined on a group G, its Fourier decomposition (analysis) is meant to be a rewriting in terms of a basis of functions that are nicely adapted to translation via group elements. It is in this sense a symmetry-guided decomposition.

In the commutative case we have eigenfunctions of translation: If $e(x) = e^{2\pi i x}$, then

$$T_y e(x) = e(x - y) = e^{-2\pi i y} e(x)$$

where T_y indicates the translation operator.

In the noncommutative case there are no simultaneous eigenfunctions for all translation operators. This is both the source of frustration as well as the spur to art for the theory and application of *the representation theory of noncommutative groups*. This forces us to look for the next best thing which is closure of some linear space under the translation action. That is, a basis of functions $e_k(x)$ on the group that have the property

$$T_y e_k(x) = e_k(xy^{-1}) = \sum_\ell T_y(k, \ell) e_\ell(x).$$

In this way we see that the eigenfunctions are naturally replaced by functions that act like, and bear the name of *matrix elements* $T_y(k, \ell)$, and it is essentially these functions which replace the sampled exponentials that create frequency space in the commutative case. When grouped together they give *matrix representations* of the group and comprise what is called *the dual* of the group (denoted $\widehat{G}$). Their study is the subject of *group representation theory*.

So in general we have, for any function f defined on a finite group G, the notion of a Fourier expansion

$$f(x) = \sum_{\rho \in \widehat{G}} c_\rho \sum_{k, \ell} \widehat{f}(k, \ell) T_x(k, \ell) \tag{3}$$

where c_ρ is some constant depending on an irreducible representation ρ, the $\widehat{f}(k, \ell)$ are the Fourier coefficients, and the matrix elements (which depend on x) now span the analogue of frequency space. The Fourier transform computes these Fourier coefficients, and it amounts to computing the discrete inner product of the function with the new basis of irreducible matrix elements.

Should G be compact, the sum is infinite (in analogy with the sum over the integers in the case of the circle) while if G is non-compact, this sum is in general some sort of integral (cf. [9] for pointers to basic representation theory references).

This new basis effects convolution in a manner akin to the commutative case:

$$\widehat{f \star g}(k, \ell) = \sum_m \widehat{f}(k, m)\widehat{g}(m, \ell) \tag{4}$$

where $f \star g(x) = \sum_{y \in G} f(xy^{-1})g(y)$.

1.4   Organization of this chapter

The majority of what follows focuses on the case of finite groups, for most of the progress has been in this area. This is the content of the next section. Included are generalizations of both the Cooley-Tukey FFT (decimation in time) in the guise of *separation of variables* group FFTs, as well as the Gentleman-Sande FFT (decimation in frequency). We also touch upon the large body of recent work devoted to the development of *quantum (finite group) FFTs*. Section three is devoted to compact group FFTs, almost exclusively compact Lie groups, while Section four discusses recent work in the difficult, but tremendously useful, noncompact case. Indeed, this raises the issue of both the utility and applicability of these algorithms, for while the abstract development of algorithms has epistemological value, it is even of greater interest when motivated by and subsequently applied to real problems. With this in mind, each section contains some indication and discussion of applications, and indeed, we hope that this chapter might inspire many new uses.

2.   Finite group FFTs

As mentioned, when G is finite and commutative, the number of operations required is bounded above by $O(|G| \log |G|)$. For arbitrary finite groups G, upper bounds of $O(|G| \log |G|)$ remain the holy grail in group FFT research. Implicit in the big-O notation is the idea that a family of groups is under consideration, with the size of the individual groups going to infinity. In 1978, A. Willsky provided the first noncommutative example by showing that certain metabelian groups had an $O(|G| \log |G|)$ Fourier transform algorithm [55].

Two of the most important algorithms in the commutative case are the Cooley-Tukey FFT and the Gentleman-Sande FFT, the former often described as decimation in time, while the latter as decimation in frequency, their similarity reflected in a natural isomorphism between the group and its dual that exists in the finite commutative case. In this section we describe in some detail the separation of variables approach [35] which generalizes the former, and an isotypic projection algorithm [32] which generalizes the latter.

2.1   Applications

While the applications of Fourier analysis on commutative groups is now legion (see the Introduction in [8] for a truly mind-boggling list!), for finite noncommutative groups the list is still short, but constantly growing.

To date, Fourier analysis on the symmetric group S_n seems to have found the most applicability. It has has been proposed and used to analyze ranked data. In this setting respondents are asked to rank a collection of n objects. As a result, each participant in effect chooses a permutation of the initially ordered list of objects. The counts of respondents choosing particular rankings then gives rise to a function on S_n for which Fourier analysis provides a natural generalization of the usual spectral analysis applied to a time series. Diaconis has used this to study voting data (cf. [14] for a discussion of this example, as well as others). More recently, Lafferty has applied this to the development of conditional probability models to analyze some partially ranked data [27].

In communications, Fourier analysis on finite matrix groups, $SL_2(p)$, the group of two-by-two matrices with determinant one with entries in a finite field, has made possible new developments in the area of low density parity check (LDPC) codes [28], and also proved instrumental in the construction of expander graphs that provide models for networks with high connectivity but relatively small numbers of links.

2.2   Cooley-Tukey revisited

The separation of variables approach generalizes the *decimation in time* FFT, which is essentially the guts of the Cooley-Tukey FFT.

Assuming that $n = pq$ (not necessarily prime), then decimation in time refers to the factorization of our time index ℓ as

$$\ell = \ell_1 q + \ell_2 \quad (0 \le \ell_1 < p, 0 \le \ell_2 < q) \tag{5}$$

which is coupled with a corresponding factorization of the frequency index k as

$$k = k_1 + k_2 p \quad (0 \le k_1 < p, 0 \le k_2 < q) \tag{6}$$

so that

$$\widehat{f}(k) = \sum_{\ell_2} e^{2\pi i \ell_2 k} \sum_{\ell_1} e^{2\pi i \ell_1 k_1 / p} f(\ell_1, \ell_2) \tag{7}$$

where $f(\ell_1, \ell_2) = f(\ell_1 q + \ell_2)$.

Notice that this rewrites a "one-dimensional" computation as a "two-dimensional" computation. The FFT organizes the calculation into two stages:

- Stage 1: For all k_1, ℓ_2 compute

$$\tilde{f}(k_1, \ell_2) = \sum_{\ell_1} e^{2\pi i \ell_1 k_1 / p} f(\ell_1, \ell_2). \tag{8}$$

This requires at most pq^2 operations.

- Stage 2: For all k_1, k_2 compute

$$\widehat{f}(k_1, k_2) = \sum_{\ell_2} e^{2\pi i \ell_2 (k_1 + k_2 p)} \tilde{f}(k_1, \ell_2). \tag{9}$$

This requires at most $p^2 q$ operations.

In toto, this gives an algorithm which requires $pq(p + q)$ operations, rather than $(pq)^2$, providing savings as long as factorization is possible.

This approach generalizes nicely. Decimation in time is naturally replaced by *group factorization*, (first generally observed by Beth [5]), but the concomitant factorization of the dual (frequency) requires a little work. For this the machinery of *Bratteli diagrams* has proved to be of immense utility. For illustration we'll revisit Cooley-Tukey in this setting.

Bratteli diagram for $\mathbb{Z}/6\mathbb{Z} > 2\mathbb{Z}/6\mathbb{Z} > 1$.

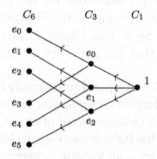

Figure 1. The Bratteli diagram for $C_6 > C_3 > C_1$.

In brief, the diagram above reflects a chain of subgroups $C_6 > C_3 > 1$, the nodes correspond to representations (frequencies) and one node is connected to another if when restricted to that subgroup it gives that corresponding representation. For example, evaluation of e_5 on the multiples of 2 (which comprise the copy of C_3 in C_6) is equivalent to simply evaluating e_2 on C_3.

A full path in the Bratteli diagram is now a frequency, and the factorization (6) gives a labeling of the legs that make up the path. Stages

1 and 2 above can now be reinterpreted diagrammatically. Stage 1 requires the computation over all *initial paths* k_1 and *subgroup elements* ℓ_1, while Stage 2 becomes a computation over all coset representatives ℓ_2 and *full paths* (k_1, k_2).

Separation of variables

As described in [35] in the noncommutative case separation of variables takes on a general form that requires more elaborate Bratteli diagrams. Once again, the initial data is a chain of subgroups $G_n > \ldots, > G_1 > G_0 = \{1\}$, but the nodes at level i now correspond to *matrix representations* of G_i. A node η at level i is connected to ρ at level $i+1$ by a number of arrows equal to the multiplicity of η in $\rho|G_i$.

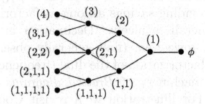

Figure 2.   The Bratteli diagram for $S_4 > S_3 > S_2 > 1$.

For example, matrix representations for the group S_n correspond to partitions of n, hence the labeling in the Bratteli diagram for the subgroup chain $S_4 > S_3 > S_2 > 1$ in Figure 2. Notice that the partition $(2, 2)$ in Figure 2 reveals that full paths are no longer uniquely described by their endpoints.

The arrows from η to ρ correspond to mutually orthogonal G_i-equivariant maps of a given irreducible vector space. In this way each full path in the diagram corresponds to a basis vector of an irreducible representation of G. Bases indexed in this fashion are called *Gel'fand-Tsetlin bases*.

Formally, this creates an isomorphism between the *path algebra* of the Bratteli diagram and the chain of semisimple algebras defined by the succession of group algebra inclusions $\mathbb{C}[G_i] \hookrightarrow \mathbb{C}[G_{i+1}]$. In this way the group algebra $\mathbb{C}[G_n]$ is realized as a *multimatrix algebra* (see e.g. [20]).

Matrix elements are now indexed by pairs of paths with a common endpoint. The beauty of the Bratteli diagram formalism lies in the convenient characterization it gives for all types of structured matrices which can arise through the use of Gel'fand-Tsetlin bases.

To begin, consider $a \in G_i \leq G_n$. According to the above explanation, the entries of $\rho(a)$ are indexed by pairs of paths from 1 to ρ in the corresponding Bratteli diagram. Since $a \in G_i$, the matrix entry $\rho_{uv}(a)$

can be nonzero only when paths u and v intersect at level i, i.e., at G_i, and agree from level i to level n. In this case the matrix coefficient $\rho_{vw}(a)$ is independent of the subpath from level i to n. This is precisely the diagrammatic realization of a block diagonal matrix with certain equal sub-blocks. It is for this reason that these are also sometimes called *adapted bases*.

For another example, consider the situation in which $a \in G_n$ centralizes G_j. Using the path algebra formalism, it is not too difficult to show that in this case $\rho_{uv}(a)$ can only be nonzero when u and v agree from level 0 to level j, then vary freely until they necessarily meet at ρ at level n. Here the matrix coefficient depends only upon the pairs of paths between levels j and n.

Finally, any factorization of the group, say into elements in subgroups of the chain as well as their centralizers, then gives a factorization of representations into sums of products of matrix elements, which by the previous discussion are only nonzero in case very particular *compatibility relations* are satisfied among the corresponding sets of contributing paths. Complexity estimates are then computed in terms of counts of compatible diagrams, and also indicate a freedom of choice among a range of possible orders of evaluation, over which the complexity estimates may vary (see e.g. [29, 36] for the case of the symmetric group). The full formalism [29, 34] phrases all of this in the language of bilinear maps and bears some resemblance to the fundamental FFT work of Winograd [56].

2.2.1 State of the art.

This separation of variables approach and its even more elaborate successors have been responsible for the fastest known algorithms for almost all classes of finite groups, including the symmetric groups [29] and their wreath products [35]. These are among the classes of groups for which $O|G|\log^c |G|$ Fourier transform algorithms are known. Other examples include the supersolvable groups [3], while the algorithms for finite matrix groups and Lie groups of finite type still have room for improvement [33, 36].

2.2.2 Finite group quantum FFTs.

By now there are many books and surveys available as introductions to quantum computing. Suffice it to say that the problem of computing a Fourier transform on a finite group in the quantum setting looks formally much like the classical setting. Using the usual bracket notation, over an arbitrary finite group G, this analogously refers to the transformation taking the state

$$\sum_{z \in G} f(z)\,|z\rangle \qquad \text{tothestate} \qquad \sum_{\omega \in \hat{G}} \hat{f}(\omega)_{ij}\,|\omega, i, j\rangle,$$

where $f : G \to \mathbb{C}$ is a function with $\|f\|_2 = 1$ and $\hat{f}(\omega)_{ij}$ denotes the i, jth entry of the Fourier transform at the representation ω. The collection $\{|v\rangle\}_v$ represent a set of basis vectors for the Hilbert space in question.

To date, the main applications of quantum computing, or more precisely, the advantages attributed to quantum computing, have relied on the use of commutative Fourier analysis for the discovery of hidden periodicities. This is similar in spirit to the applied motivations behind the implementation of classical Fourier analysis, tasked to the revelation of the periodicities whose superposition comprise the Fourier representation of a given time series.

In the quantum setting "hidden periodicity" refers to the existence of a subgroup H in a given commutative group G such that, for a particular function f defined on G, f is invariant under translation by the hidden subgroup, or equivalently, f is constant on cosets of some nontrivial subgroup H. For example, in Shor's famous quantum factoring algorithm [51] G is the cyclic group $\mathbb{Z}_n^*$ where n is the number we wish to factor, $f(x) = r^x \bmod n$ for a random $r < n$, H is the subgroup of $\mathbb{Z}_n^*$ of index order (r). His quantum solution to the discrete log problem uses $\mathbb{Z}_n \times \mathbb{Z}_n$ for G. In Simon's algorithm for the "XOR-mask" oracle problem [52] G is $\mathbb{Z}_2^n$ with H given by a subgroup of order 2^{n-1}.

Interest in *noncommutative* HSPs derives from the relation to the elusive *graph isomorphism problem*: given undirected graphs A and B, determine if they are related by a simple permutation of the vertices (which preserves the connectivity relations). It would be sufficient to solve efficiently the HSP over the permutation group S_n in order to have an efficient quantum algorithm for graph automorphism (see, e.g., Jozsa [24] for a review). This was the impetus behind the development of the first noncommutative quantum FFT [4] and is, to a large degree, the reason that the noncommutative HSP has remained such an active area of quantum algorithms research.

Most (if not all) quantum algorithms take advantage of a certain quantum parallelism by which the register (at any time a superposition of a collection of states—i.e., a particular vector in the Hilbert space) is updated via application of *local unitary transformations* which are, generally speaking, the tensor product of identity matrices with unitary matrices of bounded size. Many of these can be applied simultaneously, in essence glued together to form a single *quantum gate*, and the full transform is then effected via the application of some sequence of such gates. The efficiency of any algorithm is then measured in terms of the *quantum circuit depth*.

The various sparse factorization FFTs are, in spirit, ready-made for quantum implementations and underly efficient quantum implementations for [23] as well as some solvable groups [44]. A recent quantum adaptation of the separation of variables approach [39] provides a rederivation of Beals's original work, as well extensions to those classes of groups whose classical FFTs benefited from this framework.

2.2.3   Sparse structured factorizations.   In [17], sparse representation theoretic factorizations are put to work in helping to find factorizations of given linear transformations. In this work the goal is to describe a matrix as an element of the algebra of intertwining operators between two matrix representations. Having accomplished this, if the representations have sparse factorizations in general (e.g., of the type used in the separation of variables sorts of algorithms), then these can in turn be used to realize a sparse factorization of the original intertwining element. The paper [17] discusses optimal applications of this approach for signal transforms such as the DFT, various DCTs (Discrete Cosine Transforms) and Discrete Hartley Transforms. This approach has been partially automated and is contained in the software library AREP (Abstract REPresentations) [43], which is in turn a part of the very interesting SPIRAL Project [40, 45], a multi-university effort directed at the automatic generation of platform-optimized software.

2.3   Projection-based generalizations of the FFT

The approaches explained above rely on what is commonly known as *decimation in time*, a recursive (or, depending on your point of view, iterative) traversal of the spatial (i.e., group) domain. Decimation in time often goes by the name of *subsampling*.

An alternative, or perhaps more precisely, dual formulation is to instead recurse through the range, iteratively constructing the frequency content of the original data through successive projections which build out increasingly finer orthogonal decompositions. This is the philosophy behind *decimation in frequency*, originally due to Gentleman and Sande [19].

This idea has also found generalization in the context of computing *isotypic decompositions* of a function defined on a group or its homogeneous space. This generalization hinges on the observation that through a judicious choice of group elements and their representing matrices it can be possible to find a collection of projection operators whose application can be scheduled in such a way so as to effect the requisite projections.

2.3.1   Nested projections: The Gentleman-Sande FFT. A DFT of length n effects the projection of the data f onto the n distinct eigenvectors of the DFT operator given by the sampled exponentials. Equivalently, it is also the projection onto the eigenvectors of the *cyclic shift operator* $T_1^{(n)}$ acting on n-space via $\left(T_1^{(n)}f\right)(j) = f(\overline{j+1})$, where $\overline{j+1}$ indicates that the index is to be interpreted (mod n). The DFT eigenvectors are precisely the basis which diagonalizes the shift operator—i.e., they are also the eigenvectors for $T_1^{(n)}$.

Of course, the operator $T_1^{(n)}$ commutes with any of its powers. Suppose now that $n = pq$. Note that under the action of $T_p^{(n)} = \left(T^{(n)}\right)_1^p$, the vector space $V = \mathbb{C}^n$ decomposes into p orthogonal and $T_p^{(n)}$-invariant q-dimensional subspaces $V_j = \text{span}\{\delta_j, \delta_{j+p}, \ldots, \delta_{j+(q-1)p}\}$, where δ_ℓ denotes the standard ℓth basis vector. It is clear that the action of $T_p^{(n)}$ on any V_j is equivalent to the action of $T_1^{(q)}$ on $\mathbb{C}^q$, thus when restricted to the space V_j it is diagonalizable with eigenvectors and eigenvalues corresponding to the DFT of length q.

Thus, we see that the operator $T_p^{(n)}$ has only q distinct eigenvalues on V, one eigenspace W_j for each character of $\mathbb{Z}_q$, and by symmetry, each of these is of dimension p, and as an eigenspace is $T_p^{(n)}$-invariant. Furthermore, since $T_p^{(n)}$ commutes with $T_1^{(n)}$, there is a basis of simultaneous eigenvectors. Thus

$$W_j = W_j^0 \oplus \cdots \oplus W_j^{p-1}.$$

Note that the original DFT of length n is thus the projection of f onto the W_j^k. This suggests the following algorithm for computing the DFT:

- Stage 1: For $j = 0, \ldots, q-1$, compute $f^{(j)}$, the projection of f onto W_j.

- Stage 2: For each j and each k, compute the projection of $f^{(j)}$ onto W_j^k.

This particular fast Fourier transform is known as the Gentleman-Sande, or *decimation in frequency*, FFT (see [19]).

2.3.2   Gentleman-Sande for finite groups. The above discussion reveals that the decimation in frequency FFT can be viewed as a sequence of projections onto *isotypic subspaces*. In the commutative case these are the individual eigenspaces. For an arbitrary representation this is the decomposition into invariant subspaces each of which has an irreducible decomposition into copies of a single irreducible subspace. Thus we can attempt to generalize Gentleman-Sande by finding a collection of operators and computing (in some order) projections onto their eigenspaces of a collection of simultaneously diagonalizable linear transformations [32].

For example, suppose that $L(X)$ has three isotypic subspaces V_1, V_2, and V_3. Thus $L(X) = V_1 \oplus V_2 \oplus V_3$ and each $f \in L(X)$ may be written uniquely as $f = f_1 + f_2 + f_3$ where $f_i \in V_i$. Additionally, suppose that T and T' are diagonalizable linear transformations on $L(X)$ such that the eigenspaces of T are $V_1 \oplus V_2$ and V_3, and the eigenspaces of T' are V_1 and $V_2 \oplus V_3$. We may therefore compute the f_i by first projecting f onto the eigenspaces of T to compute $f_1 + f_2$ and f_3, and then projecting both $f_1 + f_2$ and f_3 onto the eigenspaces of T' to compute f_1, f_2 and f_3. Note that each computation is done with respect to a fixed basis of $L(X)$. This process of decomposing $L(X) = V_1 \oplus V_2 \oplus V_3$ is illustrated in Figure 3.

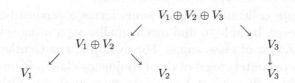

Figure 3. Decomposing $L(X) = V_1 \oplus V_2 \oplus V_3$ using T and T'.

We call the pair $\{T, T'\}$ a *separating set* for $L(X)$ because it allows us to *separate* a representation into its isotypic components.

From this point of view, the Gentleman-Sande FFT first computes a projection of the data onto the isotypic decomposition corresponding to the subgroup generated by $T_p^{(n)}$ (isomorphic to $\mathbb{Z}_q$) and then further decomposing each of these according to the decomposition of $T_1^{(n)}$ (isomorphic to the fill group G). Thus $\{T_p^{(n)}, T_1^{(n)}\}$ is a separating set for $\mathbb{Z}_{pq}$.

More generally, suppose now that $\{T_1, \ldots, T_k\}$ is a collection of diagonalizable linear transformations on a vector space V whose eigenspaces are direct sums of the isotypic subspaces of V. For each isotypic subspace V_i, let $c_i = (\mu_{i1}, \ldots, \mu_{ik})$ be the k-tuple of eigenvalues where, for $1 \le j \le k$, μ_{ij} is the eigenvalue of T_j associated to V_i. If $c_i \ne c_{i'}$ whenever $V_i \ne V_{i'}$, then we say that $\{T_1, \ldots, T_k\}$ is a *separating set* for V.

The existence of a separating set $\{T_1, \ldots, T_k\}$ for V means that the computation of the isotypic projections of $v \in V$ can be achieved through a series of eigenspace projections as follows:

Stage 1 Compute the projections of v onto each eigenspace for T_1.

Stage 2 For each $i > 1$, iteratively compute the projections of the projections previously computed for T_{i-1} onto each of the eigenspaces of T_i.

It is not difficult to see that the computed projections at stage k are precisely the isotypic projections of the vector v.

We may easily find separating sets for V by looking to the conjugacy classes $C_1, \ldots, C_h$ of G. In particular, if $T_j = \sum_{c \in C_j} \rho(c)$ is the *class sum* of C_j (with respect to ρ) and $\mu_{ij} = |C_j| \chi_i(C_j)/d_i$, then the class sum T_j is a diagonalizable linear transformation on V whose eigenspaces are direct sums of isotypic subspaces, and μ_{ij} is the eigenvalue of T_j that is associated to the isotypic subspace V_i ([32]).

The complete collection of class sums forms a separating set of V. We may, however, be able to find much smaller separating sets than the complete collection of class sums. For example, the Gentleman-Sande FFT uses approximately $\log n$ of the n conjugacy classes (since the group is commutative each element forms a conjugacy class). Other specific examples where this gives a savings include the homogeneous spaces formed from distance transitive graphs and their symmetry groups as well as quotients of the symmetric group [32].

The efficiency of this approach depends on an efficient eigenspace projection method. Since the separating sets we use consist of real symmetric matrices, in [32] *Lanczos iteration* is used. This is an algorithm that may be used to efficiently compute the eigenspace projections of a real symmetric matrix when, as in all of our examples, it has relatively few eigenspaces and when it may be applied efficiently to arbitrary vectors, either directly or through a given subroutine (see, e.g., [42]). Implicit in this iterated projection are notions of *multiresolution analysis*. See [18] for recent group theoretic interpretations of this.

2.4   Open questions for finite group FFTs

Other groups for which highly improved (but not $O(|G| \log^c |G|)$) algorithms have been discovered include the matrix groups over finite fields, and more generally, the Lie groups of finite type. See [37] for pointers to the literature. There is much work to be done finding new classes of groups which admit fast transforms, and improving on the above results. The ultimate goal is to settle or make progress on the following conjecture:

Conjecture. *There exist constants c_1 and c_2 such that for any finite group G, there is a complete set of irreducible matrix representations for which the Fourier transform of any complex function on G may be computed in fewer than $c_1|G|\log^{c_2}|G|$ scalar operations.*

Perhaps progress toward this goal will require new techniques. Indeed, it does seem as though the separation of variables approach has been pushed almost as far as it can go. One place to look, and indeed one of the more intriguing open questions in the development of FFT techniques, is for generalizations of those commutative FFT methods which are used for groups of prime order: Rader's prime FFT [46] and the "chirp-z transform" (the "chirp" here refers to radar chirp) [6, 47].

Both of these algorithms use an idea that rewrites the DFT (of prime length p) at nonzero frequencies in terms of a convolution of length $p-1$ (which, since it is composite, can be computed efficiently using other FFT methods) while computing the DFT at the zero frequency directly. Rader's prime FFT uses a generator g of $\mathbb{Z}/p\mathbb{Z}^\times$, a cyclic group (under multiplication) of order $p-1$, to write $\hat{f}(g^{-b})$ as

$$\hat{f}(g^{-b}) = f(0) + \sum_{a=0}^{p-2} f(g^a) e^{2\pi i g^{a-b}/p}. \tag{10}$$

The summation in (10) has the form of a convolution on $\mathbb{Z}/(p-1)\mathbb{Z}$, of the sequence $f'(a) = f(g^a)$, with the function $z(a) = exp^{2\pi i g^a/p}$.

Rabiner et al. [47] (see also [6]) make the change of variables $jk = (j^2 + k^2 - (j-k)^2)/2$ to obtain

$$\hat{f}(k) = \omega^{k^2/2} \sum_{j=0}^{N-1} \left(f(j)\omega^{j^2/2} \right) \omega^{(j-k)^2/2}.$$

This is a non-cyclic convolution of the sequence $\left(f(j)\omega^{j^2/2} \right)$ with the sequence $\left(\omega^{-j^2/2} \right)$, and may be performed using a cyclic convolution of any length $M \geq 2N$. Note that this gives an approach which rewrites the DFT in terms of a convolution that does not depend on N being prime. This method is commonly known as the *chirp-z transform*.

The discovery of noncommutative generalizations of these ideas would be very, very interesting.

3.   FFTs for compact groups

The DFT and FFT also have a natural extension to (continuous) compact groups as well. The terminology "discrete Fourier transform" derives from the fact that the algorithm was originally designed to compute the (possibly approximate) Fourier transform of a continuous signal from a discrete collection of sample values.

Under the simplifying assumption of periodicity a continuous function may be interpreted as a function on the unit circle, and compact commutative group, S^1. Any such function f has a *Fourier expansion* defined as

$$f(t) = \sum_{l \in \mathbf{Z}} \widehat{f}(l) e^{-2\pi i l t} \tag{11}$$

where

$$\widehat{f}(l) = \langle f, e_l \rangle = \int_0^1 f(t) e^{2\pi i l t} dt. \tag{12}$$

If $\widehat{f}(l) = 0$ for $|l| \geq N$, then f is *band-limited* with *band-limit* N and there is a *quadrature rule* or *sampling theory* for f, meaning that the Fourier coefficients of any such f can be computed as a summation using only a finite set of samples. Thus,

$$\widehat{f}(l) = \sum_{k=0}^{2N-2} \frac{1}{(2N-1)} f\left(\frac{k}{2N-1}\right) e^{2\pi i k l / (2N-1)} \tag{13}$$

where the factor $\frac{1}{2N-1}$ should be viewed as a (constant) weight function with support at the equispaced points $\{\frac{k}{2N-1}\}_{k=0}^{2N-2}$ (where the circle and unit interval have been identified). The FFT then efficiently computes these Fourier coefficients.

A more general framework capable of encompassing all continuous compact groups and their quotients is easily stated: the irreducible representations of a compact group G are all finite-dimensional, and any square-integrable function f (with respect to Haar measure) has an expansion in terms of irreducible matrix elements

$$f = \sum_{\lambda \in \Lambda} \sum_{j,k=1}^{d_\lambda} \widehat{f}(\lambda)_{jk} T_{jk}^\lambda \tag{14}$$

where Λ is some countable set, T^λ denotes an irreducible representation of degree $d_\lambda < \infty$, and the implied convergence is in the mean. The Fourier coefficients $\{\widehat{f}(\lambda)_{jk}\}$ are computed by integrals

$$\widehat{f}(\lambda)_{jk} = d_\lambda \langle f, T_{jk}^\lambda \rangle = d_\lambda \int_G f(x) T_{jk}^\lambda(x) dx \tag{15}$$

where dx denotes (the translation-invariant) Haar measure.

In turn, a general FFT schema then requires a formulation of the notion of *bandwidth*, accompanied by a corresponding *sampling theory*, and an algorithmic component for the efficient evaluation of the quadrature, or *FFT*.

3.1 Applications

To date, it is the group of rotations in three space, $SO(3)$, where most of the applications for FFTs on continuous, noncommutative compact groups have been found. Its representation theory is effectively that of the theory of *spherical harmonics*. One large source of applications comes from the climate modeling community (see e.g. [53]) where spherical harmonics are used for spectral methods approaches to solving the relevant PDEs in spherical geometry. Further applications are to be found outside the atmosphere, as spherical harmonics expansions of the CMB are the source of new information about its fine scale inhomogeneities which hope to provide new information about the shape of space and origins of the universe.

Most recently, FFTs for $SO(3)$, as applied to the development of fast convolution algorithms on $SO(3)$ [25], have been used to develop search algorithms for shape databases.

3.2 FFTs for compact groups—the work of Maslen

The first FFT for a noncommutative and continuous compact group was the efficient spherical harmonic expansion algorithm discovered by J. Driscoll and D. Healy [15]. In this case, the Fourier expansion of a function on the 2-sphere, viewed as a function on $SO(3)/SO(2)$ (with $SO(2)$ identified with the rotations that leave the north pole fixed) has a natural notion of bandwidth given by the degree of the spherical harmonic. A sampling rule on the 2-sphere, equiangular in both latitude and longitude, gives a quadrature rule, and a function of bandwidth B (and $O(B^2)$ Fourier coefficients) requires $O(B^2)$ points. The story is completed with a fast algorithm ($O(N^{3/2} \log^2 N)$ operations for $N = B^2$) that uses the three-term recurrence satisfied by the Legendre functions to produce a divide and conquer algorithm for its efficient evaluation.

Some years later, this work was extended to the full compact setting by D. Maslen: a general notion of bandwidth consistent with the commutative and spherical notions [31], a sampling rule [30], and finally an FFT which also relies on three term recurrence relations satisfied by related orthogonal polynomial systems. What follows is a brief summary of this work

There is a natural definition of band-limited in the compact case, encompassing those functions whose Fourier expansion has only a finite number of terms. The simplest version of Maslen's theory is as follows:

Definition 1. *Let $\mathcal{R}$ denote a complete set of irreducible representations of a compact group G. A system of band-limits on G is a decomposition of $\mathcal{R} = \cup_{b \geq 0} \mathcal{R}_b$ such that*

- *[1] $|\mathcal{R}_b| < \infty$ for all $b \geq 0$;*
- *[2] $b_1 \leq b_2$ implies that $\mathcal{R}_{b_1} \subseteq \mathcal{R}_{b_2}$;*
- *[3] $\mathcal{R}_{b_1} \otimes \mathcal{R}_{b_2} \subseteq span_{\mathbf{Z}} \mathcal{R}_{b_1 + b_2}$.*

Let $\{\mathcal{R}_b\}_{b \geq 0}$ be a system of band-limits on G and $f \in L^2(G)$. f is band-limited with band-limit b if $\hat{f}(T_{jk}^\lambda) = 0$ for all $\lambda \notin \mathcal{R}_b$.

The case $G = S^1$ provides the classical example. If $\mathcal{R}_b = \{\chi_j : |j| \leq b\}$ where $\chi_j(z) = z^j$, then $\chi_j \otimes \chi_k = \chi_{j+k}$ and the corresponding notion of band-limited (as per Definition 1) coincides with the usual notion.

For a noncommutative example, consider $G = SO(3)$. In this case the irreducible representations of G are indexed by the non-negative integers with V_λ the unique irreducible representation of dimension $2\lambda + 1$. Let $\mathcal{R}_b = \{V_\lambda : \lambda \leq b\}$. The *Clebsch-Gordon relations*

$$V_{\lambda_1} \otimes V_{\lambda_2} = \sum_{j=|\lambda_1 - \lambda_2|}^{\lambda_1 + \lambda_2} V_j \tag{16}$$

imply that this is a system of band-limits for $SO(3)$. When restricted to the quotient $S^2 \cong SO(3)/SO(2)$, band-limits are described in terms of the highest order spherical harmonics that appear in a given expansion.

Maslen's most general setting for a notion of band-limit develops a theory of band-limited elements for any *filtered module* over a *filtered algebra*. In the case of a connected compact Lie group G, $\mathcal{R}_s$ is defined to be the set of all matrix elements that come from representations whose highest weight is at most s. For the matrix groups $SU(r+1), Sp(r), SO(2r+1)$, it is possible to choose a norm for which $\mathcal{R}_1$ is the span of all matrix representations with highest weight given by a fundamentally analytically integral dominant weight or zero. Band-width is thus defined in terms of lengths of factorizations in sums of products of such elements expressing a given matrix element [31].

The importance of developing the band-limited theory is that in this setting there exists a sampling theory or quadrature rule that allows the

Fourier coefficients to be computed exactly as finite sums. The following is the content of [30], once the notion of bandlimit is arranged.

THEOREM 1 *Let G be compact with a system of band-limits* $\{\mathcal{R}_b\}_b$. *For any band-limit b, there exists a finite set of points* $X_b \subset G$ *such that for any function* $f \in L^2(G)$ *of band-limit b,*

$$\widehat{f}(T_{jk}^\lambda) = \sum_{x \in X_b} f(x) T_{jk}^\lambda(x) w(x) \tag{17}$$

for all $\lambda \in \mathcal{R}_b$ *and some weight function w on* X_b.

Theorem 1 reduces the integrals (15) to summations, so that efficient algorithms can now be designed to perform the computations (17). For the classical groups $U(n), SU(n), Sp(n)$, a system of band-limits $\mathcal{R}_b^n$ is chosen with respect to a particular norm on the dual of the associated Cartan subalgebra. Such a norm $\| \cdot \|$ (assuming that it is invariant under taking duals, and $\|\alpha\| \le \|\beta\| + \|\gamma\|$ for α occurring in $\beta \otimes \gamma$) defines a notion of band-limit given by all α with norm less than a fixed b. The associated sampling sets X_b^n are contained in certain one-parameter subgroups.

Implicit here are certain *discrete special function transforms*, which can often be reduced to certain *discrete polynomial transforms*

$$\widehat{f}(P_j) = \sum_{k=0}^{N-1} f(k) P_j(x_k) w_k \tag{18}$$

where $P_0, \ldots, P_{N-1}$ are a set of linearly independent polynomials with complex coefficients, $\{x_0, \ldots, x_{N-1}\}$ are a set of N distinct complex points and $\{w_0, \ldots, w_{N-1}\}$ is a set of positive weights. The case of the DFT comes from choosing equispaced roots of unity for sample points, equal weights of one, and $P_j(x) = x^j$. Direct calculation of all the $\widehat{f}(P_j)$ clearly requires N^2 operations.

If the P_j make up a family of *orthogonal polynomials*, then fast algorithms exist to speed the calculation. Here the idea is to use the *three-term recurrence* satisfied by these polynomials to create a divide-and-conquer algorithm which reduces transforms of degree n to sums of transforms of degree less than n, ultimately providing an $O(n \log^2 n)$ algorithm. (See [16] and references therein.)

By using these sorts of complexity estimates, together with a sampling theory and a careful organization of the calculation (using the diagrammatic techniques explained above), Maslen is able to derive efficient algorithms for all the classical groups.

THEOREM 2 ([31], Theorem 5)
  Assume $n \geq 2$.

(i) For $U(n)$, $T_{X_b^n}(\mathcal{R}_b^n) \leq O(b^{dim U(n)+3n-3})$

(ii) For $SU(n)$, $T_{X_b^n}(\mathcal{R}_b^n) \leq O(b^{dim SU(n)+3n-2})$

(iii) For $Sp(n)$, $T_{X_b^n}(\mathcal{R}_b^n) \leq O(b^{dim Sp(n)+6n-6})$

where $T_{X_b^n}(\mathcal{R}_b^n)$ denotes the number of operations needed for the particular sample set X_b^n and representations $\mathcal{R}_b^n$ for the associated group.

3.3   Approximate techniques.

In the bandlimited case, Maslen's techniques are exact, in the sense that if computed in exact arithmetic, they yield exact answers. Of course, in any actual implementation, errors are introduced and the utility of an algorithm will depend highly on its numerical stability.

There are also "approximate methods", approximate in the sense that they guarantee a certain specified approximation to the exact answer that depends on the running time of the algorithm. For computing Fourier transforms at non-equispaced frequencies, as well as spherical harmonic expansions, the fast *multipole method* and its variants are used [21]. Multipole-based approaches efficiently compute these quantities approximately, in such a way that the running time increases by a factor of $\log(\frac{1}{\epsilon})$ where ϵ denotes the precision of the approximation. Another approach is via the use of *quasi-classical frequency estimates* for the relevant transforms [38]. It would be interesting to generalize these sorts of techniques to compact groups and their quotients.

3.4   Open question

Maslen's work effectively creates uniform sampling grids with concomitant quadrature rules, but it may be possible that some applications may require *nonuniform grids*. In the commutative case, examples include applications in medical imaging and other forms of non-invasive testing. Noncommutative examples might include astrophysical, weather, and climate data. The corresponding measurements are rarely equidistributed (in particular, there are many large uninhabited regions in which the data is never taken) and, in fact, these two variable expansions generally use grids which evenly sample in one direction, but use Legendre points in the other [41, 53]. For example, as applied to the analysis of the cosmic microwave background, this is meant to provide a sampling that is sparse at "the center," which corresponds to avoiding our own galaxy. It would seem to be of great interest to push forward the wealth of work done in the commutative setting (see e.g. [1] and the many examples therein).

4. Noncompact groups

Much of modern signal processing relies on the understanding and implementation of Fourier analysis for $L^2(\mathbf{R})$, i.e., the noncompact commutative group $\mathbf{R}$. It is only fairly recently that noncommutative, noncompact examples have begun to attract significant attention.

In this area some of the most exciting work is being done by G. Chirikjian and his collaborators. They have been concerned primarily with the Euclidean motion group $SE(n)$. Recall that the motion groups are given as semidirect products of $\mathbb{R}^n$ with the rotation groups $SO(n)$, realized as $n+1$ by $n+1$ matrices of the form

$$\begin{pmatrix} A & v \\ \mathbf{0}_n & 1 \end{pmatrix}$$

where $A \in SO(n)$, $v \in \mathbb{R}^n$ and $\mathbf{0}_n$ denotes the all zero row vector of length n. This provides an algebraic mechanism for gluing together the group of additive translations with rotations.

Their motivation comes from a diverse collection of applications, ranging among robotics, molecular modeling and pattern matching. Applications to robotics come from the problem of *workspace determination* for *discretely actuated manipulators*. A standard example is a robot arm and a standard problem in motion planning is to determine the set of reachable configurations, as well as to plan a path to move from one configuration to another. The configuration of the end of the arm can be described with two parameters: a position in space (a vector in $\mathbb{R}^n$ for $n = 2$ or 3) as well as an orientation (an element in $SO(n)$), i.e., an element of $SE(n)$. One sort of design paradigm is to build a robot arm as an assembly of a sequence of basic modules, so that the arm takes on a worm-like or cillial form. Any single basic unit will have some finite discrete set of reachable states, defining a discrete probability density in $SE(n)$. This is the *workspace* of a single unit. The *workspace* of the full arm (defined as the linked assembly of m of these basic units) is then given as the m-fold convolution of the fundamental workspace. This is called the *workspace density*. Applications to polymer science are analogous, with similar modeling considerations used to describe the motion of a given end of a polymer (such as DNA) relative to its other end. These are but two examples. For details, as well as other applications see [9] and the many references therein.

Just as the classical FFT provides efficient computation of convolutions on the line or circle, so does an FFT for $SE(n)$ allow for efficient convolution in this setting, replacing direct convolution by FFTs, matrix multiplications and inverse FFTs.

In a collection of papers (see [9]) Chirikjian and Kyatkin create a computational framework for working with the representation theory of $SE(3)$ acting on $\mathbb{R}^3$. The matrix elements in this case are known and involve spherical harmonics, half-integer Bessel functions, glued together with Clebsch-Gordan coefficients. Explicitly, they find themselves in the position of having to compute (for a function $f(\mathbf{r}, R)$ of compact support on $SE(3)$)

$$\hat{f}^s_{\ell',m';\ell,m}(p) =$$

$$\int_{\mathbf{u} \in S^2} \int_{\mathbf{r} \in \mathbb{R}^3} \int_{R \in SO(3)} f(\mathbf{r}, R) h^s_{\ell,m}(\mathbf{u}) \times$$

$$e^{ip\mathbf{u}\cdot\mathbf{r}} \sum_{n=-\ell'}^{\ell'} \overline{U^{\ell'}_{nm'}(R) h^s_{\ell'n}(\mathbf{u})} d\mathbf{u} d^3\mathbf{r} dR. \tag{19}$$

where the h's and U's are defined in terms of generalized Legendre functions.

Computation is now effected via a host of discretizations on $\mathbb{R}^3, SO(3)$, S^2, and the dual index p, as well as some assumptions on the number of harmonics used to describe f. The exponent $p\mathbf{u}$ implies the need to convert from a rectangular to a polar $\mathbb{R}^3$ grid and so there is also an interpolation (through splines) used. The complexity of the final separation of variables style algorithm is then given by gluing together all the appropriate fast algorithms (FFTs, fast interpolation, fast spherical harmonic expansions, fast Legendre expansions).

The details of this analysis are found in [26]. This paper also contains an analogous discussion for $SE(2)$ as well as the *discrete motion groups* defined as the semidirect product of translations ($\mathbb{R}^3$) with any of the (finite number of) finite subgroups of $SO(3)$.

4.1   Open questions

To date, the techniques used here are approximate in nature and interesting open problems abound. Possibilities include the formulation of natural sampling (regular and irregular), band-limiting and time-frequency theories. The exploration of other special cases beyond the special Euclidean groups, such as semisimple Lie groups (see [2] for a beautifully written succinct survey of the Harish-Chandra theory), is also intriguing. "Fast Fourier transforms on semisimple Lie groups" has a nice ring to it!

Acknowledgments

This work was completed with partial support from the NSF under grants CCR-0219717 and F49620-00-1-0280.

References

[1] A. Aldroubi and K. Grochenig. Non-uniform sampling and reconstruction shift-invariant spaces. *SIAM Rev.*, **43** No. 4 (2001), pp. 585–620.

[2] J. Arthur, *Harmonic analysis and group representations*, Notices. Amer. Math. Soc., **47**(1) (2000), 26–34.

[3] U. Baum. Existence and efficient construction of fast Fourier transforms for supersolvable groups. *Comput. Complexity* **1** (1991), 235–256.

[4] Robert Beals. Quantum computation of Fourier transforms over symmetric groups. In ACM, editor, *Proceedings of the twenty-ninth annual ACM Symposium on the Theory of Computing: El Paso, Texas, May 4–6, 1997*, pages 48–53, New York, NY, USA, 1997. ACM Press.

[5] T. Beth. *Verfahren der schnellen Fourier-Transformation.* B. G. Teubner, Stuttgart, 1984.

[6] L. Bluestein, *A linear filtering approach to the computation of the discrete Fourier transform*, IEEE Trans. **AU-18** (1970), 451-455.

[7] O. Bratteli. Inductive limits of finite dimensional C^*-algebras. *Trans. Amer. Math. Soc.* **171** (1972), 195–234.

[8] O. Brigham. *The Fast Fourier Transform and its Applications.* Prentice Hall, NJ, 1988.

[9] G. S. Chirikjian and A. B. Kyatkin. *Engineering applications of noncommutative harmonic analysis*, CRC Press, FL, 2000.

[10] M. Clausen and U. Baum. *Fast Fourier transforms*, Bibliographisches Institut, Mannheim, 1993.

[11] W. Cochran, et. al., What is the fast Fourier transform?, *IEEE Transactions on Audio and Electroacoustics* Vol. AU-15, No. 2 (1967), 45–55.

[12] J. W. Cooley and J. W. Tukey. An algorithm for machine calculation of complex Fourier series, *Math. Comp.*, **19** (1965), 297–301.

[13] P. Diaconis. A generalization of spectral analysis with application to ranked data, *Ann. Statist.*, **17**(3), (1989), 949–979.

[14] P. Diaconis. *Group representations in probability and statistics*, IMS, Hayward, CA, 1988.

[15] J. R. Driscoll and D. Healy. Computing Fourier transforms and convolutions on the 2-sphere. *Proc. 34th IEEE FOCS*, (1989) pp. 344–349 (extended abstract); *Adv. in Appl. Math.* **15** (1994), 202–250.

[16] J. Driscoll and D. Healy and D. Rockmore. Fast discrete polynomial transforms with applications to data analysis for distance transitive graphs, *SIAM J. Comput.*, **26**, No. 4, (1997), 1066–1099.

[17] S. Egner and M. Püschel. Symmetry-based matrix factorization. *J. Symb. Comp*, to appear.

[18] R. Foote. An algebraic approach to multiresolution analysis, (2003), submitted for publication.

[19] W. Gentleman and G. Sande. Fast Fourier transform for fun and profit, *Proc. AFIPS, Joint Computer Conference*, **29**, (1966), 563-578.

[20] F. Goodman, P. de al Harpe, and V. F. R. Jones, *Coxeter graphs and towers of algebras*, Springer-Verlag, New York, 1989.

[21] L. Greengard and V. Rokhlin. A fast algorithm for particle simulations. *J. Comput. Phys.* **73** (1987) 325-348.

[22] M. T. Heideman, D. H. Johnson and C. S. Burrus. Gauss and the history of the fast Fourier transform, *Archive for History of Exact Sciences*, **34** (1985), no. 3, 265-277.

[23] Peter Höyer. Efficient quantum transforms. Technical Report quant-ph/9702028, Quantum Physics e-Print Archive, 1997.

[24] Richard Jozsa. Quantum factoring, discrete logarithms and the hidden subgroup problem. *Computing, Science, and Engineering*, **3**(2), pp. 34-43, (2001).

[25] M. Kazhdan, T. Funkhouser and S. Rusinkiewicz. Rotation invariant spherical harmonic representation of 3D-shape descriptors. *Symposium on Geometry Processing* June, (2003) pp. 167-175.

[26] A. B. Kyatkin and G. S. Chirikjian. Algorithms for fast convolutions on motion groups. *App. Comp. Harm. Anal.* **9**, 220-241 (2000).

[27] G. Lebanon and J. Lafferty. Cranking: Combining rankings using conditional probability models on permutations, in *Machine Learning: Proceedings of the Nineteenth International Conference*, San Mateo, CA: Morgan Kaufmann, 2002.

[28] J. Lafferty and D. N. Rockmore. Codes and iterative decoding on algebraic expander graphs, in *Proceedings of International Symposium on Information Theory and its Application, Honolulu, Hawaii*, 2000.

[29] D. K. Maslen. The efficient computation of Fourier transforms on the symmetric group, *Math. Comp.* **67**(223) (1998), 1121-1147.

[30] D. K. Maslen. Sampling of functions and sections for compact groups. in *Modern Signal Processing*, D. N. Rockmore and D. M. Healy, eds., Cambridge University Press, to appear.

[31] D. K. Maslen. Efficient computation of Fourier transforms on compact groups, *J. Fourier Anal. Appl.* **4**(1) (1998), 19-52.

[32] D. K. Maslen, M. Orrison, and D.N. Rockmore. Computing Isotypic Projections with the Lanczos Iteration. *SIAM J. Matrix Analysis*, to appear.

[33] D. K. Maslen and D. N. Rockmore. Separation of variables and the computation of Fourier transforms on finite groups. I. *J. Amer. Math. Soc.* **10** (1997), no. 1, 169-214.

[34] D. K. Maslen and D. N. Rockmore. Separation of variables and the computation of Fourier transforms on finite groups. II. In preparation.

[35] D. K. Maslen and D. N. Rockmore. The Cooley-Tukey FFT and group theory, *Notices Amer. Math. Soc* **48** (2001), no. 10, 1151-1160.

[36] D. Maslen and D. N. Rockmore. Double coset decompositions and computational harmonic analysis on groups.. *J. Fourier Anal. Appl.* **6**(4), 2000, pp. 349–388.

[37] D. Maslen and D. N. Rockmore. Generalized FFTs—a survey of some recent results, in *Groups and computation, II (New Brunswick, NJ, 1995)*, Amer. Math. Soc., Providence, RI, 1997, pp. 183–237.

[38] M. P. Mohlenkamp. A fast transform for spherical harmonics, *J. Fourier Anal. Appl.*, **5** no. 2–3 (1999), 159–184.

[39] C. Moore, D. N. Rockmore, and A. Russell. Generic Quantum Fourier Transforms. Technical Report quant-ph/0304064, Quantum Physics e-print Archive, 2003.

[40] J.M.F. Moura, J. Johnson, R.W. Johnson, D. Padua, V. Prasanna, M. Püschel, and M.M. Veloso, SPIRAL: Automatic Library Generation and Platform-Adaptation for DSP Algorithms, 1998, http://www.ece.cmu.edu/~spiral/

[41] S.P. Oh, D.N. Spergel, and G. Hinshaw. An Efficient Technique to Determine the Power Spectrum from the Cosmic Microwave Background Sky Maps. *Astrophysical Journal*, **510** (1999), 551–563.

[42] B. Parlett. *The symmetric eigenvalue problem*. Prentice-Hall Inc., Englewood Cliffs, N.J. 1980.

[43] M. Püschel, S. Egner, and T. Beth. In *Computer Algebra Handbook, Foundations, Applications, Systems*. Eds. J. Grabmeier, E. Kaltofen, V. Weispfenning, Springer 2003, pp. 461-462.

[44] M. Püschel, M. Rötteler, and T. Beth. Fast quantum Fourier transforms for a class of non-abelian groups. (*Proc. AAECC 99*, LNCS 1719, Springer-Verlag, pp. 148-159.

[45] M. Püschel, B. Singer, J. Xiong, J.M.F. Moura, J. Johnson, D. Padua, M.M. Veloso, and R.W. Johnson, SPIRAL. A Generator for Platform-Adapted Libraries of Signal Processing Algorithms. *J. of High Performance Computing and Applications*, accepted for publication.

[46] C. Rader. Discrete Fourier transforms when the number of data samples is prime, *IEEE Proc.* **56** (1968), 1107–1108.

[47] L. Rabiner, R. Schafer, and C. Rader, *The chirp-z transform and its applications*, Bell System Tech. J. **48** (1969), 1249-1292.

[48] D. N. Rockmore. Some applications of generalized FFTs (An appendix w/D. Healy), in *Proceedings of the DIMACS Workshop on Groups and Computation, June 7-10, 1995*. Eds. L. Finkelstein and W. Kantor, (1997), pp. 329–369

[49] D. N. Rockmore. Fast Fourier transforms for wreath products, *Appl. Comput. Harmon. Anal.*, **2**, No. 3 (1995), 279–292.

[50] J.-P. Serre. *Linear representations of finite groups*, Springer-Verlag, New York, 1977.

[51] P. W. Shor, Polynomial-time algorithms for prime factorization and discrete logarithms on a quantum computer, *SIAM J. Computing* **26** (1997), 1484–1509.

[52] Daniel R. Simon. On the power of quantum computation. *SIAM Journal on Computing*, 26(5):1474–1483, October 1997.

[53] W. F. Spotz and P. N. Swarztrauber. A Performance Comparison of Associated Legendre Projections. *Journal of Computational Physics*, 168(2), (2001) 339-355.

[54] L. Trefethen and D. Bau III. *Numerical linear algebra*. Society for Industrial and Applied Mathematics (SIAM), Philadelphia, PA, 1997.

[55] A. Willsky. On the algebraic structure of certain partially observable finite-state Markov processes. *Inform. Contr.* **38**, 179–212 (1978).

[56] S. Winograd, *Arithmetic complexity of computations*, CBMS-NSF Regional Conference Series in Applied Mathematics, Vol. 33. Society for Industrial and Applied Mathematics (SIAM), Philadelphia, Pa., 1980.

GROUP FILTERS AND IMAGE PROCESSING

Richard Tolimieri and Myoung An

Prometheus Inc.
Newport, Rhode Island U.S.A.

myoung@prometheus-inc.com

Abstract   Abelian group DSP can be completely described in terms of a special class of signals, the characters, defined by their relationship to the translations defined by abelian group multiplication. The first problem to be faced in extending classical DSP theory is to decide on what is meant by a translation. We have selected certain classes of nonabelian groups and defined translations in terms of left nonabelian group multiplications.

The main distinction between abelian and nonabelian group DSP centers around the problem of character extensions. For abelian groups the solution of the character extension problem is simple. Every character of a subgroup of an abelian group A extends to a character of A. We will see that character extensions lie at the heart of several fast Fourier transform (FFT) algorithms.

The nonabelian groups presented in this work will be among the simplest generalizations of abelian groups. A complete description of the DSP of an abelian by abelian semidirect product will be given. For such a group G, there exists an abelian normal subgroup A. A basis for signals indexed by G can be constructed by extending the characters of A and then by forming certain (left-) translations of these characters. The crucial point is that some of these characters of A cannot be extended to characters of G, but only to proper subgroups of G. In contrast to abelian group DSP expansions, expansions in nonabelian group DSP reflect local signal domain information over these proper subgroups. These expansions can be viewed as combined, local image-spectral image domain expansions in analogy to time-frequency expansions.

This DSP leads to the definition of a certain class of group filters whose properties will be explored on simulated and recorded images. Through examples we will compare nonabelian group filters with classical abelian group filters. The main observation is that in contrast to abelian group filters, nonabelian group filters can localize in the image domain. This advantage has been used for detecting and localizing the position of geometric objects in image data sets.

The works of R. Holmes M. Karpovski and E. Trachtenberg are the basis of much of the research in the application of nonabelian group

J. Byrnes (ed.) Computational Noncommutative Algebra and Applications, 255-308.

harmonic analysis to DSP, especially in the construction of group transforms and group filters. Similar ideas in coding theory have been introduced by F. J. MacWilliams.

During the last ten years considerable effort has taken place to extend
the range of applicability of nonabelian group methods to the design of
new filters and spectral analysis methodologies, as an image processing
tool. Accompanying these application efforts, there has been extensive development of algorithms for computing Fourier transforms over
nonabelian groups generalizing the classical FFT.

Keywords: finite group harmonic analysis, group filters, image domain locality.

1.   Introduction: Classical Digital Signal Processing

There are two disjoint themes in this work: the fundamental role
played by translations of data in all aspects of digital signal processing
(DSP) and the role of group structures on data indexing sets in defining
these translations.

The set of integers paired with addition modulo N is the simplest
finite abelian group. The term *abelian* refers to the fact that addition
modulo N is a commutative binary operation. We denote this group by
$\mathbf{Z}/N$ and call it the *group* of integers *modulo N*.

Denote by $\mathcal{L}(\mathbf{Z}/N)$ the space of all complex valued functions on $\mathbf{Z}/N$.
$\mathcal{L}(\mathbf{Z}/N)$ is the space of N-point data sets.

For $y \in \mathbf{Z}/N$, the mapping $T(y)$ of $\mathcal{L}(\mathbf{Z}/N)$ defined by

$$(T(y)f)(x) = f(x - y), \quad f \in \mathcal{L}(\mathbf{Z}/N), \, x \in \mathbf{Z}/N,$$

is a linear operator of $\mathcal{L}(\mathbf{Z}/N)$ called *translation* by y.

For $g \in \mathcal{L}(\mathbf{Z}/N)$, the mapping $C(g)$ of $\mathcal{L}(\mathbf{Z}/N)$ defined by

$$C(g) = \sum_{y \in \mathbf{Z}/N} g(y)T(y)$$

is a linear operator of $\mathcal{L}(\mathbf{Z}/N)$ called *convolution* by g. If $f \in \mathcal{L}(\mathbf{Z}/N)$,
then

$$(C(g)f)(x) = \sum_{y \in \mathbf{Z}/N} f(x - y)g(y), \quad x \in \mathbf{Z}/N.$$

1.1   Fourier Analysis

Fourier analysis provides the tools for the detailed analysis and construction of translation-invariant subspaces of $\mathcal{L}(\mathbf{Z}/N)$.

For $y \in \mathbf{Z}/N$, define $\chi_y \in \mathcal{L}(\mathbf{Z}/N)$ by

$$\chi_y(x) = e^{2\pi i \frac{yx}{N}}, \quad x \in \mathbf{Z}/N.$$

The collection
$$\{\chi_y : y \in \mathbf{Z}/N\}$$
is a basis of $\mathcal{L}(\mathbf{Z}/N)$ called the *exponential basis*.

A mapping $\chi : \mathbf{Z}/N \longrightarrow \mathbf{C}^\times$ is called a *character* of $\mathbf{Z}/N$ if χ satisfies
$$\chi(x + y) = \chi(x)\chi(y), \quad x, y \in \mathbf{Z}/N.$$

$\mathbf{C}^\times$ denotes the set of nonzero complex numbers. Direct computation shows that every exponential function is a character of $\mathbf{Z}/N$. Eventually, we will show that the collection of characters and exponential functions of $\mathbf{Z}/N$ coincide.

We want to show that the characters of $\mathbf{Z}/N$ can be defined in terms of the translations of $\mathcal{L}(\mathbf{Z}/N)$.

Suppose $\mathcal{M}$ is a collection of linear operators of a vector space V. For $M \in \mathcal{M}$, a nonzero vector v in V is called an *M-eigen vector* if there exists $\alpha \in \mathbf{C}$ such that
$$Mv = \alpha v.$$

We call α the *eigen value* of M corresponding to the eigen vector v. A nonzero vector v in V is called an $\mathcal{M}$-eigen vector if v is an M-eigen vector for all $M \in \mathcal{M}$. In this case, there exists a mapping $\alpha : \mathcal{M} \longrightarrow \mathbf{C}$ such that
$$Mv = \alpha(M)v, \quad M \in \mathcal{M}.$$

A basis of V is called an *M-eigen vector basis* if every vector in the basis is an $\mathcal{M}$-eigen vector. Equivalently, a basis for V is an $\mathcal{M}$-eigen vector basis if and only if the matrix of every $M \in \mathcal{M}$ with respect to the basis is a diagonal matrix.

THEOREM 1 *If χ is a character of $\mathbf{Z}/N$, then χ is a $T(\mathbf{Z}/N)$-eigen vector.*

THEOREM 2 *If f is a $T(\mathbf{Z}/N)$-eigen vector and*
$$T(y)f = \alpha(y)f, \quad y \in \mathbf{Z}/N,$$
then α is a character of $\mathbf{Z}/N$ and
$$f(y) = \alpha(-y)f(0) = \alpha(y)^* f(0), \quad y \in \mathbf{Z}/N.$$

Proof
$$f(x - y) = \alpha(y)f(x), \quad x, y \in \mathbf{Z}/N.$$
Setting $y = 0$,
$$f(x) = \alpha(0)f(x), \quad x \in \mathbf{Z}/N.$$

Since $f \neq 0$, $\alpha(0) = 1$, α is a character of $\mathbf{Z}/N$. Setting $x = 0$ and replacing $-y$ by y completes the proof.

THEOREM 3 *If χ is a $T(\mathbf{Z}/N)$-eigen vector and $\chi(0) = 1$, then χ is a character of $\mathbf{Z}/N$.*

By Theorems 1 and 3 the characters of $\mathbf{Z}/N$ are uniquely characterized as the $T(\mathbf{Z}/N)$-eigen vectors taking the value 1 at the origin of $\mathbf{Z}/N$.

We will now complete the picture and show that the characters of $\mathbf{Z}/N$ form a $T(\mathbf{Z}/N)$-eigen vector basis.

> **Linear algebra theorem** *If $\mathcal{M}$ is a commuting family of linear operators of a vector space V and each $M \in \mathcal{M}$ has finite order, then there exists an $\mathcal{M}$-eigen vector basis.*

THEOREM 4 *The set of characters of $\mathbf{Z}/N$ is a $T(\mathbf{Z}/N)$-eigen vector basis.*

Proof Choose a $T(\mathbf{Z}/N)$-eigen vector basis

$$\{\chi_0, \cdots, \chi_{N-1}\}$$

consisting of characters of $\mathbf{Z}/N$. We will show that if χ is character of $\mathbf{Z}/N$, then $\chi = \chi_n$, for some n, $0 \leq n \leq N - 1$. Write

$$\chi = \sum_{n=0}^{N-1} \alpha(n)\chi_n, \quad \alpha(n) \in \mathbf{C}, 0 \leq n \leq N - 1.$$

Without loss of generality, assume $\alpha(0) \neq 0$. For $y \in \mathbf{Z}/N$, since

$$T(y)\chi = \chi(-y)\chi = \sum_{n=0}^{N-1} \alpha(n)\chi_n(-y)\chi_n,$$

we have

$$\chi(-y)\alpha(n) = \chi_n(-y)\alpha(n), \quad 0 \leq n \leq N - 1.$$

Setting $n = 0$, for $y \in \mathbf{Z}/N$,

$$\chi(-y) = \chi_0(-y)$$

implying $\chi = \chi_0$, completing the proof.

2. Abelian Group DSP

Suppose A is an abelian group and $\mathcal{L}(A)$ is the space of all complex valued functions on A.

2.1 Translations

For $y \in A$, the mapping $T(y)$ of $\mathcal{L}(A)$ defined by

$$(T(y)f)(x) = f(y^{-1}x), \quad x \in A, \ f \in \mathcal{L}(A),$$

is a linear operator of $\mathcal{L}(A)$ called *translation* by y.

For $g \in \mathcal{L}(A)$, the mapping $C(g)$ of $\mathcal{L}(A)$ defined by

$$C(g) = \sum_{y \in A} g(y)T(y)$$

is a linear operator of $\mathcal{L}(A)$, called *convolution* by g.

By definition

$$f * g(x) = \sum_{y \in A} f(y)g(y^{-1}x), \quad x \in A.$$

THEOREM 5 *Convolution is a commutative algebra product on $\mathcal{L}(A)$.*

The algebra formed by $\mathcal{L}(A)$ paired with convolution product is called the *convolution algebra* over A.

2.2 Fourier Analysis

Suppose $\mathbf{C}A$ is the group algebra of A.

A mapping $\chi : A \longrightarrow \mathbf{C}^{\times}$ is called a *character* of A if χ satisfies

$$\chi(xy) = \chi(x)\chi(y), \quad x, y \in A.$$

A character of A is simply a group homomorphism from A into $\mathbf{C}^{\times}$.

Denote the set of all characters of A by A^*. Under the identification between $\mathcal{L}(A)$ and $\mathbf{C}A$, every character τ of A can be viewed as a formal sum in $\mathbf{C}A$

$$\tau = \sum_{x \in A} \tau(x)x.$$

Multiplication of characters will always be taken in $\mathbf{C}A$.

EXAMPLE 6 As elements in $\mathbf{C}(C_2 \times C_2)$, the characters of $C_2 \times C_2$ are the formal sums

$$
\begin{aligned}
\phi_{0,0} &= 1 + x_1 + x_2 + x_1 x_2, \\
\phi_{1,0} &= 1 - x_1 + x_2 - x_1 x_2, \\
\phi_{0,1} &= 1 + x_1 - x_2 - x_1 x_2, \\
\phi_{1,1} &= 1 - x_1 - x_2 + x_1 x_2.
\end{aligned}
$$

The mapping $\tau_0 : A \longrightarrow \mathbf{C}^\times$ defined by $\tau_0(x) = 1$, $x \in A$, is a character called the *trivial character* of A. As an element in $\mathbf{C}A$,

$$\tau_0 = \sum_{x \in A} x.$$

The importance of characters in DSP is a result of their invariance under left multiplication as stated in the following theorem.

THEOREM 7 *If $y \in A$ and τ is a character of A, then*

$$y\tau = \tau(y^{-1})\tau.$$

EXAMPLE 8 In $\mathbf{C}C_N$, if

$$\phi_n = \sum_{m=0}^{N-1} v^{nm} x^m, \quad v = e^{2\pi i \frac{1}{N}},$$

then

$$x^r \phi_n = \sum_{m=0}^{N-1} v^{nm} x^{m+r} = \sum_{m=0}^{N-1} v^{n(m-r)} x^m = v^{-nr} \phi_n.$$

COROLLARY 10.1 *If $f \in \mathbf{C}A$ and τ is a character of A, then*

$$f\tau = \hat{f}(\tau)\tau,$$

where

$$\hat{f}(\tau) = \sum_{y \in A} f(y)\tau(y^{-1}).$$

Theorem 7 implies every character τ of A is an $L(A)$-eigen vector,

$$L(y)\tau = y\tau = \tau(y^{-1})\tau, \quad y \in A.$$

The eigen value of the eigen vector τ of $L(y)$ is $\tau(y^{-1})$.

THEOREM 9 *A^* is the unique $L(A)$-eigen vector basis of $\mathbf{C}A$ satisfying the condition that the value of each basis element is 1 at the identity element 1 of A.*

The Fourier analysis over A will be developed in terms of the properties of the characters of A in $\mathbf{C}A$.

THEOREM 10 *If τ and λ are characters of A, then*

$$\tau\lambda = \left\{ \begin{array}{ll} N\tau, & \tau = \lambda, \\ 0, & \tau \neq \lambda. \end{array} \right.$$

A nonzero element e in $\mathbf{C}A$ is called an *idempotent* if $e^2 = e$. Two idempotents e_1 and e_2 in $\mathbf{C}A$ are called *orthogonal* if $e_1 e_2 = e_2 e_1 = 0$. In the language of idempotent theory Theorem 10 says that the set

$$\left\{ \frac{1}{N}\tau : \tau \in A^* \right\}$$

is a set of pairwise orthogonal idempotents.

Since A^* is a basis of the space $\mathbf{C}A$, we can write

$$1 = \sum_{\tau \in A^*} \alpha(\tau)\tau, \quad \alpha(\tau) \in \mathbf{C}.$$

For any $\lambda \in A^*$,

$$\lambda = \lambda \cdot 1 = \sum_{\tau \in A^*} \alpha(\tau)\lambda\tau = N\alpha(\lambda)\lambda$$

and $\alpha(\lambda) = \frac{1}{N}$, proving the following.

THEOREM 11

$$1 = \frac{1}{N} \sum_{\tau \in A^*} \tau.$$

A set of pairwise orthogonal idempotents $\mathcal{I}$ in $\mathbf{C}A$ is called *complete* if

$$1 = \sum_{e \in \mathcal{I}} e.$$

In the language of idempotent theory Theorems 10 and 11 say that the set

$$\left\{ \frac{1}{N}\tau : \tau \in A^* \right\}$$

is a complete set of pairwise orthogonal idempotents.

COROLLARY 10.2

$$\frac{1}{N} \sum_{\tau \in A^*} \tau(x) = \left\{ \begin{array}{ll} 1, & x = 1, \\ 0, & x \neq 1, \end{array} \right. \quad x \in A.$$

By Corollary 10.1 and completeness every $f \in \mathbf{C}A$ can be written as

$$f = \frac{1}{N} \sum_{\tau \in A^*} f\tau = \frac{1}{N} \sum_{\tau \in A^*} \widehat{f}(\tau)\tau.$$

We call this expansion of f the *Fourier expansion* of f.

$$\widehat{f}(\tau) = \sum_{y \in A} f(y)\tau(y^{-1}), \qquad \tau \in A^*,$$

is called the *Fourier coefficient* of f at τ.

2.3   Character Extensions I

In this chapter we will show that every character of a direct product $A = B \times C$ of abelian groups B and C can be represented as a product in $\mathbf{C}A$ of a character of B with a character of C.

More generally, we will relate the characters of a subgroup of an abelian group to the characters of the group through a construction called *character extensions*. The generalization of this construction to nonabelian groups will occupy a great deal of the later chapters. A is an abelian group throughout this chapter.

2.4   Direct Products

Suppose B is a subgroup of A and ρ is a character of A. The mapping $\rho_B : B \longrightarrow \mathbf{C}^{\times}$ defined by

$$\rho_B(x) = \rho(x), \qquad x \in B,$$

is a character of B called the *restriction* of ρ to B. As an element in $\mathbf{C}A$,

$$\rho_B = \sum_{x \in B} \rho(x)x.$$

THEOREM 12 *Suppose $A = B \times C$. If ρ is a character of A, then $\rho = \rho_B \rho_C$, the product in $\mathbf{C}A$ of the characters ρ_B and ρ_C of B and C defined by the restrictions of ρ.*

By Theorem 12 the character ρ of $A = B \times C$ is represented as a product in $\mathbf{C}A$ of a character of B with a character of C. The representation is unique.

Theorem 12 easily extends to direct product factorizations of A having any number of factors.

EXAMPLE 13 The character

$$1 + x_1 - x_2 - x_1 x_2 - x_3 - x_1 x_3 + x_2 x_3 + x_1 x_2 x_3$$

of $C_2(x_1) \times C_2(x_2) \times C_2(x_3)$ is the product

$$(1 + x_1)(1 - x_2)(1 - x_3)$$

of the characters $1 + x_1$, $1 - x_2$, $1 - x_3$ of $C_2(x_1)$, $C_2(x_2)$, $C_2(x_3)$.

THEOREM 14 *Suppose* $A = B \times C$. *If* τ *and* λ *are characters of* B *and* C, *then the product* $\rho = \tau\lambda$ *in* $\mathbf{C}A$ *is a character of* A.

Proof By Theorem 7 if $x = uv$, $u \in B$, $v \in C$, then

$$x\rho = (u\tau)(v\lambda) = \tau(u^{-1})\lambda(v^{-1})\tau\lambda = \rho(x^{-1})\rho,$$

implying ρ is a character of A, completing the proof.

2.5   Character Extensions

Suppose B is a subgroup of A and τ is a character of B. A character ρ of A is called an *extension* of τ if τ is the restriction of ρ to B.

Consider a complete system of B-coset representatives in A,

$$\{y_s : 1 \le s \le S\}.$$

Every $x \in A$ can be written uniquely as $x = y_s z$, $1 \le s \le S$, $z \in B$. Suppose ρ is an extension of τ. Then

$$\rho = \sum_{x \in A} \rho(x)x = \sum_{s=1}^{S} \sum_{z \in B} \rho(y_s z)y_s z.$$

Since $\rho(y_s z) = \rho(y_s)\rho(z) = \rho(y_s)\tau(z)$,

$$\rho = \left(\sum_{s=1}^{S} \rho(y_s)y_s \right) \left(\sum_{z \in B} \tau(z)z \right) = \left(\sum_{s=1}^{S} \rho(y_s)y_s \right)\tau,$$

proving the following result.

THEOREM 15 *Suppose* B *is a subgroup of* A *and* τ *is a character of* B. *If* ρ *is a character of* A *extending* τ, *then*

$$\rho = \left(\sum_{s=1}^{S} \rho(y_s)y_s \right)\tau.$$

We can use Theorem 15 to characterize the characters extending τ.

THEOREM 16 *Suppose B is a subgroup of A and τ is a character of B. A character ρ of A is an extension of τ if and only if*

$$\rho = \frac{1}{|B|}\tau\rho.$$

Assume for the rest of this section that B is a subgroup of A. For a character τ of B denote by

$$ext_\tau(A)$$

the collection of all characters of A extending the character τ of B. We will show that there always exists a character of A extending τ and describe $ext_\tau(A)$. If $A = B \times C$, then by Theorems 12 and 14 $ext_\tau(A)$ is the set of all products in $\mathbf{C}A$ of the form $\tau\lambda$, where λ is a character of C.

EXAMPLE 17 The characters of $C_{12}(x)$ extending the character

$$\tau = \sum_{n=0}^{3} v^n x^{3n}, \quad v = e^{2\pi i \frac{1}{4}},$$

of $C_{12}(x^3)$ have the form

$$\rho_w = \sum_{n=0}^{11} w^n x^n, \quad w^3 = v.$$

This will be the case if and only if

$$w = e^{2\pi i \frac{1}{12}}, \; e^{2\pi i \frac{1}{12}} e^{2\pi i \frac{1}{3}}, \; e^{2\pi i \frac{1}{12}} e^{2\pi i \frac{2}{3}}.$$

For each such w we can write the corresponding extension ρ_w as

$$\rho_w = \tau(1 + wx + w^2 x^2),$$

which is the factorization of ρ_w given in Theorem 15. In this case we also have the direct product factorization,

$$C_{12}(x) = C_{12}(x^3) \times C_{12}(x^4)$$

and can write

$$\rho_w = \tau(1 + ux^4 + u^2 x^8),$$

where

$$1 + ux^4 + u^2 x^8$$

is the character of $C_{12}(x^4)$ defined by $u = w^4$.

3. Nonabelian Groups

EXAMPLE 18 The 3×3 cyclic shift matrix

$$S_3 = \begin{bmatrix} 0 & 1 & 0 \\ 0 & 0 & 1 \\ 1 & 0 & 0 \end{bmatrix},$$

and the 3×3 time-reversal matrix

$$R_3 = \begin{bmatrix} 1 & 0 & 0 \\ 0 & 0 & 1 \\ 0 & 1 & 0 \end{bmatrix}$$

satisfy the relations

$$S_3^2 = S_3^{-1} \neq I_3, \quad , S_3^3 = I_3 = R_3^2, \quad R_3 S_3 R_3^{-1} = S_3^{-1}.$$

Denote by $\mathcal{D}_6(S_3, R_3)$ the collection of all matrices

$$\{I_3,\ S_3,\ S_3^2;\ R_3,\ S_3 R_3,\ S_3^2 R_3\}.$$

By the above relations $\mathcal{D}_6(S_3, R_3)$ is closed under matrix multiplication. For example

$$R_3 S_3 = S_3^{-1} R_3 = S_3^2 R_3.$$

Since $R_3 S_3 \neq S_3 R_3$, $\mathcal{D}_6(S_3, R_3)$ is a nonabelian group of order 6.

In general we have the following result. Assume $N \geq 3$.

THEOREM 19 *The $N \times N$ cyclic shift matrix S_N and the $N \times N$ time-reversal matrix R_N satisfy the relations $S_N^n \neq I_N$, $0 < n < N$, and*

$$S_N^N = I_N = R_N^2, \quad R_N S_N R_N = S_N^{-1}.$$

The set of all products $\mathcal{D}_{2N}(S_N, R_N)$ of the form

$$\{S_N^n R_N^j : 0 \leq n < N,\ 0 \leq j < 2\}$$

is a nonabelian group of order $2N$.

The group $\mathcal{D}_{2N}(S_N, R_N)$ is a matrix representation of the *dihedral group* of order $2N$.

3.1   Normal Subgroups

Abelian group constructions do not automatically extend to non-abelian groups. The identification of a distinguished class of subgroups, the normal subgroups, is one of the most significant insights in the development of nonabelian group theory.

The concept of a normal subgroup of a nonabelian group G is closely related to a certain class of automorphisms, the inner automorphisms of G. We will discuss these concepts in this section.

Suppose H and K are subgroups of G. We say that K *normalizes H* if

$$yHy^{-1} = H, \quad y \in K.$$

H is a *normal* subgroup of G if G normalizes H.

3.2   Quotient Groups

Suppose, to begin with, that H is any subgroup of G.

For $y \in G$, the set

$$yH = \{yx : x \in H\}$$

is called the *left H-coset in G* at y. Right H-cosets can be defined in an analogous manner.

THEOREM 20 *Two left H-cosets are either equal or disjoint.*

The collection of left H-cosets in G forms a partition of G. Denote this collection by G/H and call G/H the *quotient space* of left H-cosets in G.

THEOREM 21 *If y, $z \in G$, then $yH = zH$ if and only if $z^{-1}y \in H$.*

A set

$$\{y_1, \ldots, y_R\},$$

formed by choosing exactly one element in each left H-coset is called a *complete system of left H-coset representatives* in G.

THEOREM 22 *If H is a normal subgroup of G, then the product*

$$(yH)(zH) = yzH, \quad y, z \in G,$$

is a group product on G/H with H as identity element and

$$(yH)^{-1} = y^{-1}H, \quad y \in G.$$

The group G/H is called the *quotient group* of left H-cosets in G.

3.3 Semidirect Product

Suppose H is a normal subgroup of G. We say that G *splits* over H with *complement* K if $G = HK$ and $H \cap K = \{1\}$. If G splits over H with complement K, then we say that G is the *internal semidirect product* $H \rtimes K$. The usual argument shows that $G = H \rtimes K$ if and only if every $x \in G$ has a *unique representation* of the form $x = yz$, $y \in H$, $z \in K$. In general, the complement K is not uniquely determined.

If $G = H \rtimes K$ and K is a normal subgroup of G then $[H, K] = \{1\}$ and G is the direct product $H \times K$. What is new in the internal semidirect product is the possibility that K acts nontrivially on H.

Suppose H and K are groups and

$$\Psi : K \longrightarrow Aut(H)$$

is a group homomorphism. The *semidirect product* $H \rtimes_\Psi K$ is the pair consisting of the cartesian product $H \times K$ and the binary operation

$$(y_1, z_1)(y_2, z_2) = (y_1 \Psi_{z_1}(y_2), z_1 z_2), \quad y_1, y_2 \in H, \ z_1, z_2 \in K.$$

The semidirect product $H \rtimes_\Psi K$ is a group having identity $(1, 1)$ and inverse

$$x^{-1} = \left(\Psi_{z^{-1}}(y^{-1}), z^{-1} \right), \quad x = (y, z) \in H \rtimes_\Psi K.$$

3.4 Examples: Semidirect product constructions

DSP algorithms will be designed for nonabelian group indexing sets formed mainly by semidirect product constructions. The groups will have the form $A \rtimes H$, where A is an abelian group.

Consider the ring $\mathbf{Z}/N$ of integers modulo N. An $m \in \mathbf{Z}/N$ is called a *unit* if there exists an $n \in \mathbf{Z}/N$ such that $mn = 1$ in $\mathbf{Z}/N$. A unit in $\mathbf{Z}/N$ is an invertible element relative to multiplication in $\mathbf{Z}/N$. The unit group $U(N)$ can be characterized as the set of all integers $0 < m < N$ such that m and N are relatively prime.

THEOREM 23 *The mapping*

$$u \longrightarrow \Psi_u, \quad u \in U(N),$$

is a group isomorphism from $U(N)$ onto $Aut(C_N(x))$.

The group isomorphism in Theorem 23 establishes a one-to-one correspondence between the subgroups of $U(N)$ and those of $Aut(C_N(x))$. Under this identification, we can form $C_N(x) \rtimes K$ for any subgroup K of $U(N)$ by the group isomorphism throughout this work. A typical point in $C_N(x) \triangleleft K$ is denoted by (x^m, u), $0 \leq m < N$, $u \in K$ with multiplication given by

$$(x^m, u)(x^n, v) = (x^{m+un}, uv), \quad 0 \leq m, n < N, u, v \in K,$$

where $m + un$ is taken modulo N. Identifying $C_N(x)$ with the normal subgroup $C_N(x) \times \{1\}$ of $C_N(x) \triangleleft K$ and K with the subgroup $\{1\} \times K$ of $C_N(x) \triangleleft K$, we can write

$$x^m u = (x^m, u), \quad 0 \leq m < N, u \in K. \tag{1}$$

Groups of the form $C_N(x) \triangleleft K$ will be used in many examples as several DSP concepts are introduced in the following chapters. Although the identification in (1) is unambiguous as a description of an abstract group, we will use the modification,

$$x^m k_u = (x^m, u), \quad 0 \leq m < N, u \in U(N),$$

with $x^m k_u^j = (x^m, u^j)$.

The reason for this modification is to avoid any confusion or awkward notation in viewing $C_N(x) \triangleleft K$ as a subset of the group algebra

$$\mathbf{C}(C_N(x) \triangleleft K).$$

For a prime p the unit group $U(p)$ of $\mathbf{Z}/p$ is a cyclic group of order $p - 1$ under multiplication modulo p. Choosing a generator y for $U(p)$, we can write the elements of $U(p)$ as successive powers of y,

$$1, y, \ldots, y^{p-2},$$

with $y^{p-1} = 1$.

A typical point in $G = C_p(x) \triangleleft U(p)$ can be written as

$$x^m k_y^j, \quad 0 \leq m < p, 0 \leq j < p - 1,$$

with y a generator of $U(p)$. Multiplication in G is subject to the relations

$$x^p = 1 = k_y^{p-1}, \quad k_y x = x^y k_y.$$

Other semidirect products can be constructed replacing $U(p)$ by any subgroup of $U(p)$.

Suppose $N = p_1 p_2$, with p_1 and p_2 distinct primes. A number theoretic result asserts that $U(N) = U(p_1) \times U(p_2)$. There exist $y_1, y_2 \in U(N)$ such that every $y \in U(N)$ can be written uniquely as

$$y = y_1^{j_1} y_2^{j_2}, \quad 0 \leq j_1 < p_1 - 1, 0 \leq j_2 < p_2 - 1,$$

with $y_1^{p_1-1} = 1 = y_2^{p_2-1}$ and $y_1 y_2 = y_2 y_1$.

EXAMPLE 24 An arbitrary element in the group $C_N(x) \triangleleft U(N)$, $N = p_1 p_2$, can be written uniquely as

$$x^m k_{y_1}^{j_1} k_{y_2}^{j_2}, \quad 0 \le m < N, 0 \le j_1 < p_1 - 1, 0 \le j_2 < p_2 - 1,$$

with group multiplication subject to the relations

$$x^N = 1 = k_{y_1}^{p_1-1} = k_{y_2}^{p_2-1},$$

$$k_{y_1} x = x^{y_1} k_{y_1}, \quad k_{y_2} x = x^{y_2} k_{y_2}, \quad k_{y_1} k_{y_2} = k_{y_2} k_{y_1}.$$

Consider the abelian group $C_N(x) \times C_N(y)$. Under the usual identifications, we can write a typical point as $x^m y^n$, $0 \le m, n < N$, subject to the relations $x^N = 1 = y^N$ and $xy = yx$. We begin our discussion of $Aut(C_N(x) \times C_N(y))$ by extending the concept of a unit group to matrices.

The set $M(2, \mathbf{Z}/N)$ of all 2×2 matrices over $\mathbf{Z}/N$ is a ring with respect to matrix addition and matrix multiplication. The arithmetic required for the matrix addition and multiplication is taken in $\mathbf{Z}/N$. I_2 denotes the identity matrix in $M(2, \mathbf{Z}/N)$. A matrix $L \in M(2, \mathbf{Z}/N)$ is called a *unit* if there exists $L' \in M(2, \mathbf{Z}/N)$ such that $LL' = I_2 = L'L$. L' is called the inverse of L in $M(2, \mathbf{Z}/N)$ and is uniquely determined when it exists. If L is a unit in $M(2, \mathbf{Z}/N)$, we denote by L^{-1} the inverse of L. If L_1 and L_2 are units, then

$$(L_1 L_2)(L_2^{-1} L_1^{-1}) = I_2 = (L_2^{-1} L_1^{-1}) L_1 L_2$$

and $L_1 L_2$ is a unit with inverse $L_2^{-1} L_1^{-1}$. The set $GL(2, \mathbf{Z}/N)$ of all units in $M(2, \mathbf{Z}/N)$ is a group under matrix multiplication in $M(2, \mathbf{Z}/N)$. We call $GL(2, \mathbf{Z}/N)$ the *unit group* in $M(2, \mathbf{Z}/N)$. $GL(2, \mathbf{Z}/N)$ can be characterized as the set of all $L \in M(2, \mathbf{Z}/N)$ such that $det(L) \in U(N)$. $det(L)$ is defined by the same formula for complex matrices but with arithmetic operations taken in $\mathbf{Z}/N$.

For

$$L = \begin{bmatrix} L_1 & L_2 \\ L_3 & L_4 \end{bmatrix} \in M(2, \mathbf{Z}/N),$$

define $\Psi_L : C_N(x) \times C_N(y) \longrightarrow C_N(x) \times C_N(y)$ by

$$\Psi_L(x^m y^n) = x^{L_1 m + L_2 n} y^{L_3 m + L_4 n}, \quad 0 \le m, n < N.$$

Ψ_L is uniquely determined by its values on the generators x and y,

$$\Psi_L(x) = x^{L_1} y^{L_3}, \quad \Psi_L(y) = x^{L_2} y^{L_4}.$$

THEOREM 25 *The mapping*

$$L \longrightarrow \Psi_L, \quad L \in GL(2, \mathbf{Z}/N),$$

is a group isomorphism from $GL(2, \mathbf{Z}/N)$ *onto* $Aut(C_N(x) \times C_N(y))$.

Under the group isomorphism in Theorem 25, there is a one-to-one correspondence between the subgroups of $GL(2, \mathbf{Z}/N)$ and the subgroups of $Aut(C_N(x) \times C_N(y))$ and we can form

$$(C_N(x) \times C_N(y)) \rtimes K,$$

for any subgroup K of $GL(2, \mathbf{Z}/N)$. A typical point is denoted by $(x^m y^n, L)$, $0 \le m, n < N$, $L \in K$ with multiplication given by

$$(x^m y^n, L)(x^{m'} y^{n'}, L') = (x^{m''} y^{n''}, L''),$$

where

$$m'' = m + L_1 m' + L_2 n', \quad n'' = n + L_3 m' + L_4 n', \quad L = \begin{bmatrix} L_1 & L_2 \\ L_3 & L_4 \end{bmatrix},$$

and $L'' = LL'$. Under the usual identifications, we can write

$$x^m y^n L = (x^m y^n, L), \quad 0 \le m, n < N, L \in K.$$

As before, to avoid awkward notation in the group algebra setting, we will use the modified identification,

$$x^m y^n k_L = (x^m, y^n, L), \quad 0 \le m, n < N, L \in GL(2, \mathbf{Z}/N).$$

The discussion extends to any number of cyclic group factors as long as each factor has the same order.

3.5   Nonabelian Group DSP

The new problem raised by noncommutativity is that the characters no longer form a basis of nonabelian group algebras. As a result, there exist multidimensional irreducible ideals. Throughout G is an arbitrary nonabelian group of order N and $\mathbf{C}G$ is its group algebra.

EXAMPLE 26 The dihedral group $\mathcal{D}_{2N} = \mathcal{D}_{2N}(x, k_{N-1})$ is the set of products $x^m k_{N-1}^j$, $0 \le m < N$, $0 \le j < 2$, with multiplication subject to the relations

$$x^N = 1 = k_{N-1}^2, \quad k_{N-1}x = x^{-1}k_{N-1}.$$

The group algebra $\mathbf{C}\mathcal{D}_{2N}$ is the set of polynomials of the form

$$\sum_{m=0}^{N-1}\sum_{j=0}^{1} a_{m,j}x^m k_{N-1}^j, \qquad a_{m,j} \in \mathbf{C},$$

with polynomial multiplication subject to the relations given for the group $\mathcal{D}_{2N}$.

EXAMPLE 27 Set $A = C_N(x) \times C_N(y)$, $G = A \rtimes C_2(k_c)$, where $c = \begin{bmatrix} 1 & 0 \\ 1 & -1 \end{bmatrix}$.

In Figure 1 we display lines indexed by G and their translations under multiplications by elements from G.

Figure 1.   An image and its translates in $\mathbf{C}G_4$, $N = 64$

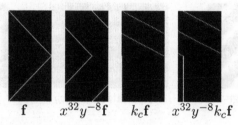

$$\mathbf{f} \qquad x^{32}y^{-8}\mathbf{f} \qquad k_c\mathbf{f} \qquad x^{32}y^{-8}k_c\mathbf{f}$$

In the following figures, we show the effects of replacing k_c by other elements of $GL(2, \mathbf{Z}/N)$ of order 2.

Set

$$b = \begin{bmatrix} 0 & 1 \\ 1 & 0 \end{bmatrix}.$$

The first image in Figure 2 is the same as in Figure 1. The remaining images are its translates by left multiplications of elements of $A \rtimes C_2(k_b)$.

Figure 2.   An image and its translates in $\mathbf{C}(A \rtimes C_2(k_b))$, $N = 64$

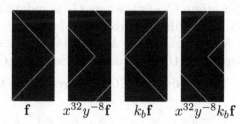

$$\mathbf{f} \qquad x^{32}y^{-8}\mathbf{f} \qquad k_b\mathbf{f} \qquad x^{32}y^{-8}k_b\mathbf{f}$$

$G_5 = A \lhd C_6(k_d)$, where $A = C_N(x) \times C_N(y)$, $N = 3 \cdot 2^K$ for an integer $K \geq 2$. Set $M = 2^K$.

$$d = \begin{bmatrix} -1 & M+1 \\ M-1 & M \end{bmatrix}.$$

Order G_5 by

$$\begin{bmatrix} A & Ak_d^2 & Ak_d^4 \\ Ak_d^3 & Ak_d^5 & Ak_d \end{bmatrix},$$

where A is ordered as before. Figure 3 is an example of translations of data indexed by G_5, $M = 16$. Starting with the top left image as an element in $\mathbf{C}G_5$, the remaining 5 images are obtained by left multiplication by successive powers of k_d.

Figure 3.   Translations by $C_6(k_d)$

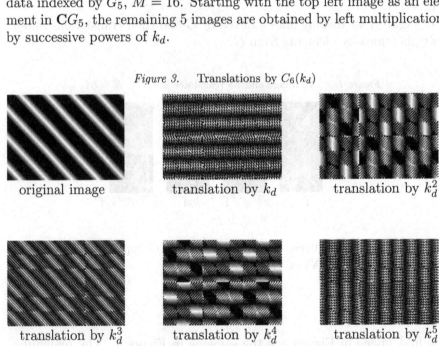

original image   translation by k_d translation by k_d^2

translation by k_d^3 translation by k_d^4 translation by k_d^5

EXAMPLE 28 Varying group structures are placed on an indexing set of data, and products are computed corresponding to the group algebra. G_1 is the abelian group $C_{2N}(x)$, where G_1 is indexed by the successive powers of x. G_2 is the dihedral group $\mathcal{D}_{2N}(x, k_{N-1})$. G_3 is constructed as follows: For $N = 2^M$, $M \in \mathbf{Z}$, $M \geq 2$, $(\frac{N}{2} + 1)^2 \equiv 1 \bmod N$; $\frac{N}{2} + 1$ generates a subgroup of $U(N)$ of order 2. Set $G_3 = C_N(x) \lhd \{1, k_{\frac{N}{2}+1}\}$, $\frac{N}{2} + 1 \in U(N)$. The product in $\mathbf{C}G_3$ is governed by the relations

$$x^N = k_{\frac{N}{2}+1} = 1, \qquad k_{\frac{N}{2}+1}x = x^{\frac{N}{2}+1}k_{\frac{N}{2}+1}.$$

Figure 4. Group algebra products

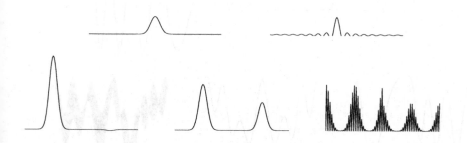

G_3 is indexed by

$$1, x, \ldots, x^{N-1}; \ k_{\frac{N}{2}+1}, xk_{\frac{N}{2}+1}, \ldots, x^{N-1}k_{\frac{N}{2}+1}.$$

Note that G_2 and G_3 are isomorphic groups.

For $N = 64$, the first two plots in Figure 4 are those of 128 points of data. The product in G_1 is the usual cyclic convolution of two sets of data and displayed in the third plot. The 4th and 5th plots are those of group algebra products in $\mathbf{C}G_2$ and $\mathbf{C}G_3$.

For $N = 128$, each of the groups determines an indexing of data of size 256. The first plots in 5 are those of 256 points of data. The product in G_1 is the usual cyclic convolution of two sets of data and displayed in the third plot. The remaining plots are those of group algebra products in $\mathbf{C}G_2$ and $\mathbf{C}G_3$.

EXAMPLE 29 Varying group structures are placed on the indexing set of two-dimensional data, and products are computed corresponding to the group algebra. For $0 \le m, \ n < N$, place the lexicographic ordering on the pair (m, n). G_1 is the abelian group $C_N(x) \times C_{2N}(y)$ with two-dimensional ordering given by

$$(x^m, y^n); \ (x^m, y^{n+N}).$$

G_2 is the abelian group $(C_N(x) \times C_N(y)) \times C_2(z)$ with two-dimensional ordering given by

$$(x^m, y^n); \ (x^m, y^n z).$$

$G_3 = (C_N(x) \times C_N(y)) \rtimes \{I_2, k_z\}, \ z = \begin{bmatrix} -1 & 0 \\ 0 & -1 \end{bmatrix}.$ The ordering of G_3 is the two-dimensional ordering given by

$$(x^m, y^n); \ (x^m, y^n k_z).$$

Figure 5.   Group algebra products

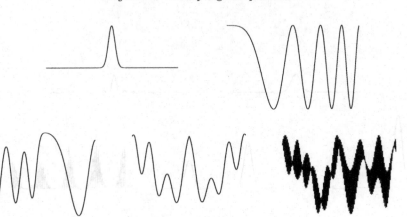

G_4 is the nonabelian group $(C_N(x) \times C_N(y)) \lhd \{I_2, k_u\}$, $u = \begin{bmatrix} 0 & 1 \\ 1 & 0 \end{bmatrix} \in$ $GL(2, \mathbf{Z}/N)$. The ordering of G_4 is the two-dimensional ordering given by

$$(x^m, y^n); (x^m, y^n k_u).$$

For $N = 32$, each of the groups determines indexing of data of size 32×64. Images in Figure 6 are log scaled intensity plots of data of size 32×64. Images in Figure 7 are convolutions of images in Figure 6 in CG_1, CG_2, CG_3 and CG_4.

3.6   Fourier Analysis

A mapping $\rho : G \longrightarrow \mathbf{C}^\times$ is called a *character* of G if it satisfies

$$\rho(xy) = \rho(x)\rho(y), \quad x, y \in G. \tag{2}$$

A character of G is simply a group homomorphism of G into $\mathbf{C}^\times$. In group representation theory the more general concept of a character of a group representation of G is defined and characters as defined in (2) are called *one-dimensional characters*. We will not use this general concept.

Figure 6. Images of size 32×64

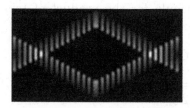 

Figure 7. Group algebra products

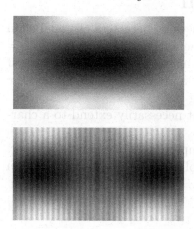 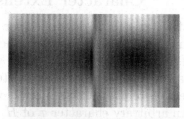

Characters of nonabelian groups are more difficult to describe than in the abelian group case. There always exists a character, the *trivial character* of G, which takes on the value 1 at all points of G.

Denote the collection of all characters of G by G^*. We can view a character ρ of G as a formal sum in $\mathbf{C}G$,

$$\rho = \sum_{x \in G} \rho(x)x,$$

and G^* as a subset of $\mathbf{C}G$.

THEOREM 30 *Suppose ρ is a character of G. If $t \in G$,*

$$t\rho = \rho t = \rho(t^{-1})\rho.$$

COROLLARY 10.3 *Suppose ρ is a character of G. If $f \in \mathbf{C}G$, then*

$$f\rho = \rho f = \widehat{f}(\rho)\rho,$$

where $\widehat{f}(\rho) = \sum_{t \in G} f(t)\rho(t^{-1})$.

Theorem 30 leads to an important formula for the product of two characters in CG.

THEOREM 31 *For characters ρ and λ of G,*

$$\rho\lambda = \begin{cases} N\rho, & \rho = \lambda, \\ 0 & \rho \neq \lambda. \end{cases}$$

3.7   Character Extensions II

Methods will be developed for constructing characters of nonabelian groups having semidirect product representations. For a group $G = H \triangleleft K$, the strategy is to begin with a character τ of H, determine if τ has an extension to a character of G and if it does, construct the collection of all characters of G extending τ.

An arbitrary character τ of H does not necessarily extend to a character of $G = H \triangleleft K$.

We assume that H is a normal subgroup of G throughout this paper. In Section 3.9, when required, we will explicitly assume that H splits in G and $G = H \triangleleft K$.

3.8   Group Action

For $y \in G$ and $f \in CG$, define $f^y \in CG$ by

$$f^y = yfy^{-1} = \sum_{x \in G} f(x)yxy^{-1}.$$

Two characters τ_1 and τ_2 of H are said to be *conjugate* in G if there exists $y \in G$ such that $\tau_2 = \tau_1^y$. Conjugacy is an equivalence relation on the set H^* of all characters of H.

Suppose τ is a character of H and $G(\tau)$ its centralizer in G. Choose a complete system of left $G(\tau)$-coset representatives in G,

$$\{y_s : 1 \leq s \leq S\}.$$

Every $y \in G$ can be uniquely written as

$$y = y_s z, \quad 1 \leq s \leq S, z \in G(\tau).$$

Define the conjugates τ_s, $1 \leq s \leq S$, of τ by

$$\tau_s = y_s \tau y_s^{-1}, \quad 1 \leq s \leq S. \tag{3}$$

THEOREM 32 *If τ is a character of H, the conjugates τ_s, $1 \leq s \leq S$, of τ defined by (3) are distinct characters of H and the set*

$$\{\tau_s : 1 \leq s \leq S\}$$

is the collection of all conjugates of τ in G.

Proof Suppose $y \in G$ and write $y = y_s z$, $1 \leq s \leq S$, $z \in G(\tau)$. Since $G(\tau)$ centralizes τ,

$$y\tau y^{-1} = y_s z \tau z^{-1} y_s^{-1} = y_s \tau y_s^{-1} = \tau_s$$

and every conjugate of τ in G is of the form τ_s, $1 \leq s \leq S$. If $\tau_s = \tau_t$, $1 \leq s, t \leq S$, then

$$\tau = y_s^{-1} y_t \tau y_t^{-1} y_s.$$

This implies $y_s^{-1} y_t \in G(\tau)$ and $y_s = y_t$, completing the proof.

By Theorem 31 we have the following corollary.

COROLLARY 10.4 *If $1 \leq s, t \leq S$ and $s \neq t$, then $\tau_s \tau_t = \tau_t \tau_s = 0$.*

The characters ϕ_n, $0 \leq n < M$, of $C_M(x)$ are defined by

$$\phi_n(x^m) = v^{nm}, \qquad 0 \leq m < M, v = e^{2\pi i \frac{1}{M}}.$$

If $0 \leq n < N$ and $m \in U(M)$, then since

$$\phi_n^{km}(x) = \phi_n(x^m) = \phi_{nm}(x),$$

the set

$$\{\phi_{nm} : m \in U(M)\} \tag{4}$$

is the set of conjugates of ϕ_n in $C_M(x) \triangleleft U(M)$. The product nm is taken modulo M. (4) is called the *conjugacy class* of ϕ_n in $C_M \triangleleft U(M)$.

3.9 Character Extension

Throughout this section H is a normal subgroup of G and K is an arbitrary subgroup of G. Following Corollary 10.5, we assume $G = H \triangleleft K$. The main results on character extensions will be proved under this assumption.

Suppose ρ is a character of K. We say that ρ *extends to a character of G* if there exists a character γ of G such that ρ is the restriction of γ to K. Generally, even if K is a normal subgroup of G, a character ρ of K will not necessarily extend to a character of G. This obstruction is a major difference between the abelian and the nonabelian group cases.

The following result establishes a necessary condition for a character τ of the normal subgroup H of G to have an extension to a character of G. For groups of the form $G = H \lhd K$, it is also a sufficient condition.

THEOREM 33 *If a character τ of H extends to a character of G, then G centralizes τ.*

THEOREM 34 *If τ is a character of H and G_1 is a subgroup of G containing H such that τ extends to a character of G_1, then $G_1 \subset G(\tau)$.*

As a special case, we have the following corollary.

COROLLARY 10.5 *If τ is a character of H and $H = G(\tau)$, then τ cannot be extended to a character of any subgroup of G strictly containing H.*

In the language introduced below, if τ is a character of H such that $H = G(\tau)$, then τ is a maximal character in G.

Assume $G = H \lhd K$ for the rest of this chapter. A typical point $x \in G$ will be written uniquely as

$$x = uv, \quad u \in H, v \in K,$$

with the understanding that $u = u \cdot 1$ and $v = 1 \cdot v$.

THEOREM 35 *If τ is a character of H centralized by K, then every product $\tau\lambda$, where λ is a character of K, is a character of G extending τ.*

Proof Choose any character λ of K and form the product $\rho = \tau\lambda$. We must show ρ is a character of G. Since $v\tau = \tau v$, $v \in K$, we have

$$x\rho = uv\tau\lambda = (u\tau)(v\lambda), \quad x = uv, u \in H, v \in K.$$

By Theorem 30

$$x\rho = \tau(u^{-1})\lambda(v^{-1})\tau\lambda = \rho(x^{-1})\rho,$$

and ρ is an $L(G)$-eigen vector. Since $\rho(1) = 1$, ρ is a character of G, completing the proof.

COROLLARY 10.6 *If τ is a character of H, then τ extends to a character of G if and only if K centralizes τ.*

Theorem 35 describes a procedure for constructing characters of G extending a character τ of H centralized by K. In fact all the characters of G extending τ are constructed in this way. For each character τ of H centralized by K denote by

$$ext_\tau(G)$$

the collection of all characters of G extending τ.

THEOREM 36 *If τ is a character of H centralized by K, then $ext_\tau(G)$ is the collection of all products $\tau\lambda$, where λ is a character of K.*

The collection G^* of all characters of G is the disjoint union of the sets

$$ext_\tau(G), \qquad \tau \in H^* \text{ and } \tau \text{ centralized by } K.$$

EXAMPLE 37 If p is prime, there are $p - 1$ characters of the group $C_p(x) \lhd U(p)$. The only character of $C_p(x)$ which extends to $C_p(x) \lhd U(p)$ is the trivial character ϕ_0. Since $U(p)$ has order $p - 1$, there are $p - 1$ characters λ_j, $0 \le j < p - 2$, of $U(p)$. The group algebra products

$$\phi_0\lambda_j, \quad 0 \le j < p - 2,$$

are the $p - 1$ characters of $C(p) \lhd U(p)$.

EXAMPLE 38 For $N \ge 2$, the characters of the normal subgroup

$$C_N(x) = \{x^m : 0 \le m < N\}$$

of the dihedral group $\mathcal{D}_{2N} = \mathcal{D}_{2N}(x, k_{N-1})$ are given by $\phi_n : C_N(x) \longrightarrow U_N$, $0 \le n < N$,

$$\phi_n(x^m) = v^{nm}, \quad 0 \le m < N, v = e^{2\pi i \frac{1}{N}}.$$

Since $\mathcal{D}_{2N} = C_N(x) \lhd K$, where $K = \{1, k_{N-1}\}$ and

$$k_{N-1}xk_{N-1}^{-1} = x^{-1},$$

ϕ_n is centralized by K if and only if $v^n = e^{2\pi i \frac{n}{N}}$ is real. For N odd, we must have $n = 0$ and the trivial character ϕ_0 is the only character of $C_N(x)$ centralized by K. If N is odd, $\mathcal{D}_{2N}$ has two characters, the trivial character and the character ρ_1 defined by

$$\rho_1(x^m k_{N-1}^j) = (-1)^j, \quad 0 \le m < N, 0 \le j < 2.$$

For N even, we must have $n = 0$ or $n = \frac{N}{2}$ and the trivial character ϕ_0 and the character $\phi_{\frac{N}{2}}$ are the only characters of $C_N(x)$ centralized by K. If N is even, $\mathcal{D}_{2N}$ has four characters, the trivial character ρ_0 and the three characters defined by

$$\rho_1(x^m k_{N-1}^j) = (-1)^j, \rho_2(x^m k_{N-1}^j) = (-1)^m, \rho_3(x^m k_{N-1}^j) = (-1)^{m+j},$$

$$0 \le m < N, 0 \le j < 2.$$

3.10   Maximal Extensions

We continue to assume that $G = H \lhd K$. If a character τ of H is *not* centralized by K, then τ cannot be extended to a character of G. Since H is a subgroup of $G(\tau)$, $G(\tau)$ has the form

$$G(\tau) = H \lhd K(\tau),$$

where $K(\tau)$ is the centralizer of τ in K,

$$K(\tau) = \{v \in K : v\tau v^{-1} = \tau\}.$$

THEOREM 39 *If τ is a character of H, then τ extends to a character of $G(\tau) = H \lhd K(\tau)$.*

Denote by

$$ext_\tau(G(\tau))$$

the collection of all characters of $G(\tau)$ extending τ.

THEOREM 40 *If τ is a character of H, then $ext_\tau(G(\tau))$ is the collection of all products in $\mathbf{C}G$ of the form $\tau\lambda$, where λ is a character of $K(\tau)$.*

For future use we have the following corollary.

COROLLARY 10.7 *If τ is a character of H, then*

$$\sum_{\rho \in ext_\tau(G(\tau))} \rho = \tau \sum_{\lambda \in K(\tau)^*} \lambda.$$

Suppose G_1 is an arbitrary group and K_1 is a subgroup of G_1. A character γ of K_1 is called a *maximal character* in G_1 if γ has no extension to a character of a subgroup of G_1 strictly containing K_1. Any extension of a character γ of K_1 to a maximal character in G_1 is called a *maximal extension* of γ in G_1.

THEOREM 41 *If τ is a character of H then $ext_\tau(G(\tau))$ is a collection of maximal extensions of τ in G.*

We will show that the left ideal generated by a maximal character is irreducible.

Suppose τ is a character of H and ρ is a character of $G(\tau)$ extending τ. $G(\tau)$ is not necessarily a normal subgroup of G. Consider the centralizer $G(\rho)$ of ρ in G. Since the support of ρ is $G(\tau)$, $G(\rho)$ normalizes $G(\tau)$,

$$yzy^{-1} \in G(\tau), \quad z \in G(\tau), \, y \in G(\rho).$$

$G(\tau)$ centralizes ρ implying $G(\tau)$ is a normal subgroup of $G(\rho)$.

THEOREM 42 *If τ is a character of H and ρ is a character of $G(\tau)$ extending τ, then $G(\rho) = G(\tau)$.*

3.11   Abelian by Abelian Semidirect Products

Groups of the form $G = A \rtimes B$ with A and B abelian groups are perhaps the simplest generalizations of abelian groups. Not surprisingly, the DSP of this class of groups closely resembles that of abelian groups.

We proved that each character τ of A extends to a character of the centralizer $G(\tau) = A \rtimes B(\tau)$ and that the set $ext_\tau(G(\tau))$ of all characters of $G(\tau)$ extending τ consists of maximal characters in G.

3.12 $EXT(G, A)$

Suppose τ is a character of A. The centralizer of τ in G is $G(\tau) = A \rtimes B(\tau)$. Denote by

$$ext_\tau(G(\tau))$$

the collection of all extensions of τ to characters of $G(\tau)$.

Denote by

$$EXT(G, A)$$

the union of the sets $ext_\tau(G(\tau))$ as τ runs over all characters of A. By Theorem 41 every character γ in $ext_\tau(G(\tau))$ is a maximal extension of τ and $EXT(G, A)$ consists of maximal characters in G.

EXAMPLE 43 If $G = C_p(x) \rtimes U(p)$, p a prime, $EXT(G, C_p(x))$ has order $2(p-1)$. The characters ϕ_n, $0 < n < p$, of $C_p(x)$ are maximal characters in G. The trivial character ϕ_0 extends to $p - 1$ characters of G.

EXAMPLE 44 If $G = C_8(x) \rtimes U(8)$, $EXT(G, C_8(x))$ has order 16. The characters ϕ_1, ϕ_3, ϕ_5 and ϕ_7 of $C_8(x)$ are maximal characters in G. As elements of $\mathbf{C}G$ they can be written explicitly as

$$
\begin{bmatrix} \phi_1 \\ \phi_3 \\ \phi_5 \\ \phi_7 \end{bmatrix}
=
\begin{bmatrix}
1 & v & v^2 & v^3 & v^4 & v^5 & v^6 & v^7 \\
1 & v^3 & v^6 & v & v^4 & v^7 & v^2 & v^5 \\
1 & v^5 & v^2 & v^7 & v^4 & v & v^6 & v^3 \\
1 & v^7 & v^6 & v^5 & v^4 & v^3 & v^2 & v
\end{bmatrix}
\begin{bmatrix} 1 \\ x \\ \vdots \\ x^7 \end{bmatrix}.
$$

The character ϕ_2 of $C_8(x)$ has two maximal extensions in G to characters of $C_8(x) \rtimes K_2$, $K_2 = \{1, 5\}$. A similar remark can be made for the character ϕ_6. The characters ϕ_0 and ϕ_4 of $C_8(x)$ extend to eight characters of G.

The first result extends, to the characters in $EXT(G, A)$, the product formula of Theorem 9 for characters of abelian groups.

For $f \in \mathbf{C}G$ set $|f|$ equal to the order of the support of f.

THEOREM 45 *If* γ_1, $\gamma_2 \in EXT(G, A)$, *then*

$$\gamma_1\gamma_2 = \begin{cases} |\gamma_1|\gamma_1, & \gamma_1 = \gamma_2, \\ 0, & \gamma_1 \neq \gamma_2. \end{cases}$$

In the language of idempotent theory a nonzero element $e \in \mathbf{C}G$ is called an *idempotent* if $e^2 = e$. Two idempotents e_1 and e_2 in $\mathbf{C}G$ are called *orthogonal* if $e_1e_2 = e_2e_1 = 0$. Theorem 45 says that

$$\left\{ \frac{1}{|\gamma|}\gamma \in EXT(G, A) \right\}$$

is a set of pairwise orthogonal idempotents. Using the language of idempotents, we call $EXT(G, A)$ a *set of pairwise orthogonal characters* in G.

The next result extends Theorem 11 to $EXT(G, A)$.

THEOREM 46

$$1 = \sum_{\gamma \in EXT(G,A)} \frac{1}{|\gamma|}\gamma.$$

A set $\mathcal{I}$ of pairwise orthogonal idempotents is said to be *complete* if

$$1 = \sum_{e \in \mathcal{I}} e.$$

Theorems 45 and 46 say that the set

$$\left\{ \frac{1}{|\gamma|}\gamma : \gamma \in EXT(G, A) \right\}$$

is a complete set of pairwise orthogonal idempotents. Using the language of idempotents, we call $EXT(G, A)$ a *complete set of pairwise orthogonal characters* in G.

Arguing as before, we have the following result.

THEOREM 47

$$\mathbf{C}G = \bigoplus_{\gamma \in EXT(G,A)} \mathbf{C}G\gamma.$$

3.13   A Basis of $\mathbf{C}G\gamma$

Suppose $\gamma \in EXT(G, A)$. We will show that $\mathbf{C}G\gamma$ can be multidimensional, determine the dimension of $\mathbf{C}G\gamma$ and construct a basis of $\mathbf{C}G\gamma$. In the next section we will show $\mathbf{C}G\gamma$ is irreducible.

Choose a complete system of left $G(\gamma)$-coset representatives in G,

$$\{y_s : 1 \le s \le S\}. \tag{5}$$

In general this system depends on γ. Since $G(\gamma)$ is a normal subgroup of G, by Theorem 32

$$\gamma_s = y_s \gamma y_s^{-1}, \quad 1 \le s \le S,$$

is the set of all conjugates of γ in G and

$$\gamma_{s_1} \gamma_{s_2} = 0, \quad 1 \le s_1, s_2 \le S, \ s_1 \ne s_2.$$

THEOREM 48   *If* $\gamma \in EXT(G, A)$, *then*

$$\{y_s \gamma : 1 \le s \le S\}$$

is a basis of the space $\mathbf{C}G\gamma$.

Proof Suppose $x \in G$ and write $x = y_s z$, $1 \le s \le S$, $z \in G(\gamma)$. By Theorem 30

$$x\gamma = y_s z\gamma = \gamma(z^{-1})y_s\gamma.$$

Since the elements $x\gamma$, $x \in G$, span $\mathbf{C}G\gamma$, the elements $y_s\gamma$, $1 \le s \le S$, span $\mathbf{C}G\gamma$.

To prove linear independence, suppose

$$0 = \sum_{s=1}^{S} \alpha(s)y_s\gamma, \quad \alpha(s) \in \mathbf{C}.$$

For any t, $1 \le t \le S$,

$$0 = \gamma_t 0 = \sum_{s=1}^{S} \alpha(s)\gamma_t \gamma_s y_s = |\gamma|\alpha(t)y_t\gamma$$

implying

$$\alpha(t) = 0, \quad 1 \le t \le S,$$

completing the proof.

The basis of $\mathbf{C}G\gamma$

$$\{y_s \gamma : 1 \le s \le S\}$$

will be denoted by $BAS(\gamma)$. It consists of the left multiplications of γ by the system (5). The basis of $\mathbf{C}G$ formed by the union of the sets $BAS(\gamma)$, $\gamma \in EXT(G, A)$, will be denoted by $BAS(G, A)$.

3.14   Irreducibility of $\mathbf{C}G\gamma$

THEOREM 49 *If* $\gamma \in EXT(G, A)$, *then* $\mathbf{C}G\gamma$ *is an irreducible left ideal.*

Proof We continue to assume that $\{y_s : 1 \leq s \leq S\}$ is a complete system of left $G(\gamma)$-coset representatives in G. Suppose V is a nonzero left ideal in $\mathbf{C}G\gamma$ and $f \in V$, $f \neq 0$. By Theorem 48

$$f = \sum_{s=1}^{S} \alpha(s) y_s \gamma,$$

where $\alpha(s_1) \neq 0$ for some $1 \leq s_1 \leq S$. Theorems 45 and 31 imply

$$\gamma_{s_1} f = \sum_{s=1}^{S} \alpha(s) \gamma_{s_1} y_s \gamma = \sum_{s=1}^{S} \alpha(s) \gamma_{s_1} \gamma_s y_s = \alpha(s_1) |\gamma| \gamma_{s_1} y_{s_1}.$$

Since V is a left ideal and $f \in V$, $\gamma_{s_1} y_{s_1} \in V$ and

$$\gamma = y_{s_1}^{-1} \gamma_{s_1} y_{s_1} \in V$$

implying $V = \mathbf{C}G\gamma$, completing the proof.

3.15   Expansion Coefficients

A formula will be derived for the expansion coefficients of $f \in \mathbf{C}G$ relative to the basis $BAS(G, A)$. An algorithm implementing this formula will be constructed by making explicit the relationship between the characters of A and $EXT(G, A)$.

Suppose $\gamma \in EXT(G, A)$. Choose a complete system of left $G(\gamma)$-coset representatives in G,

$$y_s(\gamma), \quad 1 \leq s \leq S(\gamma).$$

The dependence of the system on γ is now explicitly expressed. The basis $BAS(G, A)$ of $\mathbf{C}G$ consists of all elements of the form

$$y_s(\gamma)\gamma, \quad \gamma \in EXT(G, A), 1 \leq s \leq S(\gamma).$$

The expansion of $f \in \mathbf{C}G$ relative to this basis will be written as

$$f = \sum_{\gamma \in EXT(G,A)} \sum_{s=1}^{S(\gamma)} f_\gamma(s) y_s(\gamma)\gamma.$$

In this section we derive a formula for the expansion coefficients

$$f_\gamma(s), \quad \gamma \in EXT(G, A), 1 \leq s \leq S(\gamma).$$

In the next section we will use this formula to derive a fast Fourier-like algorithm to compute these expansion coefficients.

By Theorem 45 if $f \in \mathbf{C}G$ and $\gamma \in EXT(G, A)$, then

$$f\gamma = |\gamma| \sum_{s=1}^{S(\gamma)} f_\gamma(s) y_s(\gamma) \gamma. \tag{6}$$

The expansion coefficients of the component of f in $\mathbf{C}G\gamma$ relative to the basis $BAS(\gamma)$ are given by $f_\gamma(s)$, $1 \le s \le S(\gamma)$.

THEOREM 50 *If* $f \in \mathbf{C}G$ *and* $\gamma \in EXT(G, A)$, *then*

$$f_\gamma(s) = \frac{1}{|\gamma|} \sum_{z \in G(\gamma)} f(y_s(\gamma)z)\gamma(z^{-1}), \qquad 1 \le s \le S(\gamma).$$

Proof Set $y_s = y_s(\gamma)$ and $S = S(\gamma)$ throughout the proof. For $x \in G$, write $x = y_s z$, $1 \le s \le S$, $z \in G(\gamma)$, and

$$f = \sum_{x \in G} f(x)x = \sum_{s=1}^{S} \sum_{z \in G(\gamma)} f(y_s z) y_s z.$$

Since by Theorem 30, $y_s z\gamma = \gamma(z^{-1})y_s\gamma$, we have

$$f\gamma = \sum_{s=1}^{S} \left(\sum_{z \in G(\gamma)} f(y_s z)\gamma(z^{-1}) \right) y_s\gamma.$$

Comparing with (6) completes the proof.

4.   Examples

We will compute $EXT(G, A)$ and $BAS(G, A)$ for the groups $G = A \lhd B$, where $A = C_N(x) \times C_N(y)$ and $B = C_2(k_a) \times C_2(k_b)$ with

$$a = \begin{bmatrix} -1 & 0 \\ 0 & -1 \end{bmatrix}, \qquad b = \begin{bmatrix} 0 & 1 \\ 1 & 0 \end{bmatrix}.$$

The fixed points in A under actions by the elements of B depend on whether N is even or odd.

Fixed points

	N even	N odd
k_a	$(0,0)$, $(\frac{N}{2},0)$, $(0,\frac{N}{2})$, $(\frac{N}{2},\frac{N}{2})$	$(0,0)$
k_b	$(0,0)$, $(1,1)$, ..., $(N-1,N-1)$	$(0,0)$, $(1,1)$, ..., $(N-1,N-1)$
$k_a k_b$	$(0,0)$, $(1,N-1)$, ..., $(N-1,1)$	$(0,0)$, $(1,N-1)$, ..., $(N-1,1)$

Apart from the difference in the first row, it is important to note that for N even, $\left(\frac{N}{2}, \frac{N}{2}\right)$ is a fixed point for k_b and $k_a k_b$, but is not in the indexing set if N is odd.

Centralizer in B

Centralizer	character index	
	N even	N odd
B	$(0,0)$, $\left(\frac{N}{2}, \frac{N}{2}\right)$	$(0,0)$
B_1	$\left(\frac{N}{2}, 0\right)$, $\left(0, \frac{N}{2}\right)$	None
B_2	(l,l), $1 \le l < N$, $l \ne \frac{N}{2}$	(l,l), $1 \le l < N$
B_3	$(l, N-l)$, $1 \le l < N$, $l \ne \frac{N}{2}$	$(l, N-l)$, $1 \le l < N$

4.1 $EXT(G_1, A)$

$Ext(G_1, A)$ can be computed from the fixed points and centralizers and the characters of the subgroups of B. If the subgroup C of B is the centralizer of the point (k, l), then the collection of products in $\mathbf{C}G_1$,

$$\tau_{k,l} \lambda, \quad \lambda \in C^*$$

is the collection of all characters in $EXT(G_1, A)$ extending $\tau_{k,l}$.

The characters of B are

$$\lambda_{0,0} = 1 + k_a + k_b + k_a k_b, \quad \text{(trivial character)}.$$
$$\lambda_{0,1} = 1 + k_a - k_b - k_a k_b,$$
$$\lambda_{1,0} = 1 - k_a + k_b - k_a k_b,$$
$$\lambda_{1,1} = 1 - k_a - k_b + k_a k_b.$$

The characters of the subgroups of B are given by restriction.

Characters of subgroups of B.

Subgroups	characters
B_0	1
B_1	$1 + k_a,\ 1 - k_a$
B_2	$1 + k_b,\ 1 - k_b$
B_3	$1 + k_a k_b,\ 1 - k_a k_b$

The following table describes $EXT(G_1, A)$, for even N, by describing the subsets $ext_\tau(G_1(\tau))$ of characters in $EXT(G_1, A)$ extending $\tau \in A^*$.

$$EXT(G_1, A), \ N \text{ even.}$$

τ	$ext_\tau(G_1(\tau))$	$G_1(\tau)$
$\tau_{0,0}, \ \tau_{\frac{N}{2},\frac{N}{2}}$	$\tau\lambda_{0,0}, \ \tau\lambda_{0,1}, \ \tau\lambda_{1,0}, \ \tau\lambda_{1,1}$	G_1
$\tau_{0,\frac{N}{2}}, \ \tau_{\frac{N}{2},0}$	$\tau(1+k_a), \ \tau(1-k_a)$	$A \rtimes B_1$
$\tau_{l,l}$	$\tau(1+k_b), \ \tau(1-k_b)$ $1 \le l < N, \ l \ne \frac{N}{2}$	$A \rtimes B_2$
$\tau_{l,N-l}$	$\tau(1+k_ak_b), \ \tau(1-k_ak_b)$ $1 \le l < N, \ l \ne \frac{N}{2}$	$A \rtimes B_3$

Characters τ of A not listed above have $B_0 = \{1\}$ as their centralizer and are maximal characters in $EXT(G_1, A)$.

$$EXT(G_1, A), \ N \text{ odd.}$$

τ	$ext_\tau(G_1(\tau))$	$G_1(\tau)$
$\tau_{0,0}$	$\tau_{0,0}\lambda_{0,0}, \ \tau_{0,0}\lambda_{0,1}, \ \tau_{0,0}\lambda_{1,0}, \ \tau_{0,0}\lambda_{1,1}$	G_1
$\tau_{l,l}$	$\tau_{l,l}(1+k_b), \ \tau_{l,l}(1-k_b), \quad 1 \le l < N$	$A \rtimes B_2$
$\tau_{l,N-l}$	$\tau_{l,N-l}(1+k_ak_b), \ \tau_{l,N-l}(1-k_ak_b), \quad 1 \le l < N$	$A \rtimes B_3$

Characters τ of A not of the form given above are maximal characters in $EXT(G_1, A)$.

The basis $BAS(G_1, A)$ is the disjoint union of the bases $BAS(\gamma)$ of the irreducible left ideals $\mathbf{C}G_1\gamma, \ \gamma \in EXT(G_1, A)$. $BAS(\gamma)$ is the set given by the left multiplications of γ by the elements in the complete system of left $B(\gamma)$-coset representatives in B.

We will define an ordering on $BAS(G_1, A)$ which is compatible with the direct sum decomposition of $\mathbf{C}G_1$ into the irreducible left ideals $\mathbf{C}G_1\gamma, \ \gamma \in EXT(G_1, A)$. The elements in each basis $BAS(\gamma)$ of $\mathbf{C}G_1\gamma$ occur as contiguous elements.

$EXT(G_1, A)$ consists of all elements of the form

$$\gamma = \tau_{k,l}\lambda, \quad 0 \le k, \ l < N, \ \lambda \in B(\tau_{k,l})^*.$$

Order the characters of B as defined and order the characters of sub-groups. Order $EXT(G_1, A)$ by ordering (k, l, λ), lexicographically. In this ordering λ is the fastest running parameter, followed by l and then by k.

The ordering in $EXT(G_1, A)$ induces an ordering on the collection of all bases $BAS(\gamma), \ \gamma \in EXT(G_1, A)$. Ordering the elements in each of these bases, we have ordered $BAS(G_1, A)$ with the required compatibility condition with respect to the direct sum decomposition.

$$BAS(\gamma), \ N = 3$$

$B(\gamma)$	$BAS(\gamma)$
B	$\tau_{0,0}\lambda_{0,0}, \ \tau_{0,0}\lambda_{0,1}, \ \tau_{0,0}\lambda_{1,0}, \ \tau_{0,0}\lambda_{1,1}$
B_1	None
B_2	$\tau_{1,1}(1+k_b), \ k_a\tau_{1,1}(1+k_b), \ \tau_{1,1}(1-k_b), \ k_a\tau_{1,1}(1-k_b),$
	$\tau_{2,2}(1+k_b), \ k_a\tau_{2,2}(1+k_b), \ \tau_{2,2}(1-k_b), \ k_a\tau_{2,2}(1-k_b),$
B_3	$\tau_{1,2}(1+k_ak_b), \ k_a\tau_{1,2}(1+k_ak_b), \ \tau_{1,2}(1-k_ak_b), \ k_a\tau_{1,2}(1-k_ak_b),$
	$\tau_{2,1}(1+k_ak_b), \ k_a\tau_{2,1}(1+k_ak_b), \ \tau_{2,1}(1-k_ak_b), \ k_a\tau_{2,1}(1-k_ak_b),$
B_0	$\tau_{0,1}, \ k_b\tau_{0,1}, \ k_a\tau_{0,1}, \ k_ak_b\tau_{0,1}$
	$\tau_{0,2}, \ k_b\tau_{0,2}, \ k_a\tau_{0,2}, \ k_ak_b\tau_{0,2}$
	$\tau_{1,0}, \ k_b\tau_{1,0}, \ k_a\tau_{1,0}, \ k_ak_b\tau_{1,0}$
	$\tau_{2,0}, \ k_b\tau_{2,0}, \ k_a\tau_{2,0}, \ k_ak_b\tau_{2,0}$

Figure 8.   Normalized basis of $\mathbf{C}\left((C_3 \times C_3) \rtimes (C_2 \times C_2)\right)$

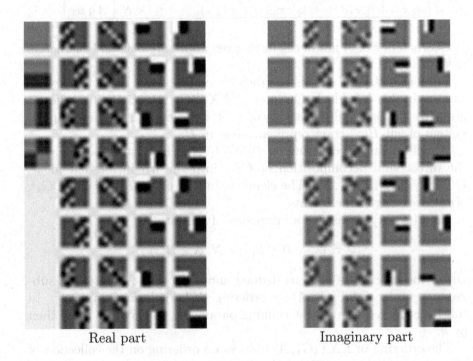

Real part   Imaginary part

Figure 9. Normalized basis of $\mathbf{C}\left((C_4 \times C_4) \updownarrow (C_2 \times C_2)\right)$

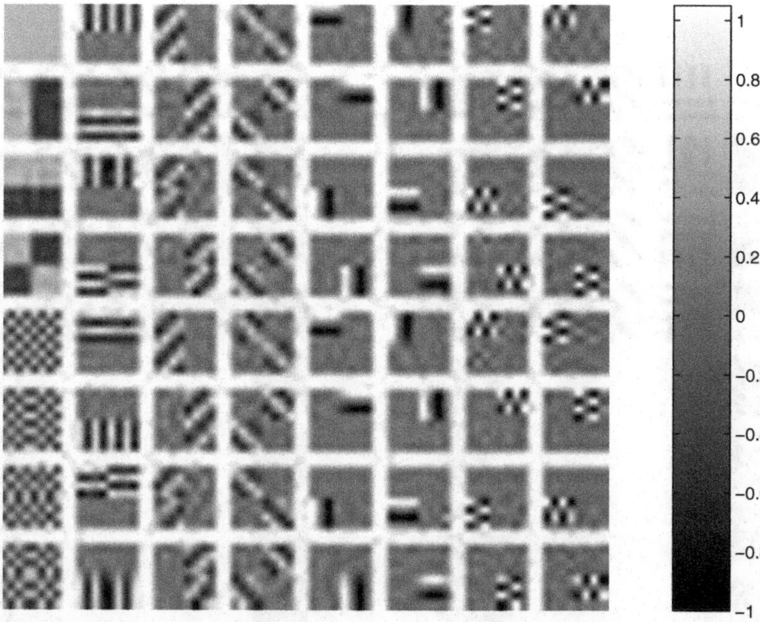

Real part

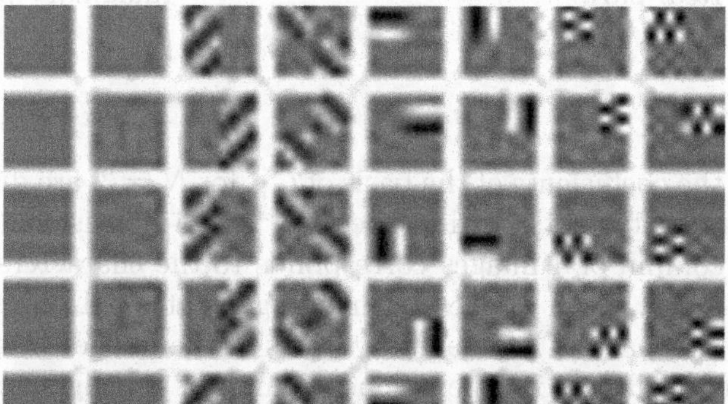

Imaginary part

Figure 10.   Generators of invariant subspaces of $\mathbf{C}\left((C_4 \times C_4) \rtimes (C_2 \times C_2)\right)$

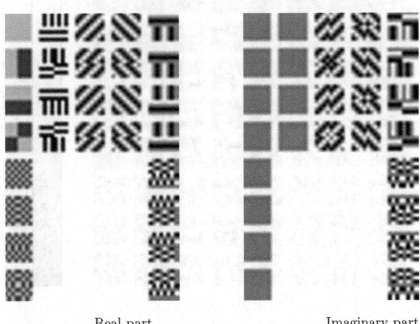

Real part   Imaginary part

In Figure 8, the coefficients of the basis of G_1, for $N = 3$, are displayed as log-scaled intensity plots, with respect to this 2-dimensional indexing. The first column of 4 images corresponds to the characters of G_1, scaled by $|G_1| = 36$. Each of these elements generates a 1-dimensional $\mathcal{L}G_1$-invariant subspace. The imaginary parts of the images in the first column are identically zero. The second column of 8 images, in pairs, corresponds to the basis of 2-dimensional invariant subspaces generated by the maximal extensions to $A \rtimes B_2$, scaled by 18. The third column of 8 images, in pairs, corresponds to the basis of 2-dimensional invariant subspaces generated by the maximal extensions to $A \rtimes B_3$, scaled by 18. The remaining two columns of images, 4 images at a time, correspond to the basis of 4-dimensional invariant subspaces generated by the characters of A which are maximal in G_1, scaled by 9.

An index matrix of G_1, for $N = 4$, is defined in an analogous way. In Figure 9, the coefficients of the basis of G_1, for $N = 4$, are displayed as log-scaled intensity plots, with respect to this 2-dimensional indexing.

An intensity scale bar is appended on the right for reference, and applies to all image plots in this chapter.

The first column of 8 images corresponds to the characters of G_1, scaled by $|G_1| = 64$. Each of these elements generates a 1-dimensional $\mathcal{L}G_1$-invariant subspace. The second column of 8 images, in pairs, corresponds to the basis of 2-dimensional invariant subspaces generated by maximal extensions to $A \lhd B_1$, scaled by $|A \lhd B_1| = 32$. The imaginary parts of the images in the first two columns are identically zero. The third column of 8 images, in pairs, corresponds to the basis of 2-dimensional invariant subspaces generated by the maximal extensions to $A \lhd B_2$, scaled by 32. The fourth column of 8 images, in pairs, corresponds to the basis of 2-dimensional invariant subspaces generated by the maximal extensions to $A \lhd B_3$, scaled by 32. The remaining columns of images, 4 images at a time, correspond to the basis of 4-dimensional invariant subspaces generated by the characters of A which are maximal in G_1, scaled by 16.

In Figure 10, the 28 generators of invariant subspaces are displayed. These are the results of adding the basis elements belonging to an invariant subspace.

5.   Group Transforms

Suppose G is an arbitrary group and Δ is a complete system of primitive, pairwise orthogonal idempotents in $\mathbf{C}G$. The *G-group transform* or simply *G-transform* is the collection of right multiplications of $\mathbf{C}G$

$$\{R(e) : e \in \Delta\}. \tag{7}$$

Reference to Δ will always be suppressed. Each right multiplication in (7) is a *G-component filter*. Any G-component filter or sum of G-component filters is a *G-filter*.

The *G-transform* of $f \in \mathbf{C}G$ is the collection

$$\{R(e)f : e \in \Delta\}. \tag{8}$$

Each product in (8) is a *G-spectral component* of f. Any G-spectral component or sum of G-spectral components of f is a *G-filtering* of f.

For an abelian group A we always take $\Delta = \Delta(A)$. Abelian group transforms are probably well-known to the reader. Since each *filter space* of the A-transform

$$\mathbf{C}Ae, \quad e \in \Delta(A),$$

is one-dimensional, the A-spectral component

$$fe = \widehat{f}(\tau)e, \quad e = \frac{1}{N}\tau,$$

can be represented by $\widehat{f}(\tau)$. The *A-spectral image* of f is the collection

$$\left\{\widehat{f}(\tau) : \tau \in A^*\right\}.$$

For any abelian by abelian semidirect product $G = A \triangleleft B$ we always take $\Delta = \Delta(G)$. The *filter spaces* of the G-transform

$$\mathbf{C}Ge, \quad e \in \Delta(G),$$

are not necessarily one-dimensional. The *G-spectral image* of f is the collection of expansion coefficients of f relative to the basis $BAS(G, A)$.

A *G-filter* is a linear operator $P : \mathbf{C}G \longrightarrow \mathbf{C}G$ satisfying

$$P(yf) = yP(f), \quad f \in \mathbf{C}G, \ y \in G.$$

By linearity if P is a G-filter, then

$$P(gf) = gP(f), \quad f, g \in \mathbf{C}G.$$

Every $f \in \mathbf{C}G$ defines a G-filter $R(f)$ by

$$R(f)g = gf, \quad g \in \mathbf{C}G.$$

$R(f)$ is called *right multiplication* by f. In general $L(f)$ is not a G-filter.

THEOREM 51 *If P is a G-filter, then there exists $f \in \mathbf{C}G$ such that $P = R(f)$.*

The image of the G-filter $R(f)$ is the left ideal $\mathbf{C}Gf$. We call $\mathbf{C}Gf$ the *filter space* of $R(f)$. Since we can have $\mathbf{C}Gf_1 = \mathbf{C}Gf_2$, for distinct f_1, $f_2 \in \mathbf{C}G$, distinct G-filters can have the same filter space.

5.1　　Matched Filtering

Suppose G is an arbitrary group and Δ is a complete system of primitive, pairwise orthogonal idempotents in $\mathbf{C}G$. For $f \in \mathbf{C}G$, the *G-matched filter* of f is the G-filter

$$e_f = \sum_{fe \neq 0} e. \tag{9}$$

The summation in (9) runs over all idempotents $e \in \Delta$ such that $fe \neq 0$. The G-matched filter of f satisfies

$$fe_f = f.$$

In a typical image processing application an image $f \in \mathbf{C}G$ is contained in noise and background. The problem is to separate the target f from the noise and background. The image data has the form

$$g = f + \eta,$$

where $\eta \in \mathbf{C}G$ represents the noise and background. The G-matched filtering

$$ge_f = f + \eta e_f$$

is usually the optimal G-filtering for separating f from η. Examples of G-matched filtering will be given in the following sections.

The concept of a G-matched filter is a major tool in both abelian and nonabelian group filter design. As defined, the formal structure of a G-matched filter does not change with different choices of group G. However, the choice of indexing group is a critical parameter affecting several important characteristics of the resulting matched filters. We have distinguished two properties which can vary as the indexing group varies: invariance under left group actions and the extent and form of image domain locality. The first will be discussed in this section. Image domain locality is the topic addressed in the next section.

For an image $f \in \mathbf{C}G$, the *G-equivalence class of images* containing f is the equivalence class of images in the set

$$\{xf : x \in G\}.$$

The G-equivalence class containing f depends on the actions of the left multiplications from G on f and can vary as G varies.

The G-matched filter of f satisfies

$$xfe_f = xf, \quad x \in G.$$

This implies that the G-matched filter of f is the G-matched filter of every image G-equivalent to f. In applications this means that the G-matched filter constructed for an image f will be the optimal filtering not only for f, but also for all G-equivalent images.

5.2   Image Domain Locality

Image domain locality is a key component of nonabelian group filters not available with abelian group filters. To understand where this property comes from and how it can be controlled at the group level, we will show that the expansion coefficients in nonabelian group DSP can encode local image space information.

The explanation is related to the character extension problem. If A is an abelian group, then this is straightforward. The basis for Fourier representation is the collection of characters. Relative to this basis, each expansion coefficient contains image information across the image plane. These coefficients are sensitive to all local changes. The placing of a geometric structure in an image data set affects all expansion coefficients.

The character extension problem is more complicated for an abelian by abelian semidirect product $G = A \rtimes B$. As we have pointed out there exist characters of A, which can not be extended to characters of G. The characters defining $\Delta(G)$ are maximal characters in G, but some have proper subgroups of G as their supports.

Suppose $e \in \Delta(G)$ and

$$e = \frac{1}{|\gamma|}\gamma,$$

where γ is a character of the subgroup $G(\gamma)$ of G. For the discussion assume $G(\gamma)$ is a proper subgroup of G. If

$$\{y_s : 1 \leq s \leq S\}$$

is a complete system of left $G(\gamma)$-coset representatives in G, then

$$\{y_s e : 1 \leq s \leq S\} \tag{10}$$

is a basis for the irreducible left ideal $\mathbf{C}Ge$. The key point is that $G(\gamma)$ is a proper subgroup of G and the basis (10) is supported on pairwise disjoint left-cosets. For each s, $1 \leq s \leq S$, the basis element $y_s e$ is localized in the image domain to the left coset $y_s G(\gamma)$.

Suppose $f \in \mathbf{C}G$. We can write

$$fe = \sum_{s=1}^{S} \alpha(s) y_s e.$$

If $g \in \mathbf{C}G$ is supported in $y_t G(\gamma)$ for some $1 \leq t \leq S$, then

$$(f + g)e = \sum_{s \neq t} \alpha(s) y_s e + \alpha'(t) y_t e.$$

The effect of the local image domain change on the expansion coefficients of the filtered image, resulting from the addition of g, is to change exactly one coefficient. The sensitivity to the local image domain depends on the relative orders of G and $G(\gamma)$.

For a fixed character τ of A, the left ideal $\mathbf{C}G\tau$ can be written as

$$\mathbf{C}G\tau = \bigoplus_{\lambda \in B(\tau)^*} \mathbf{C}G\tau\lambda.$$

$\mathbf{C}G\tau$ is not necessarily irreducible. We can write

$$f\tau = \frac{1}{|B(\tau)|} \sum_{\lambda \in B(\tau)^*} \sum_{s=1}^{S} \alpha_\lambda(s) y_s \tau \lambda.$$

This expansion shifts the basis in both the image and spectral image domains and can be viewed as a localized image-spectral image domain expansion in analogy to time-frequency expansions. Local image domain changes affect a set of prescribed coefficients and these changes describe the local image domain change in both image and spectral image domains.

6. Group Filters

In this section we study and compare properties of abelian and non-abelian group filters by illustrating their effects on lines and line segments. These examples will concentrate on the image domain locality and the left group multiplication invariance of the filters.

6.1 Abelian group filtering

For $N = 64$, set

$$A_0 = \sum_{k=0}^{N-1} \tau_{0,k} \in \mathbf{C}A.$$

A_0 is the matched filter of the vertical line and all its abelian group translates.

The following examples show detection and location by A_0 of a vertical line and several of its abelian group translates. Varying levels of noise are added to generate the images in Figure 12. The noise is modeled as Gaussian with zero-mean. Levels of noise are indicated by the standard deviation (sd). Images in Figure 13 are the results of filtering by A_0. Even in severe noise, the location of maximum intensity yields the positions of the translates of a vertical line.

Figure 11. Images containing a vertical line

Figure 12.   Noisy images containing a vertical line

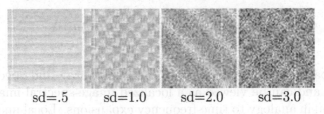

sd=.5   sd=1.0   sd=2.0   sd=3.0

Figure 13.   Results of Filtering by A_0

In Figures 14 and 15 the translates of a vertical line are replaced by line segments of one half the length. The placement of the vertical line segment is given by the coordinate of the starting position.

Figure 14.   Images containing a vertical line segment

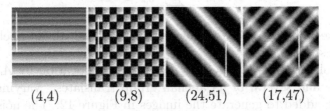

(4,4)   (9,8)   (24,51)   (17,47)

Figure 15.   Noisy images containing a vertical line segment

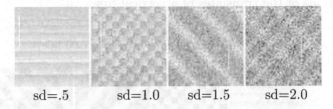

sd=.5   sd=1.0   sd=1.5   sd=2.0

The last example shows that as the length of the line diminishes, the performance of the filter A_0 for detecting line segments worsens.

Figure 16.   Results of Filtering by A_0

Figure 17.   Images of vertical line segments

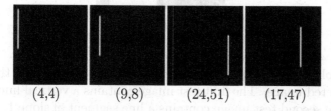

(4,4)   (9,8)   (24,51)   (17,47)

Figure 18.   Results of Filtering by A_0

Moreover, the vertical position of the line segment cannot be detected, even without any noise as shown in Figure 18.

The preceding examples show one of the major deficiencies of abelian group filtering, the lack of image domain locality.

6.2   Nonabelian group filtering

Consider the group

$$G_4 = A \rtimes C_2(k_c)$$

and the group filter defined by

$$P_0 = \frac{1}{N} \sum_{k=1}^{N-1} \tau_{0,k} + k_c \tau_{0,k}.$$

The choice of P_0 was made for its simplicity, and because it is *symmetric* or real valued. An intensity plot of the filter is shown below for $N = 32$ along with the intensity color scheme.

Figure 19.   Intensity plot of the filter P_0, $N = 32$.

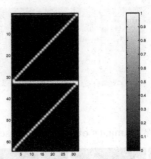

The following example shows the importance of matching the filter to the targeted image. The first test image contains a vertical line segment, while the second test image contains a line segment of slope 1. In Figure 20, the first and third images contain a vertical line segment, while the second and fourth images contain a line segment of slope 1. Noise levels are indicated by the standard deviation (sd). Figure 21 displays the results of applying P_0.

Figure 20.   Test images in noise

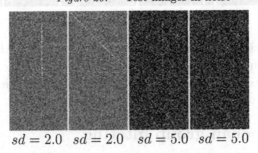

$sd = 2.0$ $sd = 2.0$ $sd = 5.0$ $sd = 5.0$

The filter P_0 is matched to the vertical line and to all left multiplications of the vertical line by the elements of G_4. Since the line segment of slope 1 is not contained in the collection of G_4-equivalent images, it cannot be detected by P_0.

Invariance of P_0 with respect to left multiplication by the elements from G_4 implies that these left actions do not affect filter performance. The previous figure shows this for left multiplications by $x^m y^n$, $0 \le m$, $n < N$.

Figure 21.   Results of applying P_0

Figure 22. k_c-translates of the images in Figure 20

Figure 23.   Results of filtering by P_0

Each frame in Figure 22 displays the k_c-translate of the corresponding frame in Figure 20. Figure 23 displays the results of filtering the images in Figure 22 by P_0.

The next pair of figures show the implications of invariance and image domain locality. In Figure 24 a vertical line segment of length 32 is placed arbitrarily, and the horizontal position is recorded. The problem is to determine the position using the filter P_0. Figure 25 shows the results of this filtering by P_0.

Figure 24. x^m-translates of an image containing line segment of length 32

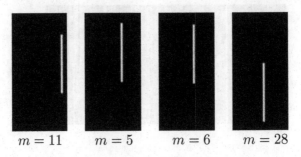

$m = 11$ $m = 5$ $m = 6$ $m = 28$

Figure 25.   Results of filtering by P_0

$(u, l) =$ $(.585, .382)$ $(.761, .207)$ $(.732, .236)$ $(.117, .851)$

The pair of numbers (u, l) in Figure 23 are the intensities of the upper and lower half of the line segments. Note the close relationship between the intensities of the line segment and its position. (The intensity of the input line segment in Figure 25 is 1.0, but this information is only relative.) The relationship between the difference of intensities and position is given by

$$m \sim N - \frac{uN}{u+l}. \tag{11}$$

Accuracy of the relationship (11) depends on N, up to $\frac{N}{16}$ pixels. Thus the smaller N is the more accurate the relationship (11).

The same experiment is repeated with varying levels of noise. The position of the line segment is estimated from the filtered image using (11), and denoted by m' in Figure 27.

In the presence of noise, the intensity is not uniform through the upper or lower half of the line segment, but the variance is very small. Estimate of location in (11) is derived using the intensity at position

Figure 26. x^m-translates of an image containing line segment of length 32

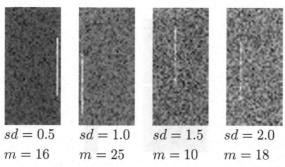

| $sd = 0.5$ | $sd = 1.0$ | $sd = 1.5$ | $sd = 2.0$ |
| $m = 16$ | $m = 25$ | $m = 10$ | $m = 18$ |

Figure 27. Results of filtering by P_0

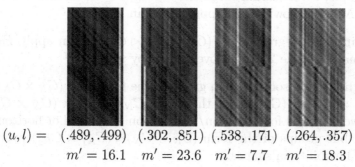

| $(u, l) =$ | $(.489, .499)$ | $(.302, .851)$ | $(.538, .171)$ | $(.264, .357)$ |
| | $m' = 16.1$ | $m' = 23.6$ | $m' = 7.7$ | $m' = 18.3$ |

$(0, y)$ and (N, y), where y is the horizontal position of the line segment. A better estimate than (11) can be derived, especially in cases where noise characteristics are known.

As these experiments show, detection and location of vertical lines of uniform intensity of known lengths can be implemented using P_0. Using similar analysis, a relationship can be derived and the same implementation can be used to locate diagonal lines shown in Figure 22. Since $k_c x^m y^n k_c^{-1} = x^m y^{m-n}$, the x-intercept is equivalent to the horizontal location of the vertical line.

Note that the length of a line segment must be at least N for exploiting the localization property consistently. More specifically, if the length of the line segment is less than N, in the case where the line segment lies entirely in lower or the upper of the two planes, we can localize the line segment only up to one of the two planes. This provides one major parameter determining the design of groups and filters in detection/localization applications: For consistent detection/localization

Figure 28.   Intensity plot of the filter P_1, $N = 32$.

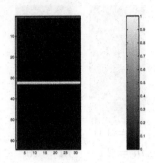

of, for example, a vertical line segment of length N in an image of size $K_1N \times K_2N$, the following two approaches can be used.

1. Starting with a group $G = (C_N \times C_N) \rtimes C_2$, we can apply P_0 to subimages of size $2N \times N$, overlapped by at least N.

2. Design filters associated to a group of the form $G = (C_N \times C_N) \rtimes (C_{K_1} \times C_{K_2})$. (Groups of the form $(C_N \rtimes C_{K_1}) \times (C_N \times C_{K_2})$ will work as well for detection/localization of vertical or horizontal lines, but groups of the form $(C_N \times C_N) \rtimes (C_{K_1} \times C_{K_2})$ provide more flexibility in shapes.)

The filter performance degrades as the noise level increases. This is particularly true with no prior knowledge of the noise characteristics. One method that will improve filter performance is to use a similar filter based on a different group of the same size. Consider the group

$$G = (C_N(x) \times C_N(y)) \rtimes C_2(k_e),$$

where $e = \begin{bmatrix} 0 & 1 \\ 1 & 0 \end{bmatrix}$. Construct the G-filter

$$P_1 = \frac{1}{N} \sum_{n=1}^{N-1} \tau_{0,k} + k_e \tau_{0,k}.$$

P_1 is also real-valued. An intensity plot of P_1 is shown in Figure 28.

In Figures 29 – 31, results of applying P_0 and P_1 are shown in the third and fourth frames along with estimates of the location of the line segment. The first frame displays the line segment and the second frame displays the line segment embedded in background and noise.

Figure 29. Results of applying P_0 and P_1

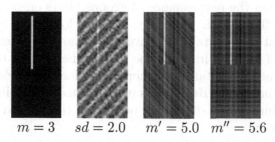

$m = 3$ $sd = 2.0$ $m' = 5.0$ $m'' = 5.6$

Figure 30. Results of applying P_0 and P_1

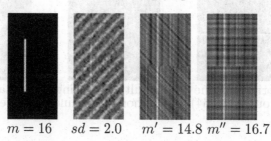

$m = 16$ $sd = 2.0$ $m' = 14.8$ $m'' = 16.7$

Figure 31. Results of applying P_0 and P_1

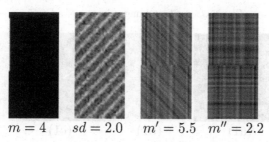

$m = 4$ $sd = 2.0$ $m' = 5.5$ $m'' = 2.2$

The accuracy of the estimate (11) differs between the filters P_0 and P_1 as the noise is *filtered out* in two distinct ways by the distinct convolutions defined by the two groups. This is another parameter determining design of a detection/location application. By applying several distinct groups, detection and location can be made more accurate. A trade-off is clearly the cost of processing.

7.   Line-like Images

In this section the performance of groups filters on non-digital, line-like objects is illustrated. Generally, filter performances are better due to the fact that non-digital lines have more than one pixel width.

In Figure 32 the first image is the digital line of slope 0. The second image is generated by rotating the horizontal line by 27° clockwise about the center of the image. Rotation is implemented using bilinear interpolation. The third image displays a portion of the rotated line near the center of rotation.

Figure 32.   Result of rotating a line

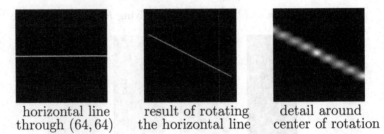

| horizontal line through (64, 64) | result of rotating the horizontal line | detail around center of rotation |

The following example compares the results of filtering a digital test image and a rotated test image in varying levels of noise.

Figure 33.   Test images in noise, $sd = .5$

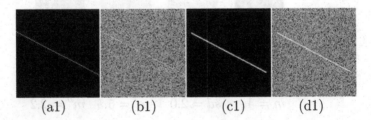

(a1)   (b1)   (c1)   (d1)

Images in Figure 33 are as follows.

(a1) digital line segment of slope $-\frac{1}{2}$,   (b1) digital line in noise,
(c1) rotated line,   (d1) rotated line in noise

The next example illustrates the performance of matched filters in detecting and locating a *line-like* object in recorded data.

Figure 34. Results of filtering in noise, $sd = .5$

Figure 35. Sonar image of size 128×128 and result of filtering

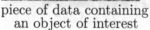
piece of data containing   result of filtering
an object of interest

The next collection of examples are included to illustrate:

1 the diversity of images which can be created by composition of parallel lines of varying orientations,

2 the sensitivity of the matched filters to changes in orientations,

3 design strategy for matched filters whose properties vary with image size.

For a positive integer K and $N = 16K$, consider the groups

$$G_6 = A \wr C_4(k_a), \quad G_7 = A \wr C_4(k_b),$$

where $A = C_N(x) \times C_N(y)$ and

$$a = \begin{bmatrix} \frac{N}{2} & \frac{N}{4}+1 \\ \frac{N}{4}+1 & 0 \end{bmatrix}, \quad b = \begin{bmatrix} \frac{N}{2} & \frac{N}{4}-1 \\ \frac{N}{4}-1 & 0 \end{bmatrix}.$$

The sum of 1-dimensional idempotents in $\mathbf{C}G_6$ and in $\mathbf{C}G_7$ define matched filters

$$P_6 = \sum_{k=0}^{N-1} \sum_{j=0}^{3} \tau_{k,(M+1)k} \lambda_j \in \mathbf{C}G_6,$$

$$P_7 = \sum_{k=0}^{N-1} \sum_{j=0}^{3} \tau_{k,(M-1)k} \mu_j \in \mathbf{C}G_7.$$

Figures 36 and 37 display the filters P_6 and P_7. Unlike the matched filters we have previously described, properties of P_6 and P_7 depend on N which determines the image size.

Figure 36.   Matched filters, $N = 16$.

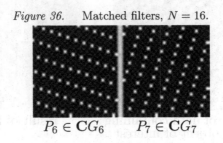

$P_6 \in \mathbf{C}G_6$ $P_7 \in \mathbf{C}G_7$

The first image in Figure 38 is a composition of two sets of intensities along parallel lines of slopes 3 and 4. The results of applying P_6 and P_7 in noise are displayed. Noise is simulated as Gaussian of zero mean with varying values of standard deviation.

The last image in Figure 38 is not an algebraic object in that it belongs neither to $\mathbf{C}G_6$ nor $\mathbf{C}G_7$. It is simply the sum of two intensities defined on an array of size $2N \times 2N$.

In Figure 40, $\mathbf{g}$ is viewed as an element of $\mathbf{C}G_7$. The last image is again simply the sum of two intensities.

Figure 37.   Matched filters, $N = 32$.

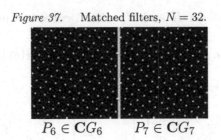

$P_6 \in \mathbf{C}G_6$ $P_7 \in \mathbf{C}G_7$

Figure 38.   Results of filtering in noise

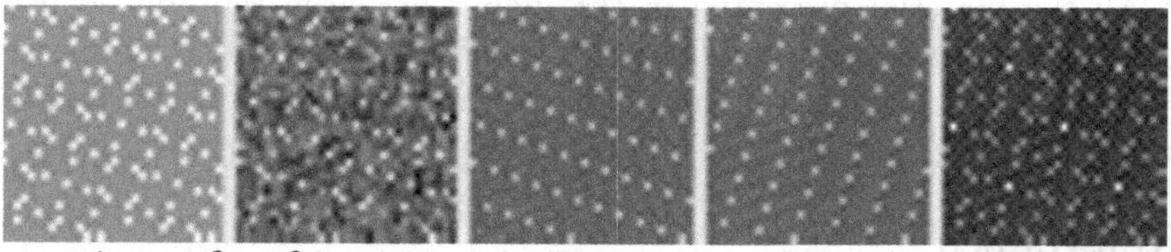

| test image **f** | **f** in noise | P_6 | P_7 | $P_6 + P_7$ |

Figure 39.   Results of filtering in noise

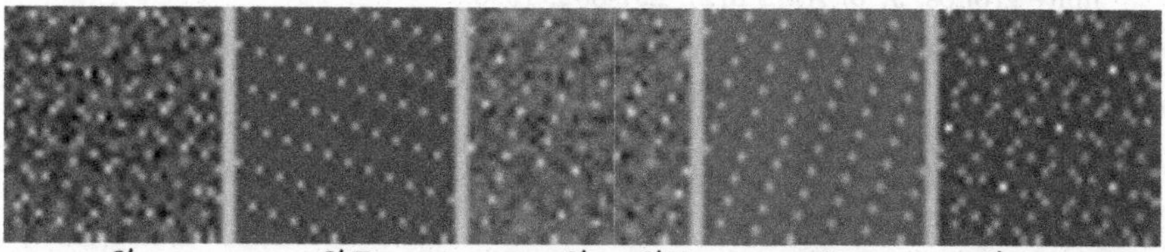

| **f**′ | **f**′P_6 | **g** = **f**′ − **f**′P_6 | **g**P_7 | **f**′P_6 + **g**P_7 |

Figure 40.   Results of filtering in noise

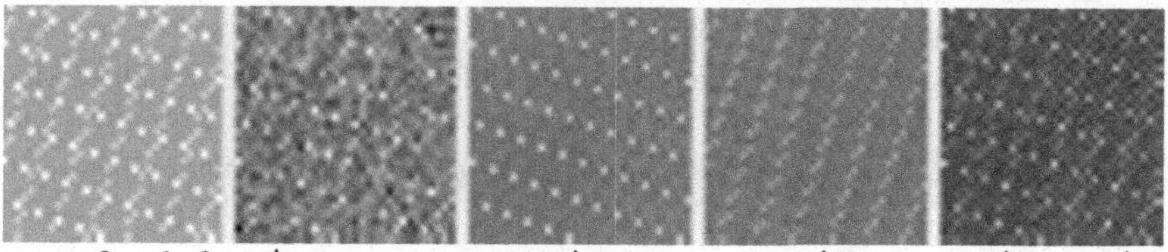

| **g** = **f** + k_b**f** | **g**′ = **g**+ noise | **g**′P_6 | **g**′P_7 | **g**′P_6 + **g**′P_7 |

The first image in Figure 40 is generated by adding to **f** its translation by k_b.

The results of applying P_6 and P_7 in noise are displayed.

Figure 41.   Results of filtering in noise, $N = 32$

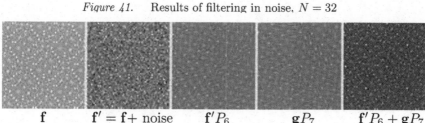

| **f** | **f**′ = **f**+ noise | **f**′P_6 | **g**P_7 | **f**′P_6 + **g**P_7 |

Acknowledgments

All the material presented in this paper has been reproduced from the text *Group Filters and Image Processing*, M. An and R. Tolimieri, Psypher Press, 2003, with explicit permissions from the authors and the publisher.

References

[1] M. An, R. Tolimieri, B. York, Technical report, AF #F49620-98-C-0043, May, 2002.

[2] P. Diaconis, D. Rockmore, Efficient computation of the Fourier transform on finite groups, *J. of AMS* **3**(2), 297-332, 1990.

[3] R. Foote, D. Healy, G. Mirchandani, T. Olson and D. Rockmore, A wreath product group approach to signal and image processing: Part I Multiresolution analysis, *IEEE Trans. in SP*, **48**(1), 2000, pp. 102–132.

[4] R. Holmes, Signal processing on finite groups, *Technical Report 873*, MIT Lincoln Laboratory, 1990.

[5] M. Karpovski and E. Trachtenberg, Filtering in a communication channel by Fourier transforms over finite groups, *Spectral techniques and fault detection*, M. Karpovski, ed., Academic Press, 1985.

[6] F.J. MacWilliams, Codes and ideals in group algebras, *Combinatorial mathematics and its applications*, 317-328, R.C. Bose and T.A. Dowlin, eds., University of North Carolina Press, 1969.

[7] D. Maslen and D. Rockmore, Generalized FFTs *DIMACS Ser. in Disc. Math. and Theor. Comp. Sci.*, **28**, 183-237, L. Finkelstein and W. Kantor, eds., 1997.

[8] R. Tolimieri, M. An, *Time-Frequency Representations*, Birkhauser, 1998.

A GEOMETRIC ALGEBRA APPROACH TO SOME PROBLEMS OF ROBOT VISION

Gerald Sommer
Institut für Informatik und Praktische Mathematik
Christian-Albrechts-Universität zu Kiel, Kiel, Germany
gs@ks.informatik.uni-kiel.de

Abstract   Geometric algebra has been proved to be a powerful mathematical language for robot vision. We give an overview of some research results in two different areas of robot vision. These are signal theory in the multidimensional case and knowledge based neural computing. In both areas a considerable extension of the modeling has been achieved by applying geometric algebra.

Keywords: Clifford algebra, Clifford spinor neuron, geometric algebra, hypersphere neuron, monogenic function, neural computing, Poisson scale-space, robot vision, signal theory.

1.   Introduction

In this tutorial paper we present a survey of some results contributed by the Kiel Cognitive Systems Group to the applications of geometric algebra in robot vision. Geometric algebra makes available a tremendous extension of modeling capabilities in comparison to the classical framework of vector algebra. It is our experience that application of geometric algebra should be strictly controlled by the geometric nature of the problem being considered. To demonstrate that tight relation between the problem at hand, its algebraic formulation and the way to find solutions will be our principle matter of concern. We i will do that by considering some key problems in robot vision. More details of the results reported here can be found in the papers and reports of the Kiel Cognitive Systems Group at our website (http://www.ks.informatik.uni-kiel.de). In addition, an extended version of this paper will be available [Sommer, 2003].

J. Byrnes (ed.) Computational Noncommutative Algebra and Applications, 309-338.
© *2004 Kluwer Academic Publishers. Printed in the Netherlands.*

1.1   Robot Vision

Robot vision is a demanding engineering discipline emerging from several contributing scientific disciplines. It aims at designing mobile technical systems which are able to take actions in its environment by using visual sensory information.

Although a matter of research for three decades, we are far away from having available seeing robots which are able to act comparable to humans in real world conditions. There are several reasons. First, many different disciplines such as image processing and signal theory, pattern recognition including learning theory, robotics, computer vision, and computing science are required for robot vision. Each of these have their own problems caused by limited modeling capabilities. Second, each of them has been developed to a great extent, isolated from the other ones, by using quite different mathematical languages of modeling. Thus, the fusion of all these disciplines within one framework is demanding by itself. Third, the most difficult problem is the design of the cognitive architecture. This concerns e.g. the gathering and use of world knowledge, controlling the interplay of perception and action, the representation of equivalence classes, invariants, and conceptions. Besides, such a system has to cope with hard real-time conditions.

The design of perception-action cycles cooperating and competing for solving a task is a demanding challenge. Of special interest is to enable the system to learn the required competence [Pauli, 2001] from experience. From a mathematical point of view the equivalence of visual and motor categories is remarkable. Both are mutually supporting [Sommer, 1997]. Of practical importance are the following two projections of a perception-action cycle. "Vision for action" means to control actions by vision and "action for vision" means the control of gaze for making vision easier.

1.2   Motivations of Geometric Algebra

Representing geometry in a general sense is a key problem of system design. But only those geometric aspects have to be represented which are of pragmatic importance. This opportunistic minimalism is tightly related to the so-called stratification of space as introduced into computer vision by Faugeras [Faugeras, 1995]. The system should be able to purposively switch from, e.g., metric to projective or kinematic aspects. We come back to that point in section 4 of [Sommer, 2003].

Another interesting topic is to ensure pragmatic completeness of the representations. But signal theory which supports image processing operations fails in that respect [Sommer, 1992, Zetzsche and Barth, 1990].

Local operations have to represent intrinsically multi-dimensional structure. The search of solutions for that problem was our original motivation for considering geometric algebra as a modeling language. We will survey our efforts in striving for a linear theory of intrinsically multi-dimensional signals in section 2.

Our aim of representing geometry in a general sense means thinking in a Kleinian sense, thus taking advantage of the tight relationship between geometry and algebra. All we have done in our work so far is based on choosing a useful geometry by embedding the problem into a certain geometric algebra. This is in fact a knowledge based approach to system design. In section 3 we will demonstrate this way of modeling in the context of neural computing. There we will profit from the chosen embedding because it results in a transformation of a non-linear problem to a linear one.

The problem of converting non-linear problems in Euclidean vector space to linear ones by embedding into a certain geometric algebra is related to another basic phenomenon which makes geometric algebra so useful in robot vision and beyond. From a geometrical point of view points are the basic geometric entities of a Euclidean vector space. Instead, a cognitive system is operating on geometric objects as a whole unique entity, e.g. a tea pot. An algebraic framework is wanted in which any object concept and transformations of it may be represented in a linear manner. Regrettably, this is an illusion. But geometric algebra enables the extension from point concepts to rather complex ones. In [Sommer, 2003] we demonstrate their linear construction and how to model their motion in a linear manner.

We will abstain from presenting a bird's eye view of geometric algebra in this contribution. Instead, we recommend the following introduction to geometric algebra [Hestenes et al., 2001]. In this paper we will use several geometric algebras as well as linear vector algebra. We will mostly write the product in the chosen algebra simply by juxtaposition of its factors. In some cases we will use a special notation for the chosen geometric product to emphasize its difference to a scalar product. Special products will be noted specially. We will use the notation $\mathbb{R}_{p,q,r}$ for the geometric algebra derived from the vector space $\mathbb{R}^{p,q,r}$ with $p+q+r = n$. These indices mark the signature of the vector space. Hence, (p, q, r) means we have $p/q/r$ basis vectors which square to $+1/-1/0$. If possible we reduce the set of signature indices, as in the case of the Euclidean vector space $\mathbb{R}^n$.

2.   Local Analysis of Multi-dimensional Signals

Image analysis is a fundamental part of robot vision. We do not understand image analysis as interpreting the whole visual data with respect to the complete scene or recognizing certain objects mapped to an image. This in fact is subject of computer vision. Instead, the aim of image analysis is to derive features from visual data for further processing and analysis. Its theoretical basis in called signal theory.

In (linear) signal theory we find the framework for handling linear shift invariant operators (LSI operators) and spectral representations of both signals and operators which are computed by Fourier transform. Both are tightly related. Although both are widely used in image analysis, there is a serious problem with respect of supporting recognition of intrinsically multi-dimensional and especially two-dimensional structure. This is a problem of incomplete representation.

We have to distinguish between two different conceptions of dimensionality of image data. The first one is the embedding dimension of data which is two in case of images and three in case of image sequences. The other one is the intrinsic dimension of local image structures. It expresses the number of degrees of freedom necessary to describe local structure. There exists structure of intrinsic dimensions zero (i0D) which is a constant signal, one (i1D) which are lines and edges and two (i2D) which are all the other possible patterns. As the meaning of "local" is depending on the considered scale, the intrinsic dimension at a chosen image position is scale-dependent too.

In this section we will show that by embedding image analysis into geometric algebra the mentioned representation gap hopefully will be closed.

2.1   Local Spectral Representations

Image analysis in robot vision means to derive locally structural hints by filtering. From the filter outputs certain useful features can be computed. A well known but naive way of filtering is template matching. Template matching filters are detectors for image structure. Because there is no way of designing templates for each possible structure, this method will get stuck soon if i2D patterns are of interest [Rohr, 1992]. Locally the only i1D patterns are lines, edges and textures constructed from these.

What filter would enable local image analysis if it is no template of the pattern in search? The alternative is to use a set of generic features as basis functions of patterns than the patterns themselves. We call this approach the split of identity.

There are many different approaches for computing the split of identity, including multi-scale analysis, local principal component analysis and variants of local Taylor series development. All of these take into consideration only the magnitude of the filters in the Fourier domain.

Our preferred approach in the split of identity is to instead use quadrature filters h_q [Granlund and Knutsson, 1995],

$$h_q = h_e + j h_o, \tag{1}$$

that is complex valued filters, where $\{h_e, h_o\}$ is a pair of real filters with even and odd symmetry, respectively. An example is the widely used Gabor filter [Granlund and Knutsson, 1995]. This kind of filtering applies evenness and oddness as a feature basis for image analysis.

Because these filters are in quadrature phase relation, that is, their phase is shifted by $-\frac{\pi}{2}$, $\{h_e, h_o\}$ is called a quadrature pair. Then convolution of the image f with such a filter h_α, $\alpha \in \{e, o\}$,

$$g_\alpha(\boldsymbol{x}) = (h_\alpha * f)(\boldsymbol{x}) \tag{2}$$

results in outputs g_e and g_o, respectively, which represent locally the even and odd symmetry of f. If h_q is a bandpass filter, then equation (2) enables the above mentioned multi-resolution analysis of local symmetry.

We restrict ourselves for the moment to 1D signals because for these the theory is well established.

Global even and odd symmetry is intrinsic to the Fourier representation of a real 1D signal. This property is called Hermite symmetry. But filtering a real function f with a quadrature filter h_q results in a complex valued function

$$g(x) = g_e(x) + j g_o(x) \tag{3}$$

no longer having Hermite symmetry in the Fourier domain. Instead, the Fourier transform $G(x)$ has the important property of approximating the analytic signal $F_A(u)$,

$$F_A(u) = F(u) + j F_H(u) = (1 + \mathrm{sign}(u)) F(u) \tag{4}$$

very well within the passband of the quadrature filter $H_q(u)$. The analytic signal in the Fourier domain is composed of the Fourier transform of the real signal $F(u)$ and the Hilbert transformed signal $F_H(u)$,

$$F_H(u) = H_I(u) F(u) = -j \,\mathrm{sign}(u) F(u). \tag{5}$$

In the spatial domain the analytic signal is

$$f_A(x) = f(x) + j f_H(x), \tag{6}$$

which is not only a complex function as in equation (4) but also has the property that f and f_H are phase shifted by $-\frac{\pi}{2}$. The functions f and f_H form a Hilbert pair. This is caused by the fact that the Hilbert transform H_I changes even signals to odd ones and vice versa. There are no amplitude changes because $|H_I| = 1$ for all frequencies. H_I, being pure imaginary, causes only phase effects. The operator of the Hilbert transform in the spatial domain reads

$$h_I(x) = \frac{1}{\pi x}. \tag{7}$$

The importance of the analytic signal results from computing the local energy $e(x)$ and the local phase $\phi(x)$,

$$e(x) = f^2(x) + f_H^2(x), \tag{8a}$$

$$\phi(x) = \arg f_A(x). \tag{8b}$$

Their use in local signal analysis decomposes the split of identity into quantitative and qualitative subtasks. If there exists a threshold ϵ of significance, then $e(x) > \epsilon$ indicates a certain local variance of data. In that case it is interesting to ask for the quality of the local structure. This is given by the local phase in the following way. An edge (odd symmetry) is indicated by $\phi(x) = \pm\frac{\pi}{2}$ and a line (even symmetry) is indicated by $\phi(x) = 0$ or $\phi(x) = \pi$. The same ideas can be used to interpret the output of a quadrature filter $g(x)$, given in equation (3).

Regrettably, this method of symmetry related split of identity which we outlined so far is not easy extendable to 2D functions, although it is used in image analysis for edge and line detection. In 2D signals it is limited to intrinsically one-dimensional functions. Hence, the aim is twofold: extension in a consistent manner to a 2D embedding and extension to intrinsically two-dimensional signals. We will present our first trials in the next subsection. These results are based on the PhD thesis of Bülow [Bülow, 1999].

2.2 Quaternionic Embedding of Line Symmetry

In this section we will first show that the algebraic nature of the Fourier transform in the complex domain causes a limited representation of symmetries for 2D functions. We propose embedding the Fourier transform into an algebraic domain with more degrees of freedom. Then we will use this idea to generalize the concept of the analytic signal for embedding dimension two to cover also the case of i2D signals.

2.2.1 Quaternionic Fourier Transform. We use the well-known technique of decomposing a 1D function $f(x)$ for any location x into an even part, $f_e(x)$, and an odd part, $f_o(x)$, according to $f(x) = f_e(x) + f_o(x)r$. Because the Fourier transform is a linear integral transform it maps this decomposition in an integral manner into the frequency domain. Let $F^c(u)$ be the complex valued Fourier transform, $F^c(u) \in \mathbb{C}$, of a one-dimensional real function $f(x)$. Then

$$F^c(u) = F_R(u) + jF_I(u) = F_e(u) + F_o(u) \tag{9}$$

as a result of all possible local decompositions $f(x) = f_e(x) + f_o(x)$. Thus, the complex Fourier transform makes explicit the only symmetry concept of a 1D function. In case of 2D functions $f(\boldsymbol{x})$, $\boldsymbol{x} = (x, y)$, the complex Fourier transform

$$F^c(\boldsymbol{u}) = F_R(\boldsymbol{u}) + jF_I(\boldsymbol{u}) , \quad \boldsymbol{u} = (u, v) \tag{10}$$

can also represent only two symmetry concepts. This contradicts the fact that in principle in two dimensions the number of different symmetries is infinite.

The 2D harmonic function

$$Q^c(\boldsymbol{u}, \boldsymbol{x}) = \exp(-j2\pi \boldsymbol{u} \cdot \boldsymbol{x}) \tag{11}$$

is the basis function of the 2D Fourier transform on a unit area

$$F^c(\boldsymbol{u}) = \iint f(\boldsymbol{x}) \, Q^c(\boldsymbol{u}, \boldsymbol{x}) \, d\boldsymbol{x}. \tag{12}$$

The 1D structure of $Q^c(\boldsymbol{u}, \boldsymbol{x})$ is the reason that in the complex Fourier domain there is no access to 2D symmetries. Looking at the decomposition

$$Q^c(\boldsymbol{u}, \boldsymbol{x}) = Q^c(u, x) \, Q^c(v, y) = \exp(-j2\pi ux) \exp(-j2\pi vy), \tag{13}$$

it is obvious that the basis functions represent a symmetry decomposition according to

$$F^c(\boldsymbol{u}) = F_{ee}(\boldsymbol{u}) + F_{oo}(\boldsymbol{u}) + F_{oe}(\boldsymbol{u}) + F_{eo}(\boldsymbol{u}) \tag{14}$$

and, thus, support in the spatial domain a decomposition

$$f(\boldsymbol{x}) = f_{ee}(\boldsymbol{x}) + f_{oo}(\boldsymbol{x}) + f_{oe}(\boldsymbol{x}) + f_{eo}(\boldsymbol{x}). \tag{15}$$

That is, the algebraic nature of equation (13) corresponds to considering products of symmetries with respect to the coordinate axes. We call

this line symmetry. But the limited degrees of freedom in the algebraic
structure of complex numbers results in a partial cover of symmetry with

$$F_R(\boldsymbol{u}) = F_{ee}(\boldsymbol{u}) + F_{oo}(\boldsymbol{u}), \quad F_I(\boldsymbol{u}) = F_{eo}(\boldsymbol{u}) + F_{oe}(\boldsymbol{u}). \tag{16}$$

Hence, although all symmetries according to equation (14) are contained
in the global spectral representation of the complex Fourier transform,
these are hidden because of equation (16) and cannot be made explicit.
Obviously this is a consequence of the complex algebra which causes the
equivalence of the basis functions according to (11) and (13), respectively. With respect to local spectral representation this results in the
incompleteness as discussed in subsection 2.1.

In [Zetzsche and Barth, 1990] Zetzsche and Barth argue from a dif-
fential geometry viewpoint that the responses of LSI filters should be
non-linearly combined to represent i2D structures. They developed in
the sequel a non-linear filter approach based on second order Volterra
series [Krieger and Zetzsche, 1996].

Instead, our approach is an algebraic extension of the degrees of freedom of a multi-dimensional Fourier transform by embedding the spectral
domain into a domain of Clifford numbers. In the case of embedding
a signal of dimension N in the spatial domain, the dimension of the
algebra has to be 2^N [Bülow and Sommer, 2001]. We call this kind
of Fourier transform a Clifford Fourier transform (CFT). The Clifford
Fourier transform has been modeled already by Brackx et al. in 1982
[Brackx et al., 1982].

In the case of signal dimension $N = 2$, the following isomorphisms
exist: $\mathbb{R}_{0,2} \simeq \mathbb{R}_{3,0}^+ \simeq \mathbb{H}$. Hence, for that case the quaternionic Fourier
transform (QFT), $F^q(\boldsymbol{u}) \in \mathbb{H}$,

$$F^q(\boldsymbol{u}) = \iint Q_i^q(u,x)\, f(\boldsymbol{x})\, Q_j^q(v,y)\, d\boldsymbol{x} \tag{17}$$

with the quaternion valued basis functions

$$Q_i^q(u,x) = \exp(-i2\pi ux)\,, \quad Q_j^q(v,y) = \exp(-j2\pi vy) \tag{18}$$

represents the symmetries of equation (14) totally uncovered in the
quaternionic domain. The algebraic decomposition of F^q is given by

$$F^q(\boldsymbol{u}) = F_R^q(\boldsymbol{u}) + iF_I^q(\boldsymbol{u}) + jF_J^q(\boldsymbol{u}) + kF_K^q(\boldsymbol{u}). \tag{19}$$

For real $f(\boldsymbol{x})$ this corresponds to the symmetry decomposition (same
order as in (19))

$$F^q(\boldsymbol{u}) = F_{ee}(\boldsymbol{u}) + F_{oe}(\boldsymbol{u}) + F_{eo}(\boldsymbol{u}) + F_{oo}(\boldsymbol{u}). \tag{20}$$

Instead of equation (13), the quaternionic basis functions

$$Q^q(\boldsymbol{u}, \boldsymbol{x}) = \exp(-i2\pi ux) \exp(-j2\pi vy) \tag{21}$$

cannot be fused as in equation (11). Those basis functions which are not positioned along the coordinate axes are indeed intrinsically 2D structures and, thus, are able to represent explicitly i2D signals.

The importance of Fourier phase as a carrier of geometric information is known, as well as the lack of reasonable phase concepts for 2D signals. The presented QFT overcomes this problem for the first time because of the representation

$$F^q(\boldsymbol{u}) = |F^q(\boldsymbol{u})| \exp(i\phi(\boldsymbol{u})) \exp(k\psi(\boldsymbol{u})) \exp(j\theta(\boldsymbol{u})) \tag{22}$$

of the QFT. Here the triple

$$(\phi, \theta, \psi) \in \left[-\pi, \pi\right[\times \left[-\frac{\pi}{2}, \frac{\pi}{2}\right[\times \left[-\frac{\pi}{4}, \frac{\pi}{4}\right]$$

of the quaternionic phase represents the 1D phases in axes directions (ϕ, θ) and the 2D phase (ψ), respectively.

2.2.2   Quaternionic Analytic Signal.   There are different approaches, with limited success, to generalizing the analytic signal in the complex domain for representation of i2D structure. This in effect means generalization of the Hilbert transform to the multi-dimensional case [Hahn, 1996]. The quaternionic analytic signal (QAS) presented in [Bülow and Sommer, 2001] takes advantage of the additional degrees of freedom in the quaternionic domain. In the spectral domain it is

$$F^q_A(\boldsymbol{u}) = (1 + \mathrm{sign}(u))(1 + \mathrm{sign}(v))F^q(\boldsymbol{u}) \tag{23}$$

and in the spatial domain

$$f^q_A(\boldsymbol{x}) = f(\boldsymbol{x}) + \boldsymbol{n}^{\mathrm{T}} \boldsymbol{f}^q_H(\boldsymbol{x}), \tag{24}$$

where $\boldsymbol{n} = (i, j, k)^{\mathrm{T}}$ is the vector of quaternionic imaginary units and $\boldsymbol{f}^q_H$ is the vector of the Hilbert transformed signal,

$$\boldsymbol{f}^q_H(\boldsymbol{x}) = f(\boldsymbol{x}) * \left(\frac{\delta(y)}{\pi x}, \frac{\delta(x)}{\pi y}, \frac{1}{\pi^2 xy}\right)^{\mathrm{T}}. \tag{25}$$

Regrettably, there is a mismatch of the chosen symmetry concept in the case of an isotropic pattern. Therefore, in the next subsection we demonstrate another way of modeling a generalization of the analytic signal.

2.3   Monogenic Embedding of Point Symmetry

The drawback of the QAS is the rigid coupling of the considered symmetry concept to the coordinate axes of the image signal. In case of deviations of the patterns from that model, e.g. in case of a rotated pattern, the quaternionic signal results in wrong local spectral representations. The alternative approach of generalizing the analytic signal which is outlined in this subsection has been developed in the PhD thesis of Felsberg [Felsberg, 2002].

The aim of this approach is to generalize the analytic signal in a rotation invariant manner. While in [Bülow et al., 2000] the identification of the isotropic multi-dimensional generalization of the Hilbert transform with the Riesz transform was presented for the first time, in [Felsberg and Sommer, 2000] the relation of the corresponding multi-dimensional generalization of the analytic signal to the monogenic functions of Clifford analysis [Stein and Weiss, 1971] has been published.

As in 2D (complex domain), in 4D (quaternion domain) no rotation invariant generalization of the Hilbert transform could be formulated. This is different for embedding a 2D function into a 3D space, or more generally an nD function into an (n+1)D space. The Riesz transform then generalizes the Hilbert transform for $n > 1$. We will give a short sketch of its derivation in the 2D case in the framework of Clifford analysis, which is an extension of the complex analysis of 1D functions to higher dimensions. Furthermore, we will discuss the extension of the analytic signal to the monogenic signal. The monogenic signal will find a natural embedding into a new scale space which is derived from the Poisson kernel. All these results are complete representations of i1D signals in 2D. We finally will present a first approach of an extension to i2D signals.

2.3.1   Solutions of the 3D Laplace Equation.

Complex analysis is mainly concerned with analytic or holomorphic functions. Such functions can be obtained by computing the holomorphic extension of a real 1D function. The resulting unique complex function over the complex plane fulfills the Cauchy-Riemann equations. There exists a mapping of a holomorphic function to a gradient field which fulfills the Dirac equation. Such a gradient field of a holomorphic function has zero divergence and zero curl everywhere and is, thus, the gradient field of a harmonic potential. The harmonic potential in turn fulfills the Laplace equation. What is known in signal theory as the Hilbert transform and analytic signal, respectively, corresponds to the 1D part of the mentioned holomorphic extension in the complex plane. For a real 2D function the corresponding framework is given in Clifford harmonic analysis [Stein and Weiss, 1971, Brackx et al., 1982].

Let $x \in \mathbb{R}^3$, $x = xe_1 + ye_2 + ze_3$, be a vector of the blades $\langle \mathbb{R}_3 \rangle_1$ of the Euclidean geometric algebra $\mathbb{R}_3$. Then a $\mathbb{R}_3$-valued analysis can be formulated, see also [Felsberg, 2002], for 3D vector fields $g(x), g = g_1 e_1 + g_2 e_2 + g_3 e_3$, which fulfill the 3D Dirac equation,

$$\nabla_3 g(x) = 0. \tag{26}$$

Here ∇_3 is the 3D nabla operator defined in $\mathbb{R}_3$. Then $g(x)$ is called a monogenic function and is a generalization of a holomorphic one. It can be shown that there exists a scalar valued function $p(x)$ called harmonic potential which is related to $g(x)$ by $g(x) = \nabla_3 p(x)$. Therefore, $g(x)$ is a gradient field called the harmonic potential field. The harmonic potential $p(x)$ fulfills the 3D Laplace equation,

$$\Delta_3 p(x) = \nabla_3 \nabla_3 p(x) = 0, \tag{27}$$

with Δ_3 as the 3D Laplace operator. If $g(x)$ is harmonic, its three components $g_i(x)$ represent a triple of harmonic conjugates. This is the important property we want to have to generalize a Hilbert pair of functions, see subsection 2.1.

The fundamental solution of the Laplace equation (27) is given by

$$p(x) = -\frac{1}{2\pi|x|} \tag{28a}$$

and its derivative,

$$g(x) = \frac{x}{2\pi|x|^3} \tag{28b}$$

is the harmonic field we started with.

Before having a closer look at the fundamental solution (28) of the Laplace equation and its derivative, we will relate it to the monogenic extension $f_m(x)$ of a real 2D function $f(x, y)$ in $\mathbb{R}_3$. The starting point is a usefull embedding of $f(x, y)$ into $\mathbb{R}_3$ as a vector field, respectively as an e_3-valued function,

$$f(x) = f(xe_1 + ye_2) = f(x, y)e_3. \tag{29}$$

Note that $z = 0$. Then the monogenic extension $f_m(x)$ of the real 2D function $f(x, y)$ can be computed by solving a boundary value problem in such a way that f_m is monogenic for $z > 0$ and its e_3-component corresponds to $f(x, y)$ for $z = 0$.
The boundary value problem reads

$$\Delta_3 p(x) = 0 \quad \text{for} \quad z > 0 \tag{30a}$$

$$e_3 \frac{\partial}{\partial z} p(\boldsymbol{x}) = f(x, y)e_3 \quad \text{for} \quad z = 0. \tag{30b}$$

Solving equation (30a) results in the fundamental solution (28) for $z > 0$. But the boundary equation (30b) introduces the contraint $g_3(x, y, 0) = f(x, y)$. Thus, the monogenic extension $\boldsymbol{f}_m(\boldsymbol{x})$ is a specific solution of (30) which in fact can be computed by convolving the function $f(x, y)$ with a set of operators derived from the fundamental solution.

The components of $\boldsymbol{g}(\boldsymbol{x})$ are derivations of $p(\boldsymbol{x})e_3$ when $z > 0$,

$$h_P(\boldsymbol{x}) = e_3 \frac{\partial}{\partial z} p(\boldsymbol{x})e_3 = \frac{z}{2\pi|\boldsymbol{x}|^3} \tag{31a}$$

$$h_{CPx}(\boldsymbol{x}) = e_1 \frac{\partial}{\partial x} p(\boldsymbol{x})e_3 = -\frac{x}{2\pi|\boldsymbol{x}|^3} e_{31} \tag{31b}$$

$$h_{CPy}(\boldsymbol{x}) = e_2 \frac{\partial}{\partial y} p(\boldsymbol{x})e_3 = \frac{y}{2\pi|\boldsymbol{x}|^3} e_{23}. \tag{31c}$$

These functions are well-known in Clifford analysis as the Poisson kernel (31a) and the conjugate Poisson kernels (31b and 31c), respectively. While $h_P = \boldsymbol{g} \cdot e_3$, the expressions (31b) and (31c) can be summarized by $\boldsymbol{h}_{CP} = h_{CPx} + h_{CPy} = \boldsymbol{g} \wedge e_3$,

$$\boldsymbol{h}_{CP}(\boldsymbol{x}) = \frac{(x e_1 + y e_2)e_3}{2\pi|\boldsymbol{x}|^3}. \tag{32}$$

Their Fourier transforms are

$$H_P(u, v, z) = \exp(-2\pi|ue_1 + ve_2|z) \tag{33a}$$

$$\boldsymbol{H}_{CP}(u, v, z) = \boldsymbol{H}_R(u, v) \exp(-2\pi|ue_1 + ve_2|z), \tag{33b}$$

where

$$\boldsymbol{H}_R(u, v) = -\left(\frac{ue_{31} - ve_{23}}{|ue_1 + ve_2|}\right)^* = \frac{ue_1 + ve_2}{|ue_1 + ve_2|} \boldsymbol{I}_2^{-1} \tag{34}$$

is the Riesz transform, which is the generalization of the Hilbert transform with respect to dimension. Note that in equation (34) $M^* = M\boldsymbol{I}_3^{-1}$ means the dual of the multivector M with respect to the pseudoscalar $\boldsymbol{I}_3$ of $\mathbb{R}_3$. Equations (31) formulate three harmonic conjugate functions which form a Riesz triple of operators in the half-space $z > 0$. Obviously the Riesz transform does not depend on the augmented coordinate

z. But equations (33) can be interpreted in such a way that z is a damping parameter in the frequency domain for all components of the Riesz triple. In fact, z must be interpreted as a scale parameter of a linear scale-space, called Poisson scale-space.

Here we must make a remark with respect to the Fourier transform used in this approach. Although the signal embedding is given in a 3D space, the Fourier transform is 2D. Actually we are using an isotropic transform to correspond to the model of point symmetry. Hence, for a given function $f(x)$ the complex valued isotropic Fourier transform is

$$F(u,v) = \iint f(x) \exp(-I_3 2\pi u \cdot x)\, dxdy. \tag{35}$$

Thus, the harmonics we consider are actually spherical ones. This important feature, together with the conception of point symmetry to realize the signal decomposition into even and odd components, overcomes the restrictions of the QFT of missing rotation invariance.

In the following we will focus on the result of the fundamental solution of the Laplace equation for $z = 0$. Besides, we will consider the scale as parameter of the representation, hence, $x = (x,y)$, and s is the scale parameter.

2.3.2 The Monogenic Signal.

Let f_M be the monogenic signal [Felsberg and Sommer, 2001] of a real 2D signal $f(x,y)$. In the plane $s = 0$ the monogenic signal f_M is composed by the e_3-valued representation $f(x) = f(x,y)e_3$ and its Riesz transform, $f_R(x)$,

$$f_M(x) = f(x) + f_R(x). \tag{36}$$

The monogenic signal in the frequency domain, $u = (u,v)$,

$$F_M(u) = F(u) + F_R(u), \tag{37}$$

is computed by

$$F_M(u) = (1 + H_R(u))\, F(u) = \left(1 + \frac{u}{|u|}I_2^{-1}\right) F(u). \tag{38}$$

In the spatial domain the Riesz transform results from convolving f with the Riesz transform kernel h_R,

$$h_R(x) = \frac{xe_3}{2\pi|x|^3}, \tag{39}$$

thus,

$$f_R(x) = \iint \frac{x'e_3}{2\pi|x'|^3} f(x - x')dx'dy'. \tag{40}$$

The Riesz transform generalizes all known Hilbert transform properties to 2D. In addition, the monogenic signal can be used to compute local spectral representations. Now the local magnitude is isotropic.

The local spectral decomposition of the monogenic signal in $\mathbb{R}^3$,

$$\boldsymbol{f}_M(\boldsymbol{x}) = |\boldsymbol{f}_M(\boldsymbol{x})| \exp(\arg(\boldsymbol{f}_M(\boldsymbol{x})e_3)) \tag{41}$$

is given by the real local amplitude

$$|\boldsymbol{f}_M(\boldsymbol{x})| = \exp(\log(|\boldsymbol{f}_M(\boldsymbol{x})|)) = \exp(\langle\log(\boldsymbol{f}_M(\boldsymbol{x})e_3)\rangle_0) \tag{42}$$

and the local rotation vector

$$r(\boldsymbol{x}) = \arg(\boldsymbol{f}_M(\boldsymbol{x}))^* = \langle\log(\boldsymbol{f}_M(\boldsymbol{x})e_3)\rangle_2^*, \tag{43}$$

see [Felsberg, 2002]. The local rotation vector $r(\boldsymbol{x})$ lies in the plane spanned by e_1 and e_2. Hence, it is orthogonal to the local amplitude vector $|\boldsymbol{f}_M(\boldsymbol{x})|e_3$. In the rotation plane the rotation vector is orthogonal to the plane spanned by $\boldsymbol{f}_M$ and e_3. The length of the rotation vector is coding the angle φ between $\boldsymbol{f}_M$ and e_3 which is the local phase of the 2D signal,

$$\varphi(\boldsymbol{x}) = \text{sign}(r \cdot e_1)|r|. \tag{44}$$

The orientation of the plane spanned by $\boldsymbol{f}_M$ and e_3 expresses the local orientation of the 2D signal in the image plane, $\theta(\boldsymbol{x})$.

Local phase and local orientation are orthogonal features expressing structural and geometric information in addition to the energy information represented by the local amplitude.

Hence, from the chosen 3D embedding of a 2D signal we obtain a more complex phase representation than in the 1D case. It includes both the local phase and the local orientation of a structure. But it is limited to the case of i1D-2D structures. Nevertheless, the result is a consistent signal theory for representing lines and edges in images which tells apart the difference between the embedding dimension and the intrinsic dimension of a signal. A practical consequence is that steering of the orientation in that approach is unnecessary for i1D signals.

The phase decomposition of a monogenic signal expresses symmetries of a local i1D structure embedded in $\mathbb{R}^3$. There is an interpretation of the local phase other than that given by a rotation vector of a $\mathbb{R}_3^+$-valued signal $\boldsymbol{f}_M e_3$ as in equation (43). Instead, the decomposition into local phase and orientation can be understood as specification of a rotation in a complex plane which is oriented in $\mathbb{R}^3$. In that case, for a given orientation θ, the well-known phase interpretation of a 1D signal (equation (8b) can be used. The embedding into $\mathbb{R}_3$ supplies the orientation of the complex plane.

2.3.3 The Poisson Scale-Space. So far we have discussed the derivation of local spectral representations, including local orientation, from the monogenic signal for the case of vanishing damping by the Poisson kernel (31a) and conjugate Poisson kernels (32), respectively. The same interpretation applies of course for all other scale parameters $s > 0$.

Because the Poisson kernel can also be derived for 1D signals by solving the Laplace equation in $\mathbb{R}_2$, there exists a similar scheme for that case [Felsberg, 2002].

The fact that $\{x; s\}$ with s being the Poisson scale parameter represents a linear scale-space [Felsberg, 2002] is surprising at first glance. There exist several axiomatics which define a set of features a scale-space should have. The first important axiomatic of scale-space theory proposed by Iijima [Iijima, 1959] excludes the existence of other scale-spaces than that one derived from solving the heat equation, that is the Gaussian scale-space. In [Felsberg and Sommer, 2003] the reason for that wrong conclusion could be identified.

As set out in subsection 2.1 the allpass characteristics of the Hilbert transform and also of the Riesz transform hinders their application in image analysis. Instead, there is a need for quadrature filters which are in fact bandpasses. By using the Poisson kernel and its conjugate counterparts (equations (33)), it is possible to design bandpasses by building the difference of Poisson filters [Felsberg, 2002]. These are abbreviated as DOP filters and $DOCP$ filters, respectively. In order to match symmetry they are either even (DOP) or odd ($DOCPs$). As h_P, h_{CPx} and h_{CPy} these three bandpasses also form a Riesz triple. The set of filters is called a spherical quadrature filter. Interestingly, $DOCP = DOCP_x + DOCP_y$ is an odd and isotropic operator which could not be designed in another way.

The Poisson scale-space is not only new, but its conception establishes a unique framework for performing phase-based image processing in scale-space. Hence, local phase and local amplitude become inherent features of a scale-space theory, in contrast to Gaussian scale-space [Felsberg and Sommer, 2003]. In [Felsberg and Sommer, 2003] there is a first study of the properties of the monogenic scale-space.

2.3.4 The Structure Multivector. The monogenic signal only copes with symmetries of i1D structure in a rotation invariant manner and enables one to estimate the orientation in addition to phase and amplitude. How can this approach based on solving the Laplace equation be extended to i2D structure? In a general sense the answer is open yet. This results from the fact that in 2D there exist infinitely many different symmetries. Because the monogenic signal is derived from first order harmonics as a transfer function of the Riesz transform, it follows that an increase of the order of harmonics to infinity would be necessary to cope with an arbitrary i2D structure. Hence, from this point of view we get stuck in a complexity problem similar to other knowledge-based approaches of designing pattern templates as filters.

Nevertheless, a first shot is given by the approach of the structure multivector [Felsberg and Sommer, 2002]. Here the first spherical harmonics of order zero to three are used to design a set of filters. This set implicitly assumes a model of two perpendicularly crossing i1D structures, thus representing in our approach a simple template of a special i2D structure [Felsberg, 2002].

Let h_S^i be the impulse response of a spherical harmonic filter of order i. Then $h_S^0(\boldsymbol{x}) \equiv \delta(\boldsymbol{x})$ and $h_S^1(\boldsymbol{x}) \equiv h_R(\boldsymbol{x})$. If $\boldsymbol{f}(\boldsymbol{x}) \in \mathbb{R}_3$, then

$$S(\boldsymbol{x}) = \boldsymbol{f}(\boldsymbol{x}) + (h_S^1 * \boldsymbol{f})(\boldsymbol{x}) + \boldsymbol{e}_3(h_S^2 * \boldsymbol{f})(\boldsymbol{x}) + \boldsymbol{e}_3(h_S^3 * \boldsymbol{f})(\boldsymbol{x}) \qquad (45)$$

is a mapping of the local structure to a 7-dimensional multivector,

$$S(\boldsymbol{x}) = s_0 + s_1\boldsymbol{e}_1 + s_2\boldsymbol{e}_2 + s_3\boldsymbol{e}_3 + s_{23}\boldsymbol{e}_{23} + s_{31}\boldsymbol{e}_{31} + s_{12}\boldsymbol{e}_{12}, \qquad (46)$$

called the structure multivector.

That response actually represents a special i2D generalization of the analytic signal. Hence, a split of identity of any i2D signal, projected to the model, can be realized in scale-space. The five independent features are local (main) orientation, two local i1D amplitudes and two local i1D phases. A major amplitude and a minor amplitude and their respective phases are distinguished. The occurrence of a minor amplitude indicates the i2D nature of the local pattern. For details the reader is referred to [Felsberg, 2002, Felsberg and Sommer, 2002].

The filter can be used to recognize both i1D and i2D structures, but in contrast to other filters which respond either to i1D or to i2D structure or mix the responses in an unspecific manner, this filter is specific to each type of structure.

3.   Knowledge Based Neural Computing

Learning the required competence in perception and action is an essential feature of designing robot vision systems within the perception-action cycle. Instead of explicitly formulating the solution, implicit representation of knowledge is used. This knowledge concerns e.g. equivalence classes of objects to be recognized, actions to be performed or visuo-motor mappings.

There are plenty of different neural architectures. But most of them have in common that they are general purpose or universal computing architectures. On the other hand we know that some architecture principles are more useful for a given task than others.

If the designer is aware of the specific features of his problem, he may integrate domain/task knowledge into the very architecture of neural computing. We will call this knowledge based neural computing (KBNC).

We will focus here on algebraic approaches which are tightly related to the geometry of data. Our choice is to take geometric algebra as an embedding framework of neural computing. We developed a general scheme of embedding neural computing into Clifford algebras in the sense of a knowledge based approach [Buchholz and Sommer, 2000b, Buchholz and Sommer, 2001b, Buchholz and Sommer, 2001a]. As an outcome we could propose several special neural processing schemes based on using geometric algebra [Banarer et al., 2003b, Buchholz and Sommer, 2000a, Buchholz and Sommer, 2000c]. In our approach we are capturing higher order information of the data within a single neuron by exploiting the special multivector structure of a chosen algebra. Because we get access to the chosen Clifford group, we are thus able to learn geometric transformation groups for the first time.

Neural learning of a model can be understood as an iterative nonlinear regression of a function to a set of training samples. By embedding the fitting of a model to given data into geometric algebra, we use algebraic knowledge on the nature of the problem to constrain the fit by the chosen algebraic rules. Furthermore, the chosen embedding causes a transformation of the function from a non-linear type in Euclidean space to a linear one in the special Clifford domain. This is a principle of minimal efforts which is called Occam's razor in statistical learning theory. Thus, learning the linear function in geometric algebra will be simpler than learning the non-linear one in vector algebra.

3.1 The Clifford Spinor Neuron

3.1.1 The Generic Neuron.
In this section we will restrict ourselves to the algebraic embedding of neurons of perceptron type arranged in feed-forward nets. Next we will summarize the basic ideas of computing with such neurons.

Let us start with a so-called generic neuron whose output, y, reads for a given input vector x and weight vector w, both of dimension n, as

$$y = g(f(x; w)). \tag{47}$$

The given neuron is defined by a propagation function $f: D^n \longrightarrow D$ and by an activation function $g: D \longrightarrow D'$. In case of a real neuron with

$D = \mathbb{R}$ and $\boldsymbol{w}, \boldsymbol{x} \in \mathbb{R}^n$, the propagation function

$$f(\boldsymbol{x}) = \sum_{i=1}^{n} w_i x_i + \theta, \tag{48}$$

where $\theta \in \mathbb{R}$ is a threshold, obviously computes the scalar product of $\boldsymbol{w}$ and $\boldsymbol{x}$. Because of the linear association of weight and input vectors, (48) is also called a linear associator. A neuron of that type with g being the identity operation is also called an adaline. By applying a non-linear activation function g to f, the neuron will become a non-linear computing unit called a perceptron. While an adaline may be interpreted as iteratively realizing a linear regression by learning, a trained perceptron enables the linear separation of two classes of samples. This classification performance results from the fact that the trained weight vector is perpendicular to a hyperplane in input space.

For the training of the generic neuron a supervised scheme can be used. That is, there is a teacher who knows the required answer, $r^i \in \mathbb{R}$, of the neuron to a given input $\boldsymbol{x}^i \in \mathbb{R}^n$. Hence, the training set is constituted by m pairs $(\boldsymbol{x}^i, r^i)$. Then the aim of learning is to find that weight vector $\boldsymbol{w}$ which minimizes the sum of squared error (SSE)

$$E = \frac{1}{2} \sum_{i=1}^{m} (r^i - y^i)^2. \tag{49}$$

This optimization is done by gradient descent because the weight correction at each step of the iterative procedure is given by

$$\Delta w_j = -\eta \frac{\partial E}{\partial w_j}, \tag{50}$$

where $\eta > 0$ is a suitable learning rate.

Because in a net of neurons the error has to be propagated back from the output to each neuron, this is also called back-propagation.

Finally, we will give a short sketch of combining several neurons to a (feed-forward) neural net. By arranging p (real) neurons in a single layer that is fully connected to the input vector $\boldsymbol{x}$, we will get a single layer perceptron network (SLP). If g is a non-linear function, the output vector $\boldsymbol{y}$ represents in each of its p components y_i a linear discriminent function in input space. By taking at least one of such layers of neurons and hiding it under an output layer, we will get a hierarchical architecture of neurons, which is called multilayer perceptron (MLP) architecture. The output layer is composed of at least one neuron which computes the superposition of the neurons of the preceding layer according to equation

(47). Hence, the MLP may be used either as a non-linear classification unit or for approximating any non-linear function.

If, on the other hand, g is the identity, the SLP computes an ordinary matrix multiplication,

$$y = Wx, \tag{51}$$

where W is the weight matrix containing the single weight vectors. If W is square ($x, y \in \mathbb{R}^n$), then it represents a linear transformation of the input vector.

3.1.2 The Clifford Neuron.

Now we will extend the model of a real valued generic neuron to that of a Clifford valued one. We will neglect for the moment the activation function. By replacing the scalar product of the linear associator by the geometric product of an algebra $\mathbb{R}_{p,q}$, $p + q = n$, we are embedding the neuron into $\mathbb{R}_{p,q}$ according to

$$f : \mathbb{R}_{p,q} \longrightarrow \mathbb{R}_{p,q}. \tag{52}$$

Hence, for $x, w, \Theta \in \mathbb{R}_{p,q}$ the propagation function is

$$f(x) = wx + \Theta \tag{53a}$$

or

$$f(x) = xw + \Theta, \tag{53b}$$

respectively. Note that the geometric product with respect to $\mathbb{R}_{p,q}$ now has to be computed. The splitting of the propagation function into the two variants of equation (53) follows obviously from the non-commutativity we have to assume for the geometric product of $\mathbb{R}_{p,q}$.

Having a training set $T := \{(x^1, r^1), \dots, (x^m, r^m) | x^i, r^i \in \mathbb{R}_{p,q}\}$, the weight corrections

$$\Delta w = \bar{x}^i (r^i - wx^i) \tag{54a}$$

for left-sided weight multiplication and

$$\Delta w = (r^i - x^i w)\bar{x}^i \tag{54b}$$

for right-sided weight multiplication enable a Clifford neuron to learn in a similar manner as a real one. Here $\bar{x}$ means the conjugate of x in the Clifford algebra. By taking this involution, the appearance of divisors of zero during the gradient descent is prevented [Buchholz and Sommer, 2001b, Buchholz and Sommer, 2001a].

What is the benefit derived from computing (53) instead of (48)? We give an illustrative example. Let $\boldsymbol{x} = x_1 + ix_2$ and $\boldsymbol{y} = y_1 + iy_2$ be two fixed complex numbers, $\boldsymbol{x}, \boldsymbol{y} \in \mathbb{C}$ with $\mathbb{C} \simeq \mathbb{R}_{0,1} \simeq \mathbb{R}_{2,0}^+$. The task of a complex neuron shall be learning the mapping $f \colon \boldsymbol{x} \longrightarrow \boldsymbol{y}$. This corresponds to learning a weight $\boldsymbol{w} \in \mathbb{C}$ so that $\boldsymbol{wx} = \boldsymbol{y}$, that is to learn a complex multiplication. This is in fact a simple task for the complex neuron. If instead real neurons should do the same, there is a need of a SLP with two neurons to compute y_1 and y_2, respectively. According to (51) the task is now to learn a weight matrix $W \in \mathbb{R}(2)$, which satisfies $W(x_1, x_2)^T = (y_1, y_2)^T$ with $\boldsymbol{x}, \boldsymbol{y} \in \mathbb{R}^2$. Here W represents a linear transformation of the vector $\boldsymbol{x}$. The SLP has to find out the constraints on W which correspond to the matrix representation of a complex number. These constraints are obviously $w_{11} = w_{22}$ and $w_{12} = -w_{21}$. As shown in [Buchholz and Sommer, 2001b] the SLP converges slower than the complex neuron. It is obviously better to use a model than to perform its simulation. Another advantage is that the complex neuron has half of the parameters (weights) to learn in comparison to the SLP.

Both of these observations can be generalized to any Clifford neuron in comparison to a corresponding SLP. Because of the $\mathbb{R}$-linearity of Clifford algebras [Porteous, 1995], any geometric product can be expressed as a special matrix multiplication. Hence, by choosing a certain Clifford algebra it is not only that a decision is made to use an algebraic model, but statistical learning will become a simpler task.

3.1.3   The Clifford Spinor Neuron.

There are additional advantages to using this approach. Next we want to extend our model. The above toy example gives us a hint. The SLP has to find out that W is indeed representing an orthogonal transformation. Because each special orthogonal matrix $W \in \mathbb{R}(2)$ is a rotation matrix, W rotates $\boldsymbol{x}$ to $\boldsymbol{y}$ in $\mathbb{R}^2$. Therefore, in the case of the complex neuron the equation $\boldsymbol{wx} = \boldsymbol{y}$ can be interpreted as mapping $\boldsymbol{x}$ to $\boldsymbol{y}$, $\boldsymbol{x}, \boldsymbol{y} \in \mathbb{C}$, by the complex number $\boldsymbol{w}$. But now $\boldsymbol{w}$ is no longer a point in the complex plane but the geometric product of $\boldsymbol{w}$ represents a special linear transformation, namely a rotation-dilation, see [Hestenes, 1993]. The representation of such an orthogonal operator in geometric algebra is given by the sum of a scalar and bivector components and is called a spinor, $\boldsymbol{S}$. In general, a spinor performs a dilation-rotation in the even subalgebra $\mathbb{R}_{p,q}^+$ of any geometric algebra $\mathbb{R}_{p,q}$.

Hence, the complex neuron learns a spinor operation, which is performing an orthogonal transformation in $\mathbb{C}$. But this is no longer true for our simple model of the Clifford neuron introduced in equation (53) if we choose vectors $x, y \in \mathbb{R}^3$. The spinors of $\mathbb{R}_3$ are quaternions because of the isomorphism $\mathbb{R}_3^+ \simeq \mathbb{H}$. To perform a rotation in $\mathbb{R}^3$, which maps vectors to vectors, the two-sided spinor product

$$\sigma(x) : x \longrightarrow Sx\widehat{S}^{-1} \tag{55}$$

has to be used in (53) instead. In this equation, the right half of the multiplication is performed with the inverse parity conjugated spinor, $\widehat{S}^{-1}$. The geometric product is now the quaternion product because of the isomorphism $\mathbb{R}_{3,0}^+ \simeq \mathbb{R}_{0,2}$. Instead of the vector $x \in \mathbb{R}^3$, we have to use its representation as a (pure) quaternion, $x \in \mathbb{R}_{0,2}$.

The greatest advantage of embedding a geometric problem into geometric algebra is to profit from such linear realization of group actions. Another advantage is that the group members, and the set of elements the group members are acting on, belong to the same algebra.

Because in our context the spinors are representing weights of the input of a neuron, we will continue using the notation w instead of S. The propagation function of a general Clifford spinor neuron, embedded into $\mathbb{R}_{p,q}^+$, is

$$f(x) = wx\widehat{w}^{-1} + \Theta. \tag{56}$$

Such a neuron computes an orthogonal transformation by using only one weight, w. Only in the case of a spinor product do we get the constraints necessary to use only one neuron with one weight for computing an orthogonal transformation. Obviously, the general Clifford spinor neuron has only half as many parameters (and half as many arithmetic operations) as the general Clifford neuron because only the even components of the weight multivector are used. But special care has to be taken to use the right coding of input and output data [Buchholz and Sommer, 2001b].

Now we will complete the neuron model by considering the activation function g, see equation (47). There is a need for generalizing an activation function on the real domain to the most popular sigmoid function

$$g_\beta(u) : \mathbb{R} \longrightarrow \mathbb{R}; \ u \longrightarrow (1 + \exp(-\beta u))^{-1} \tag{57}$$

on a Clifford valued domain. So far there is no easy way to formulate such a generalization [Georgiou and Koutsougeras, 1992]. Therefore, we use a component-wise activation function

$$g(u) = g([u]_i), \qquad i \in \{0, ..., 2^{n-1}\} \tag{58}$$

which is operating separately on all $2^n, n = p + q$, components of $\mathbb{R}_{p,q}$. This was first proposed by Arena et al. [Arena et al., 1997] for the quaternionic case.

The construction of a Clifford MLP (CMLP) or of a Clifford Spinor MLP (CSMLP) by following the principles of a real MLP is straightforward. To formulate a generalized backpropagation algorithm [Buchholz and Sommer, 2001a] is not a problem.

In principle real MLPs, CMLPs and CSMLPs have the same theoretical strength because all are universal approximators. Because they also use the same activation function, any potential advantage with respect to generalization of using the embedding of neural computation into geometric algebra should be based on the geometric product of the propagation function.

3.1.4   Learning a Euclidean 2D Similarity Transformation.

The task is to learn the plane Euclidean similarity transformation composed by a 2D rotation with $\varphi = -55^o$, a translation with $t = (+1, -0.8)$ and a dilation by the factor $\delta = 1.5$ [Buchholz and Sommer, 2001b].

Figure 1 shows both the training data (a) and test data (b), respectively. We applied a SLP with two real neurons and four weights, one complex neuron and one complex spinor neuron. While the complex neuron has two weights, the spinor neuron has only one. Figure 1c shows the convergence of the three computing units during training. As we see, the complex neuron converges faster than the real neurons. The reason for that behaviour has been discussed above. The spinor neuron needs more training steps for learning the constraints of the spinor product. But after 60 epochs its error is the lowest one of all three models. The spinor neuron learns a spinor representation, which is indicated by the successive convergence of the odd components of the propagation function to zero.

Generalization with respect to learning of a function means keeping the approximation error low if the data have not been seen during training. The behaviour of the neurons applied to the test data should be comparable to that for the training data. In the case of a KBNC approach, the learning of the model should also be robust with respect to distortions of the model in the data caused by noise. We overlayed both the training and test input data with additive median-free uniform noise up to a level of 20 percent. The mean square errors of the outputs are shown in figures 1d and 1e. Both the Clifford neuron and the Clifford spinor neuron are slightly distorted in learning the model. But their results with respect to the test data in comparison to training data indicate a good generalization. There is no significant difference in the behaviour of both Clifford neuron models with respect to generalization. This is what we expect in the case of complex algebra. The result for

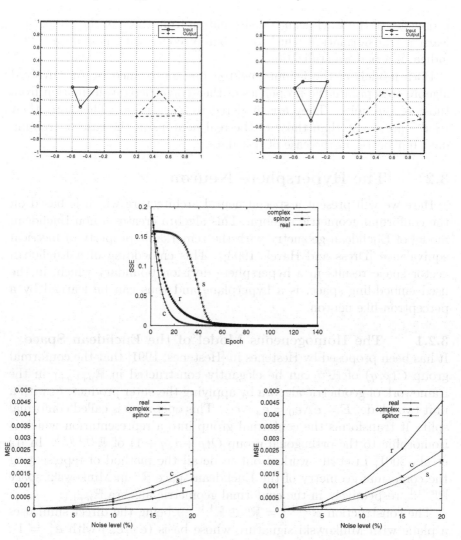

Figure 1.   Learning of a plane Euclidean similarity transformation. Upper row: left
(a): training data; right (b): test data; middle (c): convergence of the learning; lower
row: approximation error; left (d): training errors; right (e): test errors.

the Clifford neurons contrasts with the performance of the real neurons.
These are much less distorted during training but much more in the case
of test data in comparison to the knowledge based neurons.

Because real neurons are learning a general linear transformation, they
are able to learn noise better than the Clifford neurons. This causes
both the lower errors for training data and the higher ones for test data.
Because the real neurons have no constraints to separate the intrinsic

properties of the Euclidean transformation from those of the noise, they learn a transformation which only barely copes with new data. This indicates a bad generalization.

Both models of the Clifford neuron are constrained by the involved algebra to separate the properties of the geometric transformation from those of the noise. They are hindered to learn noise. Therefore, their results are worse than those of the real neurons in the case of training data but better in the case of test data.

3.2   The Hypersphere Neuron

Here we will present a special neural architecture which is based on the conformal geometric algebra. This algebra creates a non-Euclidean model of Euclidean geometry with the remarkable property of metrical equivalence [Dress and Havel, 1993]. This embedding of a Euclidean vector space results in a hypersphere decision boundary which, in the used embedding space, is a hyperplane and thus can be learned by a perceptron-like neuron.

3.2.1   The Homogeneous Model of the Euclidean Space.

It has been proposed by Hestenes in [Hestenes, 1991] that the conformal group $C(p,q)$ of $\mathbb{R}^{p,q}$ can be elegantly constructed in $\mathbb{R}_{p+1,q+1}$ in the framework of geometric algebra by applying the outer product, $\wedge$, with a unit two-blade, $\boldsymbol{E} := \boldsymbol{e} \wedge \boldsymbol{e}_0 = \boldsymbol{e}_+ \wedge \boldsymbol{e}_-$. This operation is called conformal split. It transforms the conformal group into a representation which is isomorphic to the orthogonal group $O(p+1, q+1)$ of $\mathbb{R}^{p+1,q+1}$. In [Li et al., 2001] Li et al. worked out in detail the method of representing the conformal geometry of the Euclidean space $\mathbb{R}^n$ in Minkowski space $\mathbb{R}^{n+1,1}$, respectively in the conformal geometric algebra $\mathbb{R}_{n+1,1}$.

The construction $\mathbb{R}^{n+1,1} = \mathbb{R}^n \oplus \mathbb{R}^{1,1}$, $\oplus$ being the direct sum, uses a plane with Minkowski signature whose basis $(\boldsymbol{e}_+, \boldsymbol{e}_-)$ with $e_+^2 = 1$, $e_-^2 = -1$ augments the Euclidean space $\mathbb{R}^n$ to realize a homogeneous stereographic projection of all points $\boldsymbol{x} \in \mathbb{R}^n$ to points $\underline{\boldsymbol{x}} \in \mathbb{R}^{n+1,1}$, see [Rosenhahn and Sommer, 2002]. By replacing the basis $(\boldsymbol{e}_+, \boldsymbol{e}_-)$ with the basis $(\boldsymbol{e}, \boldsymbol{e}_0)$ in the Minkowski plane, the homogeneous stereographic representation will become a representation as a space of null vectors. This is caused by the properties of the new basis vectors. These are related to the former ones by $\boldsymbol{e} = \boldsymbol{e}_- + \boldsymbol{e}_+$ and $\boldsymbol{e}_0 = \frac{1}{2}(\boldsymbol{e}_- - \boldsymbol{e}_+)$ with $e^2 = e_0^2 = 0$ and $\boldsymbol{e} \cdot \boldsymbol{e}_0 = -1$.

Any point $\boldsymbol{x} \in \mathbb{R}^{p,q}$ transforms to a point $\underline{\boldsymbol{x}} \in \mathbb{R}^{p+1,q+1}$,

$$\underline{\boldsymbol{x}} = \boldsymbol{x} + \frac{1}{2}\boldsymbol{x}^2\boldsymbol{e} + \boldsymbol{e}_0 \qquad (59)$$

with $\underline{x}^2 = 0$.

Hence, points of the Euclidean space $\mathbb{R}^3$ are represented by null vectors in the 5-dimensional space $\mathbb{R}^{4,1}$ with Minkowski signature. Actually, they are lying on a subspace of $\mathbb{R}^{4,1}$ called horosphere, N_e^3. The horosphere, which is a cut of the null cone with a hyperplane, see [Li et al., 2001], with the remarkable property of being a non-Euclidean model of Euclidean space, has been known for a long time, see [Yaglom, 1988]. But only in geometric algebra is there a practical and relevant approach to exploit this powerful non-Euclidean representation in engineering applications. The horosphere N_e^n is metrically equivalent to a Euclidean space $\mathbb{R}^n$. It is called the homogeneous model of Euclidean space, since points in $\mathbb{R}^n$ are represented in generalized homogeneous coordinates.

3.2.2 Construction and Properties of the Hypersphere Neuron.

Being metrically equivalent means that there exists a correspondence between the distance $d(x, y)$ in $\mathbb{R}^n$ and the distance $d(\underline{x}, \underline{y})$ on the horosphere N_e^n of $\mathbb{R}^{n+1,1}$. With $d(x, y) = |x - y| = \sqrt{(x - y)^2}$ this mapping reads $d(\underline{x}, \underline{y}) = -\frac{1}{2}d^2(x, y)$, see [Dress and Havel, 1993]. Given two points $x, y \in \mathbb{R}^n$, their distance $d(\underline{x}, \underline{y})$ is computed simply by the scalar product $\underline{x} \cdot \underline{y}$ in $\mathbb{R}^{n+1,1}$. Hence [Li et al., 2001],

$$\underline{x} \cdot \underline{y} = -\frac{1}{2}(x - y)^2. \tag{60}$$

Any point $\underline{c} \in N_e^n$ can be interpreted as a degenerate hypersphere $\underline{s}$ with center at $\underline{c}$ and radius $r \in \mathbb{R}$ equal to zero,

$$\underline{s} = c + \frac{1}{2}(c^2 - r^2)e + e_0. \tag{61}$$

A point $\underline{x}$ lies on a hypersphere if $|x - c| = |r|$. The general relation of a point $\underline{x}$ and a hypersphere $\underline{s}$ may be described by evaluating their distance

$$d(\underline{x}, \underline{s}) = \underline{x} \cdot \underline{s} = \underline{x} \cdot \underline{c} - \frac{1}{2}r^2 \underline{x} \cdot e = -\frac{1}{2}(x - c)^2 + \frac{1}{2}r^2. \tag{62}$$

Then the distance of a point, represented by the null vector $\underline{x}$, to a hypersphere $\underline{s} \in \mathbb{R}^{n+1,1}$ will be

$$d(\underline{x}, \underline{s}) : \begin{cases} > 0 \text{ if } \underline{x} \text{ is outside } \underline{s} \\ = 0 \text{ if } \underline{x} \text{ is on } \underline{s} \\ < 0 \text{ if } \underline{x} \text{ is inside } \underline{s} \end{cases} . \tag{63}$$

That distance measure is used for designing the propagation function of a hypersphere neuron [Banarer et al., 2003b]. If the parameters of the hypersphere are interpreted as the weights of a perceptron, then by embedding any data points into $\mathbb{R}^{n+1,1}$, the decision boundary of the perceptron will be a hypersphere. Because a vector $\underline{x} \in \mathbb{R}^{n+1,1}$ without an e_0−component represents a hypersphere with infinite radius, that is a hyperplane, the hypersphere neuron subsumes the classical perceptron as a special case.

The implementation of the hypersphere neuron uses the equivalence of the scalar products in $\mathbb{R}^{n+1,1}$ and $\mathbb{R}^{n+2}$. If $\underline{x}, \underline{y} \in \mathbb{R}^{n+1,1}$ and $\overline{x}, \overline{y} \in \mathbb{R}^{n+2}$, then $\underline{x} \cdot \underline{y} = \overline{x} \cdot \overline{y}$. Therefore, we use the following coding for a data vector $x \in \mathbb{R}^n$, $x = (x_1, ..., x_n)$, embedded in $\mathbb{R}^{n+2}$: $\overline{x} = (x_1, ..., x_n, -1, -\frac{1}{2}x^2)$. The coding of the hypersphere $\underline{s} \in \mathbb{R}^{n+1,1}$, represented in $\mathbb{R}^{n+2}$, is given by $\overline{s} = (c_1, ..., c_n, \frac{1}{2}(c^2 - r^2), 1)$. As a result of that embedding, a hypersphere in $\mathbb{R}^n$ is represented by a hyperplane in $\mathbb{R}^{n+2}$. This maps the hypersphere neuron to a perceptron with a second bias component.

In the above coding, the components of the hypersphere are considered as independent. This makes the hypersphere unnormalized and enables one control the assignment of the data either to the inner or to the outer class of a 2-class decision problem, see [Banarer et al., 2003b]. This is of special interest because the effective radius of the hypersphere will be influenced by the parameter β of the sigmoidal activation function, see equation (57), which is used to complete the neuron. If we remember the interpretation of points in $\mathbb{R}^{n+1,1}$ as a degenerate hypersphere, we are able to assign to the data points a confidence measure by extending the points to hyperspheres with imaginary radius. The confidence attributed to a data point is

$$\underline{x}_{CONF} = x + \frac{1}{2}(x^2 + r^2_{CONF})e + e_0. \tag{64}$$

From the scalar product between the hypersphere $\underline{s}$ and $\underline{x}_{CONF}$,

$$\underline{s} \cdot \underline{x}_{CONF} = \frac{1}{2}(r^2 - ((x - c)^2 + r^2_{CONF})), \tag{65}$$

follows a shift of the effective distance of the point to the hypersphere during training of the neuron. This leads to an adaption of the classification results to the confidence of the data.

3.2.3   The Performance of the Hypersphere Neuron.   The hypersphere neuron is an elemental computing unit of a hypersphere single-layer perceptron (SLHP) or multi-layer perceptron (MLHP), respectively. Its superior performance in comparison to the classical SLP or MLP, respectively, can be demonstrated with respect to several benchmark data sets [Banarer et al., 2003a].

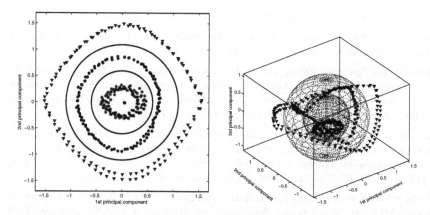

Figure 2. Classification of three toy objects. Visualization of the decision surfaces and data sets of the three first principal components.

In figure 2 we show an example from real world data. The problem is to recognize three different toy objects in a rotation invariant manner. For each of the 360 data sets (images with objects rotated each by 1°) a principal component analysis (PCA) was performed. Finally, the first three eigen-images of each class, each transformed to a data vector of length 1225, have been used for classification. Two neurons of a MLHP have been sufficient to separate the three classes. Obviously, the neuron model used is adequate for the problem.

Acknowledgments

The reported work is based on several research projects which were granted by DFG, EC and the German National Merit Foundation. Several PhD theses and diploma theses came out of that research. I have to thank all these young scientists for their enthusiasm in solving the problems I outlined in very limited space. Special thanks for the preparation of the figures to Sven Buchholz (1) and Vladimir Banarer (2).

References

[Arena et al., 1997] Arena, P., Fortuna, L., Museato, G., and Xibilia, M. (1997). Multilayer perceptrons to approximate quaternion valued functions. *Neural Networks*, 10(2):335–342.

[Banarer et al., 2003a] Banarer, V., Perwass, C., and Sommer, G. (2003a). Design of a multilayered feed-forward neural network using hypersphere neurons. In *Proc. Int. Conf. Computer Analysis of Images and Patterns, CAIP 2003, Groningen, August 2003.* accepted.

[Banarer et al., 2003b] Banarer, V., Perwass, C., and Sommer, G. (2003b). The hypersphere neuron. In *11th European Symposium on Artificial Neural Networks*,

ESANN 2003, Bruges, pages 469–474. d-side publications, Evere, Belgium.

[Brackx et al., 1982] Brackx, F., Delanghe, R., and Sommen, F. (1982). Clifford Analysis. Pitman Advanced Publ. Program, Boston.

[Buchholz and Sommer, 2000a] Buchholz, S. and Sommer, G. (2000a). A hyperbolic multilayer perceptron. In Amari, S., Giles, C., Gori, M., and Piuri, V., editors, International Joint Conference on Neural Networks, IJCNN 2000, Como, Italy, volume 2, pages 129–133. IEEE Computer Society Press.

[Buchholz and Sommer, 2000b] Buchholz, S. and Sommer, G. (2000b). Learning geometric transformations with Clifford neurons. In Sommer, G. and Zeevi, Y., editors, 2nd International Workshop on Algebraic Frames for the Perception-Action Cycle, AFPAC 2000, Kiel, volume 1888 of LNCS, pages 144–153. Springer-Verlag.

[Buchholz and Sommer, 2000c] Buchholz, S. and Sommer, G. (2000c). Quaternionic spinor MLP. In 8th European Symposium on Artificial Neural Networks, ESANN 2000, Bruges, pages 377–382.

[Buchholz and Sommer, 2001a] Buchholz, S. and Sommer, G. (2001a). Clifford algebra multilayer perceptrons. In Sommer, G., editor, Geometric Computing with Clifford Algebra, pages 315–334. Springer-Verlag, Heidelberg.

[Buchholz and Sommer, 2001b] Buchholz, S. and Sommer, G. (2001b). Introduction to neural computation in Clifford algebra. In Sommer, G., editor, Geometric Computing with Clifford Algebra, pages 291–314. Springer-Verlag, Heidelberg.

[Bülow, 1999] Bülow, T. (1999). Hypercomplex spectral signal representations for the processing and analysis of images. Technical Report Number 9903, Christian-Albrechts-Universität zu Kiel, Institut für Informatik und Praktische Mathematik.

[Bülow et al., 2000] Bülow, T., Pallek, D., and Sommer, G. (2000). Riesz transforms for the isotropic estimation of the local phase of Moiré interferograms. In Sommer, G., Krüger, N., and Perwass, C., editors, Mustererkennung 2000, pages 333–340. Springer-Verlag, Heidelberg.

[Bülow and Sommer, 2001] Bülow, T. and Sommer, G. (2001). Hypercomplex signals - a novel extension of the analytic signal to the multidimensional case. IEEE Transactions on Signal Processing, 49(11):2844–2852.

[Dress and Havel, 1993] Dress, A. and Havel, T. (1993). Distance geometry and geometric algebra. Foundations of Physics, 23(10):1357–1374.

[Faugeras, 1995] Faugeras, O. (1995). Stratification of three-dimensional vision: projective, affine and metric representations. Journal of the Optical Society of America, 12(3):465–484.

[Felsberg, 2002] Felsberg, M. (2002). Low-level image processing with the structure multivector. Technical Report Number 0203, Christian-Albrechts-Universität zu Kiel, Institut für Informatik und Praktische Mathematik.

[Felsberg and Sommer, 2000] Felsberg, M. and Sommer, G. (2000). The multidimensional isotropic generalization of quadrature filters in geometric algebra. In Sommer, G. and Zeevi, Y., editors, 2nd International Workshop on Algebraic Frames for the Perception-Action Cycle, AFPAC 2000, Kiel, volume 1888 of LNCS, pages 175–185. Springer-Verlag.

[Felsberg and Sommer, 2001] Felsberg, M. and Sommer, G. (2001). The monogenic signal. IEEE Transactions on Signal Processing, 49(12):3136–3144.

[Felsberg and Sommer, 2002] Felsberg, M. and Sommer, G. (2002). The structure multivector. In Dorst, L., Doran, C., and Lasenby, J., editors, *Applications of Geometric Algebra in Computer Science and Engineering*, pages 437–446. Proc. AGACSE 2001, Cambridge, UK, Birkhäuser Boston.

[Felsberg and Sommer, 2003] Felsberg, M. and Sommer, G. (2003). The monogenic scale-space: A unifying approach to phase-based image processing in scale-space. *Journal of Mathematical Imaging and vision*. accepted.

[Georgiou and Koutsougeras, 1992] Georgiou, G. and Koutsougeras, C. (1992). Complex domain back propagation. *IEEE Trans. Circ. and Syst. II*, 39:330–334.

[Granlund and Knutsson, 1995] Granlund, G. and Knutsson, H. (1995). *Signal Processing for Computer Vision*. Kluwer Academic Publ., Dordrecht.

[Hahn, 1996] Hahn, S. (1996). *Hilbert Transforms in Signal Processing*. Artech House, Boston, London.

[Hestenes, 1991] Hestenes, D. (1991). The design of linear algebra and geometry. *Acta Appl. Math.*, 23:65–93.

[Hestenes, 1993] Hestenes, D. (1993). *New Foundations for Classical Mechanics*. Kluwer Academic Publ., Dordrecht.

[Hestenes et al., 2001] Hestenes, D., Li, H., and Rockwood, A. (2001). New algebraic tools for classical geometry. In Sommer, G., editor, *Geometric Computing with Clifford Algebras*, pages 3–23. Springer-Verlag, Heidelberg.

[Iijima, 1959] Iijima, T. (1959). Basic theory of pattern observation (in japanese). In *Papers of Technical Group on Automation and Automatic Control*. IECE, Japan.

[Krieger and Zetzsche, 1996] Krieger, G. and Zetzsche, C. (1996). Nonlinear image operators for the evaluation of local intrinsic dimensionality. *IEEE Trans. Image Process.*, 5:1026–1042.

[Li et al., 2001] Li, H., Hestenes, D., and Rockwood, A. (2001). Generalized homogeneous coordinates for computational geometry. In Sommer, G., editor, *Geometric Computing with Clifford Algebras*, pages 27–59. Springer-Verlag, Heidelberg.

[Pauli, 2001] Pauli, J. (2001). *Learning-Based Robot Vision*, volume 2048 of *Lecture Notes in Computer Science*. Springer-Verlag, Heidelberg.

[Porteous, 1995] Porteous, I. (1995). *Clifford Algebras and the Classical Groups*. Cambridge University Press, Cambridge.

[Rohr, 1992] Rohr, K. (1992). Recognizing corners by fitting parametric models. *International Journal Computer Vision*, 9:213–230.

[Rosenhahn and Sommer, 2002] Rosenhahn, B. and Sommer, G. (2002). Pose estimation in conformal geometric algebra, part I: The stratification of mathematical spaces, part II: Real-time pose estimation using extended feature concepts. Technical Report Number 0206, Christian-Albrechts-Universität zu Kiel, Institut für Informatik und Praktische Mathematik.

[Sommer, 1992] Sommer, G. (1992). Signal theory and visual systems. In *Measurement 92*, pages 31–46. Slovac Acad. Science, Bratislava.

[Sommer, 1997] Sommer, G. (1997). Algebraic aspects of designing behavior based systems. In Sommer, G. and Koenderink, J., editors, *Algebraic Frames for the Perception and Action Cycle*, volume 1315 of *Lecture Notes in Computer Science*, pages 1–28. Proc. Int. Workshop AFPAC'97, Kiel, Springer-Verlag, Heidelberg.

[Sommer, 2003] Sommer, G. (2003). The geometric algebra approach to robot vision. Technical Report Number 0304, Christian-Albrechts-Universität zu Kiel, Institut für Informatik und Praktische Mathematik.

[Stein and Weiss, 1971] Stein, E. and Weiss, G. (1971). *Introduction to Fourier Analysis on Euclidean Spaces*. Princeton University Press, Princeton, N.J.

[Yaglom, 1988] Yaglom, M. (1988). *Felix Klein and Sophus Lie*. Birkhäuser, Boston.

[Zetzsche and Barth, 1990] Zetzsche, C. and Barth, E. (1990). Fundamental limits of linear filters in the visual processing of two-dimensional signals. *Vision Research*, 30:1111–1117.

GROUP THEORY IN RADAR AND SIGNAL PROCESSING

William Moran
Department of Electrical and Electronic Engineering
The University of Melbourne, Victoria 3010 Australia
b.moran@ee.mu.oz.au

Jonathan H. Manton
Department of Electrical and Electronic Engineering
The University of Melbourne, Victoria 3010 Australia
j.manton@ee.mu.oz.au

Keywords: ambiguity function, Heisenberg group, representation theory.

1. Introduction

This paper describes some key mathematical ideas in the theory of radar from a group theoretic perspective. The intention is to elucidate how radar theory motivates interesting ideas in representation theory and, conversely, how representation theory affords a better understanding of the inherent limitations of radar. Although most of the results presented here can be found in (Wilcox, 1991) and (Miller, 1991), there are significant differences in the selection and presentation of material. Moreover, compared with (Wilcox, 1991), (Miller, 1991) and (Moran, 2001), greater emphasis is placed here on the group theoretic approach, and in particular, its ability to arrive quickly and succinctly at basic results about radar.

Central to radar theory is the ambiguity function. Specifically, corresponding to any waveform $\mathbf{w}(t)$ is a two dimensional function $A_{\mathbf{w}}(t, f)$, called the ambiguity function, which measures the ability of that particular waveform to allow the radar system to estimate accurately the location and velocity of the target. Some waveforms perform better than others, and it is the challenge of radar engineers to design waveforms with desirable ambiguity functions while simultaneously meeting

J. Byrnes (ed.) Computational Noncommutative Algebra and Applications, 339-362.
© 2004 *Kluwer Academic Publishers. Printed in the Netherlands.*

the many other design criteria which impose constraints on the set of feasible waveforms.

A particularly elegant way of studying the ambiguity function is to write it in the form $A_{\mathbf{w}}(t,f) = \langle \mathbf{w}, \rho_{(t,f)}\mathbf{w} \rangle_{\mathsf{L}^2(\mathbf{R})}$ where $\rho_{(t,f)}$ is an operator acting on $\mathsf{L}^2(\mathbf{R})$. In fact, $\rho_{(t,f)}$ turns out to be a very special type of operator; it is an irreducible unitary multiplier representation of the additive group $\mathbf{R}^2$. It is here that group representation theory enters the picture. Functions $A : G \to \mathbf{C}$ of the form $A(g) = \langle \mathbf{w}, \rho_g \mathbf{v} \rangle$ over some group G, where ρ_g is a representation of G, are sometimes known as *special functions* in the literature. Importantly, most if not all interesting facts about special functions can be deduced from a study of the group representation ρ_g.

After defining the ambiguity function in Section 2, a brief introduction to representation theory is presented in Section 3. A feature of this presentation is that multiplier representations, along with their connections to ordinary representations and projective representations, are highlighted. Whereas it is customary to study the ambiguity function via the representation theory of the Heisenberg group, this paper studies instead the relevant multiplier representation of $\mathbf{R}^2$. Although both approaches are equivalent, the authors believe the multiplier representation approach is the more natural.

Section 4 derives fundamental facts about the multiplier representation theory of $\mathbf{R}^2$ and how it relates to the ambiguity function. A somewhat novel contribution is a generalised version of Moyal's identity (Theorem 2), whose proof has a more direct and intuitive flavour than that of previous proofs of Moyal's identity. Also covered in Section 4 are the various ways of realising the representation $\rho_{(t,f)}$ on different Hilbert spaces. These representations are equivalent to each other, but depending on the problem at hand, some spaces can be easier to work in than others.

The results in Section 4 are applied in Section 5 to answer several questions about ambiguity functions. A generating function approach is provided for finding explicit formulae for the ambiguity functions of Hermites. It is proved that the Hermite waveforms have the distinguishing property of having rotationally symmetric ambiguity functions. The potential benefits of hypothetically being able to use multiple waveforms are touched upon too.

Finally, Section 6 concludes by stating that the ambiguity function studied in this paper is the narrow band ambiguity function and is an approximation, albeit a good one in many situations, to what is known as the wide band ambiguity function. It is explained that the latter also can be studied from a group theoretic perspective, but that such a study is not undertaken here.

2. How a Radar Works

At the simplest level a radar transmits a waveform $\mathbf{w}(t)$ which is then reflected from the scene. In fact, the waveform is modulated onto (that is, multiplied by) a much higher (carrier) frequency sinusoid.

The reflection is picked up at the radar and processed to produce a picture of the scene. We assume for the purposes of this discussion that the radar is looking entirely in one direction. We can think of the scene for the moment as a collection of scatterers at various distances r_k and moving with velocities v_k relative to the radar. We measure distances in units of time — the time needed for light to travel that distance — and we measure the velocities in multiples of the speed of light. This prevents the proliferation of c's (the speed of light) in the formulae. Thus the scene can be regarded as a linear combination

$$\text{scene}(t) = \sum_k c_k\, \delta(t - r_k, v - v_k) \tag{1}$$

where $\delta(t,v)$ is the "delta" function — that is a point mass — at the origin of the "range-Doppler" (t,v) plane. The term c_k is the (complex) reflectivity of the scatterer.

Before proceeding further we remark that because radar (unlike conventional light based viewing systems) is able to keep track of the phase of the signal, or at least phase changes in it, processing in radar is done in the complex domain. A complex waveform can be modulated onto a carrier by using a phase shift to represent the argument of the waveform. The presence of the high frequency carrier and a stable oscillator producing the sinusoid means that the radar can detect phase shifts and is capable of producing a reasonably good approximation to the Hilbert transform of the return. As a result, radar engineers work in the complex domain and assume that their signals are complex, the argument corresponding to a phase shift in the carrier.

The signal returned to the receiver — and for simplicity we assume that the transmitter and receiver are collocated — is then the convolution of the transmit signal and the scene. That is,

$$\text{ret}(t) = \sum_k c_k \mathbf{w}(t - 2r_k)e^{4\pi i v_k f_c t}, \tag{2}$$

where f_c is the carrier frequency. The 2's have appeared because we are considering the round-trip of the signal rather than the one-way trip. The return is then (after stripping off the high frequency carrier) correlated with another (or the same) waveform $\mathbf{v}(t)$. The result is

$$
\begin{aligned}
\text{proc}(t) &= \int \text{ret}(\tau)\overline{\mathbf{v}(\tau - t)}\, d\tau \\
&= \sum_k c_k \int \mathbf{w}(\tau - 2r_k)e^{4\pi i v_k f_c \tau}\overline{\mathbf{v}(\tau - t)}\, d\tau,
\end{aligned}
\tag{3}
$$

which is a linear combination over the individual scatterers, as one would expect. The expression

$$
\int \mathbf{w}(\tau - 2r)e^{4\pi i v f_c \tau}\overline{\mathbf{v}(\tau - t)}\, d\tau
\tag{4}
$$

tells all about the way in which the radar behaves. Once we know this function and the scene, we can reconstruct the return. Of course this is a gross simplification but for the purposes of this paper it will be enough. It is important to mention that the key issue for the radar engineer is to do the inverse problem: given the return, how to extract the scene.

Briefly, we mention that there is scope for extraction of information from multiple waveforms rather than the single waveforms discussed here. Even this situation can be treated as that of one very long pulse and the expression (4) continues to be relevant.

By a change of variable and a renormalization of the units of measurement of the velocity, equation (4) becomes

$$
A_{\mathbf{w},\mathbf{v}}(t, f) = \int \mathbf{w}(\tau)e^{2\pi i f \tau}\overline{\mathbf{v}(\tau - t)}\, d\tau.
\tag{5}
$$

This is the *radar ambiguity function*, or at least one form of it, and is the focus of the study of this paper.

At this point we impose some mathematical constraints on the situation. The functions $\mathbf{w}$ and $\mathbf{v}$ are always assumed to be "finite energy" signals; that is, they belong to $L^2(\mathbf{R})$. It is clear then that the integrand is integrable and indeed, by the Cauchy-Schwarz Inequality, that

$$
|A_{\mathbf{w},\mathbf{v}}(t, f)| \leq \|\mathbf{w}\|_{L^2(\mathbf{R})}\|\mathbf{v}\|_{L^2(\mathbf{R})}, \qquad (t, f) \in \mathbf{R}^2,
\tag{6}
$$

for all $\mathbf{v}, \mathbf{w} \in L^2(\mathbf{R})$. The expression

$$
\rho_{(t,f)}(\mathbf{v})(\tau) = e^{-2\pi i f \tau}\mathbf{v}(\tau - t)
\tag{7}
$$

is clearly significant in the theory of the ambiguity function. We note that $\rho_{(t,f)}$ is a unitary operator in $L^2(\mathbf{R})$, and that

$$
A_{\mathbf{w},\mathbf{v}}(t, f) = \left\langle \mathbf{w}, \rho_{(t,f)}\mathbf{v} \right\rangle_{L^2(\mathbf{R})}.
\tag{8}
$$

Moreover

$$\rho_{(t_1,f_1)}\rho_{(t_2,f_2)}(\mathbf{v})(\tau) = e^{-2\pi i f_1 \tau}\rho_{(t_2,f_2)}(\mathbf{v})(\tau - t_1)$$
$$= e^{-2\pi i f_1 \tau}e^{-2\pi i f_2(\tau-t_1)}(\mathbf{v})(\tau - t_1 - t_2) \qquad (9)$$
$$= e^{2\pi i f_2 t_1}\rho_{(t_1+t_2,f_1+f_2)}(\mathbf{v})(\tau),$$

that is,

$$\rho_{(t_1,f_1)}\rho_{(t_2,f_2)} = e^{2\pi i f_2 t_1}\rho_{(t_1+t_2,f_1+f_2)}. \qquad (10)$$

This makes ρ a *multiplier representation* of $\mathbf{R}^2$ with *multiplier*

$$\sigma((t_1, f_1),(t_2, f_2)) = e^{-2\pi i t_1 f_2} \qquad (11)$$

because

$$\rho_{(t_1+t_2,f_1+f_2)} = \sigma((t_1, f_1),(t_2, f_2))\, \rho_{(t_1,f_1)}\rho_{(t_2,f_2)}. \qquad (12)$$

Our aim is to make sense of these ideas and examine their consequences.

3.   Representations

Since representations, or at least multiplier representations, play a role in the radar ambiguity function, we spend some time discussing them; first ordinary representations and then multiplier representations. The abstract theory of representations of groups goes as follows. Let G be a locally compact group. Such a group has a Haar measure m_G, which is invariant under left translation:

$$\forall h \in G, \quad \int_G F(hg)\, dm_G(g) = \int_G F(g)\, dm_G(g), \qquad (13)$$

for any integrable function F on G.

A (unitary) *representation* of G is a continuous homomorphism

$$\pi : G \to \mathcal{U}(\mathfrak{H}) \qquad (14)$$

into the unitary group of a Hilbert space $\mathfrak{H}$. It will be convenient to write the image of $g \in G$ under this map as π_g instead of the more conventional $\pi(g)$. Continuity means, in this case, that the maps

$$g \to \langle \zeta, \pi_g \xi \rangle \qquad (15)$$

are continuous for all $\zeta, \xi \in \mathfrak{H}$. The theory of unitary representations of locally compact groups, and of Lie groups in particular, is extensive. Here we focus on those (small) parts we need for the development of the theory of the radar ambiguity function.

Two representations π and θ of G on the Hilbert spaces $\mathfrak{H}$ and $\mathfrak{K}$ are *equivalent* if there is an isometry $V : \mathfrak{H} \to \mathfrak{K}$ such that

$$V\pi_g = \theta_g V \qquad (g \in G). \tag{16}$$

In pictorial form, π and θ are equivalent if there exists a V such that

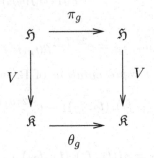

commutes. Without the isometry constraint, the operator V is called an *intertwining operator* of π with θ. If every intertwining operator of θ with itself is a scalar multiple of the identity, θ is said to be *irreducible*. This is equivalent to there being no non-trivial subspace of $\mathfrak{K}$ invariant under the action of all of the operators θ_g ($g \in G$). The connection between invariant subspaces and intertwining operators (via the spectral theorem) which makes these two definitions equivalent is called *Schur's Lemma.*

When the group G is compact every representation is a direct sum of irreducible representations and this decomposition is unique up to equivalence of representations. This is the content of the Peter-Weyl Theorem. Unfortunately, the groups of special interest in radar are not compact and it is not true for these groups that every representation is a direct sum of irreducibles. In any case, every representation is a direct integral of irreducibles; see (Mackey, 1976). In general this decomposition is not unique but, for the groups of major interest to us, uniqueness (almost everywhere) holds.

As we have already seen, it will be necessary to consider multiplier representations as well as ordinary representations. For this, we need a *multiplier*. This is a Borel map $\sigma : G \times G \to \mathbf{T}$, where $\mathbf{T}$ is the group under multiplication of complex numbers of absolute value 1, that satisfies the *cocycle* condition

$$\sigma(g_1, g_2)\sigma(g_1 g_2, g_3) = \sigma(g_1, g_2 g_3)\sigma(g_2, g_3) \qquad (g_1, g_2, g_3 \in G), \tag{17}$$

to which we add the normalization $\sigma(1, 1) = 1$. Then a multiplier representation is a map $\rho : G \to \mathcal{U}(\mathfrak{H})$ that satisfies

$$\rho_{g_1 g_2} = \sigma(g_1, g_2)\rho_{g_1}\rho_{g_2} \qquad (g_1, g_2 \in G), \tag{18}$$

and that is measurable in the same sense that an ordinary representation is continuous. We remark that while the multipliers and multiplier representations considered in this paper are all continuous, it is important for the development of the general theory to permit multipliers and representations that are only Borel. The definitions of equivalence and irreducibility for multiplier representations are unchanged from the ordinary representation case. The *projective unitary group* is

$$P(\mathfrak{H}) = \mathcal{U}(\mathfrak{H})/\{c\mathbf{I} : c \in \mathbf{T}\}. \tag{19}$$

Multiplier representations give rise to *projective representations* (that is, continuous homomorphisms from G to $P(\mathfrak{H})$) via composition with the quotient map $p : \mathcal{U}(\mathfrak{H}) \to P(\mathfrak{H})$. Conversely, since there is a Borel cross section $\eta : P(\mathfrak{H}) \to \mathcal{U}(\mathfrak{H})$ (that is, a right inverse of p) any projective representation gives rise to a multiplier representation. We note, however, that there are many such cross sections, and so there are many multiplier representations giving rise to the same projective representation — indeed different multipliers are involved in general.

The multipliers on a group G form an abelian group under multiplication. A multiplier (or *cocycle*) σ is a *coboundary* if there is a Borel map $\phi : G \to \mathbf{T}$ such that

$$\sigma(g_1, g_2) = \frac{\phi(g_1)\phi(g_2)}{\phi(g_1 g_2)}. \tag{20}$$

Coboundaries form a subgroup of the group of cocycles of G and two elements of the same coset are said to be *cohomologous*. Cohomologous multipliers have essentially the same representation theory. If σ_1 and σ_2 are cohomologous:

$$\sigma_1(g_1, g_2) = \frac{\phi(g_1)\phi(g_2)}{\phi(g_1 g_2)} \sigma_2(g_1, g_2), \tag{21}$$

and ρ is a σ_1-representation, then a simple check reveals that

$$g \to \phi(g)\rho_g \tag{22}$$

is a σ_2-representation.

As we have stated, a projective representation $\kappa : G \to P(\mathfrak{H})$ lifts to different multiplier representations using different Borel cross-sections of the quotient map p. However different lifts just produce cohomologous multipliers. If they are then made representations for the same multiplier using the trick described in equation (22), then the two representations are equivalent. Thus the theory of multiplier representations is essentially the theory of projective representations.

One final issue on the general theory of representations needs to be addressed before we return to radar theory. This is that projective/multiplier representations are really ordinary representations but for a different group. Before justifying that we point out that if projective representations did not exist (as the radar ambiguity function demonstrates they should), then we would have to invent them. There is a remarkable and beautiful theory for constructing representations of groups called "the Mackey analysis"; see (Mackey, 1976). To give a detailed exposition of this would take us well beyond the focus of this paper. We merely mention that, in order to use the Mackey analysis to construct *ordinary* representations of a group G, it is necessary to consider projective representations of subgroups of G. On the other hand, if we start trying to construct projective representations of G using the extension of the Mackey analysis to the projective case, then we still only have to consider projective representations of subgroups. Projective representations form a natural completion of the class of ordinary representations.

Now let ρ be a σ-representation of a group G. We show how to make ρ into an ordinary representation of a larger group. We can form a new group $\widetilde{G}$ whose elements are pairs $[g, z]$, where $g \in G$ and $z \in \mathbf{T}$, and with multiplication given by

$$[g_1, z_1][g_2, z_2] = [g_1 g_2, z_1 z_2 \sigma(g_1, g_2)]. \tag{23}$$

The identity element is $(1_G, 1)$ and the inverse is given by

$$[g, z]^{-1} = [g^{-1}, \overline{z\sigma(g, g^{-1})}]. \tag{24}$$

Note that a consequence of the normalization $\sigma(1, 1) = 1$ is that

$$\sigma(g, 1) = \sigma(1, g) = 1 \qquad (g \in G). \tag{25}$$

The group $\widetilde{G}$ is called a *central extension* of the group G by the circle group $\mathbf{T}$. By this we mean that $\mathbf{T}$ (as the subgroup $\{[0, z] : z \in \mathbf{T}\}$) is a subgroup of the centre of $\widetilde{G}$ and the quotient $\widetilde{G}/\mathbf{T}$ is isomorphic to G. Central extensions are classified by the group of cocycles modulo coboundaries. For details we refer the reader to (Maclane, 1975).

Thus, for any σ-representation ρ of G,

$$\pi([g, z]) = \overline{z}\rho_g \tag{26}$$

is an ordinary representation of $\widetilde{G}$. There is an exact correspondence between the σ-representation theory of G and the ordinary representation theory of those representations of $\widetilde{G}$ that restrict on the central subgroup $\mathbf{T}$ to be the homomorphism $[g, z] \to \overline{z}\mathbf{I}$.

4.   Representations and Radar

We return to the radar ambiguity function and how representation theory impinges on it. As we have seen in Section 2, the radar ambiguity function is expressible in the form

$$A_{\mathbf{w},\mathbf{v}}(t, f) = \left\langle \mathbf{w}, \rho_{(t,f)}\mathbf{v} \right\rangle, \tag{27}$$

where $(t, f) \to \rho_{(t,f)}$ is a σ-representation of $\mathbf{R}^2$ and

$$\sigma\big((t, f), (t', f')\big) = e^{-2\pi i f' t}. \tag{28}$$

We explore some of the properties of this represention over the next few subsections.

4.1   Basic Properties of the Ambiguity Function

We can quickly deduce four fairly straightforward properties:

Amb1 For all $\mathbf{u}, \mathbf{v} \in \mathsf{L}^2(\mathbf{R})$, $A_{\mathbf{u},\mathbf{v}}$ is in $\mathsf{L}^2(\mathbf{R}^2)$ and $\|A_{\mathbf{u},\mathbf{v}}\|_{\mathsf{L}^2(\mathbf{R}^2)} \leq \|\mathbf{u}\|_{\mathsf{L}^2(\mathbf{R})}\|\mathbf{v}\|_{\mathsf{L}^2(\mathbf{R})}$.

Amb2 The map $(\mathbf{u}, \mathbf{v}) \to A_{\mathbf{u},\mathbf{v}}$ is conjugate bilinear from $\mathsf{L}^2(\mathbf{R}) \times \mathsf{L}^2(\mathbf{R})$ to $\mathsf{L}^2(\mathbf{R}^2)$.

Amb3 $A_{\mathbf{u},\mathbf{v}}(t, f) = e^{2\pi i f t}\overline{A_{\mathbf{v},\mathbf{u}}(-t, -f)}$.

Amb4 For $\mathbf{u}, \mathbf{v} \in \mathsf{L}^2(\mathbf{R})$,

$$A_{\mathcal{F}\mathbf{u},\mathcal{F}\mathbf{v}}(t, f) = e^{2\pi i f t}A_{\mathbf{u},\mathbf{v}}(f, -t). \tag{29}$$

Here, $\mathcal{F}$ denotes the Fourier transform.

PROOF: The proof of **Amb1** will be dealt with as part of a more specific theorem in Section 4.3. The conjugate bilinearity is clear, hence **Amb2** follows from **Amb1**. The proof of **Amb3** is a simple calculation. First notice that

$$\rho^*_{(t,f)} = \rho^{-1}_{(t,f)} = e^{2\pi i f t}\rho_{(-t,-f)}, \tag{30}$$

in view of (10). It follows that

$$\begin{aligned}
A_{\mathbf{v},\mathbf{u}}(-t, -f) &= \langle \mathbf{v}, \rho_{(-t,-f)}\mathbf{u} \rangle \\
&= \langle \rho^*_{(-t,-f)}\mathbf{v}, \mathbf{u} \rangle \\
&= e^{2\pi i f t}\langle \rho_{(t,f)}\mathbf{v}, \mathbf{u} \rangle \\
&= e^{2\pi i f t}\overline{A_{\mathbf{u},\mathbf{v}}(t, f)}.
\end{aligned} \tag{31}$$

To deal with **Amb4**, we note first that, by the Plancherel theorem,

$$\langle \mathbf{u}, \mathbf{v} \rangle = \langle \mathcal{F}\mathbf{u}, \mathcal{F}\mathbf{v} \rangle \qquad (\mathbf{u}, \mathbf{v} \in L^2(\mathbf{R})). \tag{32}$$

Further, it is easy to check that

$$\mathcal{F}\left[\rho_{(t,f)}\mathbf{w}\right] = e^{-2\pi i f t} \rho_{(-f,t)} \mathcal{F}[\mathbf{w}]. \tag{33}$$

Now it follows that

$$\begin{aligned} A_{\mathcal{F}\mathbf{u},\mathcal{F}\mathbf{v}}(t,f) &= \langle \mathcal{F}\mathbf{u}, \rho_{(t,f)}\mathcal{F}\mathbf{v} \rangle \\ &= \langle \mathcal{F}\mathbf{u}, e^{-2\pi i f t} \mathcal{F}[\rho_{(f,-t)}\mathbf{v}] \rangle \\ &= e^{2\pi i f t} \langle \mathbf{u}, \rho_{(f,-t)}\mathbf{v} \rangle \\ &= e^{2\pi i f t} A_{\mathbf{u},\mathbf{v}}(f,-t). \end{aligned} \tag{34}$$

$\square$

It is remarked that **Amb3** is a particular case of a more general formula applying to any σ-representation ρ:

$$\langle \mathbf{u}, \rho_g \mathbf{v} \rangle = \sigma(g, g^{-1}) \overline{\langle \mathbf{v}, \rho_{g^{-1}} \mathbf{u} \rangle} \qquad (g \in G, \ \mathbf{u}, \mathbf{v} \in L^2(\mathbf{R})). \tag{35}$$

4.2 Irreducibility

The following result is really one of the two most important results in the theory of the radar ambiguity function. The other is Theorem 2.

THEOREM 1 *The σ-representation ρ is irreducible and, moreover, it is the unique irreducible σ-representation of $\mathbf{R}^2$ up to equivalence.*

PROOF: The proof of the uniqueness of ρ would take us too far from the key aim of this paper. It is called the *Stone-von Neumann Theorem* and requires some deep ideas in representations of $\mathbf{R}$. Even the proof of irreducibility relies on some results in harmonic analysis that go well beyond the scope of this paper. To prove irreducibility we use Schur's Lemma. Suppose that $B : L^2(\mathbf{R}) \to L^2(\mathbf{R})$ is an intertwining operator of ρ with itself. Then, in particular, it commutes with the translation operators

$$\rho_{(t,0)}(\mathbf{v})(\tau) = \mathbf{v}(\tau - t). \tag{36}$$

Such an object B is known as a *pseudo-measure*. The only fact we need about pseudo-measures is that they have a Fourier transform $\widehat{B}(f)$ such that

$$\widehat{B(\mathbf{v})}(f) = \widehat{B}(f)\widehat{\mathbf{v}}(f) \qquad (f \in \mathbf{R}), \tag{37}$$

and that this Fourier transform specifies a pseudo-measure uniquely. We refer the interested reader to (Katznelson, 1968) for details. Now we note that

$$\widehat{\rho_{(0,f')}(\mathbf{v})}(f) = \widehat{\mathbf{v}}(f + f'). \tag{38}$$

Since B must commute with these operators too, its Fourier transform is translation invariant and so must be constant. It follows that B itself is a scalar multiple of the identity and so ρ is irreducible. □

4.3 Moyal's Identity

Here we establish a result that is one of the key special features of the representation ρ and the ambiguity function. Before we state the result, we establish some terminology and notation. The Hilbert space tensor product of two Hilbert spaces $\mathfrak{H}$ and $\mathfrak{K}$ is the completion of the linear space of all finite formal sums

$$\sum_k \mathbf{u}_k \otimes \mathbf{v}_k, \tag{39}$$

where $\mathbf{u}_k$ and $\mathbf{v}_k$ are in $\mathfrak{H}$ and $\mathfrak{K}$ respectively. The completion is with respect to the norm obtained from the inner product

$$\left\langle \sum_k \mathbf{u}_k^{(1)} \otimes \mathbf{v}_k^{(1)}, \sum_{k'} \mathbf{u}_{k'}^{(2)} \otimes \mathbf{v}_{k'}^{(2)} \right\rangle = \sum_{k,k'} \langle \mathbf{u}_k^{(1)}, \mathbf{u}_{k'}^{(2)} \rangle_{\mathfrak{H}} \langle \mathbf{v}_k^{(1)}, \mathbf{v}_{k'}^{(2)} \rangle_{\mathfrak{H}}. \tag{40}$$

We assume there is a conjugate linear isometry $J : \mathfrak{K} \to \mathfrak{K}$ (in fact this will be conjugation in $\mathsf{L}^2(\mathbf{R})$ in our context). Evidently, given any conjugate bilinear map $B : \mathfrak{H} \times \mathfrak{K} \to \mathfrak{L}$ that satisfies

$$\left\| \sum_k B(\mathbf{u}_k, \mathbf{v}_k) \right\|_{\mathfrak{L}} \leq \left\| \sum_k \mathbf{u}_k \otimes \mathbf{v}_k \right\|_{\mathfrak{H} \otimes \mathfrak{K}} \tag{41}$$

for any finite set $\{(\mathbf{u}_k, \mathbf{v}_k)\}$ of elements of $\mathfrak{H} \times \mathfrak{K}$, there is a continuous linear map $B^{\otimes} : \mathfrak{H} \otimes \mathfrak{K} \to \mathfrak{L}$ such that the diagram

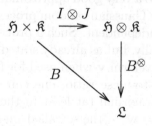

extends to a continuous linear map from $\mathfrak{H} \otimes \mathfrak{K} \to \mathfrak{L}$. We note that the tensor product of $\mathsf{L}^2(\mathbf{R})$ with itself is just $\mathsf{L}^2(\mathbf{R}^2)$. The following theorem is in essence Moyal's identity.

THEOREM 2 *The map* $(\mathbf{u}, J\mathbf{v}) \to A_{\mathbf{u},\mathbf{v}}$ *extends to an isometry from* $\mathsf{L}^2(\mathbf{R}^2)$ *to itself.*

PROOF: It is easy to see that this map (for finite sums of simple tensors in the first instance) is just

$$\Phi(F)(t, f) = \int F(\tau, \tau - t)e^{2\pi i f \tau} \, d\tau, \qquad (42)$$

which, since we know that the Hilbert space tensor product of $L^2(\mathbf{R})$ with itself is just $L^2(\mathbf{R}^2)$, clearly satisfies the condition (41). If we write $W : L^2(\mathbf{R}^2) \to L^2(\mathbf{R}^2)$ for the map

$$W(F)(\tau, t) = F(\tau, \tau - t), \qquad (43)$$

it is clear that it is an isometry of $L^2(\mathbf{R}^2)$ and that Φ is just $(I \otimes \mathcal{F}) \circ W$ where $\mathcal{F}$ is the one dimensional Fourier transform. The result is now clear.   □

COROLLARY 3 *For* $\mathbf{u}_1, \mathbf{v}_1, \mathbf{u}_2, \mathbf{v}_2 \in L^2(\mathbf{R})$,

$$\langle A_{\mathbf{u}_1,\mathbf{v}_1}, A_{\mathbf{u}_2,\mathbf{v}_2} \rangle_{L^2(\mathbf{R}^2)} = \langle \mathbf{u}_1, \mathbf{u}_2 \rangle_{L^2(\mathbf{R})} \langle \mathbf{v}_1, \mathbf{v}_2 \rangle_{L^2(\mathbf{R})}. \qquad (44)$$

It is equation (44) that is commonly referred to as *Moyal's Identity*. This has in turn another corollary.

COROLLARY 4 *For* $\mathbf{u}, \mathbf{v} \in L^2(\mathbf{R})$,

$$\|A_{\mathbf{v},\mathbf{v}}\|_{L^2(\mathbf{R}^2)} = \|\mathbf{v}\|^2. \qquad (45)$$

This is a form of the *Heisenberg uncertainty principle*. To explain its significance, we need to make a few remarks about the way radars normally operate. When the return comes back into the receiver from a scene, noise is added by the receiver. This is just thermal noise in the electronic components. To a very good approximation this is *white Gaussian noise* — a stationary Gaussian random process in which restrictions to disjoint intervals are independent. Such a process (with finite energy) cannot exist mathematically, but as already stated, this is an approximation. If one asks what waveform $\mathbf{v}$ when used for filtering in the receiver will maximize the signal-to-noise ratio, one can show that the correct choice is $\mathbf{v} = \mathbf{w}$. So it is normal (at least in theoretical discussions of radar) to assume that $\mathbf{v} = \mathbf{w}$ — the so-called *matched filter*.

We note that by (3) the processed return is just the convolution of the range-Doppler scene with the ambiguity. Evidently then, so that we can extract the scene from the return, we would like $A_{\mathbf{v},\mathbf{v}}$ (or $A_{\mathbf{v},\mathbf{w}}$) to be a delta function or close to it. Moyal's identity makes this impossible. Clearly the value of $A_{\mathbf{v},\mathbf{v}}$ at the origin is just $\|\mathbf{v}\|^2$ and this is also its L^2 norm, so that it must have considerable spread.

As we shall see later, an important collection of waveforms (in theory but not in practice) are the Hermite functions. In fact it will be convenient for us to use a renormalized form of them. We write

$$\mathbf{v}_n(t) = \pi^{-1/4}(2^n n!)^{-1/2}(-1)^n e^{-t^2/2} H_n(t), \qquad (46)$$

where H_n is the nth Hermite polynomial, defined implicitly by the generating equation

$$\sum_n \frac{1}{n!} H_n(t) z^n = e^{-z^2 + 2zt}. \qquad (47)$$

The $\mathbf{v}_n$ form an orthonormal basis of $\mathsf{L}^2(\mathbf{R})$. It follows from Moyal's identity that the collection $\{A_{\mathbf{v}_n, \mathbf{v}_m}\}_{n,m}$ therefore forms an orthonormal basis of $\mathsf{L}^2(\mathbf{R}^2)$.

4.4 The Symmetric Ambiguity Function

In practical radar systems, the location (t, f) of a target is usually estimated by maximising the magnitude of the inner product between the received waveform $\mathbf{v}(\tau)$ and the adjusted version $\rho_{(t,f)}(\mathbf{w})(\tau)$ of the transmitted waveform $\mathbf{w}(\tau)$. Here, $\rho_{(t,f)}(\mathbf{w})(\tau)$ corresponds to what would have been received in the ideal case had there been a single target at "distance" t moving at "velocity" f. Therefore, it is only the magnitude of the ambiguity function, and not its phase, which provides useful information about the performance of the system.

Let $\phi : \mathbf{R}^2 \to \mathbf{T}$ be an arbitrary Borel map. Then the magnitude of $\langle \mathbf{v}, \phi(t, f) \rho_{(t,f)} \mathbf{w} \rangle$ equals the magnitude of the ambiguity function $A_{\mathbf{v}, \mathbf{w}}(t, f)$. Moreover, the results of Section 3 imply that $\rho_{(t,f)}^{(\nu)} = \phi(t, f) \rho_{(t,f)}$ is actually a ν-representation of $\mathbf{R}^2$ for some multiplier ν cohomologous to σ. Therefore, what is of interest is not the properties of the particular ambiguity function $A_{\mathbf{v}, \mathbf{w}}(t, f)$, but rather the properties shared by all functions of the form $\langle \mathbf{v}, \rho_{(t,f)}^{(\nu)} \mathbf{w} \rangle$. Since multiplier representation theory does not distinguish between cohomologous multipliers, this shows that *radar theory really is concerned with the σ-representation theory of $\mathbf{R}^2$, no more and no less.*

Replacing σ by a cohomologous multiplier ν changes (the phase but not the magnitude of) the corresponding ambiguity function. Being able to work with different ambiguity functions can simplify calculations. The symmetric ambiguity function is now derived by considering the multiplier ν defined by

$$\nu\big((t, f), (t', f')\big) = e^{-\pi i (f't - t'f)}. \qquad (48)$$

This is cohomologous to σ because

$$\frac{\nu((t,f),(t',f'))}{\sigma((t,f),(t',f'))} = e^{\pi i(t'f+f't)} = \frac{e^{-\pi itf}e^{-\pi it'f'}}{e^{-\pi i(t+t')(f+f')}}. \tag{49}$$

In fact, one can show that up to coboundaries and automorphisms of $\mathbf{R}^2$, this is the only cocycle on $\mathbf{R}^2$. Switching to this multiplier from σ, we obtain an alternative formula for ρ, this time as a ν-representation. From the formulae (22) and (49), we obtain

$$\rho^{(\nu)}_{(t,f)}(\mathbf{v})(\tau) = e^{\pi ift}\rho_{(t,f)}(\mathbf{v})(\tau) = e^{\pi if(t-2\tau)}\mathbf{v}(\tau-t), \qquad (t,f) \in \mathbf{R}^2. \tag{50}$$

This results in the following form for the radar ambiguity function

$$A^{(\nu)}_{\mathbf{u},\mathbf{v}}(t,f) = \int_{\mathbf{R}} \mathbf{u}(\tau + \frac{t}{2})\mathbf{v}(\tau - \frac{t}{2})e^{2\pi if\tau}\, d\tau. \tag{51}$$

This is referred to as the *symmetric* form of the ambiguity function.

Note that because $\nu((t,f),(-t,-f)) = 1$, we obtain from (35) the tidier form

$$A^{(\nu)}_{\mathbf{u},\mathbf{v}}(t,f) = \overline{A_{\mathbf{v},\mathbf{u}}(-t,-f)} \tag{52}$$

of **Amb3**. In a similar vein, the new form of **Amb4** is

$$A^{(\nu)}_{\mathcal{F}\mathbf{u},\mathcal{F}\mathbf{v}}(t,f) = A^{(\nu)}_{\mathbf{u},\mathbf{v}}(f,-t). \tag{53}$$

4.5   The Heisenberg Group

Using the ideas of Section 3 and the cocycle σ corresponding to ρ, it is possible to form a central extension of $\mathbf{R}^2$ by $\mathbf{T}$. The result is a group G_0 whose centre Z_0 is (in this case) isomorphic to $\mathbf{T}$ and for which G_0/Z_0 is isomorphic to $\mathbf{R}^2$. It is customary instead to use the $\mathbf{R}$-valued cocycle $\sigma_{\mathbf{R}}$ given by

$$\sigma_{\mathbf{R}}((t,f),(t',f')) = f't, \tag{54}$$

so that $\sigma = e^{-2\pi i\sigma_{\mathbf{R}}}$. The result of applying the trick of equation (23) in this context produces a central extension of $\mathbf{R}^2$ by $\mathbf{R}$, which is called the (3-dimensional) *Heisenberg group*

$$\mathcal{H} = \begin{pmatrix} 1 & t & z \\ 0 & 1 & f \\ 0 & 0 & 1 \end{pmatrix}. \tag{55}$$

The corresponding representation of $\mathcal{H}$ is

$$\tilde{\rho}(t,f,z)(\mathbf{v})(\tau) = e^{2\pi iz}e^{-2\pi if\tau}\mathbf{v}(\tau-t). \tag{56}$$

There is another extension (but isomorphic *qua* central extensions) formed using ν, but we refrain from giving the details. In fact, we continue to present the theory of the radar ambiguity function in terms of multiplier representations of $\mathbf{R}^2$ rather than the slightly more customary ordinary representation theory of the Heisenberg group.

4.6 The Bargmann-Segal Representation

While the Stone-von Neumann Theorem tells us that there is only one irreducible σ-representation of $\mathbf{R}^2$, or equivalently that there is only one irreducible representation of the Heisenberg group that restricts to the homomorphism $z \rightarrow e^{2\pi i z}$ in the centre of $\mathcal{H}$, there are many equivalent ways of seeing this representation. The way we have discussed so far is called the *Schroedinger representation* of $\mathcal{H}$. Here and in the following section we present two other ways of looking at this representation.

It is shown in (Wilcox, 1991) that the Hermite waveforms $\mathbf{v}_n(t)$, defined in (46), have ambiguity functions whose peak at the origin is as sharp as possible (in a sense made precise there). The Hermite waveforms are also the eigenfunctions of a particular linear operator defined in Section 4.8. Since the Hermites appear to play a fundamental role, it is desirable to construct a σ-representation on a new Hilbert space $\mathfrak{F}$ such that the waveforms corresponding to the Hermites are as simple as possible. Such a representation is stated below.

Consider the space $\mathfrak{F}$ of entire functions $\mathbf{a}(z) = \sum_n a_n z^n$ satisfying $\sum_n n! |a_n|^2 < \infty$. This is a Hilbert space with the inner product

$$\langle \mathbf{a}, \mathbf{b} \rangle = \sum_n n! \, a_n \overline{b_n}. \tag{57}$$

This is actually a *reproducing kernel* Hilbert space, in that there exist elements $\mathbf{e}_u$, for any $u \in \mathbf{C}$, such that

$$\langle \mathbf{a}, \mathbf{e}_u \rangle = \mathbf{a}(u), \qquad \mathbf{a} \in \mathfrak{F}. \tag{58}$$

Furthermore, the vectors

$$\mathbf{j}_n(z) = \frac{z^n}{\sqrt{n!}}, \qquad n = 0, 1, \cdots \tag{59}$$

form an orthonormal basis for $\mathfrak{F}$.

We can explicitly write down an isometry $\Psi : L^2(\mathbf{R}) \to \mathfrak{F}$ which maps $\mathbf{v}_n(t)$ to $\mathbf{j}_n(z)$, namely

$$\Psi(\mathbf{u})(z) = \sum_{n=0}^{\infty} \langle \mathbf{u}, \mathbf{v}_n \rangle \mathbf{j}_n(z) \tag{60}$$

$$= \int_{\mathbf{R}} \frac{1}{\pi^{1/4}} e^{-(z^2+t^2)/2 - \sqrt{2}zt} u(t)\, dt.$$

Here, the second line is obtained with the help of (47).

The isometry Ψ maps the representation ρ to the equivalent representation $\rho^{\mathrm{BS}} = \Psi \circ \rho \circ \Psi^{-1}$ on $\mathfrak{F}$. This is called the *Bargmann-Segal σ-representation* of $\mathbf{R}^2$. It is given explicitly by

$$\rho^{\mathrm{BS}}_{(t,f)}(\mathbf{a})(z) = e^{-\frac{t^2 + 4\pi^2 f^2}{4} + \frac{z}{\sqrt{2}}(2\pi i f - t) - \pi i f t}\, \mathbf{a}\left(z + \frac{t + 2\pi i f}{\sqrt{2}}\right). \tag{61}$$

The derivation of (61) is postponed until Section 4.8.

4.7   The Lattice Representation

The third equivalent version of the representation ρ is the so-called *lattice representation*. It arises as an induced σ-representation from the lattice subgroup $\mathbf{Z}^2$ of $\mathbf{R}^2$ or equivalently as an induced ordinary representation from the corresponding subgroup of the Heisenberg group $\mathcal{H}$. The Hilbert space here is the space of all functions $\mathbf{r} : \mathbf{R}^2 \to \mathbf{C}$ satisfying

$$\mathbf{r}(x+n, y+m) = e^{-2\pi i n y} \mathbf{r}(x, y), \tag{62}$$

and which are square integrable in the sense that

$$\int_0^1 \int_0^1 |\mathbf{r}(x,y)|^2\, dx\, dy < \infty. \tag{63}$$

Note that, in view of (62), the integrand is periodic. This space has the obvious inner product. Now we can define a σ-representation ρ^{L} of $\mathbf{R}^2$ by

$$\rho^{\mathrm{L}}_{(t,f)}(\mathbf{r})(x,y) = e^{2\pi i f x} \mathbf{r}(x+t, y+f). \tag{64}$$

It is not hard to see that this is a σ-representation. To prove irreducibility we show directly that it is equivalent to ρ. The intertwining operator is known as the *Weil-Brezin-Zaks* transform and is given by

$$Z[\mathbf{u}](x,y) = \sum_{k=-\infty}^{\infty} e^{-2\pi i k y} \mathbf{u}(k - x) \qquad (\mathbf{u} \in L^2(\mathbf{R})). \tag{65}$$

It is straightforward to check $Z[\mathbf{u}]$ satisfies both (62) and (63). Its inverse is given by

$$Z^{-1}[\mathbf{r}](\tau) = \int_0^1 \mathbf{r}(-\tau, y) \, dy. \qquad (66)$$

To check the intertwining property, we observe

$$Z[\rho_{(t,f)}(\mathbf{u})](x,y) = e^{2\pi i f x} \sum_{k=-\infty}^{\infty} e^{-2\pi i k(f+y)} \mathbf{u}(k - x - t) \qquad (67)$$
$$= \rho_{(t,f)}^{L}(Z[\mathbf{u}])(x,y).$$

Note that this forces Z to be a scalar multiple of an isomorphism and it is straightforward to see from this that it is indeed an isomorphism.

Although not discussed here, there are links between the lattice representation and number theory. For instance, the Jacobi theta function appears in the expression for $Z[\mathbf{v}_0](x,y)$, where $\mathbf{v}_0$ is the zeroth Hermite waveform defined in (46).

4.8 The Lie Algebra Representation

Recall that the σ-representation ρ extends to a representation on the Heisenberg group $\mathcal{H}$. Since $\mathcal{H}$ is a Lie group, the representation ρ induces a Lie algebra representation on the Lie algebra $\mathfrak{h}$ associated with $\mathcal{H}$. The Hermite polynomials naturally arise in this framework as the eigenvectors of a particular linear operator.

The Heisenberg group has a Lie algebra $\mathfrak{h}$ which, as an additive group, is $\mathbf{R}^3$ and has generators T, F, Z satisfying

$$[T, F] = Z, \quad [T, Z] = 0, \quad [F, Z] = 0. \qquad (68)$$

The representation $\widetilde{\rho}$ of $\mathcal{H}$ produces a representation, which we also denote by $\widetilde{\rho}$, of the Lie algebra by unbounded operators on $\mathsf{L}^2(\mathbf{R})$. In the case of $\widetilde{\rho}$, the representations of the generators are

$$\widetilde{\rho}(F) = -2\pi i t; \quad \widetilde{\rho}(T) = -\frac{d}{dt}; \quad \widetilde{\rho}(Z) = 2\pi i. \qquad (69)$$

It is easily checked that these satisfy (68). Let

$$A = \frac{1}{\sqrt{2}}\left(T + \frac{1}{2\pi i}F\right) \qquad (70)$$

and observe that

$$\widetilde{\rho}(A) = \frac{-1}{\sqrt{2}}\left(\frac{d}{dt} + t\right). \qquad (71)$$

Its adjoint is

$$\widetilde{\rho}(A)^* = \frac{1}{\sqrt{2}}\Big(\frac{d}{dt} - t\Big). \tag{72}$$

We define $N = \widetilde{\rho}(A)^*\widetilde{\rho}(A)$ and remark that it is a self-adjoint operator. The normalized solution of $\widetilde{\rho}(A)(\mathbf{v}) = 0$ in $\mathbf{L}^2(\mathbf{R})$ is

$$\mathbf{v}_0(t) = \frac{1}{\pi^{1/4}}e^{-t^2/2}, \tag{73}$$

the zeroth Hermite function in (46). In fact, the eigenvalues of N are $\{0,\ 1,\ 2,\ \dots\}$ and the corresponding eigenvectors of N are the $\mathbf{v}_n$ as defined in (46). Indeed, one can show easily that

$$\begin{aligned}\widetilde{\rho}(A)^*(\mathbf{v}_n) &= \sqrt{n+1}\,\mathbf{v}_{n+1}\\ \widetilde{\rho}(A)(\mathbf{v}_n) &= \sqrt{n}\,\mathbf{v}_{n-1}.\end{aligned} \tag{74}$$

The Bargmann-Segal representation stated in Section 4.6 is now derived. Define the Hilbert space $\mathfrak{F}$ as in Section 4.6 and let Ψ be the isometry in (60). Deriving an expression for $\rho^{\mathrm{BS}} = \Psi \circ \rho \circ \Psi^{-1}$ directly is difficult, but it can be found indirectly by first calculating the corresponding Lie algebra representation $\widetilde{\rho^{\mathrm{BS}}}$, as follows.

Since $\rho = \Psi^{-1} \circ \rho^{\mathrm{BS}} \circ \Psi$, it follows that $\widetilde{\rho} = \Psi^{-1} \circ \widetilde{\rho^{\mathrm{BS}}} \circ \Psi$ and $\widetilde{\rho}^* = \Psi^{-1} \circ \widetilde{\rho^{\mathrm{BS}}}^* \circ \Psi$. Therefore, $\widetilde{\rho}(A)(\mathbf{v}_0) = 0$ implies $\widetilde{\rho^{\mathrm{BS}}}(A)(\mathbf{j}_0) = 0$, that is, $\widetilde{\rho^{\mathrm{BS}}}(A)(1) = 0$. Similarly, $\widetilde{\rho}(A)(\mathbf{v}_n) = \sqrt{n}\,\mathbf{v}_{n-1}$ implies $\widetilde{\rho^{\mathrm{BS}}}(A)(z^n) = nz^{n-1}$. Therefore,

$$\widetilde{\rho^{\mathrm{BS}}}(A) = \frac{d}{dz}, \tag{75}$$

or upon substituting for A,

$$\widetilde{\rho^{\mathrm{BS}}}(T) + \frac{1}{2\pi i}\widetilde{\rho^{\mathrm{BS}}}(F) = \sqrt{2}\frac{d}{dz}. \tag{76}$$

It is readily shown that $\widetilde{\rho}(T)$ and $\widetilde{\rho}(F)$ are skew-adjoint, hence so are $\widetilde{\rho^{\mathrm{BS}}}(T)$ and $\widetilde{\rho^{\mathrm{BS}}}(F)$. That is to say, $\widetilde{\rho^{\mathrm{BS}}}(T)^* = -\widetilde{\rho^{\mathrm{BS}}}(T)$ and similarly for $\widetilde{\rho^{\mathrm{BS}}}(F)$. Therefore, taking the adjoint of (76) yields the new equation

$$-\widetilde{\rho^{\mathrm{BS}}}(T) + \frac{1}{2\pi i}\widetilde{\rho^{\mathrm{BS}}}(F) = \sqrt{2}z \tag{77}$$

where use has been made of the facts that the adjoint of $\frac{d}{dz}$ is z and the adjoint of a complex number is its complex conjugate. Solving the equations (76) and (77) gives

$$\widetilde{\rho^{\mathrm{BS}}}(T) = \frac{1}{\sqrt{2}}\Big(\frac{d}{dz} - z\Big), \quad \widetilde{\rho^{\mathrm{BS}}}(F) = \sqrt{2}\pi i\Big(\frac{d}{dz} + z\Big). \tag{78}$$

Exponentiating these two operators shows that

$$\rho^{BS}_{(t,0)}(\mathbf{a})(z) = \exp\left(-\frac{t^2}{4} - \frac{tz}{\sqrt{2}}\right)\mathbf{a}\left(z + \frac{t}{\sqrt{2}}\right),$$

$$\rho^{BS}_{(0,f)}(\mathbf{a})(z) = \exp\left(-\pi^2 f^2 + \sqrt{2}\pi i f z\right)\mathbf{a}\left(z + \sqrt{2}\pi i f\right).$$

(79)

Equation (61) now follows upon noting that $\rho^{BS}_{(t,f)} = \rho^{BS}_{(0,f)}\rho^{BS}_{(t,0)}$.

4.9   The Metaplectic Representation

Here we consider automorphisms of $\mathbf{R}^2$ that preserve the structure we have discussed so far. In fact, it is more customary at this point to work with the equivalent multiplier ν, defined in (48), rather than the multiplier σ. We consider, then, the continuous automorphisms α of the group $\mathbf{R}^2$ that preserve the multiplier ν in the sense that

$$\nu(\alpha(t,f), \alpha(t',f')) = \nu((t,f),(t',f')), \qquad (t,f),(t',f') \in \mathbf{R}^2. \quad (80)$$

These are just the members of $\mathsf{SL}(2,\mathbf{R})$ — 2×2 matrices with determinant 1. Evidently, for any such automorphism, $\rho^{(\nu)} \circ \alpha$ is an irreducible ν-representation of $\mathbf{R}^2$ and so, up to equivalence, must be $\rho^{(\nu)}$ itself by the Stone-von Neumann Theorem. Thus there exists a unitary operator $U(\alpha)$ such that

$$U(\alpha)^{-1}\rho^{(\nu)}_{(t,f)}U(\alpha) = \rho^{(\nu)}_{\alpha(t,f)}, \qquad (t,f) \in \mathbf{R}^2. \quad (81)$$

Since $\rho^{(\nu)}$ is irreducible, $U(\alpha)$ is unique up to a scalar multiple.

Moreover, $U(\alpha\beta)$ must have the same effect on $\rho^{(\nu)}$ as $U(\alpha)U(\beta)$ and so, by the irreducibility of $\rho^{(\nu)}$, these two must also differ by a multiplicative constant. In other words, the map $\alpha \rightarrow U(\alpha)$ is a projective representation of $\mathsf{SL}(2,\mathbf{R})$. In fact, the multiplier is cohomologous to a two-valued one. Thus, there is a double covering $\widetilde{\mathsf{SL}(2,\mathbf{R})}$ of $\mathsf{SL}(2,\mathbf{R})$ and U lifts to a unitary representation of $\widetilde{\mathsf{SL}(2,\mathbf{R})}$ on $\mathsf{L}^2(\mathbf{R})$. This is called the *metaplectic representation* of $\mathsf{SL}(2,\mathbf{R})$. (The representation is not irreducible, but rather, $\mathsf{L}^2(\mathbf{R})$ decomposes as the direct sum of two subspaces, where the representation restricted to either of the two subspaces is irreducible.)

5.   Ambiguity Functions

This section uses the earlier results on the σ-representation theory of $\mathbf{R}^2$ to establish several facts about ambiguity functions.

5.1   Ambiguity Functions of the Hermites

Let $\rho_{nm}(t, f) = \langle \mathbf{v}_n, \rho_{(t,f)} \mathbf{v}_m \rangle$ denote the ambiguity functions associated with the Hermite waveforms $\mathbf{v}_n$ defined in (46). This section outlines how a closed form expression for $\rho_{nm}(t, f)$ can be obtained.

The Hermite waveforms are simpler in the Bargmann-Segal space, so the first step is to notice that $\rho_{nm}(t, f) = \langle \mathbf{j}_n, \rho_{(t,f)}^{\mathrm{BS}} \mathbf{j}_m \rangle$ because the isometry Ψ maps $\mathbf{v}_n$ to $\mathbf{j}_n$; see Section 4.6 for notation. Next, write down the generating function

$$G(t, f; a, b) = \left\langle \mathbf{e}_{\bar{a}}, \rho_{(t,f)}^{\mathrm{BS}}(\mathbf{e}_b) \right\rangle$$
$$= \sum_{n,m=0}^{\infty} \left\langle \mathbf{j}_n, \rho_{(t,f)}^{\mathrm{BS}}(\mathbf{j}_m) \right\rangle \frac{a^n b^m}{\sqrt{n!\, m!}} \tag{82}$$

where the second equality follows from the fact that $\mathbf{e}_u$, defined implicitly by (58), is given explicitly by

$$\mathbf{e}_u = \sum_{k=0}^{\infty} \frac{\bar{u}^k}{\sqrt{k!}} \mathbf{j}_k. \tag{83}$$

Using the reproducing kernel property (58) shows that

$$G(t, f; a, b) = \overline{\rho_{(t,f)}^{\mathrm{BS}}(\mathbf{e}_b)(\bar{a})}$$
$$= e^{-\frac{t^2+4\pi^2 f^2}{4} - \frac{a}{\sqrt{2}}(t+2\pi i f)+\pi i f t} e^{b\left(a+(t-2\pi i f)/\sqrt{2}\right)}. \tag{84}$$

Equating coefficients of $a^n b^m$ in (82) and (84) results in an explicit expression for $\rho_{nm}(t, f) = \langle \mathbf{j}_n, \rho_{(t,f)}^{\mathrm{BS}} \mathbf{j}_m \rangle$. Moreover, making the substitution $(t, f) = (r \cos\theta, r \sin\theta)$ allows this expression to be written in terms of the Laguerre polynomials. See (Wilcox, 1991) or (Miller, 1991) for details.

5.2   Symmetries of Ambiguity Functions

This section introduces the basic machinery for studying the possible symmetries of an ambiguity function. Although only the magnitude $|A(t, f)|$ of an ambiguity function $A(t, f)$ is of interest in general, it is significantly simpler to study symmetries of $A(t, f)$ rather than of $|A(t, f)|$. Changing multipliers changes the phase of the ambiguity function, thereby potentially altering the symmetry. It is therefore important to note that this section chooses to work with the symmetric ambiguity function defined in Section 4.4.

Recall from Section 4.9 that the elements of $\mathrm{SL}(2, \mathbf{R})$ preserve the multiplier ν. Moreover, from (81), it follows that if $A^{(\nu)}(t, f)$ is an ambiguity function then so too is $A^{(\nu)}\big(\alpha(t, f)\big)$ for any $\alpha \in \mathsf{SL}(2, \mathbf{R})$. Indeed,

$$
\begin{aligned}
A^{(\nu)}_{\mathbf{u},\mathbf{v}}\big(\alpha(t, f)\big) &= \big\langle \mathbf{u}, \rho^{(\nu)}_{\alpha(t,f)} \mathbf{v} \big\rangle \\
&= \big\langle \mathbf{u}, U(\alpha)^{-1} \rho^{(\nu)}_{(t,f)} U(\alpha) \mathbf{v} \big\rangle \qquad (85) \\
&= A^{(\nu)}_{U(\alpha)\mathbf{u}, U(\alpha)\mathbf{v}}(t, f).
\end{aligned}
$$

It follows that if $\mathcal{S}$ is a subgroup of $\mathsf{SL}(2, \mathbf{R})$ and the waveform $\mathbf{u}$ satisfies $U(\alpha)\mathbf{u} = \mathbf{u}$ for all $\alpha \in \mathcal{S}$ then the ambiguity function $A^{(\nu)}_{\mathbf{u},\mathbf{u}}$ is symmetric with respect to $\mathcal{S}$. In fact, since it is shown in (Miller, 1991) that $A^{(\nu)}_{\mathbf{u},\mathbf{u}} = A^{(\nu)}_{\mathbf{u}',\mathbf{u}'}$ if and only if $\mathbf{u} = \lambda \mathbf{u}'$ for some $\lambda \in \mathbf{T}$, we say that $\mathbf{u}$ has an $\mathcal{S}$-symmetric ambiguity function if and only if there exists a function $\lambda : \mathcal{S} \to \mathbf{T}$ such that $U(\alpha)\mathbf{u} = \lambda(\alpha)\mathbf{u}$ for all $\alpha \in \mathcal{S}$.

One interesting subgroup of $\mathsf{SL}(2, \mathbf{R})$ is the rotation group. However, it is more natural to rotate a dilated version of the ambiguity function so that the units of time and frequency are compatible. Let $\mathcal{S}$ denote the group whose elements $S(\theta)$ are dilated rotations, namely

$$
\begin{aligned}
S(\theta) &= \begin{pmatrix} \sqrt{2\pi} & 0 \\ 0 & \frac{1}{\sqrt{2\pi}} \end{pmatrix} \begin{pmatrix} \cos\theta & -\sin\theta \\ \sin\theta & \cos\theta \end{pmatrix} \begin{pmatrix} \frac{1}{\sqrt{2\pi}} & 0 \\ 0 & \sqrt{2\pi} \end{pmatrix} \\
&= \begin{pmatrix} \cos\theta & -2\pi\sin\theta \\ \frac{1}{2\pi}\sin\theta & \cos\theta \end{pmatrix}, \quad \theta \in [0, 2\pi).
\end{aligned} \qquad (86)
$$

Note that $\mathcal{S}$ is a subgroup of $\mathsf{SL}(2, \mathbf{R})$ and that $S(-\theta)$ is the inverse of $S(\theta)$. It is now shown that the only waveforms with rotationally symmetric (that is, $\mathcal{S}$-symmetric) ambiguity functions are the Hermites.

Anticipating the involvement of the Hermites, we choose to work in Bargmann-Segal space. Define $U^{\mathrm{BS}}(\alpha) = \Psi \circ U(\alpha) \circ \Psi^{-1}$ so that $\rho^{\mathrm{BS}(\nu)}_{\alpha(t,f)} = U^{\mathrm{BS}}(\alpha)^{-1} \rho^{\mathrm{BS}(\nu)}_{(t,f)} U^{\mathrm{BS}}(\alpha)$. Here, Ψ is the isomorphism defined in (60) and $\rho^{\mathrm{BS}(\nu)}$ is the ν-representation analogue of (61), namely

$$
\rho^{\mathrm{BS}(\nu)}_{(t,f)}(\mathbf{a})(z) = e^{-\frac{t^2 + 4\pi^2 f^2}{4} + \frac{z}{\sqrt{2}}(2\pi i f - t)} \mathbf{a}\left(z + \frac{t + 2\pi i f}{\sqrt{2}}\right). \qquad (87)
$$

It is claimed

$$
U^{\mathrm{BS}}\big(S(\theta)\big)(\mathbf{a})(z) = \mathbf{a}(e^{i\theta} z). \qquad (88)
$$

Indeed, $S(\theta)(t, f) = \big(t\cos\theta - 2\pi f\sin\theta, (2\pi)^{-1}t\sin\theta + f\cos\theta\big)$, hence

$$
\begin{aligned}
\rho^{\mathrm{BS}(\nu)}_{S(\theta)(t,f)}(\mathbf{a})(z) &= e^{-\frac{t^2 + 4\pi^2 f^2}{4} + \frac{z}{\sqrt{2}}(2\pi i f - t)e^{-i\theta}} \mathbf{a}\left(z + \frac{t + 2\pi i f}{\sqrt{2}}e^{i\theta}\right) \\
&= U^{\mathrm{BS}}\big(S(\theta)\big)^{-1} \rho^{\mathrm{BS}(\nu)}_{(t,f)} U^{\mathrm{BS}}\big(S(\theta)\big)(\mathbf{a})(z),
\end{aligned} \qquad (89)
$$

as required. The rotationally invariant waveforms in the Bargmann-Segal domain are thus those waveforms $\mathbf{a}(z)$ which satisfy

$$\forall \theta \in [0, 2\pi), \quad \mathbf{a}(e^{i\theta}z) = \lambda(\theta)\mathbf{a}(z) \tag{90}$$

for some λ mapping θ into $\mathbf{T}$. The only solutions are the Hermite waveforms $\mathbf{a}(z) = \mathbf{j}_n(z)$, and scalar multiples of them, for $n = 0, 1, \cdots$.

It is remarked that if the ordinary rotations $R(\theta)$ are considered instead of the dilated rotations $S(\theta)$, then $U(R(\theta))$ turns out to be the fractional Fourier transform. Indeed, we know from (53) that $U(R(\theta))$ is the Fourier transform if $\theta = -\pi/2$. Thus, $U(R(\theta))$ embeds the Fourier transform in a one parameter group.

5.3   Multiple Waveforms

This section presents an intriguing observation about how an ambiguity function hypothetically can be made to resemble the ideal delta function, thus overcoming the Heisenberg uncertainty principle.

Assume it was somehow possible to transmit simultaneously but separately many different waveforms $\{\mathbf{u}_n\}_{n=1}^N$ and to receive the returns separately too. What we have in mind is not using different portions of the electromagnetic spectrum, which does not separate the waveforms in the strict sense we require, but rather to imagine the hypothetical situation of multiple universes where the target exists in each one but a different waveform can be used to detect it in each universe.

If the waveforms are orthogonal and have equal energy with total energy one, so that $\|\mathbf{u}_n\|^2 = N^{-1}$, then it follows from Moyal's identity that the ambiguity functions $A_{\mathbf{u}_n,\mathbf{u}_n}(t, f)$ are orthogonal, and in particular, the ambiguity function $A(t, f) = \sum_{n=1}^N A_{\mathbf{u}_n,\mathbf{u}_n}(t, f)$ of the whole system has norm

$$\|A(t, f)\|^2 = \sum_{n=1}^N \|A_{\mathbf{u}_n,\mathbf{u}_n}\|^2 = \sum_{n=1}^N \|\mathbf{u}_n\|^4 = \frac{1}{N}. \tag{91}$$

This shows that there is less volume under the ambiguity surface for a given total energy as N increases. Therefore, since $A(0, 0) = 1$, the total ambiguity function can be made to approach the ideal delta function as N approaches infinity.

6.   The Wide Band Case

Our description of the Doppler effect is actually only an approximation that works when the radial velocities of moving objects in the scene are much smaller than the speed of light, and the transmitted signal has a spectrum in a narrow band around its carrier frequency. This is, of course, a valid assumption on most circumstances in radar. However, in sonar, which works on similar principles, it can often be the case that objects move at a non-trivial proportion of the speed of sound in water (around 1500 metres per second). Here it is appropriate to replace the "narrow band approximation" by the so-called "wide band theory". As the name implies, this theory is also appropriate when the spectrum of the transmit signal is broad, as is the case, for example, where the transmit signal is a very short pulse. Such signals are not typical in conventional radar systems, but again are in sonar, where the pulse is often created by a small explosive charge.

The effect of Doppler is not, as we have suggested in (2), a shift in frequency, but is a dilation of the signal. Thus the return from an object moving at a radial velocity (towards the transmitter/receiver) of v is, leaving aside the magnitude term,

$$\text{ret}(t) = \mathbf{w}\left(\alpha t - \frac{2r}{c+v}\right),\tag{92}$$

where c is the speed of the wave and

$$\alpha = \frac{\left(1 - \frac{v}{c}\right)}{\left(1 + \frac{v}{c}\right)}.\tag{93}$$

This leads, after a rescaling, to an ambiguity function of the form

$$\mathbf{W_{u,v}}(r, \alpha) = \sqrt{\alpha} \int_{-\infty}^{\infty} \mathbf{u}(t)\overline{\mathbf{v}\big(\alpha(t+r)\big)}\, dt.\tag{94}$$

The relevant representation in the wide band case is the representation

$$\rho_{(a,b)}(\mathbf{v})(t) = \sqrt{a}\,\mathbf{v}(at+b), \quad \mathbf{v} \in \mathsf{L}^2(\mathbf{R}),\tag{95}$$

of the affine group

$$\mathcal{A} = \left\{ \begin{pmatrix} a & b \\ 0 & 1 \end{pmatrix} : a > 0,\ b \in \mathbf{R} \right\}\tag{96}$$

with group multiplication given by matrix multiplication. Note that $\rho_{(a,b)}$ is actually a representation *on the right*, meaning $\rho_{g_1 g_2} = \rho_{g_2} \rho_{g_1}$ for all $g_1, g_2 \in \mathcal{A}$. Analogous to the narrow band case, the wide band ambiguity function can be expressed (after a change of coordinates) as

$$\mathbf{W_{u,v}}(a, b) = \langle \mathbf{u}, \rho_{(a,b)} \mathbf{v} \rangle\tag{97}$$

and its properties are studied by investigating the representation $\rho_{(a,b)}$ of the affine group.

This investigation is not undertaken here for reasons of space. The interested reader is, however, referred to (Miller, 1991) for details. It is remarked though that the representation theory of the affine group is complicated by the affine group not being unimodular. That is to say, whereas the left invariant Haar measure defined in (13) of the Heisenberg group is also the right invariant measure of the Heisenberg group, the left and right invariant measures of the affine group differ.

References

Katznelson, Y. (1968). *An Introduction to Harmonic Analysis.* Wiley.

Mackey, G. W. (1976). *The Theory of Unitary Group Representations.* Chicago Lectures in Mathematics. The University of Chicago Press.

Maclane, S. (1975). *Homology.* Springer-Verlag.

Miller, W., Jr. (1991). *Topics in Harmonic Analysis with Applications to Radar and Sonar,* volume 32 of *I.M.A. series in Mathematics and its Applications,* pages 66–168. Springer-Verlag.

Moran, W. (2001). The mathematics of radar. In Byrnes, J. S., editor, *Twentieth Century Harmonic Analysis.* Kluwer.

Wilcox, C. H. (1991). *The synthesis problem for radar ambiguity functions,* volume 32 of *I.M.A. series in Mathematics and its Applications,* pages 229–260. Springer-Verlag.

GEOMETRY OF PARAVECTOR SPACE WITH APPLICATIONS TO RELATIVISTIC PHYSICS

William E. Baylis
Physics Dept., University of Windsor
Windsor, ON, Canada N9B 3P4

Abstract   Clifford's geometric algebra, in particular the algebra of physical space (APS), lubricates the paradigm shifts from the Newtonian worldview to the post-Newtonian theories of relativity and quantum mechanics. APS is an algebra of vectors in physical space, and its linear subspaces include a 4-dimensional space of paravectors (scalars plus vectors). The metric of the latter has the pseudo-Euclidean form of Minkowski spacetime, with which APS facilitates the transition from Newtonian mechanics to relativity without the need of tensors or matrices. APS also provides tools, such as spinors and projectors, for solving classical problems and for smoothing the transition to quantum theory. This lecture concentrates on paravectors and applications to relativity and electromagnetic waves. A following lecture will extend the treatment to the quantum/classical interface.

Keywords: algebra of physical space, Clifford algebra, instruction in the quantum age, paravectors, quantum/classical interface, relativity.

Introduction

Devices such as single-electron transistors and switches, single photon masers, and quantum computers are hot topics. Not only is it difficult to ignore quantum and relativistic effects in such devices; these effects may often be dominant. Indeed, the theory of electromagnetic phenomena is inherently relativistic, and impetus behind quantum computation is to employ quantum superposition to make a new breed of computer vastly superior to current ones for certain tasks.

Relativity and quantum theory were introduced roughly a century ago. They both entail paradigm shifts from the assumptions made by

J. Byrnes (ed.) Computational Noncommutative Algebra and Applications, 363-387.
© 2004 *Kluwer Academic Publishers. Printed in the Netherlands.*

Newton, but our teaching has not fully adapted. Undergraduate instruction still begins with Newtonian mechanics, and only after Newton's worldview is ingrained does it progress to the more difficult and abstract mathematics of quantum theory and relativity. In this lecture, I suggest that much of the apparent dichotomy between the mathematics of Newtonian mechanics and that of both quantum theory and relativity arises because we have not been using the best formulation to describe the physics.

Clifford's geometric algebra, in particular the *algebra of physical space (APS)*, empowers classical physics with geometric tools that lead to a covariant formulation of relativity and are strikingly similar to tools common in quantum theory. [1, 2] With APS, quantum theory and relativity can be taught with the same mathematics as Newtonian physics, and this permits an earlier, smoother introduction to post-Newtonian physics. This, in turn, encourages students to build intuition consistent with relativistic and quantum phenomena and properly prepares them for the quantum age of the 21st century.

The principal purpose of this lecture is to demonstrate that the structure and geometry of APS make it a natural and minimal model for both Newtonian and relativistic mechanics. It is natural in that it associates quantities much in the same way that humans usually do, and it is minimal in that it avoids the assumption of additional structure that is not relevant to the physics. I start by reviewing the Clifford algebras commonly used in physics and their relation to APS, which I introduce as an algebra of spatial vectors. However, we quickly note that APS contains a 4-dimensional paravector space with the metric of spacetime, and it can be used to formulate a covariant approach to relativity. Multiparavectors and their Lorentz transformations are also discussed and interpreted. The relation of APS to the spacetime algebra (STA) is discussed in detail with an emphasis on the difference between absolute and relative formulations of relativity. The quantum-like tools of eigenspinors and projectors are introduced, along with applications to the electrodynamics of Maxwell and Lorentz, including a study of Stokes parameters and light polarization.

1.   Clifford Algebras in Physics

The importance of Clifford algebras to physics and engineering is increasingly recognized, [3, 4, 5, 6, 7, 8] and most physicists have encountered Clifford algebras in some guise. The three most commonly employed in physics are the quaternion algebra $\mathbb{H} = C\ell_{0,2}$, the algebra of physical space (APS) $C\ell_3$, and the spacetime algebra (STA) $C\ell_{1,3}$. They are closely related. We use the common notation [9] $C\ell_{p,q}$ for the Clifford algebra of a vector space of metric signature (p, q), and $C\ell_p \equiv C\ell_{p,0}$.

When Hamilton introduced vectors in 1843, they were part of an *algebra of quaternions.* [10] The superiority of $\mathbb{H}$ for matrix-free and coordinate-free computations of rotations in physical space has been recently rediscovered by space programs, the computer-games industry, and robotics engineering. Furthermore, $\mathbb{H}$ has been investigated as a replacement of the complex field in an extension of Dirac theory. [11] Quaternions were used by Maxwell and Tait to express Maxwell's equations of electromagnetism in compact form, and they motivated Clifford to find generalizations based on Grassmann theory.

Hamilton's biquaternions (complex quaternions) are isomorphic to APS: $\mathbb{H} \otimes \mathbb{C} \simeq C\ell_3$, familiar to physicists as the algebra of the Pauli spin matrices. The even subalgebra $C\ell_3^+$ is isomorphic to $\mathbb{H}$ over the reals, and the correspondences $\mathbf{i} \leftrightarrow \mathbf{e}_3\mathbf{e}_2$, $\mathbf{j} \leftrightarrow \mathbf{e}_1\mathbf{e}_3$, $\mathbf{k} \leftrightarrow \mathbf{e}_2\mathbf{e}_1$ identify pure quaternions with bivectors in APS. APS distinguishes cleanly between vectors and bivectors, in contrast to most approaches with complex quaternions. The identification of the volume element $\mathbf{e}_1\mathbf{e}_2\mathbf{e}_3 = i$ (see next section) endows the unit imaginary with geometrical significance and helps explain the widespread use of complex numbers in physics. [12] The sign of i is reversed under parity inversion, and imaginary scalars and vectors correspond to pseudoscalars and pseudovectors, respectively.

APS is also isomorphic to the even part of STA: $C\ell_3 \simeq C\ell_{1,3}^+$. STA is familiar as the algebra of Dirac's gamma matrices, where each matrix γ_μ, $\mu = 0, 1, 2, 3$, represents a unit vector in spacetime. To be sure, Dirac's electron theory (1928) was based on a matrix representation of $C\ell_{1,3}$ over the *complex field*, whereas STA, pioneered by Hestenes [13, 14, 15] for use in many areas of physics, is $C\ell_{1,3}$ over the *reals*.

Clifford algebras of higher-dimensional spaces have also been used in robotics [16], many-electron systems, and elementary-particle theory [17]. This lecture focuses on APS, although generalizations to $C\ell_n$ are made where convenient, and one section is devoted to the relation of APS to STA. A full study of APS is beyond the scope of this lecture and can be found elsewhere [6], but the algebra is sufficiently simple that we can easily present its foundation and structure.

1.1   APS: an Algebra of Vectors

To form any algebra, we need elements and an associative product among them. The elements of APS are the vectors of physical space $\mathbf{u}, \mathbf{v}, \mathbf{w}$, and all their products $\mathbf{uv}, \mathbf{uvw}, \mathbf{uu}, \ldots$. If we start with vectors in an n-dimensional Euclidean space, then only one axiom is needed to define the algebraic product: the square of any vector $\mathbf{u}$ is its square length (a real number, a scalar):

$$\mathbf{u}\mathbf{u} \equiv \mathbf{u}^2 = \mathbf{u} \cdot \mathbf{u} . \tag{1}$$

That's it. This axiom, together with the usual rules for adding and multiplying square matrices, determines the entire algebra.

Let's put $\mathbf{u} = \mathbf{v} + \mathbf{w}$. The axiom implies that

$$\mathbf{v}\mathbf{w} + \mathbf{w}\mathbf{v} = 2\mathbf{v} \cdot \mathbf{w} . \tag{2}$$

Evidently the algebra is not commutative. If $\mathbf{v}$ and $\mathbf{w}$ are perpendicular, they anticommute. Let $\{\mathbf{e}_1, \mathbf{e}_2, \cdots\}$ be a basis of orthogonal unit vectors in the n-dimensional Euclidean space. Then $\mathbf{e}_1^2 = 1$ and $\mathbf{e}_1\mathbf{e}_2 = -\mathbf{e}_2\mathbf{e}_1$. We can be sure that $\mathbf{e}_1\mathbf{e}_2$ doesn't vanish because it squares to -1 : $\mathbf{e}_1\mathbf{e}_2\mathbf{e}_1\mathbf{e}_2 = -1$. The product of perpendicular vectors is a new element called a *bivector*. It represents a directed area in the plane of the vectors. The "direction" corresponds to circulation in the plane: if the circulation is reversed, the sign of the bivector is reversed. The bivector replaces the vector cross product of polar vectors, but unlike the usual cross product, it is intrinsic to the plane and can be applied to planes in spaces of more than 3 dimensions.

Bivectors can also be viewed as operators on vectors. They generate rotations and reflections in the plane. To rotate any vector $\mathbf{u} = u^1\mathbf{e}_1 + u^2\mathbf{e}_2$ in the $\mathbf{e}_1\mathbf{e}_2$ plane by a right angle, multiply it by the unit bivector $\mathbf{e}_1\mathbf{e}_2$: $\mathbf{u}\mathbf{e}_1\mathbf{e}_2 = u^1\mathbf{e}_2 - u^2\mathbf{e}_1$. The counterclockwise sense of the rotation when $\mathbf{u}$ is multiplied from the right corresponds to the circulation used to define the "direction" of $\mathbf{e}_1\mathbf{e}_2$. Multiplication from the left reverses the rotation. To rotate $\mathbf{u}$ in the plane by an arbitrary angle ϕ multiply it by a linear combination of 1 (no rotation) and $\mathbf{e}_1\mathbf{e}_2$:

$$\mathbf{u} \left(\cos\phi + \mathbf{e}_1\mathbf{e}_2 \sin\phi \right) = \mathbf{u} \exp\left(\mathbf{e}_1\mathbf{e}_2\phi\right)$$

Note the exponential function of the bivector $\mathbf{e}_1\mathbf{e}_2\phi$. It can be defined by its power-series expansion because all powers of bivectors can be calculated in the algebra. The Euler-type relation for $\exp\left(\mathbf{e}_1\mathbf{e}_2\phi\right)$ follows from the fact that $\mathbf{e}_1\mathbf{e}_2$ squares to -1.

A general vector $\mathbf{v}$, with components both in the plane and perpendicular to it, is rotated by the angle ϕ in the $\mathbf{e}_1\mathbf{e}_2$ plane by

$$\mathbf{v} \rightarrow R\mathbf{v}R^{\dagger}, \tag{3}$$

where the *rotors* $R, R^\dagger$ are

$$R = \exp\left(-\mathbf{e}_1\mathbf{e}_2\phi/2\right) = \cos\frac{\phi}{2} - \mathbf{e}_1\mathbf{e}_2\sin\frac{\phi}{2} \tag{4}$$

$$R^\dagger = \cos\frac{\phi}{2} - (\mathbf{e}_1\mathbf{e}_2)^\dagger\sin\frac{\phi}{2} = \cos\frac{\phi}{2} - \mathbf{e}_2\mathbf{e}_1\sin\frac{\phi}{2} = R^{-1}. \tag{5}$$

The dagger † denotes a conjugation called *reversion*,[1] in which the order of vectors in products is reversed. Thus, for any vectors $\mathbf{v}, \mathbf{w}$,, $(\mathbf{vw})^\dagger = \mathbf{wv}$. The reversion of other elements, say $\mathbf{AB}$, can then be found from the rule $(\mathbf{AB})^\dagger = \mathbf{B}^\dagger\mathbf{A}^\dagger$. An element equal to its reversion is said to be *real*, whereas one equal to minus its reversion is *imaginary*. The two-sided *spinorial form* of (3) preserves the reality of the transformed vector. From (5), R is unitary and consequently all products of vectors transform in the same way (3). In particular, the bivector $\mathbf{e}_1\mathbf{e}_2$ commutes with the rotors $\exp\left(\pm\mathbf{e}_1\mathbf{e}_2\phi/2\right)$ and is therefore invariant under rotations in the $\mathbf{e}_1\mathbf{e}_2$ plane. It is equally well expressed as the ordered product of any pair of orthonormal vectors in the plane.

The trivector $\mathbf{e}_1\mathbf{e}_2\mathbf{e}_3$ squares to -1 and commutes with all vectors that are linear combinations of $\mathbf{e}_1, \mathbf{e}_2$, and $\mathbf{e}_3$. More generally, products of k orthonormal basis vectors $\mathbf{e}_j$ can be reduced if two of them are the same, but if they are all distinct, their product is a basis k-vector. In an n-dimensional space, the algebra contains $\binom{n}{k}$ such linearly independent k-vectors, and any real linear combination of them is said to be an element of *grade* k. Thus, scalars have grade 0, vectors grade 1, bivectors grade 2, trivectors grade 3, and so on.

In APS, where the number of dimensions is $n = 3$, the $\mathbf{e}_1\mathbf{e}_2\mathbf{e}_3$ is the highest-grade element, namely the *volume element*, and it commutes with every vector and hence with all elements. It can be identified with the unit imaginary:

$$\mathbf{e}_1\mathbf{e}_2\mathbf{e}_3 = i. \tag{6}$$

Note that i changes sign under spatial inversion: $\mathbf{e}_k \to -\mathbf{e}_k$, $k = 1, 2, 3$. Imaginary scalars are called *pseudoscalars* because of this sign change. Any bivector can be expressed as an imaginary vector, called a *pseudovector*. For example,

$$\mathbf{e}_1\mathbf{e}_2 = \mathbf{e}_1\mathbf{e}_2\left(\mathbf{e}_1\mathbf{e}_2\mathbf{e}_3\right)/i = i\mathbf{e}_3. \tag{7}$$

The center of APS (the part that commutes with all elements) is spanned by $\{1, i\}$ and is identified with the complex field. Every element of APS

---

[1] A tilde ~ is often used to indicate reversal, but in spaces of definite metric such as Euclidean spaces, the dagger is common since it corresponds to Hermitian conjugation in any matrix representation in which the matrices representing the basis vectors are Hermitian.

is a linear combination of $1, \mathbf{e}_1, \mathbf{e}_2, \mathbf{e}_3, i\mathbf{e}_1, i\mathbf{e}_2, i\mathbf{e}_3, i$ over the reals or, equivalently, of $1, \mathbf{e}_1, \mathbf{e}_2, \mathbf{e}_3$ over the complex field. The element $^*x = -ix$ is said to be the *Clifford-Hodge dual* of x.

1.2 Existence

We should check that our algebra exists. It is possible to define structures that are not self-consistent. The existence of a *matrix representation* is sufficient to prove existence. The canonical one replaces unit vectors by Pauli spin matrices. There are an infinite number of valid representations. They share the same *algebra* and that is all that matters.

2. Paravector Space as Spacetime

APS includes not only the 3-dimensional linear space of physical vectors, but also other linear spaces, in particular the 4-dimensional linear space of scalars plus vectors. In mathematical terms, this 4-dimensional linear space is also a vector space and its elements are vectors, but to distinguish them from spatial vectors, we call them *paravectors*. A paravector p can generally be written

$$p = p^0 + \mathbf{p} \tag{8}$$

where p^0 is a scalar and $\mathbf{p} = p^k \mathbf{e}_k$, $k = 1, 2, 3$, a physical vector (the summation convention for repeated indices is used). It is convenient to put $\mathbf{e}_0 = 1$ so that we can write

$$p = p^\mu \mathbf{e}_\mu, \quad \mu = 0, 1, 2, 3. \tag{9}$$

The *metric* of the 4-dimensional paravector space is determined by the quadratic form ("square length") of paravectors. Since p^2 is not generally a pure scalar, we need the *Clifford conjugate* of p, $\bar{p} = p^\mu \bar{\mathbf{e}}_\mu = p^0 - \mathbf{p}$ since the product $p\bar{p} = \bar{p}p = (p^0)^2 - \mathbf{p}^2$ is always a scalar. If we take $p\bar{p}$ to be the quadratic form of p and use

$$\langle x \rangle_S \equiv \frac{1}{2}(x + \bar{x}) = \langle \bar{x} \rangle_S \tag{10}$$

to denote the scalarlike (that is scalar plus pseudoscalar) part of any element x, then

$$p\bar{p} = \langle p\bar{p} \rangle_S = p^\mu p^\nu \eta_{\mu\nu} \tag{11}$$

and the *Minkowski metric* of spacetime $\eta_{\mu\nu} = \langle \mathbf{e}_\mu \bar{\mathbf{e}}_\nu \rangle_S$ arises automatically. Spacetime can be viewed as paravector space, and spacetime vectors are (real) paravectors in APS. The vector part of the paravector is the usual spatial vector, and the scalar part is the time component.

EXAMPLE 1 *In units with the speed of light* $c = 1$, *the quadratic form (or "square length") of the spacetime displacement* $dr = dt + d\mathbf{r}$ *is*

$$dr \, d\bar{r} = dt^2 - d\mathbf{r}^2 = dt^2 \left(1 - \mathbf{v}^2\right),$$

where $\mathbf{v}$ *is the velocity* $d\mathbf{r}/dt$. *If we define the dimensionless scalar known as the proper time* τ *by* $dr \, d\bar{r} = d\tau^2$, *then we see that* $\gamma d\tau = dt$ *with*

$$\gamma = \frac{dt}{d\tau} = \left[1 - \mathbf{v}^2\right]^{-1/2}.$$

In particular, $d\tau = dt$ *in a rest frame of the displacement. The dimensionless proper velocity is*

$$u = \frac{dr}{d\tau} = \frac{dt}{d\tau} \left(1 + \frac{d\mathbf{r}}{dt}\right) = \gamma \left(1 + \mathbf{v}\right),$$

and by definition it is unimodular: $u\bar{u} = 1$. *Other spacetime vectors can be similarly represented as paravectors in APS. For example, the energy-momentum paravector of a particle is* $p = mu = E + \mathbf{p}$.

In his article [18] of 1905 on special relativity, Einstein did not mention a spacetime continuum. That was a construction proposed three years later by Minkowski. I like to think that Einstein, had he seen it, would have appreciated the natural appearance of the spacetime geometry in paravector space.

2.1   Multiparavectors

From the quadratic form of the sum $p + q$ of paravectors, we obtain an expression for the *scalar product* of paravectors p and q :

$$\langle p\bar{q}\rangle_S = \frac{1}{2}\left(p\bar{q} + q\bar{p}\right). \tag{12}$$

One says that paravectors p and q are *orthogonal* if and only if $\langle p\bar{q}\rangle_S = 0$. The *vectorlike* (vector plus pseudovector) part of $p\bar{q}$ is

$$\langle p\bar{q}\rangle_V \equiv \frac{1}{2}\left(p\bar{q} - q\bar{p}\right) = p\bar{q} - \langle p\bar{q}\rangle_S \tag{13}$$

and represents the directed plane in paravector space that contains paravectors p and q. It is called a *biparavector* and can be expanded in a basis of unit biparavectors $\langle \mathbf{e}_\mu \bar{\mathbf{e}}_\nu\rangle_V$ with $\langle \mathbf{e}_\mu \bar{\mathbf{e}}_\nu\rangle_V^2 = \pm 1$:

$$\langle p\bar{q}\rangle_V = p^\mu q^\nu \langle \mathbf{e}_\mu \bar{\mathbf{e}}_\nu\rangle_V$$

The biparavectors form a six-dimensional linear subspace of APS equal to the direct sum of vector and bivector spaces. Since bivectors of APS are also pseudovectors, any biparavector is also a complex vector.

Biparavectors arise most frequently in APS as operators on paravectors. Thus, the unit biparavector $\langle \mathbf{e}_\mu \bar{\mathbf{e}}_\nu \rangle_V$, $\mu \neq \nu$, rotates any paravector $p = a\mathbf{e}_\mu + b\mathbf{e}_\nu$ in the plane to an orthogonal direction:

$$\langle \mathbf{e}_\mu \bar{\mathbf{e}}_\nu \rangle_V \, p = \frac{1}{2} \left(a\mathbf{e}_\mu \bar{\mathbf{e}}_\nu \mathbf{e}_\mu + b\mathbf{e}_\mu \bar{\mathbf{e}}_\nu \mathbf{e}_\nu - a\mathbf{e}_\nu \bar{\mathbf{e}}_\mu \mathbf{e}_\mu - b\mathbf{e}_\nu \bar{\mathbf{e}}_\mu \mathbf{e}_\nu \right)$$

$$= -a\eta_{\mu\mu}\mathbf{e}_\nu + b\eta_{\nu\nu}\mathbf{e}_\mu.$$

In analogy with bivectors, biparavectors generate rotations in paravector space. One of the most important biparavectors in physics is the electromagnetic field (or "Faraday") which in SI units with $c = 1$ can be written $\mathbf{F} = \langle \partial \bar{A} \rangle_V = \mathbf{E} + i\mathbf{B}$, where $\partial = \mathbf{e}^\mu \partial / \partial x^\mu$ is the paravector gradient operator and $A = \phi + \mathbf{A}$ is the paravector potential. We will see below how to define $\mathbf{F}$ in terms of rotation rate.

Triparavectors can also be formed. The triparavector subspace of APS is four-dimensional, the direct sum of pseudovector and pseudoscalar spaces. The volume element of paravector space is the same as that of the underlying vector space:

$$\mathbf{e}_0 \bar{\mathbf{e}}_1 \mathbf{e}_2 \bar{\mathbf{e}}_3 = \mathbf{e}_1 \mathbf{e}_2 \mathbf{e}_3 = i \ . \tag{14}$$

As seen before, it commutes with all elements of APS.

2.2   Paravector Rotations and Lorentz Transformations

Rotations and reflections in paravector space preserve the scalar product $\langle p\bar{q} \rangle_S$ of any two paravectors. Paravector rotations have the same spinorial form as vector rotations,

$$p \to LpL^\dagger, \tag{15}$$

where L is a unimodular element ($L\bar{L} = 1$) known as a *Lorentz rotor*. Lorentz rotations are the physical Lorentz transformations of relativity: boosts, spatial rotations, and their products. In APS they can be calculated algebraically without matrices or tensors.

Lorentz rotors for spatial rotations are just the same rotors (4) introduced above, and those for boosts are similar except that the rotation plane in paravector space includes the time axis $\mathbf{e}_0$. For a boost along $\mathbf{e}_1$ for example, the Lorentz rotor L has the real form

$$L = \exp\left(\mathbf{e}_1 \bar{\mathbf{e}}_0 \frac{w}{2} \right) = \cosh\frac{w}{2} + \mathbf{e}_1 \sinh\frac{w}{2} = L^\dagger, \tag{16}$$

where w is a scalar parameter called the *rapidity*. The Lorentz rotation (15) can be calculated directly with two algebraic products, but because it is linear in p, it is sufficient to determine the transformed basis paravectors

$$\mathbf{u}_\mu \equiv L\mathbf{e}_\mu L^\dagger. \tag{17}$$

Since the proper velocity in the rest frame is unity, the Lorentz rotation of $\mathbf{e}_0 = 1$ must give the proper velocity u induced by the boost on objects initially at rest:

$$u = L\mathbf{e}_0 L^\dagger = LL^\dagger = \gamma\,(1 + v\mathbf{e}_1) \tag{18}$$

$$\gamma = \cosh w, \ \ \gamma v = \sinh w\ . \tag{19}$$

Since $\mathbf{e}_1$ also commutes with the biparavector $\mathbf{e}_1\bar{\mathbf{e}}_0$ of the rotation plane whereas $\mathbf{e}_2$ and $\mathbf{e}_3$ anticommute with it, the boost of the paravector basis elements gives

$$\mathbf{u}_0 = u\mathbf{e}_0, \ \mathbf{u}_1 = u\mathbf{e}_1 \tag{20}$$

$$\mathbf{u}_2 = \mathbf{e}_2, \ \mathbf{u}_3 = \mathbf{e}_3\ .$$

The set $\{\mathbf{u}_0, \mathbf{u}_1, \mathbf{u}_2, \mathbf{u}_3\} \equiv \{\mathbf{u}_\mu\}$ is an orthonormal basis of paravectors for the boosted system. It follows that the boost of any paravector $p = p^\mu \mathbf{e}_\mu$ produces

$$LpL^\dagger = p^\mu \mathbf{u}_\mu = u\left(p^0\mathbf{e}_0 + p^1\mathbf{e}_1\right) + p^2\mathbf{e}_2 + p^3\mathbf{e}_3 \tag{21}$$

$$= \gamma\left(p^0 + vp^1\right)\mathbf{e}_0 + \gamma\left(p^1 + vp^0\right)\mathbf{e}_1 + p^2\mathbf{e}_2 + p^3\mathbf{e}_3\ .$$

We can eliminate the dependence on the paravector basis by introducing components of p coplanar with the rotation plane

$$p^\triangle = p^0\mathbf{e}_0 + p^1\mathbf{e}_1$$

and perpendicular to it

$$p^\perp = p^2\mathbf{e}_2 + p^3\mathbf{e}_3 = p - p^\triangle.$$

Then the boost of p is

$$p \to LpL^\dagger = up^\triangle + p^\perp. \tag{22}$$

No matrices or tensors are required, and the algebra is trivial. Algebraic calculation of the boost is sufficiently simple to be taught at an early stage of a student's study of physics. Students can perform most calculations in introductory relativity texts using no more than the transformation (22) and the basic axiom (2) of the algebra. A couple of examples will illustrate the simplicity of the approach.

EXAMPLE 2 *Let Carol have proper velocity u_{BC} as seen by Bob, and let Bob have proper velocity u_{AB} with respect to Alice. To find Carol's proper velocity u_{AC} with respect to Alice, we boost the paravector $p = u_{BC}$ by $L = L^\dagger = u_{AB}^{1/2}$ as in (22). If the vector parts of u_{AB} and u_{BC} are collinear, u_{AB} and u_{BC} lie in the same spacetime plane and commute. Then $p^\perp = 0$ and the transformation (22) reduces to the product*

$$u_{AC} = u_{AB} u_{BC} \ .$$

By writing each proper velocity in the form $u = \gamma(1 + \mathbf{v})$, one easily extracts the usual result for collinear velocity composition[2]

$$\mathbf{v}_{AC} = \frac{\langle u_{AC} \rangle_V}{\langle u_{AC} \rangle_S} = \frac{\mathbf{v}_{AB} + \mathbf{v}_{BC}}{1 + \mathbf{v}_{AB} \cdot \mathbf{v}_{BC}}.$$

EXAMPLE 3 *Consider the change in the wave paravector k of a photon when its source is boosted from rest to proper velocity u. The wave paravector, like the momentum $\hbar k$, is null, $k\bar{k} = 0$, and can be written $k = \omega + \mathbf{k} = \omega \left(1 + \hat{\mathbf{k}}\right)$, where the unit vector $\hat{\mathbf{k}}$ gives the direction and ω the frequency of k. From (22), writing out the coplanar part of k as $k^\triangle = \omega + \mathbf{k}^\parallel$ and noting that $\mathbf{k}^\perp$ is a vector, we find*

$$k = \omega \left(1 + \hat{\mathbf{k}}\right) \rightarrow k' = LkL^\dagger = u \left(\omega + \mathbf{k}^\parallel\right) + \mathbf{k}^\perp$$

with $u = \gamma(1 + \mathbf{v})$ and $\mathbf{k}^\parallel = \mathbf{k} \cdot \hat{\mathbf{v}} \, \hat{\mathbf{v}} = \mathbf{k} - \mathbf{k}^\perp$, where $\hat{\mathbf{v}}$ is the unit vector along $\mathbf{v}$. This transformation describes what happens to the photon momentum when the light source is boosted. Evidently $\mathbf{k}^\perp$ is unchanged, but there is a Doppler shift in ω and a change in $\mathbf{k} \cdot \hat{\mathbf{v}}$:

$$\omega' = \left\langle u \left(\omega + \mathbf{k}^\parallel\right)\right\rangle_S = \gamma\omega\left(1 + \hat{\mathbf{k}} \cdot \mathbf{v}\right)$$

$$\mathbf{k}' \cdot \hat{\mathbf{v}} = \left\langle u\hat{\mathbf{v}}\left(\omega + \mathbf{k}^\parallel\right)\right\rangle_S = \gamma\omega\left(v + \hat{\mathbf{k}} \cdot \hat{\mathbf{v}}\right) = \omega' \cos\theta'.$$

The ratio $\mathbf{k}' \cdot \hat{\mathbf{v}}/\omega'$ shows how the photons are thrown forward

$$\cos\theta' = \frac{v + \cos\theta}{1 + v\cos\theta} \ .$$

in what is called the "headlight" effect.

EXAMPLE 4 *The results of the previous example can be combined with a qualitative description of Thomson scattering to explain how high-energy*

---

[2]Note that the cleanest way to compose general Lorentz rotations is to multiply rotors: $L_{AC} = L_{AB} L_{BC}$. See also the section below on eigenspinors.

gamma-ray photons are produced near sources of energetic electrons. In Thomson scattering, a electron initially at rest scatters an unpolarized beam of radiation into all directions. It is the limit of Compton scattering when $\omega \ll m$, the rest energy of the electron. In the Lab frame, the electrons are ultra-relativistic with energies $\gamma m \gg m$, and they collide with photons of the 2.7 K blackbody radiation that permeates space. Let ω_0 be Lab frequency of the background radiation. In the rest frame of the electron, this gets Doppler shifted to roughly $\gamma\omega_0$ (within a factor between 0 and 2, depending on angle), and Thomson scattering occurs, redistributing the photons in all directions. Transforming the scattered photons back to the Lab gives a collimated beam of photons in the direction of the electron velocity with energies of order $\gamma^2\omega_0$. Thus, 5 GeV electrons ($\gamma = 10^4$) can raise 10^{-2} eV photons to MeV energies.

An attractive feature of this simple algebraic approach to introductory relativity is that it is not restricted to such simple cases. Rather it is part of an algebra that simplifies computations for all relativistic phenomena. Lorentz rotations of multiparavectors, in particular, are readily found by putting together those for paravectors. Thus, for the boost along $\mathbf{e}_1$ considered above, the biparavectors $\mathbf{e}_1\bar{\mathbf{e}}_0$ and $\mathbf{e}_2\bar{\mathbf{e}}_3$ are seen to be invariant whereas $\mathbf{e}_1\bar{\mathbf{e}}_2, \mathbf{e}_1\bar{\mathbf{e}}_3, \mathbf{e}_0\bar{\mathbf{e}}_2, \mathbf{e}_0\bar{\mathbf{e}}_3$ are multiplied from the left by u. More generally, the paravector product $p\bar{q}$ transforms to $LpL^\dagger\overline{(LqL^\dagger)} = Lp\bar{q}\bar{L}$. From this we can confirm that scalar products (12) of paravectors are Lorentz invariant and that any biparavector, say $\mathbf{F}$, transforms as

$$\mathbf{F} \to L\mathbf{F}\bar{L} \, . \tag{23}$$

The power of APS allows us to generalize expression (22) to an arbitrary Lorentz rotation. We start with an arbitrary *simple Lorentz rotation*, that is, a rotation in a single paravector plane. Simple rotations include all spatial rotations, pure boosts, and many boost-rotation combinations. We again split p into one component $p^\triangle$ coplanar with the rotation plane plus another component $p^\perp$ perpendicular to it: $p = p^\triangle + p^\perp$. Consider a biparavector $\langle u\bar{v}\rangle_V$, which represents the plane containing the independent paravectors u and v. By expansion it is easy to see that

$$\langle u\bar{v}\rangle_V \, u = \frac{u\bar{v}u - v\bar{u}u}{2} = u\left(\frac{\bar{v}u - \bar{u}v}{2}\right) = u\langle u\bar{v}\rangle_V^\dagger \, .$$

Similarly, one finds $\langle u\bar{v}\rangle_V \, v = v\langle u\bar{v}\rangle_V^\dagger$. The component $p^\triangle$ coplanar with $\langle u\bar{v}\rangle_V$ is a linear combination of u and v and therefore obeys the same relation

$$\langle u\bar{v}\rangle_V \, p^\triangle = p^\triangle \langle u\bar{v}\rangle_V^\dagger \, .$$

On the other hand, the component $p^\perp$ is orthogonal to u and v, $\left\langle \bar{u}p^\perp \right\rangle_S = 0 = \left\langle u\bar{p}^\perp \right\rangle_S$ and similarly for u replaced by v. Consequently,

$$\left\langle u\bar{v} \right\rangle_V p^\perp = \frac{u\bar{v}p^\perp - v\bar{u}p^\perp}{2} = -p^\perp \left\langle u\bar{v} \right\rangle_V^\dagger .$$

It follows that if L is any Lorentz rotation in the $\left\langle u\bar{v} \right\rangle_V$ plane, then

$$p^\triangle L^\dagger = Lp^\triangle, \ \ p^\perp L^\dagger = \bar{L}p^\perp ,$$

and the rotation (15) reduces to

$$p \rightarrow Lp^\triangle L^\dagger + Lp^\perp L^\dagger = L^2 p^\triangle + p^\perp . \tag{24}$$

Expression (22) is a special case of this relation for a pure boost. We can now generalize the result further to include compound Lorentz rotations. Any rotor can be expressed as

$$L = \pm \exp\left(\frac{1}{2}\mathbf{W}\right), \tag{25}$$

and the arbitrary biparavector $\mathbf{W}$ can always be expanded as a sum of simple biparavectors $\mathbf{W} = \mathbf{W}_1 + \mathbf{W}_2 = (1 + i\alpha)\mathbf{W}_1$, where α is a real scalar. The relation $\mathbf{W}_2 = i\alpha\mathbf{W}_1$ means that $\mathbf{W}_2$ is proportional to the dual of $\mathbf{W}_1$ so that $\mathbf{W}_1$ and $\mathbf{W}_2$ represent orthogonal planes: every paravector in $\mathbf{W}_1$ is orthogonal to every paravector in $\mathbf{W}_2$. Since $\mathbf{W}_1$ and $\mathbf{W}_2$ commute, L can be written as a product of commuting simple Lorentz rotations:

$$L = L_1 L_2, \ \ L_1 = e^{\mathbf{W}_1/2}, \ \ L_2 = e^{\mathbf{W}_2/2}.$$

Every paravector p can be split into one component, p_1, coplanar with $\mathbf{W}_1$ and another, p_2, coplanar with $\mathbf{W}_2$. Two applications of the simple transformation result (24) gives its generalization

$$p \rightarrow LpL^\dagger = L_1 L_2 (p_1 + p_2) L_2^\dagger L_1^\dagger = L_1^2 p_1 + L_2^2 p_2 . \tag{26}$$

Note that the Lorentz transformations of paravectors p and $p^\dagger$ are the same, as are those of biparavectors $\mathbf{F}$ and $\bar{\mathbf{F}}$. However, p and $\bar{p}$ have distinct transformations as do $\mathbf{F}$ and $\mathbf{F}^\dagger$. As a consequence, Lorentz transformations mix time and space components of p, but not its real and imaginary parts, whereas such transformations of $\mathbf{F}$ leave $\mathbf{F}$ vectorlike but mix vector (real) and bivector (imaginary) parts.

3.   Interpretation

Note that in APS, the Lorentz rotation acts directly on paravectors and their products, and not on scalar coefficients. Contrast this with matrix or tensor formulations where transformations are given only for the coefficients. Of course, the two approaches are easily related. For example, in the boost (21) of p, if p and the transformed $p' = LpL^\dagger$ are expanded in the same basis $\{\mathbf{e}_\mu\}$, we can simply read off the transformation of coefficients:

$$p'^0 = \gamma \left(p^0 + vp^1 \right)$$
$$p'^1 = \gamma \left(p^1 + vp^0 \right)$$
$$p'^2 = p^2, \; p'^3 = p^3,$$

and this is easily cast into standard matrix form. The two approaches are therefore equivalent, but the APS approach of transforming paravectors is more geometric and does not require Lorentz transformations that are aligned with basis elements. Indeed, the algebraic transformation $p \rightarrow p' = LpL^\dagger$ is independent of basis.

A basis is needed only to compare measured coefficients. In general, the transformed paravector p' is $p' = p'^\mu \mathbf{e}_\mu = p^\nu \mathbf{u}_\nu$, where as above $\mathbf{u}_\mu = L\mathbf{e}_\mu L^\dagger$ are the transformed basis paravectors. To isolate individual coefficients from an expansion, we introduce reciprocal basis paravectors $\mathbf{e}^\mu$ defined by the scalar products

$$\left\langle \mathbf{e}_\mu \bar{\mathbf{e}}^\nu \right\rangle_S = \delta_\mu^\nu \, ,$$

where δ_μ^ν is the usual Kronecker delta. The reciprocal paravectors of the standard basis elements $\mathbf{e}_\mu$ are $\mathbf{e}^0 = \mathbf{e}_0, \mathbf{e}^k = -\mathbf{e}_k$, $k = 1, 2, 3$. With their help, we obtain

$$p'^\mu = p^\nu \left\langle \mathbf{u}_\nu \bar{\mathbf{e}}^\mu \right\rangle_S = p^\nu \left\langle L\mathbf{e}_\nu L^\dagger \bar{\mathbf{e}}^\mu \right\rangle_S \equiv \mathcal{L}_\nu^\mu p^\nu. \qquad (27)$$

3.1   Active, Passive, and Relative Transformations

It is common to distinguish active transformations from passive ones. In active transformations, a single observer compares objects in different inertial frames, whereas in passive transformations a single object is observed by inertial observers in relative motion to each other. The transformations described above were active. However, APS accommodates both active and passive interpretations. The mathematics is the same, as seen from expression (27) for the transformation elements

$$\mathcal{L}_\nu^\mu = \left\langle \left(L\mathbf{e}_\nu L^\dagger \right) \bar{\mathbf{e}}^\mu \right\rangle_S = \left\langle \mathbf{e}_\nu \left(L^\dagger \bar{\mathbf{e}}^\mu L \right) \right\rangle_S = \left\langle \mathbf{e}_\nu \overline{\left(\bar{L} \mathbf{e}^\mu \bar{L}^\dagger \right)} \right\rangle_S . \qquad (28)$$

Here we used the property that the scalar part of a product is independent of the order $\langle xy \rangle_S = \langle yx \rangle_S$ for any elements x, y of APS. The first equality of (28) finds the components of the transformed basis paravectors on the original basis, whereas the last finds components of the original basis on the inversely transformed basis. These are alternative ways to interpret the same expression.

To see the relation between active and passive transformations more explicitly, consider the passive transformation of a fixed paravector from one observer, say Alice, to another, say Bob. Let $p_A = p_A^\mu \mathbf{e}_\mu$ be the paravector as seen by Alice in terms of Alice's standard (rest) basis. Bob, who moves at proper velocity u with respect to Alice, will measure a different paravector, namely $p_B = p_B^\mu \mathbf{e}_\mu$, with respect to his rest frame. Note that the paravector basis in expansions of both p_A and p_B is the same, namely the standard basis $\{\mathbf{e}_\mu\}$, even though p_A is expressed relative to Alice and p_B relative to Bob. The reason is that the standard basis $\{\mathbf{e}_\mu\}$ is *relative*; it is at rest relative to the observer. To relate p_A and p_B, both must be expressed relative to the same observer. Bob's frame as seen by Alice is

$$\{\mathbf{u}_\mu\} = \left\{ L\mathbf{e}_\mu L^\dagger \right\} \tag{29}$$

with $\mathbf{u}_0 = u$. Thus p_A can be written

$$p_A = p_A^\mu \mathbf{e}_\mu = p_B^\nu \mathbf{u}_\nu = L p_B L^\dagger , \tag{30}$$

which can be inverted to give the passive transformation

$$p_A \rightarrow p_B = \bar{L} p_A \bar{L}^\dagger. \tag{31}$$

This has the same form as the active transformation, but with the inverse Lorentz rotor.

APS uses the same mathematics to find not only passive and active Lorentz rotations, but also any mixture of passive and active rotations: all that counts is the *relative* Lorentz rotation of the observed object with respect to the observer. This property means that the basis paravectors themselves represent not an absolute frame, but rather a frame *relative* to the observer (or Lab). The proper basis $\{\mathbf{e}_\mu\}$ with $\mathbf{e}_0 = 1$ represents a frame at rest with respect to the observer. In APS, as in experiments, it is only the *relative* motion and orientation of the observed object with respect to the observer that is significant.

3.2 Covariant Elements and Invariant Properties

Experiments generally measure real scalars such as the size of paravector components on a given basis. The most meaningful geometric quantities in relativity, however, are spacetime vectors and products thereof that simply rotate and reflect in paravector space under the action of Lorentz transformations. Such quantities are said to be *covariant*. In APS, covariant spacetime vectors are real paravectors, and the biparavectors and triparavectors formed from them are also covariant, but one can move back and forth easily between covariant quantities and their components. Individual components are not generally covariant. Some properties, such as the scalar product of covariant paravectors and the square of simple covariant biparavectors, are *invariant*, unchanged by Lorentz rotations. Such properties are known as Lorentz scalars.

3.3   Relation of APS to STA

An alternative to the paravector model of spacetime is the spacetime algebra (STA) introduced by David Hestenes [13, 15]. APS and STA are closely related, and it is the purpose of this section show how.

STA is the geometric algebra $\mathcal{C}\ell_{1,3}$ of Minkowski spacetime. Whereas the Minkowski spacetime metric appears automatically in APS, it is imposed in STA. In each frame STA starts with a 4-dimensional orthonormal basis $\{\gamma_0, \gamma_1, \gamma_2, \gamma_3\} \equiv \{\gamma_\mu\}$ satisfying

$$\gamma_\mu \gamma_\nu + \gamma_\nu \gamma_\mu = 2\eta_{\mu\nu},$$

where as previously, $\eta_{\mu\nu}$ are elements of the metric tensor $(\eta_{\mu\nu}) = \text{diag}(1, -1, -1, -1)$. The volume element in STA is $\mathbf{I} = \gamma_0 \gamma_1 \gamma_2 \gamma_3$. Although it is referred to as the unit pseudoscalar and squares to -1, it anticommutes with vectors, thus behaving more like an additional spatial dimension than a scalar.

The frame chosen for a system can be independent of the observer and her frame $\{\hat{\gamma}_\mu\}$. Any spacetime vector $p = p^\mu \gamma_\mu$ expanded in the system frame $\{\gamma_\mu\}$ can be multiplied by observer's time axis $\hat{\gamma}_0$ to give

$$p\hat{\gamma}_0 = p \cdot \hat{\gamma}_0 + p \wedge \hat{\gamma}_0 = p^\mu \mathbf{u}_\mu, \qquad (32)$$

where[3] $\mathbf{u}_\mu = \gamma_\mu \hat{\gamma}_0$. In particular, $\mathbf{u}_0 = \gamma_0 \hat{\gamma}_0$ is the proper velocity of the system frame $\{\gamma_\mu\}$ with respect to the observer frame $\{\hat{\gamma}_\mu\}$. If the system frame is at rest with respect to the observer, then $\gamma_0 = \hat{\gamma}_0$ and

---

[3]A double arrow might be thought more appropriate than an equality here, because $\mathbf{u}_\mu$ and $\gamma_\mu, \hat{\gamma}_0$ act in different algebras. However, we are identifying $\mathcal{C}\ell_3$ with the even subalgebra of $\mathcal{C}\ell_{1,3}$, so that the one algebra is embedded in the other. Caution is still needed to avoid statements such as

$$i = \mathbf{e}_1 \mathbf{e}_2 \mathbf{e}_3 = \mathbf{e}_1 \wedge \mathbf{e}_2 \wedge \mathbf{e}_3 \overset{\text{wrong!}}{=} \hat{\gamma}_1 \wedge \hat{\gamma}_0 \wedge \hat{\gamma}_2 \wedge \hat{\gamma}_0 \wedge \hat{\gamma}_3 \wedge \hat{\gamma}_0 = 0.$$

This is not valid because the wedge products on either side of the third equality refer to different algebras and are not equivalent.

the relative basis paravectors $\mathbf{u}_\mu$ are replaced by proper basis elements $\mathbf{e}_\mu$. The result

$$p\hat{\gamma}_0 = p^\mu \mathbf{e}_\mu = p^0 + \mathbf{p},$$

where $\mathbf{e}_0 \equiv 1$ and $\mathbf{e}_k \equiv \gamma_k \hat{\gamma}_0$, is called a *space/time split*. The association $\gamma_\mu \hat{\gamma}_0 = \mathbf{e}_\mu$ when $\gamma_0 = \hat{\gamma}_0$ establishes the previously mentioned isomorphism between the even subalgebra of STA and APS. Together with $\mathbf{u}_\mu = \gamma_\mu \hat{\gamma}_0$ for a more general system basis, it emphasizes that the basis vectors in APS are *relative*: they always relate two frames, the system frame and the observer frame, each of which has its own basis in STA.

Clifford conjugation in APS corresponds to reversion in STA, indicated by a tilde. For example, $\bar{\mathbf{u}}_\mu = (\gamma_\mu \hat{\gamma}_0)^{\tilde{}} = \hat{\gamma}_0 \gamma_\mu$. In particular, the proper velocity of the observer frame $\{\hat{\gamma}_\mu\}$ with respect to γ_0 is $\bar{\mathbf{u}}_0 = \hat{\gamma}_0 \gamma_0$, the inverse of $\mathbf{u}_0 = \gamma_0 \hat{\gamma}_0$. It is not possible to make all of the basis vectors in any STA frame Hermitian, but one usually takes $\hat{\gamma}_0^\dagger = \hat{\gamma}_0$ and $\hat{\gamma}_k^\dagger = -\hat{\gamma}_k$ in the observer's frame $\{\hat{\gamma}_\mu\}$. More generally, Hermitian conjugation of an arbitrary element Γ in STA combines reversion with reflection in the observer's time axis $\hat{\gamma}_0$: $\Gamma^\dagger = \hat{\gamma}_0 \tilde{\Gamma} \hat{\gamma}_0$. For example, the relation

$$\mathbf{u}_\mu^\dagger = \left[\hat{\gamma}_0 \, (\gamma_\mu \hat{\gamma}_0)^{\tilde{}} \, \hat{\gamma}_0 \right] = \gamma_\mu \hat{\gamma}_0 = \mathbf{u}_\mu$$

shows that all the paravector basis vectors $\mathbf{u}_\mu$ are Hermitian. It is important to note that Hermitian conjugation is frame dependent in STA just as Clifford conjugation of paravectors is in APS.

EXAMPLE 5 *The Lorentz-invariant scalar part of the paravector product $p\bar{q}$ in APS has the same expansion as in STA:*

$$\langle p\bar{q} \rangle_S = \frac{1}{2} p^\mu q^\nu \left(\mathbf{e}_\mu \bar{\mathbf{e}}_\nu + \mathbf{e}_\nu \bar{\mathbf{e}}_\mu \right)$$

$$= \frac{1}{2} p^\mu q^\nu \left(\gamma_\mu \hat{\gamma}_0 \hat{\gamma}_0 \gamma_\nu + \gamma_\nu \hat{\gamma}_0 \hat{\gamma}_0 \gamma_\mu \right)$$

$$= p^\mu q^\nu \eta_{\mu\nu}.$$

Basis biparavectors in APS become basis bivectors in STA:

$$\frac{1}{2} \left(\mathbf{e}_\mu \bar{\mathbf{e}}_\nu - \mathbf{e}_\nu \bar{\mathbf{e}}_\mu \right) = \frac{1}{2} \left(\gamma_\mu \hat{\gamma}_0 \hat{\gamma}_0 \gamma_\nu - \gamma_\nu \hat{\gamma}_0 \hat{\gamma}_0 \gamma_\mu \right)$$

$$= \frac{1}{2} \left(\gamma_\mu \gamma_\nu - \gamma_\nu \gamma_\mu \right).$$

Lorentz transformations in STA are effected by $\gamma_\mu \to L\gamma_\mu \tilde{L}$, with $L\tilde{L} = 1$. Every product of basis vectors transforms the same way. An *active transformation* keeps the observer frame fixed and transforms only the system frame. Suppose the system frame is related to the observer frame by the Lorentz rotor $L : \gamma_\mu = L\hat{\gamma}_\mu \tilde{L}$. Then

$$\mathbf{u}_\mu = \gamma_\mu \hat{\gamma}_0 = L\hat{\gamma}_\mu \tilde{L}\hat{\gamma}_0 = L\hat{\gamma}_\mu \hat{\gamma}_0 \left(\hat{\gamma}_0 \tilde{L}\hat{\gamma}_0 \right) = L\mathbf{e}_\mu L^\dagger,$$

which coincides with the relation in APS. In a *passive transformation*, it is the system frame that stays the same and the observer's frame that changes. Let us suppose that the observer moves from frame $\{\gamma_\mu\}$ to frame $\{\hat{\gamma}_\mu\}$ where $\gamma_\mu = L\hat{\gamma}_\mu \tilde{L}$. Then

$$\mathbf{e}_\mu = \gamma_\mu \gamma_0 \rightarrow \mathbf{u}_\mu = \gamma_\mu \hat{\gamma}_0 .$$

To re-express the transformed relative coordinates $\mathbf{u}_\mu$ in terms of the original $\mathbf{e}_\mu$, we must expand the system frame vectors γ_μ in terms of the observer's transformed basis vectors $\hat{\gamma}_\mu$. Thus

$$\mathbf{u}_\mu = L\hat{\gamma}_\mu \tilde{L}\hat{\gamma}_0 = L\mathbf{e}_\mu L^\dagger.$$

The mathematics is identical to that for the active transformation, but the interpretation is different. It is important to stress that the system and observer frames are distinct in STA, and under active or passive transformations, only one of them changes. Confusion about this point can easily lead to errors in the transformation properties of elements and their relation space/time splits.

Since the transformations can be realized by changing the observer frame and keeping the system frame constant, the physical objects can be taken to be fixed in STA, giving what is sometimes referred to as an *invariant* formulation of relativity. Note, however, that the name "invariant" for covariant objects is consistent only if no active Lorentz transformations are needed. In order to avoid inconsistency and to prevent confusion of covariant expressions with Lorentz scalars such as scalar products of spacetime vectors, I prefer to call STA an *absolute-frame* formulation of relativity.

We have seen that a Lorentz rotation has the same physical effect whether we rotate the object forward or the observer backward or some combination. This is trivially incorporated in APS where only the object frame relative to the observer enters. The absolute frames of STA, while sometimes convenient, impose an added structure not required by experiment.

The space/time split of a property in APS is simply a result of expanding into vector grades in the observer's proper basis $\{\mathbf{e}_\mu\}$:

$$p = p^0 + \mathbf{p}$$
$$\mathbf{F} = \mathbf{E} + i\mathbf{B} .$$

Although p and $\mathbf{F}$ are covariant, the split is not; it is valid only in a rest frame of the observer. To relate this to the split as seen in a frame moving with proper velocity u with respect to the observer, we expand p in the paravector basis $\left\{\mathbf{u}_\mu = Le_\mu L^\dagger\right\}$ and $\mathbf{F}$ in the corresponding biparavector basis $\left\{\langle \mathbf{u}_\mu \bar{\mathbf{u}}_\nu \rangle_V\right\}$, where $u = \mathbf{u}_0$. The passive transformation of the observer to the moving frame replaces $\mathbf{u}_\mu$ by $\mathbf{e}_\mu$, and the new space/time split simply separates the result into vector grades. The physical fields, momenta, etc. are transformed and are not invariant in APS but *covariant,* that is, under a Lorentz transformation, the form of equations relating them remains the same even though the vectors and multivectors themselves change.[4]

STA uses the metric of signature $(1,3)$, but the pseudoEuclidean metric of signature $(3,1)$ could equally well have been used. This alternative uses the real Clifford algebra $Cl_{3,1}$ in place of STA's $Cl_{1,3}$. Because the two algebras are inequivalent, there is no simple transformation between them, and some authors have debated the relative merits of the two choices. APS easily accommodates both possibilities. Our formulation above gives a paravector metric of signature $(1,3)$, but simply by changing the overall sign on the definition of the quadratic form (or, equivalently, of the Clifford conjugate), we obtain a paravector space of the other signature, namely $(3,1)$.

STA and APS seem equally adept at modeling relativistic phenomena. This at first is surprising since STA has $2^4 = 16$ linearly independent elements whereas APS has only half that many. To understand how APS achieves its compactness, note that Lorentz scalars are grade 0 objects in both STA and APS, but spacetime vectors in STA are homogeneous elements of grade 1 whereas in APS they are paravectors, which mix grades 0 and 1. APS maintains a formal grade distinction between an observer's proper time axis and spatial directions by making time a scalar and spatial direction a vector. Furthermore, spacetime planes are represented by elements of grade 2 in STA and by biparavectors, which combine elements of grades 1 and 2, in APS. Elements of a given grade evidently play a double role in APS. Rather than being a disadvantage, however, the double roles mirror common usage.

For example, the spacetime momentum of a particle at rest has only one nonvanishing component, namely a time component equal to its mass $(c = 1)$, but the mass is also a Lorentz scalar giving the invariant "length" of the momentum. In APS, the mass is simply a scalar that can fill both roles. In STA, the two roles are represented by expressions

---

[4]You can have *absolute frames* in APS, if you want them for use in passive transformations, by introducing an *absolute observer*.

of different grades: the Lorentz-invariant mass is a scalar while the rest-frame momentum is a vector. They are not equal but are instead related by a space/time split, which requires multiplication by $\hat{\gamma}_0$.

Another example is provided by the electromagnetic field **F**, which for a given observer reduces to the electric field **E** if there is no magnetic part. In APS, **F** is a biparavector with a vector part **E** and a bivector part $ic\mathbf{B}$ for any given observer. There is no problem in identifying **E** both as a spatial vector and a spacetime plane that includes the time axis. In STA, on the other hand, the two choices require different notation. Since **F** is an element of grade 2, we either specify **E** as $\mathbf{F}\cdot\hat{\gamma}_0$ or as this times $\hat{\gamma}_0$. The expression $\mathbf{F}\cdot\hat{\gamma}_0$ has the form of a spacetime vector in STA, but of course **E** transforms differently. The correct transformation behavior of $\mathbf{F}\cdot\hat{\gamma}_0$ is obtained in STA if, as discussed above, one distinguishes between the observer frame $\{\hat{\gamma}_\mu\}$ and the system frame $\{\gamma_\mu\}$ and applies the Lorentz rotation to only one of them.

The double-role playing of vector grades in APS is responsible for its efficiency in modeling spacetime. STA requires twice as many degrees of freedom to model the same phenomena. This appears to be the cost of having an absolute-frame formulation of relativity. Both STA and APS easily relate a covariant representation to observer-dependent measurements, although the connection is more direct in APS.

4.   Eigenspinors

The motion of a particle is described by the special Lorentz rotor $L = \Lambda$ that transforms the particle from rest to the lab. Any property known in the rest frame can be transformed to the lab by Λ, which is known as the *eigenspinor* of the particle and is generally a function of its proper time τ. For example, the spacetime momentum in the lab is $p = \Lambda m \Lambda^\dagger$. The term "spinor" refers to the form of a Lorentz rotation of Λ, namely

$$\Lambda \to L\Lambda, \tag{33}$$

which is the form for the composition of Lorentz rotations but is distinct from Lorentz rotations of paravectors and their products.

The eigenspinor $\Lambda(\tau)$ is the solution of a time evolution equation in the simple linear form

$$\frac{d\Lambda}{d\tau} \equiv \dot{\Lambda} = \frac{1}{2}\Omega\Lambda(\tau), \tag{34}$$

where Ω is the *spacetime rotation rate* of the particle in the lab. This approach offers new tools for classical physics. It implies that

$$\dot{p} = \dot{\Lambda}m\Lambda^\dagger + \Lambda m\dot{\Lambda}^\dagger = \langle\Omega p\rangle_\Re . \tag{35}$$

For the motion of a charge e in the electromagnetic field $\mathbf{F} = \mathbf{E} + i\mathbf{B}$, the identification

$$\Omega = \frac{e}{m}\mathbf{F}, \tag{36}$$

when substituted into (35), gives the covariant Lorentz-force equation. Note that (36) can be taken as a covariant definition of $\mathbf{F}$, valid in any inertial frame. It is trivial to find Λ for any uniform field $\mathbf{F}$:

$$\Lambda(\tau) = \exp\left(\frac{e\mathbf{F}\tau}{2m}\right)\Lambda(0). \tag{37}$$

Solutions can also be found for relativistic charge motion in plane waves, plane-wave pulses, or plane waves superimposed on static longitudinal electric or magnetic fields. [19]

5.   Maxwell's Equation

Maxwell's famous equations are inherently relativistic, and it is a shame that so many texts treat much of electrodynamics in nonrelativistic approximation. One reason given for introducing relativity only late in an electrodynamics course is that tensors and or matrices are required which makes the presentation more abstract and harder to interpret. In APS we can easily display and exploit relativistic symmetries in simple vector and paravector terms without the need of tensors or matrices.

Maxwell's equations were written as a single quaternionic equation by Conway (1911), Silberstein (1912), and others. In APS we can write

$$\bar{\partial}\mathbf{F} = \mu_0\bar{j}, \tag{38}$$

where $\mu_0 = \varepsilon_0^{-1} = 4\pi \times \dot{3}0$ Ohm is the impedance of the vacuum, with $\dot{3} \equiv 2.99792458$. The usual four equations are simply the four vector grades of this relation, extracted as the real and imaginary, scalarlike and vectorlike parts. It is also seen as the *necessary covariant extension* of Coulomb's law $\nabla \cdot \mathbf{E} = \rho/\varepsilon_0$. The covariant field is not $\mathbf{E}$ but $\mathbf{F} = \mathbf{E} + i\mathbf{B}$, the divergence is part of the covariant gradient $\bar{\partial}$, and ρ must be part of $\bar{j} = \rho - \mathbf{j}$. The combination is Maxwell's equation.[5]

It is a simple exercise to derive the continuity equation $\langle\partial\bar{j}\rangle_S = 0$ in one step from Maxwell's equation. One need only note that the D'Alembertian $\partial\bar{\partial}$ is a scalar operator and that $\langle\mathbf{F}\rangle_S = 0$.

---

[5] We have assumed that the source is a real paravector current and that there are no contributing pseudoparavector currents. Known currents are of the real paravector type, and a pseudoparavector current would behave counter-intuitively under parity inversion. Our assumption is supported experimentally by the apparent lack of magnetic monopoles.

5.1 Directed Plane Waves

In source-free space ($\bar{j} = 0$), there are solutions $\mathbf{F}(s)$ that depend on spacetime position only through the Lorentz invariant $s = \langle k\bar{x} \rangle_S = \omega t - \mathbf{k} \cdot \mathbf{x}$, where $k = \omega + \mathbf{k} \neq 0$ is a constant propagation paravector. Since $\partial \langle k\bar{x} \rangle_S = k$, Maxwell's equation gives

$$\bar{\partial}\mathbf{F} = \bar{k}\mathbf{F}'(s) = 0 . \tag{39}$$

In a division algebra, we could divide by $\bar{k}$ and conclude that $\mathbf{F}'(s) = 0$, a rather uninteresting solution. There is another possibility here because APS is not a division algebra: $\bar{k}$ may have no inverse. Then k has the form $k = \omega \left(1 + \hat{\mathbf{k}}\right)$, and after integrating (39) from some s_0 at which $\mathbf{F}$ is presumed to vanish, we get $\left(1 - \hat{\mathbf{k}}\right)\mathbf{F}(s) = 0$, which means $\mathbf{F}(s) = \hat{\mathbf{k}}\mathbf{F}(s)$. The scalar part of $\mathbf{F}$ vanishes and consequently $\left\langle \hat{\mathbf{k}}\mathbf{F}(s) \right\rangle_S = \hat{\mathbf{k}} \cdot \mathbf{F}(s) = 0$ so that the fields $\mathbf{E}$ and $\mathbf{B}$ are perpendicular to $\hat{\mathbf{k}}$ and thus anticommute with it. Furthermore, equating imaginary parts gives $i\mathbf{B} = \hat{\mathbf{k}}\mathbf{E}$ and it follows that

$$\mathbf{F} = \mathbf{E} + i\mathbf{B} = \left(1 + \hat{\mathbf{k}}\right)\mathbf{E}(s) \tag{40}$$

with $\mathbf{E} = \langle\mathbf{F}\rangle_{\Re}$ real. This is a plane-wave solution with $\mathbf{F}$ constant on spatial planes perpendicular to $\hat{\mathbf{k}}$. Such planes propagate at the speed of light along $\hat{\mathbf{k}}$. In spacetime, $\mathbf{F}$ is constant on the light cone $\hat{\mathbf{k}} \cdot \mathbf{x} = t$. However, $\mathbf{F}$ is not necessarily monochromatic, since $\mathbf{E}(s)$ can have any functional form, including a pulse, and the scale factor ω, although it has dimensions of frequency, may have nothing to do with any physical oscillation. The structure of the plane wave $\mathbf{F}$ is that of a simple biparavector representing the spacetime plane containing both the null paravector $1 + \hat{\mathbf{k}}$ and the orthogonal direction $\mathbf{E}$. This structure ensures that $\mathbf{F}$ itself is *null*: $\mathbf{F}^2 = 0$. In fact, $\mathbf{F}$ is what Penrose calls a *null flag*. The *flagpole* $1 + \hat{\mathbf{k}}$ lies on the light cone in the plane of the flag but is orthogonal to both itself and the flag. This is the basis of an important symmetry that is critical for determining charge dynamics in plane waves. [19] The null-flag structure is beautiful and powerful, but you miss it entirely if you only write out separate electric and magnetic fields. The electric and magnetic fields are simply components of the null flag; $\mathbf{E}$ gives the extent of the flag perpendicular to the flagpole, and $\mathbf{B}$ represents the spatial plane swept out by $\mathbf{E}$ as it propagates along $\hat{\mathbf{k}}$.

The electric field $\mathbf{E}(s)$ determines the polarization of the wave. If the direction of $\mathbf{E}$ is constant, for example $\mathbf{E}(s) = \mathbf{E}(0)\cos s$, the wave is linearly polarized along $\mathbf{E}(0)$. If $\mathbf{E}$ rotates around $\hat{\mathbf{k}}$, for example

$\mathbf{E}(s) = \mathbf{E}(0)\exp\left(i\kappa\hat{\mathbf{k}}s\right)$, $\kappa = \pm 1$, the wave is circularly polarized with helicity κ. Note that the flagpole can gobble the unit vector $\hat{\mathbf{k}}$:

$$\mathbf{F}(s) = \left(1+\hat{\mathbf{k}}\right)\mathbf{E}(0)\exp\left(i\kappa\hat{\mathbf{k}}s\right) = \left(1+\hat{\mathbf{k}}\right)\exp\left(-i\kappa\hat{\mathbf{k}}s\right)\mathbf{E}(0)$$

$$= \left(1+\hat{\mathbf{k}}\right)\mathbf{E}(0)\exp\left(-i\kappa s\right). \tag{41}$$

This establishes an equivalence for null flags between rotations about the spatial direction $\hat{\mathbf{k}}$ of the flagpole and multiplication by a phase factor. Since the dual of $\mathbf{F}$ is $-i\mathbf{F}$, the phase factor is said to induce duality rotations. The energy density $\mathcal{E} = \frac{1}{2}\left(\varepsilon_0\mathbf{E}^2 + \mathbf{B}^2/\mu_0\right)$ and the Poynting vector $\mathbf{S} = \mathbf{E}\times\mathbf{B}/\mu_0$ for the plane wave are combined in

$$\frac{1}{2}\varepsilon_0\mathbf{F}\mathbf{F}^\dagger = \mathcal{E} + \mathbf{S} = \varepsilon_0\mathbf{E}^2\left(1+\hat{\mathbf{k}}\right).$$

5.2   Polarization Basis

The application of APS to the polarization of a beam of light or other electromagnetic radiation gives a formulation vastly simpler than the usual approach with Mueller matrices, and it demonstrates spinorial type transformations and additional uses of paravectors. Furthermore, the mathematics is the same as used to describe electron polarization, which I discuss in my next lecture.

The field $\mathbf{F}$ of a beam of monochromatic radiation can be expressed as a linear combination of two independent polarization types. Both linear and circular polarization bases are common, but a circular basis is most convenient, partially because of the relation noted above between spatial and duality rotations. Circularly polarized waves also have the simple form used popularly by R. P. Feynman [20] to discuss light propagation in terms of a rotating pointer that we can take to be $\mathbf{E}(s)$. A linear combination of both helicities of a directed plane wave gives

$$\mathbf{F} = \left(1+\hat{\mathbf{k}}\right)\hat{\mathbf{E}}_0 e^{i\delta\hat{\mathbf{k}}}\left(E_+ e^{is\hat{\mathbf{k}}} + E_- e^{-is\hat{\mathbf{k}}}\right)$$

$$= \left(1+\hat{\mathbf{k}}\right)\hat{\mathbf{E}}_0 e^{-i\delta}\left(E_+ e^{-is} + E_- e^{is}\right),$$

where $E_\pm$ are the real field amplitudes, δ gives the rotation of $\mathbf{E}$ about $\hat{\mathbf{k}}$ at $s = 0$ from the unit vector $\hat{\mathbf{E}}_0$, and in the second line, we gobbled $\hat{\mathbf{k}}$'s. Because every directed plane wave can be expressed in the form $\mathbf{F} = \left(1+\hat{\mathbf{k}}\right)\mathbf{E}(s)$, it is sufficient to determine the electric field $\mathbf{E}(s) =$

$\langle \mathbf{F} \rangle_{\Re}$:

$$\begin{aligned}
\mathbf{E} &= \left\langle \left(1 + \hat{\mathbf{k}} \right) \hat{\mathbf{E}}_0 E_+ e^{-i\delta} e^{-is} + \left(1 + \hat{\mathbf{k}} \right) \hat{\mathbf{E}}_0 E_- e^{-i\delta} e^{is} \right\rangle_{\Re} \\
&= \left\langle \left[\left(1 + \hat{\mathbf{k}} \right) \hat{\mathbf{E}}_0 E_+ e^{-i\delta} + \hat{\mathbf{E}}_0 \left(1 + \hat{\mathbf{k}} \right) E_- e^{i\delta} \right] e^{-is} \right\rangle_{\Re} \\
&= \left\langle \left(\epsilon_+, \epsilon_- \right) \Phi e^{-is} \right\rangle_{\Re} ,
\end{aligned}$$

where the complex polarization basis vectors $\epsilon_\pm = 2^{-1/2} \left(1 \pm \hat{\mathbf{k}} \right) \hat{\mathbf{E}}_0$ are null flags satisfying $\epsilon_- = \epsilon_+^\dagger$, $\epsilon_+ \cdot \epsilon_+^\dagger = 1 = \epsilon_- \cdot \epsilon_-^\dagger$, and the *Poincaré spinor*

$$\Phi = \sqrt{2} \begin{pmatrix} E_+ e^{-i\delta} \\ E_- e^{i\delta} \end{pmatrix} \tag{42}$$

gives the (real) electric-field amplitudes and their phases, and it contains all the information needed to determine the polarization and intensity of the wave. The spinor (42) is related by unitary transformation to the Jones vector, which uses a basis of linear polarization.

5.2.1   Stokes Parameters.   To describe partially polarized light, we can use the *coherency density*, [6] which in the case of a single Poincaré spinor is

$$\rho = \varepsilon_0 \Phi \Phi^\dagger = \rho^\mu \sigma_\mu , \tag{43}$$

where the σ_μ are the Pauli spin matrices. The normalization factor ε_0 has been chosen to make ρ^0 the time-averaged energy density. The coefficients $\rho^\mu = \langle \rho \sigma_\mu \rangle_S$ are the *Stokes parameters*. The coherency density $\rho = \rho^0 + \boldsymbol{\rho}$ is a paravector in the algebra for the space spanned by $\{ \sigma_1, \sigma_2, \sigma_3 \}$. This space, called *Stokes space*, is a 3-D Euclidean space analogous to physical space. It is not physical space, but its geometric algebra is isomorphic to APS, and it illustrates how Clifford algebras can arise in physics for spaces other than physical space.

As defined for a single Φ, ρ is null ($\det \rho = \rho \bar{\rho} = 0$) and can be written $\rho = \rho^0 \left(1 + \mathbf{n} \right)$, where $\mathbf{n}$, a unit vector in the direction of $\boldsymbol{\rho}$, specifies the type of polarization. In particular, for positive helicity light, $\boldsymbol{n} = \sigma_3$, for negative helicity polarization $\boldsymbol{n} = -\sigma_3$, and for linear polarization at an angle $\delta = \phi/2$ with respect to $\mathbf{E}_0$, $\boldsymbol{n} = \sigma_1 \cos \phi + \sigma_2 \sin \phi$. Other directions correspond to elliptical polarization.

5.2.2   Polarizers and Phase Shifters.   The action of ideal polarizers and phase shifters on the wave is modeled mathematically by transformations on the Poincaré spinor Φ of the form $\Phi \rightarrow T\Phi$. For *polarizers* T that pass polarization of type **n**, T is the *projector* $T = \mathsf{P_n} =$

$\frac{1}{2}(1+\mathbf{n})$, which is proportional to the pure state of that polarization. Projectors are idempotent ($\mathrm{P}_\mathbf{n}^2 = \mathrm{P}_\mathbf{n}$), just as we would expect for ideal polarizers since a second application of $\mathrm{P}_\mathbf{n}$ changes nothing further. The polarizer represented by the complementary projector $\bar{\mathrm{P}}_\mathbf{n}$ annihilates $\mathrm{P}_\mathbf{n}$: $\bar{\mathrm{P}}_\mathbf{n}\mathrm{P}_\mathbf{n} = \mathrm{P}_{-\mathbf{n}}\mathrm{P}_\mathbf{n} = 0$, and in general, opposite directions in Stokes space correspond to orthogonal polarizations.

If the wave is split into orthogonal polarization components ($\pm\mathbf{n}$) and the two components are given a relative phase shift of α, the result is equivalent to rotating ρ by α about $\mathbf{n}$ in Stokes subspace: $T = \mathrm{P}_\mathbf{n}e^{i\alpha/2} + \bar{\mathrm{P}}_\mathbf{n}e^{-i\alpha/2} = e^{i\mathbf{n}\alpha/2}$. Depending on $\mathbf{n}$, this operator can represent both the Faraday effect and the effect of a birefringent medium with polarization types $\mathbf{n}$ and $-\mathbf{n}$ corresponding to the slow and fast axes, respectively.

5.2.3   Coherent Superpositions and Incoherent Mixtures.

A superposition of two waves of the same frequency is *coherent* because their relative phase is fixed. Mathematically, one adds spinors in such cases: $\Phi = \Phi_1 + \Phi_2$, where the subscripts refer to the two waves, not to spinor components. Waves of different frequencies have a continually changing relative phase, and when averaged over periods large relative to their beat period, combine *incoherently*: $\rho = \rho_1 + \rho_2$. The degree of polarization is given by the length of ρ relative to ρ_0 and can vary from 0 to 100%. Any transformation T of spinors, $\Phi \to T\Phi$, transforms the coherency density by $\rho \to T\rho T^\dagger$, and transformations that do not preserve the polarization can also be applied to ρ. [6]

6.   Conclusions

The multiparavector structure of APS makes the algebra ideal for modeling relativistic phenomena. It presents a covariant formulation based on relative motion and orientation but provides a simple path to the spatial vectors for any given observer as well as to the operators that act on the vectors. The formulation is simple enough to be used in introductory physics courses, and it holds the promise of becoming a key factor in any curriculum revision designed to train students to contribute significantly to the quantum age of the 21st century.

Since APS is isomorphic to complex quaternions, any calculation in APS can be repeated with quaternions taken over the complex field. However, the geometry is considerably clearer in APS. In its descriptive and computational power for relativistic physics, APS seems as capable as STA. However, STA has twice the size of APS in order to add the non-observable structure of absolute frames to its formulation.

Acknowledgement

Support from the Natural Sciences and Engineering Research Council of Canada is gratefully acknowledged.

References

[1] R. Abłamowicz and G. Sobczyk, eds., *Lectures on Clifford Geometric Algebras,* Birkhäuser, Boston, 2003.

[2] D. Hestenes, *Am. J. Phys.* **71**:104–121, 2003.

[3] W. E. Baylis, editor, *Clifford (Geometric) Algebra with Applications to Physics, Mathematics, and Engineering,* Birkhäuser, Boston 1996.

[4] J. Snygg, *Clifford Algebra, a Computational Tool for Physicists,* Oxford U. Press, Oxford, 1997.

[5] K. Gürlebeck and W. Sprössig, *Quaternions and Clifford Calculus for Physicists and Engineers,* J. Wiley and Sons, New York, 1997.

[6] W. E. Baylis, *Electrodynamics: A Modern Geometric Approach,* Birkhäuser, Boston, 1999.

[7] R. Abłamowicz and B. Fauser, eds., *Clifford Algebras and their Applications in Mathematical Physics, Vol. 1: Algebra and Physics,* Birkhäuser, Boston, 2000.

[8] R. Abłamowicz and J. Ryan, editors, *Proceedings of the 6th International Conference on Applied Clifford Algebras and Their Applications in Mathematical Physics,* Birkhäuser Boston, 2003.

[9] P. Lounesto, *Clifford Algebras and Spinors,* second edition, Cambridge University Press, Cambridge (UK) 2001.

[10] W. R. Hamilton, *Elements of Quaternions,* Vols. I and II, a reprint of the 1866 edition published by Longmans Green (London) with corrections by C. J. Jolly, Chelsea, New York, 1969.

[11] S. Adler, *Quaternion Quantum Mechanics and Quantum Fields,* Oxford University Press, Oxford (UK), 1995.

[12] W. E. Baylis, J. Wei, and J. Huschilt, *Am. J. Phys.* **60**:788–797, 1992.

[13] D. Hestenes, *Spacetime Algebra,* Gordon and Breach, New York 1966.

[14] D. Hestenes, *New Foundations for Classical Mechanics,* 2nd edn., Kluwer Academic, Dordrecht, 1999.

[15] C. Doran and A. Lasenby, *Geometric Algebra for Physicists,* Cambridge University Press, Cambridge (UK), 2003.

[16] G. Somer, ed., *Geometric Computing with Clifford Algebras: Theoretical Foundations and Applications in Computer Vision and Robotics,* Springer-Verlag, Berlin, 2001.

[17] G. Trayling and W. E. Baylis, *J. Phys. A* **34**:3309-3324, 2001.

[18] A. Einstein, *Annalen der Physik* **17**:891, 1905.

[19] W. E. Baylis and Y. Yao, *Phys. Rev.* **A60**:785–795, 1999.

[20] R. P. Feynman, *QED: The Strange Story of Light and Matter,* Princeton Science, 1985.

A UNIFIED APPROACH TO FOURIER-CLIFFORD-PROMETHEUS SEQUENCES, TRANSFORMS AND FILTER BANKS

Ekaterina L.-Rundblad, Valeriy Labunets, and Ilya Nikitin
Ural State Technical University
Ekaterinburg, Russia
lab@rtf.ustu.ru

Abstract   In this paper we develop a new unified approach to the so-called generalized *Fourier-Clifford-Prometheus sequences, transforms* (FCPTs) and *M-channel Filter Banks*. It is based on a new generalized FCPT-generating construction. This construction has a rich algebraic structure that supports a wide range of fast algorithms.

Keywords: Clifford algebra, filter banks, Golay-Shapiro sequences, Fourier-Clifford-Prometheus transforms.

1.   Introduction

The basis which has come to be known as the *Prometheus Orthonormal Set* (PONS) was introduced in [1] to prove the H.S. SHAPIRO global uncertainty principle conjecture. Each function in PONS is called a *Golay-Shapiro sequence*. They are defined on $[0, 1]$, piecewise ± 1 and can change sign only at points of the form $j/2^n$, $j = 0, 1, \ldots, 2^n - 1$, $n = 1, 2, \ldots$. These basis functions satisfy almost all standard properties of the Walsh functions. Discrete classical Fourier-Prometheus Transforms (FPT) in bases of different Golay-Shapiro sequences can be used in many signal processing applications: multiresolution by discrete orthogonal wavelet decomposition, digital audio, digital video broadcasting, communication systems (Orthogonal Frequency Division Multiplexing, Multi-Code Code-Division Multiple Access), radar, and cryptographic systems.

Golay-Shapiro (GS) 2-complementary (± 1)-valued sequences associated with the cyclic group $\mathbf{Z}_2$ were introduced by SHAPIRO and GOLAY

J. Byrnes (ed.) Computational Noncommutative Algebra and Applications, 389-400.
© 2004 *Kluwer Academic Publishers. Printed in the Netherlands.*

in 1949–1951 [2]–[7]. In 1961, Golay [3] gave an explicit construction for binary Golay complementary pairs of length 2^m and later noted [4] that the construction implies the existence of at least $2^m m!/2$ binary Golay sequences of this length. They are known to exist for all lengths $N = 2^\alpha 10^\beta 26^\gamma$, where α, β, γ are integers and $\alpha, \beta, \gamma \geq 0$ [8], but do not exist for any length N having a prime factor congruent to 3 modulo 4 [9]. BUDISIN [10] using the earlier work of SIVASWAMY [11], gave a more general recursive construction for Golay complementary pairs and showed that the set of all binary Golay complementary pairs of length 2^m obtainable from it concides with those given explicitly by Golay [3]. For a survey of results on nonbinary Golay Complementary pairs, see [12]–[13]. Recently, DAVIS and JEDWAB [14], combining results appearing in the work of Golay and Shapiro cited above, gave an explicit description of a large class of Golay complementary sequences in terms of certain cosets of the first order Reed-Muller codes. The following general elements are used for building the classical Fourier-Prometheus transforms in bases of classical Golay-Shapiro sequences: 1) the Abelian group $\mathbf{Z}_2^n$, 2) the 2-point Fourier transform $\mathcal{F}_2$, and 3) the complex field $\mathbf{C}$; i.e., these transforms are associated with the triple $(\mathbf{Z}_2^n, \mathcal{F}_2, \mathbf{C})$.

The multiresolution analysis (MRA) operates upon a discrete signal $x(l)$ of length 2^n, where n is an integer. The sequence $x(l)$ is convolved with two filters L and H. Each convolution results in a sequence half the length of the original sequence. The result from the convolution with the low-pass filter is again transformed. Each re-transformed sequence of the low-pass output is referred to as a dilation. For a sequence $x(l)$ of length 2^n, a maximum of n dilations can be performed. MRA applied to a real-valued sequence $x(l)$ is defined recursively by the equations:

$$c^{(p)}(l) = L\left\{c^{(p-1)}(l)\right\}, \quad d^{(p)} = H\left\{c^{(p-1)(l)}\right\},$$

where $p = n, n-1, \ldots, 1, 0$, $c^n(l) = x(l)$, and

$$c(l) = (Lx)(l) = \sum_{l=0}^{2^n-1} k_{lp}(l - 2k)x(l),$$

$$d(l) = (Hx)(l) = \sum_{l=0}^{2^n-1} k_{hp}(l - 2k)x(l)$$

are low-pass and high-pass filters, respectively.

The sequences $c^{(p)}(l)$ and $d^{(p)}(l)$ are called the "averages" and "differences" of the original signal. The inverse discrete wavelet transform reconstructs $c^{(n)}(l) = x(l)$ using the recursive algorithm

$$c^{(p+1)}(l) = L^*\{c^{(p)}(l)\} + H^*\{d^{(p)}(l)\},$$

where L^* and H^* are the inverse filters of L and H, respectively. All filters L, H, and L^*, H^*, satisfy the following equation $LL^* = I$, $\quad HH^* = I$, and

$$LL^* + HH^* = 2I, \quad LH^* = H^*L = 0, \tag{1}$$

where I and 0 denote the identity and zero operators. Note that a pair of filters having these properties required of the transformations L and H are known as quadrature mirror filters, having the perfect reconstruction property.

The conditions (1) can be rewritten in terms of the $\mathcal{Z}$-transform as

$$|k_{lp}(z)|^2 + |k_{hp}(z)|^2 = 2, \quad \overline{k}_{lp}(z)k_{lp}(-z) + \overline{k}_{hp}(z)k_{hp}(-z) = 0, \quad \forall z \in \mathbb{T}_1$$

where $\mathbb{T}_1$ is the unit circle of the complex field $\mathbb{C}$. These conditions mean that impulse responses $k_{lp}(l)$ and $k_{hp}(l)$ form a Golay-Shapiro (GS) 2-complementary pair.

In this paper we develop a new unified approach to the so-called generalized *Fourier-Clifford-Prometheus* (FCP) *sequences*, FCP *transforms* (FCPTs), and *M-channel Filter Banks*. We describe the precise theoretical and computational relationship between M-band wavelets, M-channel filterbanks and generalized Golay-Shapiro sequences. The approach is based on a new generalized FCPT-generating construction. This construction has a rich algebraic structure that supports a wide range of fast algorithms. This construction is associated not with the triple $\left(\mathbf{Z}_2^n, \mathcal{F}_2, \mathbf{C}\right)$, but rather with other groups instead of $\mathbf{Z}_2^n$, other unitary transforms instead of $\mathcal{F}_2$, and other algebras (Clifford algebras) instead of the complex field $\mathbf{C}$.

2. New construction of classical and multiparametric Prometheus transforms

We begin by describing the original Golay 2-complementary (± 1)-valued sequences.

DEFINITION 1 *Let* $\mathbf{p}(t) := (p_0, p_1, \ldots, p_{N-1})$, $\mathbf{q}(t) := (q_0, q_1, \ldots, q_{N-1})$, *where* $p_i, q_i \in \{\pm 1\}$. *The sequences* $\mathbf{p}(t), \mathbf{q}(t)$ *are called a 2-complementary (± 1)-valued or Golay complementary pair over* $\{\pm 1\}$ *if*

$$\mathrm{COR}[\mathbf{p}, \mathbf{p}](\tau) + \mathrm{COR}[\mathbf{q}, \mathbf{q}](\tau) = N\delta(\tau),$$

or

$$|\mathbf{p}(z)|^2 + |\mathbf{q}(z)|^2 = N, \quad \forall z \in \mathbb{T}_1,$$

where $\mathrm{COR}[\mathbf{f}, \mathbf{f}](\tau)$ *is the periodic correlation function of* $\mathbf{f}(t)$; $\mathbf{p}(z)$ *and* $\mathbf{q}(z)$ *are* $\mathcal{Z}$-transforms of $\mathbf{p}(t)$ *and* $\mathbf{q}(t)$, *respectively. Any sequence which is a member of a Golay complementary pair is called a Golay sequence.*

The Fourier-Prometheus matrix of depth n has size $2^n \times 2^n$: $\mathcal{FP}_{2^n} = [\mathrm{Pr}_\alpha(t)]_{\alpha,t=0}^{2^n-1}$. For α and t we shall use binary representations $\alpha = \alpha_{[n]} := (\alpha_1, \alpha_2, \ldots, \alpha_n)$, $t = t_{[n]} := (t_1, t_2, \ldots, t_n)$, where $\alpha_i, t_i \in \{0,1\}$, $i = 1, 2, \ldots, n$. Obviously, $\alpha_{[1]} = (\alpha_1)$, $\alpha_{[2]} = (\alpha_1, \alpha_2)$, $\alpha_{[3]} = (\alpha_1, \alpha_2, \alpha_3)$, $\ldots$ $t_{[1]} = (t_1)$, $t_{[2]} = (t_1, t_2)$, $t_{[n]} = (t_1, t_2, \ldots, t_n), \ldots$. For this reason,

$$2^n\mathcal{FP}_{(\alpha_{[n-1]},\alpha_n)} = \begin{bmatrix} \mathrm{Pr}_{(0,0,\ldots,0,0)}(t_1,\ldots,t_n) \\ \mathrm{Pr}_{(0,0,\ldots,0,1)}(t_1,\ldots,t_n) \\ \hline \mathrm{Pr}_{(0,0,\ldots,1,0)}(t_1,\ldots,t_n) \\ \mathrm{Pr}_{(0,0,\ldots,1,1)}(t_1,\ldots,t_n) \\ \hline \cdots \\ \cdots \\ \hline \mathrm{Pr}_{(1,1,\ldots,1,0)}(t_1,\ldots,t_n) \\ \mathrm{Pr}_{(1,1,\ldots,1,1)}(t_1,\ldots,t_n) \end{bmatrix} = \boxplus_{\alpha_{[n-1]}=0}^{2^{n-1}-1} \begin{bmatrix} \mathrm{Pr}_{(\alpha_{[n-1]},0)}(t) \\ \mathrm{Pr}_{(\alpha_{[n-1]},1)}(t) \end{bmatrix},$$

where $\mathrm{Pr}_{(\alpha_{[n-1]},0)}(t)$ and $\mathrm{Pr}_{(\alpha_{[n-1]},1)}(t)$ are a pair of GS 2-complementary sequences and $\boxplus$ represents the vertical concatenation of matrices.

The classical matrix $\mathcal{FP}_{2^n}$ is formed by starting with the (2×2)-matrix $2^1\mathcal{FP} = \begin{bmatrix} \mathrm{Pr}_0(t) \\ \mathrm{Pr}_1(t) \end{bmatrix} = \begin{bmatrix} 1 & 1 \\ 1 & -1 \end{bmatrix}$ and by repeated application of the **PONS**-iteration construction to pairs of rows in the matrix. In the $(n+1)$st iteration this construction takes each pair $\begin{bmatrix} \mathbf{p} \\ \mathbf{q} \end{bmatrix} = \begin{bmatrix} \mathrm{Pr}_{(\alpha_{[n-1]},0)}(t) \\ \mathrm{Pr}_{(\alpha_{[n-1]},1)}(t) \end{bmatrix}$ of

$$2^n\mathcal{FP}_{(\alpha_{[n-1]},\alpha_n)} = \boxplus_{\alpha_{[n-1]}=0}^{2^{n-1}-1} \begin{bmatrix} \mathrm{Pr}_{(\alpha_{[n-1]},0)}(t) \\ \mathrm{Pr}_{(\alpha_{[n-1]},1)}(t) \end{bmatrix}$$

and constructs four rows of twice the length

$$\mathbf{PONS(p,q)} = \begin{bmatrix} \mathbf{p} & \mathbf{q} \\ \mathbf{p} & -\mathbf{q} \\ \hline \mathbf{q} & \mathbf{p} \\ -\mathbf{q} & -\mathbf{p} \end{bmatrix} = \begin{bmatrix} \mathbf{p} & \mathbf{q} \\ \mathbf{p} & -\mathbf{q} \end{bmatrix} \boxplus \begin{bmatrix} \mathbf{q} & \mathbf{p} \\ -\mathbf{q} & -\mathbf{p} \end{bmatrix}$$

$$= \left(\begin{bmatrix} 1 & 1 \\ 1 & -1 \end{bmatrix} \begin{bmatrix} \mathbf{p} & \\ \hline & \mathbf{q} \end{bmatrix} \right) \boxplus \left(\begin{bmatrix} 1 & 1 \\ 1 & -1 \end{bmatrix} \begin{bmatrix} & \mathbf{p} \\ \hline \mathbf{q} & \end{bmatrix} \right)$$

$$= \left(\begin{bmatrix} 1 & 1 \\ 1 & -1 \end{bmatrix} \begin{bmatrix} \mathbf{p} & \\ \hline & \mathbf{q} \end{bmatrix} \begin{bmatrix} 1 & \\ & 1 \end{bmatrix} \right) \boxplus \left(\begin{bmatrix} 1 & 1 \\ 1 & -1 \end{bmatrix} \begin{bmatrix} \mathbf{p} & \\ \hline & \mathbf{q} \end{bmatrix} \begin{bmatrix} & 1 \\ 1 & \end{bmatrix} \right)$$

$$= \left(\mathcal{F}_2 \begin{bmatrix} \mathbf{p} & \\ \hline & \mathbf{q} \end{bmatrix} T_2^0 \right) \boxplus \left(\mathcal{F}_2 \begin{bmatrix} \mathbf{p} & \\ \hline & \mathbf{q} \end{bmatrix} T_2^1 \right),$$

where $\{T^{\alpha_1}\}_{\alpha_1=0}^1$ are dyadic shifts. Using this construction for all 2^{k-2} complementary pairs $(\alpha_{[k-2]} = 0, 1, \ldots, 2^{k-2} - 1)$, we obtain

$$2^{n+1}\mathcal{FP}_{(\alpha_{[n]}, \alpha_{n+1})} = \boxplus_{\alpha_{[n]}=0}^{2^n-1} \left(\mathcal{F}_2 \left[\begin{array}{c|c} \mathbf{p} \\ \hline & \mathbf{q} \end{array} \right] T_2^{\alpha_n} \right)$$

$$= \boxplus_{\alpha_{[n]}=0}^{2^n-1} \left(\mathcal{F}_2 \left[\begin{array}{c|c} \mathrm{Pr}_{(\alpha_{[n-1]},0)}(t) & \\ \hline & \mathrm{Pr}_{(\alpha_{[n-1]},1)}(t) \end{array} \right] T_2^{\alpha_n} \right). \qquad (2)$$

Repetition of this construction yields the Fourier-Prometheus matrix $2^{n+1}\mathcal{FP}$ of size $2^{n+1} \times 2^{n+1}$.

Our new PONS construction uses in (2) three parametric unitary matrices

$$\mathcal{U}_2(\beta, \varphi, \gamma) = \left[\begin{array}{cc} e^{i(\beta+\gamma)} \cos\varphi & e^{i(\beta-\gamma)} \sin\varphi \\ e^{-i(\beta-\gamma)} \sin\varphi & -e^{-i(\beta+\gamma)} \cos\varphi \end{array} \right]$$

instead of $\mathcal{F}_2$:

$$2^{n+1}\mathcal{FP}_{(\alpha_{[n]}, \alpha_{n+1})}(\vec{\beta}_{n+1}, \vec{\varphi}_{n+1}, \vec{\gamma}_{n+1}) = \boxplus_{\alpha_{[n]}=0}^{2^n-1} \Big(\mathcal{U}(\beta_{n+1}, \varphi_{n+1}, \gamma_{n+1})$$

$$* \left[\begin{array}{c|c} \mathrm{Pr}_{(\alpha_{[n-1]},0)}(t|\vec{\beta}_n, \vec{\varphi}_n, \vec{\gamma}_n) & \\ \hline & \mathrm{Pr}_{(\alpha_{[n-1]},1)}(t|\vec{\beta}_n, \vec{\varphi}_n, \vec{\gamma}_n) \end{array} \right] T_2^{\alpha_k} \Big), \qquad (3)$$

where

$$\vec{\beta}_{n+1} = (\beta_1, \ldots, \beta_{n+1}), \quad \vec{\varphi}_{n+1} = (\varphi_1, \ldots, \varphi_{n+1}), \quad \vec{\gamma}_{n+1} = (\gamma_1, \ldots, \gamma_{n+1})$$

are three $(n+1)$D vectors of parameters. Extra parameters $\beta_k, \varphi_k, \gamma_k$ $(k = 1, 2, \ldots, n+1)$ are changed from stage to stage in this construction. The resulting matrix still has orthogonal rows and every pair is 2-complementary in the Golay-Shapiro sense.

3. PONS associated with Abelian groups

3.1 Abelian groups Z_N^n

A natural generalization of a 2-complementary Golay pair is an N-complementary Golay N-member orthogonal set of Clifford-valued sequences $\mathbf{p}_0(t), \ldots, \mathbf{p}_{N-1}(t)$, where $t = 0, 1, \ldots, N^n - 1$.

DEFINITION 2 *Let* $\mathbf{p}_0(t), \mathbf{p}_1(t), \ldots, \mathbf{p}_{N-1}(t)$ *be an N-member orthogonal set of Clifford-valued sequences, where* $\mathbf{p}_i(t) \in \{\varepsilon_N^k\}_{k=0}^{N-1}$, $\varepsilon_N :=$

$e^{2\pi\mathbf{u}/N} \in Cla$, Cla is a Clifford algebra, and $\mathbf{u}$ is an appropriate bivector with the property $\mathbf{u}^2 = -1$. The sequences $\{\mathbf{p}_i(t)\}_{i=0}^{N-1}$ are called N-complementary $\{\varepsilon_N^k\}_{k=0}^{N-1}$-valued sequences of length N^n if

$$\mathrm{COR}[\mathbf{p}_0, \mathbf{p}_0](\tau) + \ldots + \mathrm{COR}[\mathbf{p}_{N-1}, \mathbf{p}_{N-1}](\tau) = N^n \delta(\tau),$$

or $|\mathbf{p}_0(z)|^2 + |\mathbf{p}_1(z)|^2 + \ldots + |\mathbf{p}_{N-1}(z)|^2 = N^n$, $\forall z \in \mathbb{T}_1$, where $\mathbf{p}_i(z)$ are $\mathcal{Z}$-transforms of $\mathbf{p}_i(t)$, $i = 0, 1, \ldots, N-1$, respectively.

Let, for example, $N = 3$. Then for the group $\mathbf{Z}_3$ we define the Fourier-Clifford-Prometheus transform as the Fourier-Clifford transform

$$^{3^1}\mathcal{FCP} := {}^{3^1}\mathcal{FC} = \begin{bmatrix} \mathrm{Pr}_0(t) \\ \mathrm{Pr}_1(t) \\ \mathrm{Pr}_2(t) \end{bmatrix} = \begin{bmatrix} 1 & 1 & 1 \\ 1 & \varepsilon_3 & \varepsilon_3^2 \\ 1 & \varepsilon_3^2 & \varepsilon_3 \end{bmatrix}.$$

For the group $\mathbf{Z}_3^2$ we define the Fourier-Clifford-Prometheus transform using the classical PONS-construction (2) by

$$^{3^2}\mathcal{FCP}_{\alpha_1,\alpha_2} = \boxplus_{\alpha_1=0}^2 \left(\mathcal{FC}_3 \begin{bmatrix} \mathrm{Pr}_0(t) & & \\ & \mathrm{Pr}_1(t) & \\ & & \mathrm{Pr}_2(t) \end{bmatrix} T_3^{\alpha_1} \right)$$

$$= \begin{bmatrix} \begin{array}{ccc|ccc|ccc} 1 & 1 & 1 & & & & 1 & 1 & 1 & & & \\ 1 & \varepsilon_3 & \varepsilon_3^2 & & & & & & & 1 & \varepsilon_3 & \varepsilon_3^2 \\ 1 & \varepsilon_3^2 & \varepsilon_3 & & & & & & & & & 1 & \varepsilon_3^2 & \varepsilon_3 \end{array} \end{bmatrix}$$

$$= \begin{bmatrix} \begin{array}{ccc|ccc|ccc} 1 & 1 & 1 & 1 & \varepsilon_3 & \varepsilon_3^2 & 1 & \varepsilon_3^2 & \varepsilon_3 \\ 1 & 1 & 1 & \varepsilon_3 & \varepsilon_3^2 & 1 & \varepsilon_3^2 & \varepsilon_3 & 1 \\ 1 & 1 & 1 & \varepsilon_3^2 & 1 & \varepsilon_3 & \varepsilon_3 & 1 & \varepsilon_3^2 \\ \hline 1 & \varepsilon_3^2 & \varepsilon_3 & 1 & 1 & 1 & 1 & \varepsilon_3 & \varepsilon_3^2 \\ \varepsilon_3^2 & \varepsilon_3 & 1 & 1 & 1 & 1 & \varepsilon_3 & \varepsilon_3^2 & 1 \\ \varepsilon_3 & 1 & \varepsilon_3^2 & 1 & 1 & 1 & \varepsilon_3^2 & 1 & \varepsilon_3 \\ \hline 1 & \varepsilon_3 & \varepsilon_3^2 & 1 & \varepsilon_3^2 & \varepsilon_3 & 1 & 1 & 1 \\ \varepsilon_3 & \varepsilon_3^2 & 1 & \varepsilon_3^2 & \varepsilon_3 & 1 & 1 & 1 & 1 \\ \varepsilon_3^2 & 1 & \varepsilon_3 & \varepsilon_3 & 1 & \varepsilon_3^2 & 1 & 1 & 1 \end{array} \end{bmatrix} = \begin{bmatrix} \mathrm{Pr}_{(0,0)}(t) \\ \mathrm{Pr}_{(0,1)}(t) \\ \mathrm{Pr}_{(0,2)}(t) \\ \hline \mathrm{Pr}_{(1,0)}(t) \\ \mathrm{Pr}_{(1,1)}(t) \\ \mathrm{Pr}_{(1,2)}(t) \\ \hline \mathrm{Pr}_{(2,0)}(t) \\ \mathrm{Pr}_{(2,1)}(t) \\ \mathrm{Pr}_{(2,2)}(t) \end{bmatrix},$$

where $\{T^{\alpha_1}\}_{\alpha_1=0}^2$ are 3-cyclic shift operators. After $n+1$ iterations we obtain the following Fourier-Clifford-Prometheus transform on the group

$\mathbf{Z}_3^{n+1}$:

$$^{3^{n+1}}\mathcal{FCP}_{(\alpha_{[n]},\alpha_{n+1})} = \boxplus_{\alpha_{[n]}=0}^{3^n-1} \Big(\mathcal{FC}_3$$

$$* \begin{bmatrix} \mathrm{Pr}_{(\alpha_{[n-1]},0)}(t) & & \\ & \mathrm{Pr}_{(\alpha_{[n-1]},1)}(t) & \\ & & \mathrm{Pr}_{(\alpha_{[n-1]},2)}(t) \end{bmatrix} T_3^{\alpha_n}\Big).$$

The same expression is true for the Fourier-Clifford-Prometheus transform on the group $\mathbf{Z}_N^n$:

$$^{N^{n+1}}\mathcal{FCP}_{(\alpha_{[n]},\alpha_{n+1})} = \boxplus_{\alpha_{[n]}=0}^{N^{[n]}-1} \Big(\mathcal{FC}_N$$

$$* \begin{bmatrix} \mathrm{Pr}_{(\alpha_{[n-1]},0)}(t) & & & \\ & \mathrm{Pr}_{(\alpha_{[n-1]},1)}(t) & & \\ & & \ddots & \\ & & & \mathrm{Pr}_{(\alpha_{[n-1]},N-1)}(t) \end{bmatrix} T_N^{\alpha_n}\Big),$$

where $\mathcal{FC}_N$ is the Fourier-Clifford transform on the group $\mathbf{Z}_N$,

$$\{T^{\alpha_1}\}_{\alpha_1=0}^{N-1}$$

are N-cyclic shift operators.

3.2   Abelian groups $\mathbf{Z}_{N_1} \oplus \mathbf{Z}_{N_2} \oplus \ldots \oplus \mathbf{Z}_{N_n}$

Let $\mathbf{Z}_{N_1} \oplus \mathbf{Z}_{N_2} \oplus \ldots \oplus \mathbf{Z}_{N_n}$ be an Abelian group, where $N_1, N_2, \ldots, N_n$ are positive integers. The classical Fourier-Prometheus transforms are generated by the Fourier-Walsh transform $\mathcal{F}_2$ and by dyadic shifts. Fourier-Clifford-Prometheus transforms associated with $\mathbf{Z}_N^n$ are generated by the Fourier-Clifford transform $\mathcal{FC}_N$ of the group $\mathbf{Z}_N$ and by N-ary shifts. We shall generate new Fourier-Clifford-Prometheus transforms associated with Abelian groups $\mathbf{Z}_{N_1} \oplus \mathbf{Z}_{N_2} \oplus \ldots \oplus \mathbf{Z}_{N_n} \oplus \mathbf{Z}_{N_{n+1}}$ by using the set of Fourier-Clifford transforms $\mathcal{FC}_{N_1}, \mathcal{FC}_{N_2}, \ldots, \mathcal{FC}_{N_n}, \mathcal{FC}_{N_{n+1}}$. For example, the group $\mathbf{Z}_2 \oplus \mathbf{Z}_3 \oplus \mathbf{Z}_4$ requires three Fourier-Clifford transforms

$$\mathcal{F}_2 = \begin{bmatrix} 1 & 1 \\ 1 & -1 \end{bmatrix}, \quad \mathcal{F}_3 = \begin{bmatrix} 1 & 1 & 1 \\ 1 & \varepsilon_3 & \varepsilon_3^2 \\ 1 & \varepsilon_3^2 & \varepsilon_3 \end{bmatrix}, \quad \mathcal{F}_4 = \begin{bmatrix} 1 & 1 & 1 & 1 \\ 1 & \varepsilon_4^1 & \varepsilon_4^2 & \varepsilon_4^3 \\ 1 & \varepsilon_4^2 & 1 & \varepsilon_4^2 \\ 1 & \varepsilon_4^3 & \varepsilon_4^2 & \varepsilon_4^1 \end{bmatrix}.$$

Let us consider the group $\mathbf{Z}_2 \oplus \mathbf{Z}_3$. $\mathcal{FCP}_2 = \mathcal{FC}_2 = \begin{bmatrix} \mathrm{Pr}_0(t) \\ \mathrm{Pr}_1(t) \end{bmatrix} = \begin{bmatrix} 1 & 1 \\ 1 & -1 \end{bmatrix}$. We define the Fourier-Clifford-Prometheus transform associated with the Abelian group $\mathbf{Z}_2 \oplus \mathbf{Z}_3$ by using the classical PONS construction

$$^{2 \cdot 3}\mathcal{FCP}_{(\alpha_1, \alpha_2)}$$

$$= \boxplus_{\alpha_1=0}^{1} \left(\mathcal{FC}_3 \begin{bmatrix} \mathrm{Pr}_{\langle(\alpha_1,0)\rangle_2}(t) & & \\ \hline & \mathrm{Pr}_{\langle(\alpha_1,1)\rangle_2}(t) & \\ \hline & & \mathrm{Pr}_{\langle(\alpha_1,2)\rangle_2}(t) \end{bmatrix} T_3^{\alpha_1} \right),$$

where $\langle(\alpha_1, \beta_2)\rangle_2 := (\alpha_1, \beta_2) \bmod 2$. Therefore,

$$\mathcal{FCP}_{2 \cdot 3} = \begin{bmatrix} \begin{bmatrix} 1 & 1 & 1 \\ 1 & \varepsilon_3 & \varepsilon_3^2 \\ 1 & \varepsilon_3^2 & \varepsilon_3 \end{bmatrix} & \\ & \begin{bmatrix} 1 & 1 & 1 \\ 1 & \varepsilon_3 & \varepsilon_3^2 \\ 1 & \varepsilon_3^2 & \varepsilon_3 \end{bmatrix} \end{bmatrix} \begin{bmatrix} \begin{bmatrix} 1 & 1 \\ 1 & -1 \end{bmatrix} & \\ & \begin{bmatrix} 1 & 1 \\ 1 & -1 \end{bmatrix} \\ \begin{bmatrix} 1 & 1 \\ 1 & -1 \end{bmatrix} & \end{bmatrix}$$

$$= \begin{bmatrix} 1 & 1 & 1 & -1 & 1 & 1 \\ 1 & 1 & \varepsilon_3 & \varepsilon_3 & \varepsilon_3^2 & \varepsilon_3^2 \\ 1 & 1 & \varepsilon_3 & -\varepsilon_3 & \varepsilon_3^2 & \varepsilon_3^2 \\ \hline 1 & -1 & 1 & 1 & 1 & -1 \\ \varepsilon_3^2 & -\varepsilon_3^2 & 1 & 1 & \varepsilon_3 & -\varepsilon_3 \\ \varepsilon_3 & -\varepsilon_3 & 1 & 1 & \varepsilon_3^2 & -\varepsilon_3^2 \end{bmatrix} = \begin{bmatrix} \mathrm{Pr}_{(0,0)}(t) \\ \mathrm{Pr}_{(0,1)}(t) \\ \mathrm{Pr}_{(0,2)}(t) \\ \hline \mathrm{Pr}_{(1,0)}(t) \\ \mathrm{Pr}_{(1,1)}(t) \\ \mathrm{Pr}_{(1,2)}(t) \end{bmatrix}. \quad (4)$$

We design Fourier-Clifford-Prometheus transforms associated with the Abelian groups $\mathbf{Z}_{N_1} \oplus \mathbf{Z}_{N_2} \oplus \ldots \oplus \mathbf{Z}_{N_{n+1}}$ by the same classical PONS construction

$$^{N^{[n+1]}}\mathcal{FCP}_{(\alpha_{[n]}\alpha_{n+1})}$$

$$= \boxplus_{\alpha_{[n]}=0}^{N^{[n]}-1} \left(\mathcal{FC}_{N_{n+1}} \begin{bmatrix} \mathrm{Pr}_{(\alpha_{[n-1]}, \langle 0 \rangle_{N_n})} & & \\ \hline & \ddots & \\ \hline & & \mathrm{Pr}_{(\alpha^{(n-1)}, \langle N_{n+1}-1 \rangle_{N_n})} \end{bmatrix} T_{N_{n+1}}^{\alpha_n} \right)$$

where $\mathcal{F}_{N_{n+1}}$ is the Fourier-Clifford transform on the group $\mathbf{Z}_{N_{n+1}}$, $\langle \alpha_n + \beta_{n+1} \rangle_{n+1} := (\alpha_n + \beta_{n+1}) \bmod N_{n+1}$, $\alpha_{[n]} := (\alpha_1, \alpha_2, \ldots, \alpha_n)$, $N^{[n]} := N_1 N_2 \cdots N_n$, $(\alpha_{[n]}, \beta_{n+1}) := (\alpha_1, \ldots, \alpha_n, \beta_{n+1})$, and, hence,

$$\langle(\alpha_{[n]}, \beta_{n+1})\rangle_n := (\alpha_1, \ldots, \alpha_n, \beta_{n+1}) \bmod N_n.$$

4. Fast Fourier-Prometheus Transforms

4.1 Radix-2 Fast Transforms

Let us return to the Fourier-Clifford-Prometheus transform

$$\mathcal{FP}_{2^2} = \begin{bmatrix} \mathrm{Pr}_{(0,0)}(t) \\ \mathrm{Pr}_{(0,1)}(t) \\ \mathrm{Pr}_{(1,0)}(t) \\ \mathrm{Pr}_{(1,1)}(t) \end{bmatrix} = \left[\begin{array}{cc|cc} 1 & 1 & 1 & -1 \\ 1 & 1 & -1 & 1 \\ 1 & -1 & 1 & 1 \\ -1 & 1 & 1 & 1 \end{array} \right]$$

$$= \left[\begin{array}{c|c} \begin{matrix} 1 & \\ & 1 \end{matrix} & \\ \hline & \begin{matrix} 1 & \\ & -1 \end{matrix} \end{array} \right] \left[\begin{array}{c|c} \begin{matrix} 1 & \\ & 1 \end{matrix} & \\ \hline & \begin{matrix} 1 & \\ & 1 \end{matrix} \end{array} \right] \left[\begin{array}{cc|cc} 1 & 1 & 1 & 1 \\ 1 & -1 & 1 & -1 \\ 1 & 1 & -1 & -1 \\ 1 & -1 & -1 & 1 \end{array} \right] \left[\begin{array}{c|c} \begin{matrix} 1 & \\ & 1 \end{matrix} & \\ \hline & \begin{matrix} 1 & \\ & -1 \end{matrix} \end{array} \right]$$

$$= \Delta_0 \Pi_4 (\mathcal{F}_2 \otimes \mathcal{F}_2) \Delta_0, \tag{5}$$

where $\Delta_0 := \mathbf{diag}(\mathrm{Pr}_0(t)) = \mathbf{diag}(\mathrm{Pr}_\alpha(0))$ is a diagonal matrix and Π_4 is a special permutation matrix. From this expression we see that Prometheus functions up to constant factor are modulated Walsh functions:

$$\mathrm{Pr'}_{(\alpha_1,\alpha_2)}(t_1, t_2) = (-1)^{\alpha_1\alpha_2} \left[\mathrm{Wal}_{(\alpha_1,\alpha_2)}(t_1, t_2)(-1)^{t_1 t_2} \right],$$

where $(-1)^{\alpha_1\alpha_2}$ and $(-1)^{t_1 t_2}$ are the so-called Shapiro multipliers, and $\mathrm{Wal}_{(\alpha_1,\alpha_1)}(t_1, t_2) = (-1)^{\alpha_1 t_1 \oplus_2 \alpha_2 t_2}$. The same result is true in the general case for the Fourier-Clifford-Prometheus $(2^n \times 2^n)$-transform $\mathcal{FP}_{2^n} = \Delta_0 \Pi_{2^n} (\mathcal{F}_2 \otimes \mathcal{F}_2 \otimes \cdots \otimes \mathcal{F}_2) \Delta_0$, where Π_{2^n} is a special permutation matrix and $\Delta_0 = \mathbf{diag}(\mathrm{Pr}_0(t)) = \mathbf{diag}(\mathrm{Pr}_\alpha(0))$ is the diagonal matrix whose diagonal elements form the Shapiro (± 1)-multipliers. If $\alpha = (\alpha_1, \alpha_2, \ldots, \alpha_n)$ is the binary representation of the number in the αth row of Δ_0, where $\alpha_i \in \mathbf{Z}_2$, then for diagonal elements $\Delta_{\alpha,\alpha}$ we have the expression $\Delta_{\alpha,\alpha} = (-1)^{\sum_{i=1}^{n-1} \alpha_i\alpha_{i+1}}$. The quantity $b(\alpha) = \sum_{i=1}^{n-1} \alpha_i\alpha_{i+1}$ is the number of occurrences of the block $B = (11)$ in the binary representation of α, $(\alpha_1, \alpha_2, \ldots, \alpha_n)$. For this reason the Fourier-Clifford-Prometheus transform has the Cooley-Tukey fast algorithm

$$\mathcal{FP}_{2^n} = \Delta_0 \Pi_{2^n} \left[CT_{2^n}^1 CT_{2^n}^2 \cdots CT_{2^n}^n \right] \Delta_0, \tag{6}$$

where $CT_{2^n}^i := I_2 \otimes \ldots \otimes \mathcal{F}_2 \otimes \ldots \otimes I_2$ for $i = 1, 2, \ldots, n$ are the so-called *Cooley-Tukey sparse matrices*.

Now we can prove that an analogous result is true for Davis-Jedwab Clifford-valued sequences. Let $\mathbf{MC}_{2^h} = \{\varepsilon_{2^h}^k\}_{k=0}^{2^h-1}$ be the multiplicative cyclic group of 2^hth roots of unity and ε_{2^h} be a 2^hth primitive root in a

Clifford algebra $\mathcal{C}la$. Let $(c_1, c_2, \ldots, c_n) \in \mathbf{Z}_{2^h}^n = \mathbf{Z}_{2^h} \oplus \mathbf{Z}_{2^h} \oplus \ldots \oplus \mathbf{Z}_{2^h}$, be an nD vector of parameters over $\mathbf{Z}_{2^h}$, where $\mathbf{Z}_{2^h}^n$ is a set of nD vectors (labels). Let

$$\mathcal{F}C_2(\varepsilon_{2^h}^{c_k}) := \begin{bmatrix} 1 & \varepsilon_{2^h}^{c_k} \\ 1 & -\varepsilon_{2^h}^{c_k} \end{bmatrix} = \begin{bmatrix} 1 & 1 \\ 1 & -1 \end{bmatrix} \begin{bmatrix} 1 & \\ & \varepsilon_{2^h}^{c_k} \end{bmatrix}, \quad k = 1, 2, \ldots, n, \tag{7}$$

be a set of (2×2)-matrices. Then the tensor product of these matrices

$$\mathcal{F}CP\mathcal{D}\mathcal{J}_{2^n}^{(c_1, c_2, \ldots, c_n)} := \Delta_0 \Pi_{2^n} \Big(\mathcal{F}_2(\varepsilon_{2^h}^{c_1}) \otimes \mathcal{F}_2(\varepsilon_{2^h}^{c_2}) \otimes \cdots \otimes \mathcal{F}_2(\varepsilon_{2^h}^{c_n}) \Big) \Delta_0 \tag{8}$$

gives us new multi-parametric Fourier-Prometheus transforms with fast Cooley-Tukey algorithm:

$$\mathcal{F}CP\mathcal{D}\mathcal{J}_{2^n}^{(c_1, c_2, \ldots, c_n)} = \Delta_0 \Pi_{2^n} \Big[CT_{2^n}^1(\varepsilon_{2^h}^{c_1}) CT_{2^n}^2(\varepsilon_{2^h}^{c_2}) \cdots CT_{2^n}^n(\varepsilon_{2^h}^{c_n}) \Big] \Delta_0, \tag{9}$$

where $CT_{2^n}^k(\varepsilon_{2^h}^{c_k}) := \Big[I_2 \otimes \ldots \otimes \mathcal{F}_2(\varepsilon_{2^h}^{c_k}) \otimes \ldots \otimes I_2 \Big]$, and $k = 1, 2, \ldots, n$.

4.2   Radix-N Fourier-Prometheus transforms

Let us consider the case of $\mathbf{Z}_3^2$. In this case $\mathcal{F}CP_{3^2} = \Delta_0 \Pi_9 (\mathcal{F}_3 \otimes \mathcal{F}_3) \Delta_0$, where $\Delta_0 = \mathbf{diag}\{\mathrm{Pr}_{(0,0)}(t_1, t_2)\} = \mathbf{diag}\{\mathrm{Pr}_{(t_1, t_2)}(0, 0)\}$ and Π_9 is a special permutation matrix. From this expression we see that Prometheus functions up to constant factor are modulated Chrestenson-Clifford sequences (i.e., Clifford-valued characters of the group $\mathbf{Z}_3^2$):

$$\mathrm{Pr'}_{(\alpha_1, \alpha_2)}(t_1, t_2) = \mathrm{Pr}_{(\alpha_1, \alpha_2)}(0, 0) \Big[\mathrm{Ch}_{(\alpha_1, \alpha_2)}(t_1, t_2) \cdot \mathrm{Pr}_{(0,0)}(t_1, t_2) \Big]$$

$$= \varepsilon_3^{\alpha_1 \alpha_2} \Big[\mathrm{Ch}_{(\alpha_1, \alpha_2)}(t_1, t_2) \varepsilon_3^{t_1 t_2} \Big] = \varepsilon_3^{\alpha_1 \alpha_2} \Big[\varepsilon_3^{\alpha_1 t_1 \oplus 2\alpha_2 t_2} \cdot \varepsilon_3^{t_1 t_2} \Big],$$

where

$$\mathrm{Pr}_{(\alpha_1, \alpha_2)}(0, 0) = \varepsilon_3^{\alpha_1 \alpha_2}$$

$$\mathrm{Pr}_{(0,0)}(t_1, t_2) = \varepsilon_3^{t_1 t_2},$$

and

$$\mathrm{Ch}_{(\alpha_1, \alpha_1)}(t_1, t_2) = \varepsilon_3^{\alpha_1 t_1 \oplus 2\alpha_2 t_2}.$$

For this reason, this Fourier-Clifford Prometheus transform has the Cooley-Tukey fast algorithm $\mathcal{F}P_{3^2} = \Delta_0 \Pi_9 \Big[CT_9^1 \cdot CT_9^2 \Big] \Delta_0$, where $CT_9^1 := \mathcal{F}_3 \otimes I_3$, $CT_9^2 = I_3 \otimes \mathcal{F}_3$. The same result is true in the general case for Fourier-Clifford-Prometheus $(3^n \times 3^n)$-transforms

$$\mathcal{F}P_{3^n} = \Delta_0 \Pi_{3^n} \Big(\mathcal{F}_3 \otimes \mathcal{F}_3 \otimes \cdots \otimes \mathcal{F}_3 \Big) \Delta_0 = \Delta_0 \Pi_{3^n} \Big[CT_{3^n}^1 CT_{3^n}^2 \cdots CT_{3^n}^n \Big] \Delta_0,$$

where $CT_{3^n}^i := I_3 \otimes \ldots \otimes \mathcal{F}_3 \otimes \ldots \otimes I_3$ for $i = 1, 2, \ldots, n$ are the so-called *Cooley-Tukey sparse matrices*. Now we are ready to write the analogous expression for Fourier-Clifford-Prometheus $(N^n \times N^n)$-transforms

$$\mathcal{FP}_{N^n} = \Delta_0 \Pi_{N^n} \left(\mathcal{F}_N \otimes \mathcal{F}_N \otimes \cdots \otimes \mathcal{F}_N \right) \Delta_0$$

$$= \Delta_0 \Pi_{N^n} \left[CT_{N^n}^1 CT_{N^n}^2 \cdots CT_{N^n}^n \right] \Delta_0, \qquad (10)$$

where $CT_{N^n}^i := I_N \otimes \ldots \otimes \mathcal{F}_N \otimes \ldots \otimes I_N$ for $i = 1, 2, \ldots, n$ are the so-called *Cooley-Tukey sparse matrices*. The same result is true in the general case for the Fourier-Clifford-Prometheus $(N^{[n]} \times N^{[n]})$-transform

$$\mathcal{FP}_{N^{[n]}} = \Delta_0 \Pi_{N^n} \left(\mathcal{F}_{N_1} \otimes \mathcal{F}_{N_2} \otimes \cdots \otimes \mathcal{F}_{N_n} \right) \Delta_0$$

$$= \Delta_0 \Pi_{N^n} \left[CT_{N^{[n]}}^1 CT_{N^{[n]}}^2 \cdots CT_{N^{[n]}}^n \right] \Delta_0, \qquad (11)$$

where $\Delta_0 := \mathbf{diag}(\mathrm{Pr}_0(t)) = \mathbf{diag}(\mathrm{Pr}_\alpha(0))$ is a diagonal matrix and Π_{N^n} a permutation matrix, and $CT_{N^{[n]}}^i := I_{N_1} \otimes \ldots \otimes \mathcal{F}_{N_i} \otimes \ldots \otimes I_{N_n}$ for $i = 1, 2, \ldots, n$ are the so-called *Cooley-Tukey sparse matrices*.

5. Conclusions

We have shown how Clifford algebras can be used to formulate a new unified approach to so-called generalized *Fourier-Clifford-Prometheus transforms*. It is based on a new generalized FCPT-generating construction. This construction has a rich algebraic structure that supports a wide range of fast algorithms. This construction is associated not with the triple $\left(\mathbf{Z}_2^n, \mathcal{F}_2, \mathbf{C} \right)$, but rather with other groups instead of $\mathbf{Z}_2^n$, other unitary transforms instead of $\mathcal{F}_2$, and other algebras (Clifford algebras) instead of the complex field $\mathbf{C}$.

Acknowledgments

The work was supported by the Russian Foundation for Basic Research, research project no. 03–01–00735. The paper contains some results obtained in the project no. 3258 of the Ural State Technical University.

We thank the Organizing Committee of the NATO Advanced Study Institute "Computational Noncommutative Algebra and Applications". The authors are grateful for their NATO support.

References

[1] Byrnes, J.S. (1994). Quadrature mirror filter, low crest factor arrays, functions achieving optimal uncertainty principle bounds, and complete orthonormal sequences - a unified approach. *Applied and Computational Harmonic Analysis*, pp. 261–264.

[2] Golay, M.J.E. (1949). Multislit spectrometry. *J.Optical Society Am.*, 39:437.

[3] Golay, M.J.E. (1951). Complementary series. *IRE Trans. Inform. Theory*, IT-7, pp. 19–61.

[4] Golay, M.J.E. (1977). Sieves for low autocorrelation binary sequences. *IEEE Trans. Inform. Theory*, IT-23, pp. 43–51.

[5] Shapiro, H.S. (1951). *Extremal problems for polynomials and power series*. ScM.Thesis, Massachusetts Institute of Technology.

[6] Shapiro, H.S. (1958). A power series with small partial sums. *Notices of the AMS*, 6(3):366.

[7] Rudin, W. (1959). Some theorems on Fourier coefficients. *Proc. Amer. Math. Soc.*, 10, pp. 855–859.

[8] Turan, R.J. (1974). Hadamard matrices, Baumert-Hall units, four-symbol sequences, pulse compression, and surface wave encodings. *J.Combin. Theory (A)*. vol. 16, pp. 313–333.

[9] Eliahou, S., Kervaire, M., and Saffari, B. (1990). A new restriction on the lengths of Golay complementary sequences. *J. Combin. Theory (A)*, vol. 55, pp. 49–59.

[10] Budisin, S.Z. (1990). New complementary pairs of sequences. *Electron. Lett.*, Vol. 26, pp. 881–883.

[11] Sivaswamy, R. (1978). Multiphase complementary codes. *IEEE Trans. Inform. Theory*, vol. IT-24, pp. 546–552.

[12] Saffari, B.: History of Shapiro polynomials and Golay sequences. In preparation.

[13] Fan, P. and Darnell, M. (1996). *Sequence Design for Communications Applications* (Communications Systems, Technologies and Applications). Taunton, U.K.: Res. Studies,

[14] Davis, A. and Jedwab, J. (1999). Peak-to-Mean Power control in OFDM, Golay complementary Sequences, Reed-Muller codes. *IEEE Trans. Inform. Theory*, vol. IT-45, No. 7, pp. 2397–2417

FAST COLOR WAVELET-HAAR-HARTLEY-PROMETHEUS TRANSFORMS FOR IMAGE PROCESSING

Ekaterina L.-Rundblad, Alexei Maidan, Peter Novak, Valeriy Labunets
Urals State Technical University
Ekaterinburg, Russia
lab@rtf.ustu.ru

Abstract   This paper present a new approach to the Color Fourier Transformation. Color image processing is investigated in this paper using an algebraic approach based on triplet (color) numbers. In the algebraic approach, each image color pixel is considered not as a 3D vector, but as a triplet (color) number. The so-called orthounitary transforms are introduced and used for color image processing. These transforms are similar to a fast orthogonal and unitary transforms. Simulations using the color Wavelet-Haar-Prometheus transforms on color image compression have also been performed.

Keywords: Clifford algebra, color images, color wavelet, edge detection, orthounitary transforms.

1.   Introduction

Fourier analysis based on orthogonal and unitary transforms plays an important role in digital image processing. Transforms, notably the classical discrete Fourier transform, are extensively used in digital image filtering and in power spectrum estimation. Other Fourier transforms—e.g., the discrete cosine/sine transforms, wavelet transforms—are often employed in digital image compression. All the above-mentioned transforms are used in digital grey-level image processing. However, in recent years an increasing interest in color processing has been observed. Our approach to color image processing is in using so-called color triplet numbers [1]–[7] for color images and to operate directly on three-channel (RGB-valued) images as on single-channel triplet-valued images. In the classical approach every color is associated to a point of the 3D color

J. Byrnes (ed.) Computational Noncommutative Algebra and Applications, 401-411.
© 2004 *Kluwer Academic Publishers. Printed in the Netherlands.*

RGB vector space. In our approach, each image color pixel is considered not as a 3D RGB vector, but as a triplet (color) number.

A natural question that arises in our approach is the definition of color (RGB-channel) transforms that can be used efficiently in edge detection and digital image compression. The so-called orthounitary (triplet-valued or color-valued) Fourier transforms are introduced and are used for color image processing. These transforms are similar to fast orthogonal and unitary transforms. Therefore, fast algorithms for their computation can be easily constructed. Simulations of application of color transforms to color image compression have also been performed. The main contributions of this paper are: a) the definition and analysis of properties of the orthounitary (color) Fourier transforms (in particular, color Wavelet-Haar-Prometheus transforms); b) showing that the triplet (color) algebra can be used to solve color image processing problems in a natural and effective manner.

2.   Color images

The aim of this section is to present algebraic models of the subjective perceptual color space. The color representation we are using is based on Young's theory (1802), asserting that any color can be visually reproduced by a proper combination of three colors, referred to as primary colors. The color image appears on the retina as a 3D vector-valued $((R, G, B)$-valued) function

$$\mathbf{f}_{col}(\mathbf{x}) = \Big(f_R(\mathbf{x}), f_G(\mathbf{x}), f_B(\mathbf{x})\Big) = f_R(\mathbf{x})\mathbf{i}_R + f_G(\mathbf{x})\mathbf{i}_G + f_B(\mathbf{x})\mathbf{i}_B,$$

where $f_R(\mathbf{x}) = \int_\lambda s^{obj}(\mathbf{x}, \lambda)H_R(\lambda)d\lambda$, $f_G(\mathbf{x}) = \int_\lambda s^{obj}(\mathbf{x}, \lambda)H_G(\lambda)d\lambda$, and $f_B(\mathbf{x}) = \int_\lambda s^{obj}(\mathbf{x}, \lambda)H_B(\lambda)d\lambda$, $s^{obj}(\mathbf{x}, \lambda)$ is the color spectrum received from the object, $H_R(\lambda)$, $H_B(\lambda)$, $H_R(\lambda)$ are three photoreceptor (cone or sensor) sensitivity functions, λ is the wavelength and $\mathbf{i}_R := (1, 0, 0)$, $\mathbf{i}_G := (0, 1, 0)$, $\mathbf{i}_B := (0, 0, 1)$.

In our approach, each image color pixel is considered not as a 3D RGB vector, but as a triplet number in the following two forms (see [1] in this book): $\mathbf{f}_{col}(\mathbf{x}) = f_R(\mathbf{x})1_{col} + f_G(\mathbf{x})\varepsilon_{col} + f_B(\mathbf{x})\varepsilon_{col}^2$, $\mathbf{f}_{col}(\mathbf{x}) = f_{lu}(\mathbf{x})\mathbf{e}_{lu} + \mathbf{f}_{Ch}(\mathbf{x})\mathbf{E}_{Ch}$, where $\varepsilon^3 = 1$. The first and the second expressions are called the $\mathcal{A}_3(\mathbf{R}|1, \varepsilon_{col}, \varepsilon_{col}^2)$- and $\mathcal{A}_3(\mathbf{R}, \mathbf{C})$- *representations of color image*, respectively. Numbers of the form $\mathcal{C} = x + y\varepsilon_{col} + z\varepsilon_{col}^2$ are called the *triplet, 3-cycle,* or *color numbers*. Every color number $\mathcal{C} = x + y\varepsilon^1 + z\varepsilon^2$ is a linear combination $\mathcal{C} = x + y\varepsilon^1 + z\varepsilon^2 = a_{lu}\mathbf{e}_{lu} + \mathbf{z}_{Ch}\mathbf{E}_{Ch} = (a_{lu}, \mathbf{z}_{Ch})$ of the "scalar" $a_{lu}\mathbf{e}_{lu}$ and "complex" parts $\mathbf{z}_{Ch}\mathbf{E}_{Ch}$ in the idempotent basis $\{\mathbf{e}_{lu}, \mathbf{E}_{Ch}\}$. Real numbers $a_{lu} \in \mathbf{R}$ we will call *intensity (luminance) numbers*, and complex numbers $\mathbf{z}_{Ch} =$

$b + jc \in \mathbf{C}$ are called the *chromaticity numbers*. Thus $f_{lu}(\mathbf{x})$ is a real-valued (grey-level) image and $\mathbf{f}_{Ch}$ is a complex-valued (chromatic-valued) image.

A 2D discrete color image can be defined as a 2D array $\mathbf{f}_{col} :=$ $[\mathbf{f}_{col}(i,j)]_{i,j=1}^{N}$ i.e., as a 2D discrete $\mathcal{A}_3^{col}$-valued function in one of the following forms $\mathbf{f}_{col}(i,j) : \mathbf{Z}_N^2 \longrightarrow \mathcal{A}_3^{col}(1,\varepsilon^1,\varepsilon^2)$, $\mathbf{f}_{col}(i,j) : \mathbf{Z}_N^2 \longrightarrow$ $\mathcal{A}_3^{col}(\mathbf{R},\mathbf{C})$, $(i,j) \in \mathbf{Z}_N^2$. Here, every color pixel $\mathbf{f}_{col}(i,j)$ at position (i,j) is a color number of the type

$$\mathbf{f}_{col}(i,j) := f_r(i,j) + f_g(i,j)\varepsilon^1 + f_b(i,j)\varepsilon^2$$

or of the type

$$\mathbf{f}_{col}(i,j) := f_{lu}(i,j)\mathbf{e}_{lu} + f_{Ch}(i,j)\mathbf{E}_{Ch}.$$

The set of all such images forms N^2D Greaves-Hilbert space $\mathbb{L}(\mathbf{Z}_N^2,$ $\mathcal{A}_3^{col}) = (\mathcal{A}_3^{col})^{N^2} = \mathbf{R}^{N^2}1 + \mathbb{R}^{N^2}\varepsilon^1 + \mathbf{R}^{N^2}\varepsilon^2 = \mathbf{R}^{N^2}\mathbf{e}_{lu} + \mathbf{C}^{N^2}\mathbf{E}_{Ch} =$ $\mathbf{R}^{N^2} \oplus \mathbf{C}^{N^2}$, where $\mathbf{R}^{N^2}1$, $\mathbf{R}^{N^2}\varepsilon^1$, $\mathbb{R}^{N^2}\varepsilon^2$ are real N^2D Hilbert spaces of red, green, and blue images, respectively, $\mathbf{R}^{N^2}$ is the N^2D real space of gray-level images, and $\mathbf{C}^{N^2}$ is the N^2D complex space of chromaticity images.

A color linear operator $\mathcal{L}_{2D} : \left(\mathcal{A}_3^{col}\right)^{N^2} \longrightarrow \left(\mathcal{A}_3^{col}\right)^{N^2}$, $\mathcal{L}_{2D}[\mathbf{f}_{col}] =$ $\mathbf{F}_{col}$ is said to be *orthounitary* if $\mathcal{L}_{2D}^{-1} = \mathcal{L}_{2D}^*$. Orthounitary operators preserve scalar product $\langle .|. \rangle$ and form orthounitary group transforms $\mathbb{OU}(\mathcal{A}_3^{col})$. This group is isomorphic to the direct sum of orthogonal and unitary groups $\mathbb{O}(\mathbb{R})\mathbf{e}_{lu} + \mathbb{U}(\mathbb{C})\mathbf{E}_{Ch}$ and every element has the representation $\mathcal{L}_{2D} = \mathbf{O}_{2D}\mathbf{e}_{lu} + \mathcal{U}_{2D}\mathbf{E}_{ch}$, where $\mathbf{O}_{2D} \in \mathbb{O}(\mathbf{R})$ and $\mathcal{U}_{2D} \in \mathbb{U}(\mathbf{C})$ are orthogonal and unitary transforms, respectively. For color image processing we shall use separable $2D$ transforms. The orthounitary transform $\mathcal{L}_{2D}[\mathbf{f}_{col}] = \mathbf{F}_{col}$ is called separable if it can be represented as $\mathbf{F}_{col} = \mathcal{L}_{2D}[\mathbf{f}_{col}] = \mathcal{L}_{1D}[\mathbf{f}_{col}]\mathfrak{M}_{2D}$, i.e. $\mathcal{L}_{2D} = \mathcal{L}_{1D} \otimes \mathfrak{M}_{1D}$ is the tensor product of two $1D$ orthounitary transforms of the form $\mathcal{L}_{2D} = \mathcal{L}_{1D} \otimes \mathcal{L}_{1D} = (\mathbf{O}_1 \otimes \mathbf{O}_2)\mathbf{e}_{lu} + (\mathcal{U}_1 \otimes \mathcal{U}_2)\mathbf{E}_{ch}$, where $\mathbf{O}_1, \mathbf{O}_2$ and $\mathcal{U}_1, \mathcal{U}_2$ are 1D orthogonal and unitary transforms, respectively. Thus, we can obtain any orthounitary transform, using any two pairs of orthogonal $\mathbf{O}_1, \mathbf{O}_2$ and unitary transforms $\mathcal{U}_1, \mathcal{U}_2$. In this work we shall use one pair of orthogonal and unitary transforms where $\mathbf{O}_1 = \mathbf{O}_2 =$ $\mathbf{O}$ and $\mathcal{U}_1 = \mathcal{U}_2 = \mathcal{U}$. In this case we obtain a wide family of orthounitary transforms of the form $\mathcal{L}_{2D} = (\mathbf{O} \otimes \mathbf{O})\mathbf{e}_{lu} + (\mathcal{U} \otimes \mathcal{U})\mathbf{E}_{ch}$ using different 1D orthogonal transforms. In this work we shall use the more simple orthounitary transforms $\mathcal{L} = \mathbf{O} \cdot \mathbf{e}_{lu} + \mathcal{U} \cdot \mathbf{E}_{Ch}$ with $\mathcal{U} = \mathbf{O} \cdot \mathbf{diag}(z_0, z_1, \ldots, z_{N-1})$, where $\mathbf{diag}(z_0, z_1, \ldots, z_{N-1})$ is a diagonal matrix of chromatic (complex) numbers. In this case we have

$\mathfrak{L} = \mathbf{O} \cdot \mathbf{e}_{lu} + \mathbf{O} \cdot \mathbf{diag}(z_0, z_1, \ldots, z_{N-1}) \cdot \mathbf{E}_{Ch} = \mathbf{O} \cdot \mathbf{diag}(\mathcal{C}_0, \mathcal{C}_1, \ldots, \mathcal{C}_{N-1})$, where $\mathcal{C}_k := 1\mathbf{e}_{lu} + z_k \mathbf{E}_{Ch}$, $k = 0, 1, \ldots, N-1$, are color (triplet) numbers. Hence, our orthounitary transforms are represented as the product of an orthogonal transform $\mathbf{O}$ and a triplet-valued (color) diagonal matrix $\mathbf{diag}(\mathcal{C}_0, \mathcal{C}_1, \ldots, \mathcal{C}_{N-1})$. A large number of orthounitary transforms can be devised by appropriate choice of the orthogonal transform $\mathbf{O}$ and the diagonal transform parameters $\mathcal{C}_0, \mathcal{C}_1, \ldots, \mathcal{C}_{N-1}$.

3. Color Wavelet-Haar-Prometheus transforms

One of the aims of this paper is to define three-channel transforms (the so-called color Fourier transforms) that could eventually be used in digital color image compression. We have experimented with the so-called *color Wavelet-Haar* $\mathbf{WH}_{2^n}$, *color Wavelet-Haar-Prometheus* $\mathbf{WHP}_{2^n}$, *Wavelet-Haar-Hartley* $\mathbf{WHH}_{3^n}$, and *Wavelet-Haar-Hartley-Prometheus* $\mathbf{WHHP}_{3^n}$, *transforms* [8]–[9].

Orthogonal and unitary Wavelet-Haar and Wavelet-Haar-Prometheus transforms have the factorizations: $\mathbf{WHP}_{2^n} = \mathbf{WH}_{2^n} \Delta_{2^n}$, $\mathcal{WHP}_{2^n} = \mathcal{WH}_{2^n} \Delta_{2^n}$, and

$$\mathbf{WH}_{2^n} = \prod_{i=1}^{n} \left[(\mathbf{I}_{2^{n-i}} \overset{\circ}{\otimes} \mathbf{F}_2) \oplus \mathbf{I}_{2^n - 2^{n-i+1}} \right], \tag{1}$$

$$\mathcal{WH}_{2^n} = \prod_{i=1}^{n} \left[(\mathbf{I}_{2^{n-i}} \overset{\circ}{\otimes} \mathcal{F}_2) \oplus \mathbf{I}_{2^n - 2^{n-i+1}} \right], \tag{2}$$

respectively, where

$$\mathbf{I}_{2^{n-i}} \overset{\circ}{\otimes} \mathbf{F}_2 = C \left[\frac{\mathbf{I}_{2^{n-i}} \otimes [1 \quad 1]}{\mathbf{I}_{2^{n-i}} \otimes [1 - 1]} \right], \quad \mathbf{I}_{2^{n-i}} \overset{\circ}{\otimes} \mathcal{F}_2 = C \left[\frac{\mathbf{I}_{2^{n-i}} \otimes [1 \quad \omega_3]}{\mathbf{I}_{2^{n-i}} \otimes [1 - \omega_3]} \right]$$

are the generalized tensor products of the identity matrix $\mathbf{I}_{2^{n-i}}$ with $\mathbf{F}_2 = \begin{bmatrix} 1 & 1 \\ 1 & -1 \end{bmatrix}$ and $\mathcal{F}_2 = \begin{bmatrix} 1 & \omega_3 \\ 1 & -\omega_3 \end{bmatrix} = \frac{\sqrt{2}}{2} \begin{bmatrix} 1 & 1 \\ 1 & -1 \end{bmatrix} \begin{bmatrix} 1 & \\ & \omega_3 \end{bmatrix}$, respectively. Here, $c = \frac{\sqrt{2}}{2}$, $\mathbf{F}_2$ is the classical Walsh transform and $\mathcal{F}_2$ is the complex ω_3-deformed Walsh transform, where $\omega_3 = \sqrt[3]{1} = e^{2\pi i/3}$ and Δ_{2^n} is a diagonal matrix, whose diagonal elements form the Shapiro (± 1)-sequence. If $\alpha = (\alpha_1, \alpha_2, \ldots, \alpha_n)$ is the binary representation of the number of the αth row of Δ_{2^n}, where $\alpha_i \in \mathbf{Z}_2$, then for diagonal elements $\Delta_{\alpha,\alpha}$ we have the expression $\Delta_{\alpha,\alpha} = (-1)^{\sum_{i=1}^{n-1} \alpha_i \alpha_{i+1}}$. The quantity $b(\alpha) = \sum_{i=1}^{n-1} \alpha_i \alpha_{i+1}$ is the number of occurrences of the digital block $B = (11)$ in the binary representation $(\alpha_1, \alpha_2, \ldots, \alpha_n)$ of the number α.

Using two pairs $(\mathbf{WH}_{2^n}, \mathcal{WH}_{2^n})$ and $(\mathbf{WHP}_{2^n}, \mathcal{WHP}_{2^n})$ of fast orthogonal and unitary Haar-Wavelet transforms (1–2), we construct 1D color Wavelet-Haar and Wavelet-Haar-Prometheus 2^n-point fast transforms as

$$\mathfrak{WH}_{2^n} = \mathbf{WH}_{2^n} \cdot \mathbf{e}_{lu} + \mathcal{WH}_{2^n} \cdot \mathbf{E}_{Ch} = \left(\prod_{i=1}^{n} \left[(\mathbf{I}_{2^{n-i}} \overset{\circ}{\otimes} \mathfrak{F}_2) \oplus \mathbf{I}_{2^n - 2^{n-i+1}} \right] \right),$$
(3)

$$\mathfrak{WHP}_{2^n} = \mathfrak{WH}_{2^n} \Delta_{2^n} = \left(\mathbf{WHP}_{2^n} \cdot \mathbf{e}_{lu} + \mathcal{WHP}_{2^n} \cdot \mathbf{E}_{Ch} \right) \Delta_{2^n} =$$

$$= \left(\prod_{i=1}^{n} \left[(\mathbf{I}_{2^{n-i}} \overset{\circ}{\otimes} \mathfrak{F}_2) \oplus \mathbf{I}_{2^n - 2^{n-i+1}} \right] \right) \Delta_{2^n},$$
(4)

where $\mathfrak{F}_2 = \mathbf{F}_2 \cdot \mathbf{e}_{lu} + \mathcal{F}_2 \cdot \mathbf{E}_{Ch} =$

$$\frac{\sqrt{2}}{2} \left(\begin{bmatrix} 1 & 1 \\ 1 & -1 \end{bmatrix} \cdot \mathbf{e}_{lu} + \begin{bmatrix} 1 & \omega_3 \\ i & -\omega_3 \end{bmatrix} \cdot \mathbf{E}_{Ch} \right) = \frac{\sqrt{2}}{2} \left[\begin{array}{c|c} 1 & \varepsilon \\ \hline 1 & -\varepsilon \end{array} \right].$$

We see that the color Haar-Wavelet transform has the same fast algorithm as the orthogonal and unitary transforms in Eqs. (1). Note that the product of ε with a color pixel $\mathbf{f}_{col} = (f_R, f_G, f_B) = f_R 1 + f_G \varepsilon^1 + f_B \varepsilon^2$ is realized without multiplications as the right shift of color components $\varepsilon \mathbf{f}_{col} = \varepsilon(f_R, f_G, f_B) = \varepsilon(f_R 1 + f_G \varepsilon^1 + f_B \varepsilon^2) = (f_B 1 + f_R \varepsilon^1 + f_B \varepsilon^2) = (f_B, f_R, f_G)$. In this case the situation is the same as for Number Theoretical Transforms.

The next example of color (orthounitary) Haar-like wavelet transforms is based on Haar-Hartley $\mathbf{H}_3$ (3×3)-transforms. Using these transforms we construct an "elementary" three-point color transform of the following form in the $\mathcal{A}_3(\mathbf{R}, \mathbf{C})$-algebra: $\mathfrak{HF}_3 = \mathbf{H}_3 \cdot \mathbf{e}_{lu} + \mathcal{F}_3 \cdot \mathbf{E}_{Ch} =$

$$\frac{1}{\sqrt{3}} \begin{bmatrix} 1 & 1 & 1 \\ 1 & h_1 & h_2 \\ 1 & h_2 & h_1 \end{bmatrix} \mathbf{e}_{lu} + \frac{1}{\sqrt{3}} \begin{bmatrix} 1 & 1 & 1 \\ 1 & h_1\omega_3^1 & h_2\omega_3^2 \\ 1 & h_2\omega_3^1 & h_1\omega_3^2 \end{bmatrix} \mathbf{E}_{Ch} =$$

$$\frac{1}{\sqrt{3}} \begin{bmatrix} 1 & 1 & 1 \\ 1 & h_1\varepsilon^1 & h_2\varepsilon^2 \\ 1 & h_2\varepsilon^1 & h_1\varepsilon^2 \end{bmatrix} = \frac{1}{\sqrt{3}} \begin{bmatrix} 1 & 1 & 1 \\ 1 & h_1 & h_2 \\ 1 & h_2 & h_1 \end{bmatrix} \begin{bmatrix} 1 & & \\ & \varepsilon^1 & \\ & & \varepsilon^2 \end{bmatrix},$$
(5)

where $h_1 := \mathrm{cas}\left(\frac{2\pi}{3}\right) = \cos\left(\frac{2\pi}{3}\right) + \sin\left(\frac{2\pi}{3}\right)$, $h_2 := \mathrm{cas}\left(\frac{2\pi 2}{3}\right) = \cos\left(\frac{2\pi 2}{3}\right) + \sin\left(\frac{2\pi 2}{3}\right)$ and $\omega_3^1 := \mathrm{cis}\left(\frac{2\pi}{3}\right) = \cos\left(\frac{2\pi}{3}\right) + i\sin\left(\frac{2\pi}{3}\right)$, $\omega_3^2 := \mathrm{cis}\left(\frac{2\pi 2}{3}\right) = \cos\left(\frac{2\pi 2}{3}\right) + i\sin\left(\frac{2\pi 2}{3}\right)$. We use the orthogonal and unitary Wavelet-Haar-Hartley and Wavelet-Haar-Hartley-Prometheus 3^n-point transforms that follow: $\mathbf{WHP}_{3^n} = \mathbf{WH}_{3^n}\Delta_{3^n}$, $\mathcal{WHP}_{3^n} = \mathcal{WH}_{3^n}\Delta_{3^n}$, and

$$\mathbf{WH}_{3^n} = \prod_{i=1}^{n}\left[(\mathbf{I}_{3^{n-i}} \overset{\circ}{\otimes} \mathbf{F}_3) \oplus \mathbf{I}_{3^n-3^{n-i+1}}\right], \tag{6}$$

$$\mathcal{WH}_{3^n} = \prod_{i=1}^{n}\left[(\mathbf{I}_{3^{n-i}} \overset{\circ}{\otimes} \mathcal{F}_3) \oplus \mathbf{I}_{3^n-2^{n-i+1}}\right], \tag{7}$$

respectively, where

$$\mathbf{I}_{3^{n-i}} \overset{\circ}{\otimes} \mathbf{F}_3 = \begin{bmatrix} \mathbf{I}_{3^{n-i}}\otimes[1 \ 1 \ 1 \ ] \\ \mathbf{I}_{3^{n-i}}\otimes[1 \ h_1 \ h_2] \\ \mathbf{I}_{3^{n-i}}\otimes[1 \ h_2 \ h_1] \end{bmatrix}, \quad \mathbf{I}_{3^{n-i}} \overset{\circ}{\otimes} \mathcal{F}_3 = \begin{bmatrix} \mathbf{I}_{3^{n-i}}\otimes[1 \ \ \omega_3^1 \ \ \omega_3^2 \ ] \\ \mathbf{I}_{3^{n-i}}\otimes[1 \ h_1\omega_3^1 \ h_2\omega_3^2] \\ \mathbf{I}_{3^{n-i}}\otimes[1 \ h_2\omega_3^1 \ h_1\omega_3^2] \end{bmatrix}$$

are the generalized tensor products of the identity matrix $\mathbf{I}_{3^{n-i}}$ with

$$\mathbf{H}_3 = \frac{1}{\sqrt{3}}\begin{bmatrix} 1 & 1 & 1 \\ 1 & h_1 & h_2 \\ 1 & h_2 & h_1 \end{bmatrix} \quad \text{and} \quad \mathcal{H}_3 = \frac{1}{\sqrt{3}}\begin{bmatrix} 1 & 1 & 1 \\ 1 & h_1\omega_3^1 & h_2\omega_3^2 \\ 1 & h_2\omega_3^1 & h_1\omega_3^2 \end{bmatrix},$$

respectively, and Δ_{3^n} is a diagonal matrix whose diagonal elements form the 3-point Shapiro ω_3-sequence. If $\alpha = (\alpha_1, \alpha_2, \ldots, \alpha_n)$ is the 3-ary representation of the number in the αth row of Δ_{3^n}, where $\alpha_i \in \mathbf{Z}_2$, then for diagonal elements $\Delta_{\alpha,\alpha}$ we have the expression $\Delta_{\alpha,\alpha} = \omega_3^{\sum_{i=1}^{n-1}\alpha_i\alpha_{i+1}}$.

Using these fast orthogonal and unitary Wavelet-Haar-Hartley-Prometheus transforms (6–7) we construct 1D fast color Wavelet-Haar-Hartley-Prometheus 3^n-point transforms by $\mathfrak{WHP}_{3^n} = \mathfrak{WH}_{3^n}\Delta_{3^n}$ and

$$\mathfrak{WH}_{3^n} = \left(\prod_{i=1}^{n}\left[(\mathbf{I}_{3^{n-i}} \overset{\circ}{\otimes} \mathfrak{HF}_3) \oplus \mathbf{I}_{3^n-3^{n-i+1}}\right]\right), \tag{8}$$

where $\mathfrak{HF}_3$ is the color 3-point Hartley transform (5).

4. Edge detection and compression of color images

One of the primary applications of this work could be in edge detection and color image compression. For edge detection, we convolve the color (3×3)-masks $\mathbf{m}_{col}(i,j)$ with a color image $\mathbf{f}_{col}(i,j)$ of size $N \times N$:

$$\widehat{f}(i,j) = \sum_{(i,j)\in\mathbf{Z}_N^2} m_{col}(k,l)\mathbf{f}_{col}(i-k,j-l).$$

We use color Prewitt's-like masks for detection of horizontal, vertical, and diagonal edges. As entries instead of real numbers these masks have

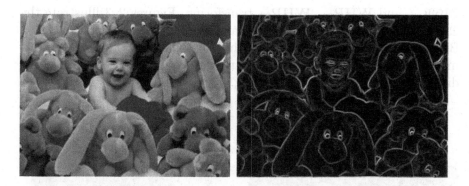

Figure 1. Color edge detector. Left: original image, right: detected edges.

triplet numbers:

$$\mathbf{m}_{col}^{H} = \begin{bmatrix} 1 & \varepsilon & \varepsilon^2 \\ 0 & 0 & 0 \\ -1 & -\varepsilon & -\varepsilon^2 \end{bmatrix}, \quad \mathbf{m}_{col}^{V} = \begin{bmatrix} 1 & 0 & -1 \\ \varepsilon & 0 & -\varepsilon \\ \varepsilon^2 & 0 & -\varepsilon^2 \end{bmatrix},$$

$$\mathbf{m}_{col}^{LD} = \begin{bmatrix} \varepsilon & \varepsilon^2 & 0 \\ 1 & 0 & -\varepsilon^2 \\ 0 & -1 & -\varepsilon \end{bmatrix}, \quad \mathbf{m}_{col}^{RD} = \begin{bmatrix} 0 & 1 & \varepsilon \\ -1 & 0 & \varepsilon^2 \\ -\varepsilon & -\varepsilon^2 & 0 \end{bmatrix}.$$

The effect of the masks in homogenous and nonhomogenous color regions differs substantially. Let us analyze both cases in detail. At any position (i, j) of a homogenous color region after convolution we get

$$\widehat{\mathbf{f}}_{col} = \begin{bmatrix} f_R^{11} + f_B^{12} + f_G^{13} \\ f_G^{11} + f_R^{12} + f_B^{13} \\ f_B^{11} + f_G^{12} + f_R^{13} \end{bmatrix} - \begin{bmatrix} f_R^{31} + f_B^{32} + f_G^{33} \\ f_G^{31} + f_R^{32} + f_B^{33} \\ f_B^{31} + f_G^{32} + f_R^{33} \end{bmatrix} = 0,$$

since, for all 9 pixels we have $\mathbf{f}_{col}^{11} = \mathbf{f}_{col}^{12} = \mathbf{f}_{col}^{13} = \mathbf{f}_{col}^{21} = \mathbf{f}_{col}^{22} = \mathbf{f}_{col}^{23} = \mathbf{f}_{col}^{31} = \mathbf{f}_{col}^{32} = \mathbf{f}_{col}^{33} = \text{color const} = \mathcal{C}$. For horizontal nonhomogenous color regions we have $\mathbf{f}_{col}^{11} = \mathbf{f}_{col}^{12} = \mathbf{f}_{col}^{13} = \mathcal{C}_1 = a_{lu}^1 \mathbf{e}_{lu} + bf z_{ch}^1 \mathbf{E}_{ch}$, and $\mathbf{f}_{col}^{31} = \mathbf{f}_{col}^{32} = \mathbf{f}_{col}^{33} = \mathcal{C}_2 = a_{lu}^2 \mathbf{e}_{lu} + z_{ch}^2 \mathbf{E}_{ch}$. Hence, $\widehat{\mathbf{f}}_{col} =$

$$= \begin{bmatrix} f_R^{11} + f_B^{12} + f_G^{13} \\ f_G^{11} + f_R^{12} + f_B^{13} \\ f_B^{11} + f_G^{12} + f_R^{13} \end{bmatrix} - \begin{bmatrix} f_R^{31} + f_B^{32} + f_G^{33} \\ f_G^{31} + f_R^{32} + f_B^{33} \\ f_B^{31} + f_G^{32} + f_R^{33} \end{bmatrix} = \begin{bmatrix} a_{lu}^1 \\ a_{lu}^1 \\ a_{lu}^1 \end{bmatrix} - \begin{bmatrix} a_{lu}^2 \\ a_{lu}^2 \\ a_{lu}^2 \end{bmatrix} = \Delta a_{lu} \mathbf{e}_{lu}.$$

Fig. 1 shows the result of color edge detection.

We have performed a number of simulations on the use of orthounitary transforms in the area of color image compression. We have experimented both with $2^n \times 2^n$ and $3^n \times 3^n$-pixel images by using $\mathfrak{WHP}_{2^n}$,

$\mathfrak{WHP}_{3^n}$, and $\mathbf{WHP}_{2^n}$, $\mathbf{WHP}_{3^n}$ transforms. Figures 2–3 illustrate the $\mathfrak{WHP}_{3^n}$ and $\mathbf{WHP}_{2^n}$, $\mathbf{WHP}_{3^n}$ transforms of the color image "BABOON" after the first iterations of the fast algorithms of these transformations. Examples of 2D color histrograms of chromaticity planes for different spectrums are shown in Fig. 4.

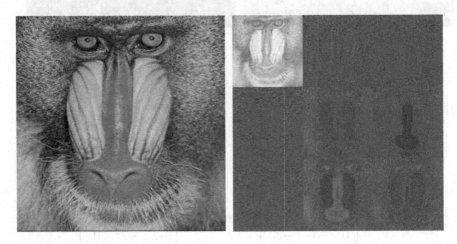

Figure 2. Left: original "BABOON" image, right: wavelet spectrum of "BABOON" after the first iterations of $\mathfrak{WHP}_{3^n}$.

5. Conclusion

A system of color-valued 2D basis functions has been defined in this paper. This system can be used to obtain color orthounitary Fourier transforms and series analyses of color images. Properties of the color Fourier transforms are presented. It is shown that such color series have properties similar to the classical orthogonal Fourier series. A family of discrete color orthounitary 2D Fourier transforms has also been presented that can be used in color image compression. In particular, the color Wavelet-Haar-Prometheus transforms are defined and used to obtain the color Wavelet-Haar-Prometheus series.

The analysis presented in this paper provides a very general framework for the definition of other multicolor transforms based on multiplet hypercomplex numbers. The derivation of such multicolor transforms is the subject of ongoing research. The motivation of this ongoing re-

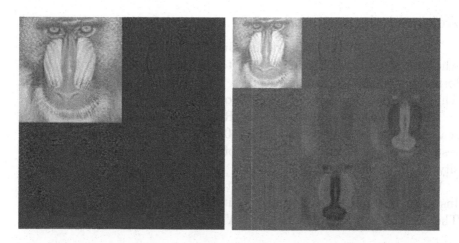

Figure 3. Wavelet spectrum of "BABOON" after the first iterations. Left: $\mathbf{WH}_{2^n}$, right: $\mathfrak{W}\mathfrak{H}_{3^n}$.

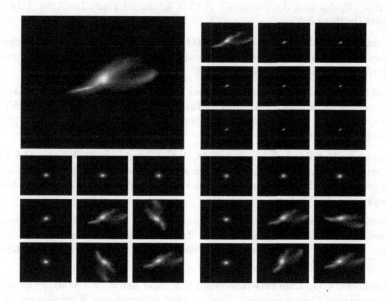

Figure 4. Examples of 2D histograms of the chromaticity plane. From top to bottom from left to right: original "BABOON" image, $\mathbf{WH}_3$ spectrum, $\mathfrak{W}\mathfrak{H}_3$ spectrum, $\mathfrak{W}\mathfrak{H}\mathfrak{P}_3$ spectrum.

search is to define multicolor transforms that can be used efficiently in multicolor satellite image compression.

Acknowledgments

The authors would like to thank Jim Byrnes from Prometheus Inc., who first stimulated our interest in the Prometheus orthogonal transforms and Shapiro-Golay sequences. Part of this work was performed while the first author was visiting Prometheus Inc.

The work was supported by the Russian Foundation for Basic Research, research project no. 03–01–00735. The paper contains some results obtained in project no. 3258 of the Ural State Technical University.

We thank the Organizing Committee of the NATO Advanced Study Institute "Computational Noncommutative Algebra and Applications". The authors are grateful for their NATO support.

References

[1] Labunets, Valery G. (2003). "Clifford Algebras as a Unified Language for Multi-color Image Processing and Pattern Recognition", *"Computational Noncommutative Algebra and Applications"*, NATO Advanced Study Institute, July 6–19, 2003 (in this book)

[2] Labunets-Rundblad, Ekaterina V. (2000). *Fast Fourier-Clifford Transforms Design and Application in Invariant Recognition.* PhD thesis, Tampere University Technology, Tampere, Finland, P. 262

[3] Labunets-Rundblad, Ekaterina V. and Labunets, Valery G. (2001). Chapter 7. "Spatial-Colour Clifford Algebra for Invariant Image Recognition", pp. 155–185. (In: *Geometric Computing with Clifford Algebra),* Edt. G. Sommer. Springer, Berlin Heideberg, 452 p.

[4] Labunets-Rundblad, E.V., Labunets, V.G., Astola, J. (2001). "Is the Visual Cortex a "Fast Clifford algebra quantum computer"? *Clifford Analysis and Its Applications,* II. Mathematics, Physics and Chemistry, Vol.25, NATO Science Series, pp. 173–183.

[5] Rundblad-Labunets, E.V., and Labunets, V.G. (1999). "Fast invariant recognition of multicolour images based on Triplet-Fourier-Gauss transform," *Second Int. Workshop on Transforms and Filter Banks, Tampere, Finland, TICSP Series,* 4, pp. 405–438.

[6] Labunets, V.G., Labunets-Rundblad, E.V., and Astola, J. (2000). "Algebra and Geometry of Colour Images," *Proc. of First International Workshop on Spectral Techniques and Logic Design for Future Digital Systems, Tampere, Finland,* pp. 231–361.

[7] L.-Rundblad, E.V., Labunets, V.G., and Astoia, J. (2000). "Fast Calculation Algorithms of Invariants for Color and Multispectral Image Recognition," *Algebraic Frames for the Perception Action Cycle. Second Inter. Workshop, AFPAC 2000.*

[8] L.-Rundblad E., Nikitin I., and Labunets V. (2003). "Unified Approach to Fourier-Clifford-Prometheus Sequences, Transforms and Filter Banks", *"Com-*

putational Noncommutative Algebra and Applications", NATO Advanced Study Institute, July 6–19, 2003.

[9] L.-Rundblad E., Nikitin I., Novak P., Labunets M., (2003). "Wavelet Fourier-Clifford-Prometheus Transforms", *"Computational Noncommutative Algebra and Applications"*, NATO Advanced Study Institute, July 6–19, 2003.

SELECTED PROBLEMS

Various Authors

1. Transformations of Euclidean Space and Clifford Geometric Algebra

Georgi H. Georgiev

Shumen University
Shumen 9712, Bulgaria
g.georgiev@shu-bg.net

1.1 Similarity transformations

Any direct similarity transformation of Euclidean affine space $\mathcal{E}^n$ is a product of an orientation-preserving homothety and a proper rigid motion. In other words any direct similarity transformation of $\mathcal{E}^n$ preserves the orientation and the angles (see [1] for details). Let $Sim^+(\mathcal{E}^n)$ denote the group of all direct similarities of $\mathcal{E}^n$.

If we identify the Euclidean plane $\mathcal{E}^2$ with the field of complex numbers $\mathbb{C}$, then $f \in Sim^+(\mathcal{E}^2)$ has a representation

$$f(\mathbf{z}) = \alpha\, \mathbf{z} + \beta, \qquad \mathbf{z} \in \mathbb{C} \cong \mathcal{E}^2,$$

where $\alpha \in \mathbb{C}\backslash\{0\}$ and $\beta \in \mathbb{C}$ are constant.

In the case $n = 3$, we may identify three-dimensional Euclidean space $\mathcal{E}^3$ with the space of pure quaternions $\mathrm{Im}\,\mathbb{H}$. Then an arbitrary similarity $f \in Sim^+(\mathcal{E}^3)$ has a representation

$$f(\mathbf{z}) = \lambda\, \mathbf{n}\, \mathbf{z}\, \mathbf{n}^{-1} + \mathbf{a}, \qquad \mathbf{z} \in \mathrm{Im}\,\mathbb{H} \cong \mathcal{E}^3,$$

where $\lambda \in \mathbb{R}^+$, $\mathbf{n}$ is a fixed unit quaternion and $\mathbf{a}$ a fixed pure quaternion. Clearly, f is a homothety whenever $\lambda \neq 1$ and $\mathbf{n} = \pm 1$, f is a translation whenever $\lambda = 1$ and $\mathbf{n} = \pm 1$, f is a rotation whenever $\lambda = 1$ and $\mathbf{a} = 0$.

The next step is the discovery of a common representation of the similarities by one equation in any dimension.

Problem 1. Using Clifford geometric algebra find a representation of $f \in Sim^+(\mathcal{E}^n)$ for any $n > 1$ in terms of the geometric product.

Note that the dilations, rotations and translations can be represent by rotors (see [3]).

1.2   Quadratic transformations

Let Q be a non-degenerate quadric hypersurface in the real projective space $\mathbb{P}^n$, $n > 1$, let p be a point lying on Q, and let τ be the tangent hyperplane to Q at p. Then the pair (Q, p) determines a quadratic involutory transformation

$$\varphi : \mathbb{P}^n \backslash \tau \longrightarrow \mathbb{P}^n \backslash \tau$$

with the set of fixed points $Q \backslash p$. In fact for $x \in \mathbb{P}^n \backslash \tau$,

$$\varphi(x) = x' = \mathrm{Pol}_Q(x) \cap \langle px \rangle,$$

where $\mathrm{Pol}_Q(x)$ is the polar hyperplane of x and $\langle px \rangle$ is the straight line passing through p and x. If $\mathbb{P}^n$ is the projective extension of $\mathcal{E}^n$ and τ is the hyperplane at infinity, then φ is a one-to-one transformation of $\mathcal{E}^n$. The three-dimensional Clifford algebra $\mathcal{G}_3$ and techniques from [4] and [5] are applied in [2] for a description of the case $n = 2$.

If $Q = p + \lambda\, q + \lambda^2\, r$ $(\lambda \in \mathbb{R})$ is a conic in $\mathbb{P}^2$, where $p \in Q$, q and r are non-collinear points in the projective plane, the line $L = p \wedge x = \langle px \rangle$ passes through p and x, and meets the conic Q at a second point x^0 (see [2]).

Then φ has the equation

$$x' = 2p \cdot x \left(x^0 \cdot x - x \cdot x \right) p + \left(p \cdot x\, p \cdot x^0 - p \cdot p\, x^0 \cdot x \right) x,$$

where the second intersection point $x^0 = [rpx]\,^2 p - [qpx]\,[rpx]\,q + [qpx]\,^2 r$, $[p\, q\, r] = (p \wedge q \wedge r).I^{-1}$, and I is the unit pseudoscalar in $\mathcal{G}_3$.

Problem 2. Using Clifford geometric algebra obtain an explicit equation for φ for $n \geq 3$.

The techniques stated in [6] can be used for the solution of the above problems.

References

[1] M. Berger. *Geometry I*. Springer, Berlin, 1998.

[2] G. Georgiev. Quadratic transformations of the projective plane. In L. Dorst, C. Doran and J. Lasenby Eds. *Applications of Geometric Algebra in Computer Science and Engineering* pp. 187–191, Birkhäuser, Boston, 2002.

[3] J. Lasenby. Using Clifford geometric algebra in robotics. In *Computational Non-commutative Algebra and Applications,* Kluwer Academic Publishers, Dordrecht, 2003.

[4] J. Lasenby, E. Bayro-Corrochano. Analysis and computation of projective invariants from multiple views in the geometric algebra framework. *International Journal of Pattern Recognition and Artificial Intelligence* **13**:1105–1119, 1999.

[5] D. Hestenes, R. Ziegler. Projective geometry with Clifford algebra. *Acta Applicandae Mathematicae* **23**:25–63, 1991.

[6] G. Sobczyk. Clifford geometric algebras in multilinear algebra and non-Euclidean geometries. In *Computational Noncommutative Algebra and Applications,* Kluwer Academic Publishers, Dordrecht, 2003.

2. On the Distribution of Kloosterman Sums on Polynomials over Quaternions

N.M. Glazunov

Glushkov Institute of Cybernetics NAS
Kiev 03187, Ukraine

glanm@yahoo.com

Abstract

We formulate two problems concerning the distribution of angles of Kloosterman sums and the algebraic structure of curves over quaternions.

Keywords: density of a distribution, elliptic curve, Kloosterman sum, quaternions.

2.1 The Problem of the distribution of Kloosterman sums

Let

$$T_p(c,d) = \sum_{x=1}^{p-1} e^{2\pi i(\frac{cx+\frac{d}{x}}{p})}$$

$$1 \le c, d \le p - 1; \; x, c, d \in \mathbf{F}_p^*$$

be the Kloosterman sum.
By the A. Weil [1] estimate

$$T_p(c,d) = 2\sqrt{p}\cos\theta_p(c,d),$$

$$\theta_p(c,d) = \arccos\left(\frac{T_p(c,d)}{2\sqrt{p}}\right)$$

on the interval $[0, \pi)$:

Problem 1. *Let c and d be fixed and p vary over all primes not dividing c and d. What is the distribution of angles $\theta_p(c,d)$ as $p \to \infty$?*

Motivation. Kloosterman sums of different types have in recent years played an increasingly important role in the study of many problems in analytic number theory [2].
During 1983 and 1989 the presenter of this problem implemented computations with Kloosterman sums $T_p(c,d)$ for primes p, $2 \leq p \leq 13499$, $c = d = 1$, and for $1 \leq c \leq p-1$, $1 \leq d \leq p-1$, for some primes p [3, 4].

Conjecture. *Under the conditions of Problem 1 the angles $\theta_p(c,d)$ are distributed on $[0,\pi)$ with Sato-Tate density $\frac{2}{\pi}\sin^2 t$.*

2.2 On algebraic structure of curves over quaternions

Let E be the elliptic curve over a field K, $E(K)$ the group of rational points of E. $E(k)$ is an abelian group with a finite number of generators (Poincaré, Mordell, Weil).
Let $End(E) = Hom(E, E)$ be the ring of endomorphisms of E. In some cases the ring $End(E)$ is noncommutative (Deuring).
Example. Let $K = \mathbf{F}_{2^2}$ and E be defined by the equation

$$y^2 + y = x^3.$$

Let $\alpha(x,y) = (\varepsilon x, y)$ where $\varepsilon^2 + \varepsilon + 1$, $\varepsilon \in \mathbf{F}_{2^2}$ and Frobenius $\phi(x,y) = (x^2, y^2)$. Then, since $\varepsilon^2 \neq \varepsilon$, $\alpha \circ \phi(x,y) \neq \phi \circ \alpha(x,y)$. Moreover the ring $End(E)$ is isomorphic to the set of integer quaternions $End(E) = \{m_1 + m_2i + m_3j + m_4k, \ m_i \in \mathbf{Z}$ or $m_i + 1/2 \in \mathbf{Z}\}$, so E is a supersingular elliptic curve.

Let $\mathbf{H} = \{a_1 + a_2i + a_3j + a_4k, \ a_i \in \mathbf{R}\}$ be the ring of quaternions and

$$EH : y^2 = x^3 + ax + b$$

be the equation with $a, b \in \mathbf{H}$.

Problem 2. *What can be said about the algebraic properties and structure of the set*

$$EH(\mathbf{H}) = \{(x, y) \in \mathbf{H} \times \mathbf{H} \ \& \ y^2 = x^3 + ax + b\}$$

and the set $Mor(EH(\mathbf{H}))$ of maps $EH(\mathbf{H}) \to EH(\mathbf{H})$?

References

[1] A. Weil. On some exponential sums. *Proc. N.A.S.* **34**:204–207, 1948.

[2] W. Duke, J. Friedlander, H. Iwaniec. Bilinear forms with Kloosterman fractions. *Invent. Math.* **128**:23-43, 1997.

[3] N. Glazunov. On Equidistribution of Values of Kloosterman Sums. *Doklady Acad. Nauk Ukr. SSR, ser. A* No.2:9–12, 1983.

[4] N. Glazunov. On moduli spaces, equidistribution, bounds and rational points of algebraic curves. *Ukrainian Math. Journal.* No.9:1174–1183, 2001.

3. Harmonic Sliding Analysis Problems

Vladimir Ya. Krakovsky

International Research-Training Centre
for Information Technologies and Systems
40 Academician Hlushkov Avenue, Kyiv 03680 Ukraine
vladimir_krakovsky@uasoiro.org.ua

Harmonic sliding analysis (HSA) is a dynamic spectrum analysis [1] in which the next analysis interval differs from the previous one by including the next signal sample and excluding the first one from the previous analysis interval. Such a harmonic analysis is necessary for time-frequency localization [2] of the analysed signal given peculiarities. Using the well-known Fast Fourier transform (FFT) is not effective in this context. More effective are known recursive algorithms which use only one complex multiplication for computing one harmonic during each analysis interval. The presenter of this problem improved one of those algorithms, so that it became possible to use one complex multiplication for computing two, four and even eight (for complex signals) harmonics simultaneously [3], [4], [5]. One problem of HSA was mentioned in [6]. In [7] there is a short review of the presenter's papers devoted to HSA. A basic problem of HSA is: are there methods to further increase the speed of the response using up-to-date nanotechnology? If this is possible, how does one do it?

References

[1] V.N. Plotnikov, A.V. Belinsky, V.A. Sukhanov and Yu.N. Zhigulevtsev, "Spectrum digital analyzers," Moscow: Radio i svjaz', 1990. - 184 p. (In Russian).

[2] R. Tolimieri, M. An, Time-Frequency Representations, Birkhauser, 1998.

[3] V.Ya. Krakovsky, "Algorithms for increase of speed of response for digital analyzers of the instant spectrum," Kibernetika, No.5, pp.113-115, 1990. (In Russian).

[4] V.Ya. Krakovskii. "Generalized representation and implementation of speed-improvement algorithms for instantaneous spectrum digital analyzers." Cybernetics and System Analysis, Vol.32, No.4, pp.592-597, March 1997.

[5] V.Ya. Krakovsky. "Spectrum Sliding Analysis Algorithms & Devices." Proceedings of the 1st International Conference on Digital Signal Processing and Its Application, June 30 - July 3, 1998, Moscow, Russia, Volume I, pp.104-107.

[6] V.Ya. Krakovsky. "Harmonic Sliding Analysis Problem." Twentieth Century Harmonic Analysis - A Celebration, Edited by James S. Byrnes, NATO Science Series, II. Mathematics, Physics and Chemistry - Vol. 33, 2000, p. 375-377.

[7] V.Ya. Krakovsky, "Harmonic Sliding Analysis Problems." - http://www.prometheus-inc.com/asi/algebra2003/posters.html

4.   Spectral Analysis under Conditions of Uncertainty

Sergey I. Kumkov

IMM Ural Branch of RAS
Ekaterinburg, Russia
kumkov@imm.uran.ru

In the classic spectral analysis, a sample $\{x_n, t_n, \ n = \overline{0, N-1}\}$ of the discrete time periodic process (x_n is the value of the process, t_n is the argument; N is the sample length) is expanded in the spectrum $C = \{C_k = (\sum_{n=0}^{N-1} x_n \bar{\phi}_k(t_n))/N, \ k = \overline{0, N-1}\}$ by using some basis system $\{\phi_k(t_n), k = \overline{0, N-1}; n = \overline{1, N-1}\}$. The reverse representation of the process is $\{\tilde{x}_n(C) = \sum_{k=0}^{N-1} C_k \phi_k(t_n)\}$, respectively.

Usually the sample is composed of the disturbed values with some model, for example, the model of disturbance can have the form $x_n = x_n^* + \varepsilon_n, \ n = \overline{0, N-1}$, where x_n^* is the unknown true value of the process, $\vec{\varepsilon} = \{\varepsilon_n\}$ is the additive disturbance.

If the disturbance has a stochastic nature and is described by some probability characteristics, there are classical approaches for description and computation of the spectrum for this stochastic process.

But in practice, often the statistical properties of disturbance are unknown, the researcher can show only the geometric constraint $|\varepsilon_n| \le \varepsilon_{max}$, and the sample gives the collection of the *uncertainty sets* $\{H_n =$

$[x_n - \varepsilon_{max}, x_n + \varepsilon_{max}]\}$. Additionally, the type of the process can be given, i.e., $x^*(t) = f(t, \vec{P})$, where $f(\cdot)$ is the describing function, $\vec{P}$ is the vector of parameters of dimension M, and $M < N$.

Now, each component C_k mentioned above becomes a set $C_k = C_k + (\sum_{n=0}^{N-1} \varepsilon_n \bar{\phi}_k(t_n))/N$, where all $\varepsilon_n \in [-\varepsilon_{max}, \varepsilon_{max}]$, and the process spectrum $\mathcal{C}$ becomes the body in the space generated by the taken basis $\{\phi_k\}$. The researcher is interested only in the vectors $C \in \mathcal{C}$ consistent with the shown collection of the uncertainty sets, i.e., $\tilde{x}_n(C \in \mathcal{C}) \in H_n$ for all $n = \overline{0, N-1}$.

Problem. *Since the sets C_k depend on the vector of the disturbance $\vec{\varepsilon}$, they are not mutually independent and the spectrum $\mathcal{C}$ is not a rectangular parallellotop. How to describe and to compute the frontier of the body-spectrum $\mathcal{C}$ consistent with the collection of the uncertainty sets and the given type of the descriptive function?*

5. A Canonical Basis for Maximal Tori of the Reductive Centralizer of a Nilpotent Element

Alfred G. Noël

Mathematics Department
The University of Massachusetts
Boston, Massachusetts 02125-3393, USA

anoel@math.umb.edu

Abstract   We describe a problem that we partially solved in *[1], [2] and [3]* using a computational scheme. However a more conceptual argument might be very enlightening

5.1   Problem Description.

Let $\mathfrak{g}$ be a real semisimple Lie algebra with adjoint group G and $\mathfrak{g}_{\mathbb{C}}$ its complexification. Also let $\mathfrak{g} = \mathfrak{k} \oplus \mathfrak{p}$ be the Cartan decomposition of $\mathfrak{g}$. Finally, let θ be the corresponding Cartan involution of $\mathfrak{g}$ and σ the conjugation of $\mathfrak{g}_{\mathbb{C}}$ with regard to $\mathfrak{g}$. Then $\mathfrak{g}_{\mathbb{C}} = \mathfrak{k}_{\mathbb{C}} \oplus \mathfrak{p}_{\mathbb{C}}$ where $\mathfrak{k}_{\mathbb{C}}$ and $\mathfrak{p}_{\mathbb{C}}$ are obtained by complexifying $\mathfrak{k}$ and $\mathfrak{p}$ respectively. Denote by $K_{\mathbb{C}} \subseteq G_{\mathbb{C}}$ the connected subgroup of the adjoint group $\mathfrak{G}_{\mathbb{C}}$ of $\mathfrak{g}_{\mathbb{C}}$, with Lie algebra $\mathfrak{k}_{\mathbb{C}}$.

A triple (x, e, f) in $\mathfrak{g}_{\mathbb{C}}$ is called a standard triple if $[x, e] = 2e$, $[x, f] = -2f$ and $[e, f] = x$. If $x \in \mathfrak{k}_{\mathbb{C}}$, e and $f \in \mathfrak{p}_{\mathbb{C}}$ then (x, e, f) is said to be normal. It is a result of Kostant and Rallis that any nilpotent e of $\mathfrak{p}_{\mathbb{C}}$

can be embedded in a standard normal triple (x, e, f). Let $\mathfrak{k}_{\mathfrak{c}}^{(x,e,f)}$ be the centralizer of (x, e, f) in $\mathfrak{k}_{\mathfrak{c}}$.

Maintaining the above notations we would like to solve the following problem:

5.2

PROBLEM: Let $\mathfrak{t}$ be a Cartan subalgebra of $\mathfrak{k}_{\mathfrak{c}}$ such that $x \in \mathfrak{t}$. Then find two nilpotent elements e and f in $\mathfrak{p}_{\mathfrak{c}}$ such that (x, e, f) is a standard triple and $\mathfrak{t}_1 = \mathfrak{t} \cap \mathfrak{k}_{\mathfrak{c}}^{(x,e,f)}$ is a maximal torus in $\mathfrak{k}_{\mathfrak{c}}^{(x,ef)}$. More precisely give a natural basis for $\mathfrak{t}_1$.

The reader should be aware that in general $\mathfrak{t}_1 \neq \mathfrak{t}^{(x,e,f)}$ for an arbitrary e. A counterexample can be found in *[1]*. Furthermore, there is currently no good characterization of such a torus in the literature. Our conversation with several experts led us to believe that such a characterization may be quite technical. We wish to thank Prof. Andreas Dress from Bielefeld University for simplifying the statement of the problem.

5.3   Partial Results.

When $\mathfrak{g}$ is a real form of an exceptional non-compact simple complex Lie algebra we have developed a computational scheme to compute $\mathfrak{t}_1$. These results are found in *[1], [2], and [3]*. Here is a concise version of the algorithm:

Algorithm

Input: $\mathfrak{g}$ is a real exceptional simple Lie algebra, $\Delta_k = \{\beta_1, \ldots, \beta_l\}$ a set of fundamental roots of $\mathfrak{k}_{\mathfrak{c}}$, and t is a Cartan subalgebra of $\mathfrak{g}_{\mathfrak{c}}$ define by Δ_k.

Computation

- Compute x using the values of $\beta_i(x)$.

- Using x, express $e = \sum_{i=1}^{r} c_{\gamma_i} X_{\gamma_i}$, where γ_i is a non compact root. X_{γ_i} a non zero root vector and c_{γ_i} a complex number, in a regular semisimple subalgebra l_e of minimal rank r. Create the normal triple (x, e, f).

- Compute the intersection of the kernels of γ_i on t. *Observe that the complex span of such an intersection is a maximal torus in* $\mathfrak{k}_{\mathfrak{c}}^{(x,e,f)}$.

Output: $\mathfrak{t}_1$ is the complex span of the intersection computed above.

We do not have at this moment a good way of describing a natural basis for $\mathfrak{t}_1$ in general. The above choice of e suggests that e must be given as a linear combination of a minimal set of root vectors. In designing the above algorithm we used some information about the type of $\mathfrak{k}_{\mathbb{C}}^{(x,e,f)}$ from the work of Dragomir Djokovič. We anticipate that we will need such information for the classical groups where the nilpotent orbits are parametrized by partitions.

References

[1] A. G. Noël *Classification of Admissible Nilpotent Orbits In simple Exceptional real Lie algebras of Inner type.* AMS Journal of RepresentationTheory 5 (2001) 455-493.

[2] A. G. Noël *Classification of Admissible Nilpotent Orbits In simple real Lie algebras $E_{6(6)}$ and $E_{6(-26)}$.* AMS Journal of Representation Theory 5 (2001) 494A-502.

[3] A. G. Noël *Maximal Tori of Reductive Centralizers of Nilpotents in Exceptional Complex Symmetric Spaces.* (submitted)

6. The Quantum Chaos Conjecture

L.D. Pustyl'nikov

Keldysh Institute of Applied Mathematics
Russian Academy of Sciences, Moscow, Russia
pustylni@physik.uni-bielefeld.de

According to the general formulation of the quantum chaos conjecture introduced in [1] (see also [2]), the distribution of the distances between adjacent energy levels of a quantum system should be close to a Poisson distribution with density e^{-x} provided such a system is a quantum analogue of a classical integrable system. The quantum chaos conjecture has not been proved previously for any system. The quantum system is described by the Schrödinger equation and obtained from the classical system with the help of some standard procedures. We present a class of quantum systems containing as a special case the well-known and popular model of a rotating particle subject to δ-shocks ("kicked rotator"), which in many papers has been considered as one of the main objects in connection with confirming the conjecture.

The proof of the quantum chaos conjecture for this model makes essential use of results on the distribution of distances between adjacent fractional parts of the values of a polynomial, and the estimate of the remainder is based on a new theory of generalized continued fractions for number vectors ([3]-[6]). We consider a one-dimensional nonlinear oscillator determined by the Hamilton function $H = H(\phi, I, t) = H_0(I) + H_1(\phi, t)$:

$$\frac{d\phi}{dt} = \frac{\partial H}{\partial I} = \frac{dH_0}{dI}, \quad \frac{dI}{dt} = -\frac{\partial H}{\partial \phi} = -\frac{\partial H_1}{\partial \phi}, \tag{1}$$

where I, ϕ are "action-angle" variables, t is an independent variable and the function $H_1(\phi, t)$ has period 2π with respect to ϕ and period $T > 0$ with respect to t and can be represented as

$$H_1(\phi, t) = F(\phi) \sum_{k=-\infty}^{\infty} \delta(t - kT), \tag{2}$$

with $F(\phi)$ a smooth 2π-periodic function and $\delta = \delta(t)$ the delta-function.

We assume that $H_0(I) = \sum_{s=0}^{n} b_s I^s$ is a polynomial of degree $n \geq 2$ with coefficients $b_s = a_s/\hbar^s$ $(s = 0, \ldots, n)$, where $\hbar$ is the Planck constant and the a_s are real numbers. In the special case when $n = 2$, $a_0 = a_1 = 0$, and $F(\phi) = \gamma \cos \phi$ $(\gamma$ is a constant), the system (1) represents a "kicked rotator". We introduce the Hilbert space L^2 of complex 2π-periodic functions as the state space of the quantum system and we define the momentum operator by $\hat{I} = \frac{\hbar}{i}\frac{\partial}{\partial \phi}$. The evolution of the wave function $\Psi = \Psi(\phi, t) \in L^2$ is described by the Schrödinger equation

$$i\hbar\frac{\partial}{\partial t}\Psi(\phi, t) = \hat{H}(t)\Psi(\phi, t) ,$$

where $\hat{H}(t) = \hat{H}_0 + \hat{H}_1(t)$, $\hat{H}_0 = \sum_{s=0}^{n} b_s \hat{I}^s$, and $\hat{H}_1(t)$ is the limit as $\varepsilon \to 0$ $(\varepsilon > 0)$ of the operators of multiplication by the function $H_1^{(\varepsilon)}$ obtained from the function H_1 in (2) when the delta-function δ is replaced by a smooth positive function δ_ε supported on the interval $[0, \varepsilon]$ and having integral 1.

References

[1] M.V.Berry and M.Tabor. *Level clustering in the regular spectrum.* Proc. R. Soc. Lond. A (1977), 356,375-394.

[2] A.Knauf and Y.G.Sinai. *Classical Nonintegrability, Quantum Chaos.* Birkhäser Verlag (1977), 1-98.

[3] L.D.Pustyl'nikov. *Proof of the quantum chaos conjecture and generalized continued fractions.* Russian Math. Surveys (2002), 1, 159-160.

[4] L.D.Pustyl'nikov. *The quantum chaos conjecture and generalized continued fractions.* Sb.Math.(2003), 194, 4, 575-587.

[5] L.D.Pustyl'nikov. *Probabilistic laws in the distribution of the fractional parts of the values of polynomials and evidence for the quantum chaos conjecture.* Russian Math. Surveys (1999), 6, 1259-1260.

[6] L.D.Pustyl'nikov *Generalized continued fractions and ergodic theory.* Russian Math. Surveys (2003), 58:1, 109-159.

7.   Four Problems in Radar

Michael C. Wicks and Braham Himed

Air Force Research Laboratory
Sensors Directorate
26 Electronic Parkway
Rome, New York 13441-4514
Michael.Wicks@rl.af.mil, Braham.Himed@rl.af.mil

For more context on these four problems, please consult the chapter in this book by the same authors, from which this section is excerpted.

As analog hardware performance matures to a steady plateau, and Moore's Law provides for a predictable improvement in throughput and memory, it is only the advances in signal and data processing algorithms that offer potential for performance improvements in fielded sensor systems. However, it requires a revolution in system design and signal processing algorithms to dramatically alter the traditional architectures and concepts of operation. One important aspect of our current research emphasizes new and innovative sensors that are electrically small (on the order of 10 wavelengths or less), and operate in concert with a number of other electrically small sensor systems within a wide field of view (FOV). Our objective is to distribute the power and the aperture of the conventional wide area surveillance radar among a number of widely disbursed assets throughout the battlefield environment. Of course, we must have an algorithm for distributing those assets in real time as the dynamically changing surveillance demands. The mathematical challenge here relates to the traveling salesman problem. Classically, the traveling salesman must select his route judiciously in order to maximize potential sales. Recent analysis in the literature addresses multiples salesmen covering the same territory. This is analogous to our problem, where multiple unmanned aerial vehicle (UAV) Ð based sensors are charged with the mission of detecting, tracking, and identifying all targets (friend or foe). Not only must these sensors detect and identify threat targets, they must also process data coherently across multiple platforms. Our mathematical challenge problem reduces to one in which the position and velocity all UAV-based sensors are selected to maximize detection performance and coverage area, and minimize revisit rate.

Enhancing one of the sensors described above, to be more like a classical radar with a large power-aperture product, leads to the second mathematical challenge problem to be addressed by this community. With a larger aperture and more precise estimates of target parameters (angle, Doppler), an opportunity to expand the hypothesis testing problem to include both detection and estimation emerges. Here, conventional wisdom dictates that we perform filtering and false alarm rate control as part of the detection process, yet perform track processing as a post-detection analysis, where the parameter estimation is focused upon target position and velocity history. Clearly, parameter estimation need not be accomplished as a post-detection process. Since this segmented approach to detection and track processing has been in effect for decades, it will require a dramatic demonstration of improvement before it will be embraced by the radar community.

A third challenge problem arises in the formulation of the Generalized Likelihood Ratio Test (GLRT). In Kelly's formulation of a GLRT, conditioning on finite sample support is incorporated into the basic test. As such, a statistical method developed under the assumption that only finite training data are available for sample covariance matrix formulation was made available. The next generalization to be made, in an extension of Kelly's GLRT, is to incorporate prior knowledge of the structure of the sample covariance matrix into the mathematical development of a statistical test. This mathematical structure arises due to the fact that the phase spectra of ground clutter as seen by an airborne radar is determined only by geometry, and remains independent of the underlying clutter statistics (except for initial phase). The effect of this geometric dependence is to localize the interference along a contour in the transform domain (Fourier analysis). Our objective is to formulate a single GLRT which incorporates the effects of finite training data as well as geometric dependence.

The fourth mathematical challenge facing the modern radar engineer is to incorporate adaptivity on transmit into the basic formulation of the signal processing algorithm. Since this is a new research topic, an opportunity exists to formulate the basic mathematical framework for fully adaptive radar on both transmit and receive. Further extensions arise by incorporating the above challenge problems into this analysis.

Topic Index

Author Index